International
REVIEW OF
Neurobiology
Volume 88

International

R E V I E W   O F

Neurobiology

Volume 88

New Concepts of Psychostimulants Induced Neurotoxicity

EDITED BY

HARI SHANKER SHARMA

Department of Surgical Sciences
Anaesthesiology and Intensive Care Medicine
University Hospital, Uppsala University
SE-751 85 Uppsala
Sweden

AMSTERDAM • BOSTON • HEIDELBERG • LONDON
NEW YORK • OXFORD • PARIS • SAN DIEGO
SAN FRANCISCO • SINGAPORE • SYDNEY • TOKYO
Academic Press is an imprint of Elsevier

ELSEVIER

Academic Press is an imprint of Elsevier
360 Park Avenue South, New York, NY 10010-1700
525 B Street, Suite 1900, San Diego, California 92101-4495, USA
32 Jamestown Road, London NW1 7BY, UK

This book is printed on acid-free paper.

For information on all Academic Press publications
visit our Web site at www.elsevierdirect.com

ISBN-13: 978-0-12-374504-0

PRINTED AND BOUND IN THE UNITED STATES OF AMERICA
09 10 11 12   8 7 6 5 4 3 2 1

Working together to grow
libraries in developing countries

www.elsevier.com | www.bookaid.org | www.sabre.org

ELSEVIER   BOOK AID
             International   Sabre Foundation

CONTENTS

Acute Methamphetamine Intoxication: Brain Hyperthermia, Blood–brain Barrier, Brain Edema, and Morphological Cell Abnormalities

EUGENE A. KIYATKIN AND HARI S. SHARMA

Molecular Bases of Methamphetamine-Induced Neurodegeneration

JEAN LUD CADET AND IRINA N. KRASNOVA

Involvement of Nicotinic Receptors in Methamphetamine- and Mdma-Induced Neurotoxicity: Pharmacological Implications

E. ESCUBEDO, J. CAMARASA, C. CHIPANA, S. GARCÍA-RATÉS, AND D. PUBILL

Ethanol Alters the Physiology of Neuron–Glia Communication

ANTONIO GONZÁLEZ AND GINÉS M. SALIDO

Therapeutic Targeting of "DARPP-32": A Key Signaling Molecule in the Dopiminergic Pathway for the Treatment of Opiate Addiction

SUPRIYA D. MAHAJAN, RAVIKUMAR AALINKEEL, JESSICA L. REYNOLDS, BINDUKUMAR B. NAIR, DONALD E. SYKES, ZIHUA HU, ADELA BONOIU, HONG DING, PARAS N. PRASAD, AND STANLEY A. SCHWARTZ

Pharmacological and Neurotoxicological Actions Mediated By Bupropion and Diethylpropion

HUGO R. ARIAS, ABEL SANTAMARÍA, AND SYED F. ALI

Neural and Cardiac Toxicities Associated With 3, 4-Methylenedioxymethamphetamine (MDMA)

MICHAEL H. BAUMANN AND RICHARD B. ROTHMAN

Cocaine-Induced Breakdown of the Blood–Brain Barrier and Neurotoxicity

HARI S. SHARMA, DAFIN MURESANU, ARUNA SHARMA, AND RANJANA PATNAIK

Cannabinoid Receptors in Brain: Pharmacogenetics, Neuropharmacology, Neurotoxicology, and Potential Therapeutic Applications

EMMANUEL S. ONAIVI

Intermittent Dopaminergic Stimulation causes Behavioral Sensitization in the Addicted Brain and Parkinsonism

FRANCESCO FORNAI, FRANCESCA BIAGIONI, FEDERICA FULCERI, LUIGI MURRI, STEFANO RUGGIERI, AND ANTONIO PAPARELLI

The Role of the Somatotrophic Axis in Neuroprotection and Neuroregeneration of the Addictive Brain

FRED NYBERG

CONTRIBUTORS

Numbers in parentheses indicate the pages on which the authors' contributions begin.

Ravikumar Aalinkeel (199), Department of Medicine, Division of Allergy, Immunology, and Rheumatology, Buffalo General Hospital, State University of New York, Buffalo, New York 14203, USA

Syed F. Ali (223), Neurochemistry Laboratory, Division of Neurotoxicology, National Center of Toxicological Research, Food and Drug Administration, Jefferson, Arkansas 72079, USA

Francesco Angelucci (1), Division of Psychiatry, Karolinska Institutet, Institution of Clinical Neuroscience, Huddinge University Hospital, S-141 86 Huddinge, Sweden; and Department of Clinical and Behavioural Neurology, IRCCS Santa Lucia Foundation, Rome, Italy

Hugo R. Arias (223), Department of Pharmaceutical Sciences, College of Pharmacy, Midwestern University, Glendale, Arizona 85308, USA

Masato Asanuma (43), Department of Brain Science, Okayama University Graduate School of Medicine, Dentistry and Pharmaceutical Sciences, Okayama 700–8558, Japan

Michael H. Baumann (257), Clinical Psychopharmacology Section, Intramural Research Program (IRP), National Institute on Drug Abuse (NIDA), National Institutes of Health (NIH), Baltimore, Maryland 21224, USA

Francesca Biagioni (371), Laboratory of Neurobiology of Movement Disorders I.R.C.C.S., I.N.M. Neuromed (IS), Pozzilli, Isernia, Italy

Adela Bonoiu (199), Institute for Lasers Photonics and Biophotonics, State University of New York (SUNY), Buffalo, New York 14260, USA

Pietro Bria (1), Institute of Psychiatry, Catholic University, Rome, Italy

Jean Lud Cadet (101), Molecular Neuropsychiatry Branch, NIDA-Intramural Research Program, NIH/DHHS, Baltimore, Maryland 21224, USA

Carlo Caltagirone (1), Department of Neuroscience, Tor Vergata University; and Department of Clinical and Behavioural Neurology, IRCCS Santa Lucia Foundation, Rome, Italy

J. Camarasa (121), Unitat de Farmacologia i Farmacognòsia, Facultat de Farmàcia, Universitat de Barcelona, Barcelona 08028, Spain

C. Chipana (121), Unitat de Farmacologia i Farmacognòsia, Facultat de Farmàcia, Universitat de Barcelona, Barcelona 08028, Spain

Hong Ding (199), Institute for Lasers Photonics and Biophotonics, State University of New York (SUNY), Buffalo, New York 14260, USA

E. Escubedo (121), Unitat de Farmacologia i Farmacognòsia, Facultat de Farmàcia, Universitat de Barcelona, Barcelona 08028, Spain

Francesco Fornai (371), Laboratory of Neurobiology of Movement Disorders I.R.C.C.S., I.N.M. Neuromed (IS), Pozzilli, Isernia, Italy; and Department of Human Morphology and Applied Biology, University of Pisa, Pisa, Italy

Federica Fulceri (371), Department of Human Morphology and Applied Biology, University of Pisa, Pisa, Italy

S. García-Ratés (121), Unitat de Farmacologia i Farmacognòsia, Facultat de Farmàcia, Universitat de Barcelona, Barcelona 08028, Spain

Antonio González (167), Department of Physiology (Cell Physiology Research Group), University of Extremadura, 10071, Cáceres, Spain

Zihua Hu (199), Department of Biostatistics, Center for Computational Research, State University of New York (SUNY), Buffalo, New York 14260, USA

Taizo Kita (43), Laboratory of Pharmacology, Kyushu Nutrition Welfare University Graduate School of Food and Nutrition, Kitakyushu, Fukuoka 803–8511, Japan

Eugene A. Kiyatkin (65), Behavioral Neuroscience Branch, National Institute on Drug Abuse–Intramural Research Program, National Institutes of Health, Baltimore, Maryland 21224, USA

Irina N. Krasnova (101), Molecular Neuropsychiatry Branch, NIDA-Intramural Research Program, NIH/DHHS, Baltimore, Maryland 21224, USA

Supriya D. Mahajan (199), Department of Medicine, Division of Allergy, Immunology, and Rheumatology, Buffalo General Hospital, State University of New York, Buffalo, New York 14203, USA

H. Manev (25), Department of Psychiatry, The Psychiatric Institute, University of Illinois at Chicago, Chicago, Illinois 60612, USA

Aleksander A. Mathé (1), Division of Psychiatry, Karolinska Institutet, Institution of Clinical Neuroscience, Huddinge University Hospital, S-141 86 Huddinge, Sweden

Ikuko Miyazaki (43), Department of Brain Science, Okayama University Graduate School of Medicine, Dentistry and Pharmaceutical Sciences, Okayama 700–8558, Japan

Dafin Muresanu (297), Department of Neurology, Medical University of Cluj-Napoca, Cluj-Napoca, Romania

Luigi Murri (371), Department of Neurosciences, Clinical Neurology, University of Pisa, Pisa, Italy

Bindukumar B. Nair (199), Department of Medicine, Division of Allergy, Immunology, and Rheumatology, Buffalo General Hospital, State University of New York, Buffalo, New York 14203, USA

Fred Nyberg (399), Department of Pharmaceutical Biosciences, Division of Biological Research on Drug Dependence, Uppsala University, S-75124 Uppsala, Sweden

Emmanuel S. Onaivi (335), Department of Biology, William Paterson University, Wayne, New Jersey 07470, USA

Antonio Paparelli (371), Department of Human Morphology and Applied Biology, University of Pisa, Pisa, Italy

Ranjana Patnaik (297), Department of Biomaterials, School of Biomedical Engineering, Institute of Technology, Banaras Hindu University, Varanasi 2210005, India

Paras N. Prasad (199), Institute for Lasers Photonics and Biophotonics, State University of New York (SUNY), Buffalo, New York 14260, USA

D. Pubill (121), Unitat de Farmacologia i Farmacognòsia, Facultat de Farmàcia, Universitat de Barcelona, Barcelona 08028, Spain

Jessica L. Reynolds (199), Department of Medicine, Division of Allergy, Immunology, and Rheumatology, Buffalo General Hospital, State University of New York, Buffalo, New York 14203, USA

Valerio Ricci (1), Institute of Psychiatry, Catholic University, Rome, Italy; and Division of Psychiatry, Karolinska Institutet, Institution of Clinical Neuroscience, Huddinge University Hospital, S-141 86 Huddinge, Sweden

Richard B. Rothman (257), Clinical Psychopharmacology Section, Intramural Research Program (IRP), National Institute on Drug Abuse (NIDA), National Institutes of Health (NIH), Baltimore, Maryland 21224, USA

Stefano Ruggieri (371), Laboratory of Neurobiology of Movement Disorders I.R.C.C.S., I.N.M. Neuromed (IS), Pozzilli, Isernia, Italy

Ginés M. Salido (167), Department of Physiology (Cell Physiology Research Group), University of Extremadura, 10071, Cáceres, Spain

Abel Santamaría (223), Laboratorio de Aminoácidos Excitadores, Instituto Nacional de Neurología y Neurocirugía, México City, Mexico

Stanley A. Schwartz (199), Department of Medicine, Division of Allergy, Immunology, and Rheumatology, Buffalo General Hospital, State University of New York, Buffalo, New York 14203, USA

Hari S. Sharma (65, 297), Laboratory of Cerebrovascular Research & Pain Research Laboratory, Department of Surgical Sciences, Anesthesiology & Intensive Care Medicine, University Hospital, Uppsala University, SE-75185 Uppsala, Sweden; and Behavioral Neuroscience Branch, National Institute on Drug Abuse–Intramural Research Program, National Institutes of Health, Baltimore, Maryland 21224, USA

Aruna Sharma (297), Laboratory of Cerebrovascular Research & Pain Research Laboratory, Department of Surgical Sciences, Anesthesiology & Intensive Care Medicine, University Hospital, Uppsala University, SE-75185 Uppsala, Sweden

Gianfranco Spalletta (1), Department of Neuroscience, Tor Vergata University; and Department of Clinical and Behavioural Neurology, IRCCS Santa Lucia Foundation, Rome, Italy

Donald E. Sykes (199), Department of Medicine, Division of Allergy, Immunology, and Rheumatology, Buffalo General Hospital, State University of New York, Buffalo, New York 14203, USA

Mika Takeshima (43), Laboratory of Pharmacology, Kyushu Nutrition Welfare University Graduate School of Food and Nutrition, Kitakyushu, Fukuoka 803–8511, Japan

T. Uz (25), Department of Psychiatry, The Psychiatric Institute, University of Illinois at Chicago, Chicago, Illinois 60612, USA

George C. Wagner (43), Department of Psychology, Rutgers University, New Brunswick, New Jersey 08854, USA

PREFACE

NEW CONCEPTS OF PSYCHOSTIMULANTS INDUCED NEUROTOXICITY

With recent advancements in new knowledge, it has become evident that psychostimulants and related drugs of abuse are influencing our central nervous system (CNS) remarkably and could alter their function for long time. Experiments carried out in our laboratory recently demonstrated that psychostimulants, such as methamphetamine, MDMA, and related substance of abuse (i.e., morphine and GHB), are able to induce breakdown of the blood–brain barrier (BBB). Since most of the psychostimulants induce hypertehrmia that depends on the surrounding environmental temperatures, it appears that drugs of abuse induced rise in body and brain temperatures could be instrumental in breakdown of the BBB and precipitating mental abnormities and/or brain damage. This volume is the first to focus on substance abuse-induced brain pathology in widest sense as it covers alterations in neuronal, glial, and endothelial cell functions under the influence of acute or chronic usage of substance abuse.

This volume is a refereed collection of invited papers from leading experts in the world to highlight the latest developments in this newly emerging field on psychostimulants-induced neurotoxicity. It appears that substance abuse could lead to premature neurodegeneration by modifying not only the BBB but by altering the permeability of the blood–CSF barrier (BCSFB) and the blood–spinal cord barrier (BSCB) to proteins and other harmful agents as well. Obviously, breakdown of the blood–CNS barriers will allow many harmful agents to enter into the CNS microenvironment from the vascular compartments leading to development of vasogenic brain edema, abnormal gene expression in brain, and neuronal and nonneuronal cell death. These cascades initiated by drugs of abuse could lead to permanent alterations in the CNS structure and function causing mental anomalies and long-term functional disability of the victims.

I hope that these newly emerging concepts discussed in this volume for the first time will help in expanding our current understanding in the field of psychostimulants and brain pathology. In addition, the volume will provide new insights for further research to tackle these problems worldwide in minimizing the CNS damage that is the utmost need of our society currently. The ideas and facts presented in this volume will further help in stimulating new interest in the field to newcomers and provide the state-of-the-art knowledge to students and open new avenues for research and development in the subject to professionals. The volume will be a precious collection for the students, researchers, teachers, education and health providers, policy makers, and science planners in the field of neuroanatomy, neurosurgery, neurochemistry, neurophysiology, neuropharmacology, neuropsychiatry, neuropathology, and neuropsychopharmacology in the areas of substance abuse and its prevention in the future.

HARI SHANKER SHARMA
Uppsala University, Sweden

EFFECTS OF PSYCHOSTIMULANTS ON NEUROTROPHINS: IMPLICATIONS FOR PSYCHOSTIMULANT-INDUCED NEUROTOXICITY

Francesco Angelucci,*,§ Valerio Ricci,*,† Gianfranco Spalletta,*,‡
Carlo Caltagirone,*,‡ Aleksander A. Mathé,§ and Pietro Bria†

*Department of Clinical and Behavioural Neurology, IRCCS Santa Lucia Foundation,
Rome, Italy
†Institute of Psychiatry, Catholic University, Rome, Italy
‡Department of Neuroscience, Tor Vergata University, Rome, Italy
§Division of Psychiatry, Karolinska Institutet, Institution of Clinical Neuroscience,
Huddinge University Hospital, S-141 86 Huddinge, Sweden

It is well documented that psychostimulants may alter neuronal function and neurotransmission in the brain. Although the mechanism of psychostimulants is still unknown, it is known that these substances increase extracellular level of several neurotransmitters including dopamine (DA), serotonin, and norepinephrine by competing with monoamine transporters and can induce physical tolerance and dependence. In addition to this, recent findings also suggest that psychostimulants may damage brain neurons through mechanisms that are still under investigation.

In the recent years, it has been demonstrated that almost all psychostimulants are able to affect a class of proteins, called neurotrophins, in the peripheral and

INTERNATIONAL REVIEW OF
NEUROBIOLOGY, VOL. 88
DOI: 10.1016/S0074-7742(09)88001-X

1

central nervous system (CNS). Neurotrophins, such as nerve growth factor (NGF) and brain-derived neurotrophic factor (BDNF), have relevant action on neurons involved in psychostimulant action, such as DA and serotonergic neurons, and can play dual roles: first, in neuronal survival and death, and, second, in activity-dependent plasticity.

In this review, we will focalize on the effects of psychostimulants on this class of proteins, which may be implicated, at least in part, in the mechanism of the psychostimulant-induced neurotoxicity. Moreover, since altered neurotrophins may participate in the pathogenesis of psychiatric disorders and psychiatric disorders are common in drug users, one plausible hypothesis is that psychostimulants can cause psychosis through interfering with neurotrophins synthesis and utilization by CNS neurons.

I. Introduction

A psychostimulant is a substance that enhances locomotor behavior. The most widely used psychostimulant drugs include amphetamines, methamphetamine, cocaine, Methylphenidate (MPH) and other recreational drugs such as 3,4-Methylenedioxymethanphetamine (MDMA, "ecstasy"). Other possible psychostimulant compounds are cannabis, alcohol, morphine, and *gamma*-Hydroxybutyric acid (GBH). Although the exact molecular mechanism of psychostimulants remains unknown, it is known that these substances increase the extracellular level of several neurotransmitters including dopamine (DA), serotonin, and norepinephrine through distinct molecular mechanisms and can induce physical tolerance and dependence (Aldrich and Barker, 1976). In addition to this, recent findings in the field suggest that psychostimulants may damage our brain and spinal cord through mechanisms that are still under investigation.

In this review, we will focalize on the effects of psychostimulants on a class of proteins, called neurotrophins, which are involved in survival of neurons and may be implicated, at least in part, in the mechanism of the psychostimulant-induced neurotoxicity.

II. Psychostimulants

Many psychostimulants can be classified as releasers (i.e., amphetamine analogs) or uptake blockers (i.e., cocaine-like drugs) based on the mechanism of their acute effects on neurotransmitter flux through the dopamine transporter (DAT) and the vesicular monoamine transporter-2 (VMAT-2), a phenomenon likely related to the neurotoxic potential of these drugs to DAergic neurons.

A. AMPHETAMINES

"Amphetamines" is the generic term for amphetamine, methamphetamine, methylenedioxy congeners, and a number of other less commonly known substances. One of the best known is MDMA or ecstasy. The action of these substances in the central nervous system (CNS) is complex, with several molecular sites of action.

The molecular effects of amphetamines are mainly attributed to their binding to the DAT, resulting in both reuptake inhibition and increased release of DA at the mesocorticolimbic dopaminergic nerve terminals. Amphetamine and methamphetamine treatment causes long-term decreases in tyrosine hydroxylase activity, DAT function, decreases concentrations of associated neurotransmitters and metabolites (Buening and Gibb, 1974; Eisch et al., 1992; Hotchkiss and Gibb, 1980; Morgan and Gibb, 1980; Seiden et al., 1976). High-dose amphetamine administration causes persistent DA deficits (Ricaurte et al., 1983), whereas MDMA-induced deficits are typically selective against the serotonin system (Insel et al., 1989; Johnson et al., 1988). These deficits are thought to represent a neurotoxic response reflecting the destruction of corresponding monoamine axons terminals.

It is unclear whether increases in extraneuronal DA levels or a redistribution of intraneuronal DA are responsible for amphetamine effects. Early data suggested that extraneuronal DA might be of importance because pretreatment with DA postsynaptic receptor antagonists prevents the long-term dopaminergic deficits resulting from methamphetamine treatment (Sonsalla et al., 1986). In contrast, other evidence implicates changes in the disposition of intraneuronal DA. LaVoie and Hastings (1999) demonstrated a dissociation between extracellular DA concentrations, formation of DA oxidation products and long-term dopaminergic deficits suggesting that the toxicity of methamphetamine is not exclusively the result of increases in extracellular DA concentrations. In any case, changes in DA disposition resulting from administration of amphetamine analogs appear necessary for long-term toxicities. For example, DA depletion resulting from pretreatment with methyl-p-tyrosine attenuates the persistent DA deficits caused by methamphetamine treatment (Gibb and Kogan, 1979; Schmidt et al., 1985; Wagner et al., 1983).

B. Cocaine

Cocaine (benzoylmethyl ecgonine) is a crystalline tropane alkaloid obtained from the leaves of the coca plant. Cocaine causes an increase in DA by preventing the reuptake of newly released DA into striatal neurons via DAT (Heikkila *et al.*, 1975; Nicolaysen and Justice, 1988). Chronic cocaine administration can lead to a number of persisting alterations in behavior such as the long-lasting sensitization to DA agonists (Robinson and Becker, 1986) and development of paranoid psychoses (Satel *et al.*, 1991), as well as enduring alterations in brain chemistry, including alterations in a number of neurotransmitters and receptor populations (Kuhar and Pilotte, 1996).

The interaction of cocaine with DAT dramatically elevates extracellular DA, and this action is thought to be the initiating molecular event that reinforces drugseeking behaviors and addiction (Koob *et al.*, 2004). Other effects of cocaine are caused by the blockade of the serotonin and norepinephrine transporters that give sympathomimetics effects, whereas the blockade of sodium channels produces local anesthetic actions (Gissen *et al.*, 1980). Recent studies in rats also suggest that the serotonergic neurons from the raphe nuclei to limbic areas modulate the mesocorticolimbic DAergic-glutamatergic system (Bortolozzi *et al.*, 2005) and may have a role in neuropsychotoxicity induced by psychostimulants (Ago *et al.*, 2006; Müller *et al.*, 2007).

In conclusion, cocaine administration results in an increase in DA neurotransmission, which may be related to stimulant-induced psychosis, behavioral hyperactivity, and the reinforcing properties of the drug (Clarke *et al.*, 1988; Kelley and Iverson, 1975). These effects appear to be mediated primarily by the DA associated with the mesocorticolimbic circuitry.

C. Heroin and Opioid System

Heroin is a semisynthetic drug, obtained from acetylation of morphine and pertaining to the class of opioid drugs: natural opioids (morphine, codeine, tebanine), semisynthetic opioids (heroine oximorfon, idrossimorfon) and synthetic opioids such as methadone (Hutchinson and Somogyi, 2002; White and Irvine, 1999). Opiates exert their effects by binding with one or more of several specific types of ORs (Gold, 1993; Nestler, 1993): μ-receptors mediate analgesia, euphoria, physical dependence, κ-receptors for sedation analgesia and mitosis, δ-receptors for analgesia release of growth hormones and σ-receptor for hallucination, midriasis and respiratory and vasomotor stimulation.

Heroin possesses little or no opioid activity but its metabolism generates 6-monoacetylmorphine (6-MAM) and morphine, two μ-opioid receptor (MOR) agonists (White and Irvine, 1999). Stimulation of MORs localized on inhibitory

GABAergic neurons in the midbrain ventral tegmental area (VTA) leads to enhanced DAergic activity in rats (Johnson and North, 1992) and subsequent elevation of DA levels in projection terminals such as the nucleus accumbens (NAc) (Pontieri *et al.*, 1995; Spanagel *et al.*, 1990). In addition to the direct stimulation of MORs in the ventral striatum (NAc), indirect regulation of mesocorticolimbic DA neurotransmission has also been implicated in the pathophysiology of opioid addiction.

Heroin use may also have toxic effects. Computed tomography (CT) scans of chronic heroin abusers revealed cerebral atrophy (Cala and Mastaglia,1980; Pezawas *et al.*, 1988) with significantly smaller sulci and ventricle/brain indices larger than normal (Wolf and Mikhael, 1979). However, other studies showed no gross anatomical abnormalities (Rose *et al.*, 1996). On magnetic resonance imaging (MRI) scans, areas of demyelination in the white matter were described by Volkow *et al.* (1998) while other quantitative MRI studies were not able to detect specific differences between drug-dependent men and age-matched controls (Amass *et al.*, 1992; Rose *et al.*, 1996). Apoptotic cell death appears to be involved in the loss of neuronal function and citotoxicity induced by heroin and opioids. In fact, heroin induces apoptosis in neuronal-like rat pheochromocytoma (PC12) cells (Oliveira *et al.*, 2002) and morphine was described to induce apoptosis in neurons (Mao *et al.*, 2002) and microglia (Hu *et al.*, 2002).

D. Cannabis

The term cannabinoid comes from the chemical characterization of the major psychoactive component of marijuana, Δ-9-tetrahydrocannabinol or Δ-9-THC (Gaoni and Mechoulam, 1964). Δ-9-THC produces a variety of effects in the brain by acting on several neurotransmitters including DA, GABA, glutamate, serotonine, and noradrenaline (Howlett *et al.*, 2002, 2004).

It is known that cannabis action in the CNS is mediated by cannabinoid receptors abundant in the so-called "reward pathways" which is implicated in the neuropathophysiology of addiction. A specific receptor in the mammalian brain, identified as CB1, is localized in the substantia nigra pars reticulata, globus pallidus, cerebellum, and hippocampus (Devane *et al.*, 1988; Matsuda *et al.*, 1990). A second type of cannabinoid receptor, CB2 (Munro *et al.*, 1993), is present mainly in immune tissues and cells (Galiegue *et al.*, 1995). Studies performed in rats have demonstrated that CB-1R activation increases mesolimbic DA activity thus providing one explanation for the positive psychotic symptom induced by Δ-9-THC (Chen *et al.*, 1990, 1991; French, 1997; Melis *et al.*, 2000; Pistis *et al.*, 2002; Tanda *et al.*, 1997). In addition, cannabinoids have also been shown to influence glutamatergic synaptic transmission and plasticity in the prefrontal cortex favoring long-term depression (LTD) (Auclair *et al.*, 2000).

Cannabis use may also induce long-lasting changes in brain structures. Chronic exposure to Δ-9-THC or marijuana extracts persistently alters the structure and function of the rat hippocampus (Ameri, 1999), a brain region involved with learning and memory processes. In humans brain imaging studies evidenced that chronic cannabis use alters neuronal and axonal integrity in the dorsolateral prefrontal cortex (Hermann *et al.*, 2007).

III. Neurotrophins

Nerve growth factor (NGF), brain-derived neurotrophic factor (BDNF), neurotrophin-3 (NT-3), and neurotrophin-4/5 (NT-4/5) are derived from a common ancestral gene and are therefore collectively named neurotrophins (Hallbook, 1999). Neurotrophins regulate development, maintenance, and function of peripheral (PNS) and central (CNS) nervous systems. These proteins also regulate many aspects of neural function. In the mature nervous system, they control synaptic function and synaptic plasticity, while continuing to modulate neuronal survival. All neurotrophins have a similar overall structure, with a small number of amino acid differences between them, determining the specificity of receptor binding. The mature active forms of neurotrophins are very stable noncovalently associated homodimers, with molecular weights around 28 kDa. Dimerization seems to be an essential requisite for NT receptor activation, and is a feature characterizing other growth factors as well (Ibanez, 1998).

A. Nerve Growth Factor

NGF, the first member of the neurotrophin family to be characterized, was discovered during a search for such survival factors (cf review Levi-Montalcini, 1987). It is considered the prototypical growth factor and is a dimer of two identical polypeptide chains, each of 118 amino acid residues (McDonald and Blundell, 1991). NGF was initially purified as a factor able to support survival of sympathetic and sensory spinal neurons in culture (Levi-Montalcini, 1987). However, successively other relevant functions of NGF in the PNS and CNS were evidenced (Levi-Montalcini, 2004; Levi-Montalcini *et al.*, 1996).

Anti-NGF injections demonstrated that this factor is important in maintaining survival of sympathetic neurons *in vivo* as well as *in vitro*. In the PNS, NGF is synthesized and secreted by sympathetic and sensory target organs (Korsching, 1993). From these sources, it is captured in nerve terminals by receptor-mediated endocytosis and is transported through axons to neuronal cell bodies where it acts to promote neuronal survival and differentiation. Within the target organs, synthesis of NGF and of other neurotrophins is associated with peripheral tissues such as cutaneous tissues, internal organs, and hair follicles, which become innervated by sensory and sympathetic neurons.

In contrast to the periphery, NGF expression in the CNS is much more restricted; NGF mRNA and protein are expressed in a number of brain regions, with the hippocampus providing the single largest source in the entire CNS (Korsching et al., 1985). In the hippocampus, NGF RNA and protein are expressed by the principal excitatory (glutamate) neurons, as well as by a subset of T-aminobutyric acid (GABA)-containing inhibitory neurons (Rocamora et al., 1996). These hippocampal target cells receive rich innervation from ascending neurons with their cell bodies in the basal forebrain.

B. Brain-Derived Neurotrophic Factor

BDNF was purified from pig brain, due to its survival-promoting action on a subpopulation of dorsal root ganglion neurons (Barde et al., 1982). The amino acid sequence of mature BDNF has a strong homology with that of NGF (Leibrock et al., 1989; Rosenthal et al., 1991). As is the case for NGF, BDNF is necessary for survival of some peripheral sensory neurons, notably those in the vestibular ganglia and nodose-petrosum ganglia. Interestingly, part of the trophic effects of BDNF in the PNS seems to depend on autocrine loops and paracrine interactions between adjacent neurons, since sensory neurons can express both BDNF and its high-affinity receptor TrkB.

BDNF is more highly expressed and widely distributed than NGF in the CNS, and has survival-promoting actions on a variety of CNS neurons including hippocampal and cortical neurons (Ghosh et al., 1994; Lindholm et al., 1996), cholinergic neurons (Alderson et al., 1990), and nigral dopaminergic neurons (Hyman et al., 1991). These facts raised keen interest in BDNF as a potential therapeutic agent for Parkinson's disease and Alzheimer's disease, among other neurodegenerative disorders and nondegenerative pathologies. BDNF can be anterogradely transported in the CNS, a fact that has considerably expanded the concept of neuronal-derived trophic support, and sustains the hypothesis that BDNF can have actions at the synaptic level (Altar and Di Stefano, 1998).

C. Neurotrophin Receptors

The biological actions of neurotrophins are mediated by two classes of membrane receptors (Barbacid, 1995; Bothwell, 1996; Chao et al., 1998; Dechant, 2001): three highly related receptor protein-tyrosine kinases, known as high-affinity receptors (TrkA, TrkB, and TrkC), and a protein named as neurotrophin receptor p75 (p75NTR), a member of the TNF receptor superfamily.

The mature form of neurotrophins binds with distinct selectivities to the Trk family of receptor tyrosine kinases. NGF specifically binds to TrkA, while BDNF to TrkB, and NT-3 binds preferentially to TrkC, but also to TrkB and TrkA with lower efficacy (Kaplan and Miller, 1997; Klein, 1994). Through this receptors, neurotrophins activate many signaling pathways, including those mediated by RAS and members of the cdc-42/ras/rho G protein families, and the MAP kinase, PI-3 kinase, and Jun kinase cascades. Within neural precursors and neurons, the pathways regulated by Trk receptors include survival and differentiation, axonal and dendritic growth, and remodeling, assembly of the cytoskeleton, membrane trafficking and fusion, and synapse formation and function.

The mechanisms of transduction mediating the biological effects of p75NTR in neurons are poorly understood. On the one hand, p75NTR can modulate cellular responses to neurotrophins, by interacting with their high-affinity TrkA. Modulation of TrkA interaction with NGF has been considered as the main p75NTR mechanism of action since the discovery of Trk receptors (Barbacid, 1995; Bothwell, 1996; Chao and Hempstead, 1995). However, it has been now clearly established that p75NTR can induce cellular responses in the absence of Trk receptors, such as apoptotic cell death, when it is activated by the immature form of neurotrophins (Pro-neurotrophins) (Beattie *et al.*, 2002; Harrington *et al.*, 2004).

IV. Neurotrophins and Psychostimulants

A. NGF AND PSYCHOSTIMULANTS

Table I includes all the data reported so far in human and animal models on NGF and psychostimulant drugs.

The first experimental evidence came in 1979 when it was demonstrated that NGF may modulate the response to D-amphetamine in adult male rats with 6-hydroxydopamine-induced lesions of the NAc (Lewis *et al.*, 1979). More recently, it has been shown that amphetamine administration may directly reduce NGF synthesis in selected regions of the rat brains, such as hippocampus, occipital cortex, and hypothalamus (Angelucci *et al.*, 2007a). Conversely, NGF is able to rescue the damage induced by amphetamine administration in dopaminergic areas of the rat brain (Chaturvedi *et al.*, 2006). In rat PC12 cells, which differentiate in presence of NGF, it has been shown that repeated, intermittent treatment with amphetamine may also have a neurotrophic action since it induces neurite outgrowth and this effect is not blocked by pretreatment with NGF antibody or DA receptor antagonists (Park *et al.*, 2002).

TABLE I
DATA REPORTED IN HUMAN AND ANIMAL MODELS ON NGF AND PSYCHOSTIMULANT DRUGS

Psychostimulant	Human data	Animal data
Amphetamines		Modulation of amphetamine response by NGF in rats with with 6-hydroxydopamine lesions (Lewis *et al.*, 1979) Reduced NGF levels in the hippocampus, occipital cortex, and hypothalamus (Angelucci *et al.*, 2007a) NGF rescues amphetamine-induced DA lesions (Chaturvedi *et al.*, 2006) Amphetamine-induced stimulation of neurite outgrowth in PC12 cells (Park *et al.*, 2002)
Cocaine	Reduced NGF serum levels (Angelucci *et al.*, 2007b)	Inhibition of neuronal differentiation in NGF-stimulated PC12 cells (Zachor *et al.*, 2000a,b) NGF reverses cocaine-induced inhibition of PC12 proliferation (Tosk *et al.*, 1996)
Heroin and opioids	Reduced NGF serum levels (Angelucci *et al.*, 2007b)	Disruption of normal development of striatal cholinergic neurons and reduction of striatal NGF in rats (Wu *et al.*, 2001)
Cannabis	Reduced NGF serum levels (Angelucci *et al.*, 2008) Increased NGF serum levels in schizophrenic patients (Jockers-Scherübl *et al.*, 2003)	

DA = dopamine; PC12 = pheochromocytoma cells. Further description of experimental data is given in the text.

Few data on cocaine and heroin effects on NGF are available, so far. It is known that cocaine inhibits neuronal differentiation in a dose-dependent fashion in the NGF-stimulated PC12 cells through modulation of D(1)-type DA receptors (Zachor *et al.*, 2000a,b) while NGF itself may partially reverse this cocaine-induced inhibition of proliferation (Tosk *et al.*, 1996). In humans, chronic cocaine abuse may lead to a consistent reduction in NGF peripheral serum levels (Angelucci *et al.*, 2007b). Heroin also reduces serum NGF levels in humans (Angelucci *et al.*, 2007b) while administration of the synthetic opioid methadone, a standard method of managing heroin addicted patients, has been shown to cause disruption of normal development of striatal cholinergic neurons and reduction of striatal NGF content in the rat brain (Wu *et al.*, 2001).

Cannabis use also raised NGF serum levels in a population of drug free schizophrenic patients (Jockers-Scherübl *et al.*, 2003) and, at present, we have evidence that chronic cannabis abuse produces a reduction of NGF serum levels in normal nonpsychotic group of patients as compared to healthy subjects (Angelucci *et al.*, 2008).

B. BDNF AND PSYCHOSTIMULANTS

Data on the effects of psychostimulant drugs on BDNF in human and animal models are reported in Table II.

High concentrations of BDNF have been found in plasma of chronic methamphetamine users (Kim *et al.*, 2005). Moreover, an association between the BDNF polymorphism Val66Met and amphetamine response has been reported: amphetamine produced less pronounced self-ratings of arousal and energy in the val/met and met/met compared to the val/val group (Flanagin *et al.*, 2006). Another study reported no differences in allele frequency among Japanese amphetamine users and healthy subjects (Itoh *et al.*, 2005). Amphetamine i.p. administration in rats may result in either reduced BDNF protein (Angelucci *et al.*, 2007a) or increased BDNF mRNA (Meredith *et al.*, 2002) in the cortex, hypothalamus, and amigdala. In addition, after cortical contusion injury in rats, amphetamine treatment increases hippocampal BDNF (Griesbach *et al.*, 2008). Moreover, the release of DA and DA-related behaviors induced by methamphetamine are blocked by pretreatment with intra-NAc injection of either BDNF or TrkB antibody, providing evidence that BDNF/TrkB pathway is implicated in the mechanism of action of amphetamines (Martin-Iverson *et al.*, 1994; Narita *et al.*, 2003).

In adult rats, repeated cocaine injection causes an upregulation of BDNF mRNA and protein in the prefrontal cortex (Fumagalli *et al.*, 2007) while BDNF infusion in the same region suppresses cocaine-seeking indicating that this neurotrophin regulates the cortical pathway implicated in drug seeking (Berglind *et al.*, 2007; Pu *et al.*, 2006). BDNF mRNA in the hippocampus increases in rats subjected to cocaine withdrawal (Filip *et al.*, 2006). For these reasons a role of BDNF in cocaine reward has been postulated (Schoenbaum *et al.*, 2007).

Cocaine exposure during pregnancy causes a reduction of BDNF in the rat hippocampus (Yan *et al.*, 2004). This reduction of BDNF may make neurons more vulnerable to cocaine's toxic effects and precipitate cocaine-induced CNS damages, as also demonstrated by RNA-interference studies in cultured cells (Yan *et al.*, 2007).

Chronic heroin abusers are characterized by low levels of circulating BDNF (Angelucci *et al.*, 2007b). Furthermore, significant differences in BDNF Val66Met genotype distribution were found between subjects dependent on methamphetamine or heroin and controls. In the heroin-dependent group, the Val/Val homozygotes had a later onset of substance abuse compared with the Met allele carriers, suggesting that the BDNF Val66Met polymorphism is associated with substance abuse (Cheng *et al.*, 2005).

TABLE II
DATA REPORTED IN HUMAN AND ANIMAL MODELS ON BDNF AND PSYCHOSTIMULANT DRUGS

Psychostimulant	Human data	Animal data
Amphetamines	Increased BDNF plasma levels (Kim *et al.*, 2005) Association between Val66Met BDNF polymorphism and amphetamine response (Flanagin *et al.*, 2006)	Reduced BDNF levels in the cortex and hypothalamus (Angelucci *et al.*, 2007a) Increased BDNF mRNA in the cortex, hypothalamus and amigdala (Meredith *et al.*, 2002) Blockade of methamphetamine-induced striatal DA release by BDNF antibody (Martin-Iverson *et al.*, 1994; Narita *et al.*, 2003) Increased hippocampal BDNF after cortical contusion injury in rats (Griesbach *et al.*, 2008)
Cocaine		Upregulation of BDNF mRNA in the prefrontal cortex after repeated cocaine injections (Fumagalli *et al.*, 2007) Suppression of cocaine seeking by BDNF infusion in the prefrontal cortex (Berglind *et al.*, 2007; Pu *et al.*, 2006) Increased hippocampal BDNF mRNA in rats during cocaine withdrawal (Filip *et al.*, 2006) Reduced BDNF levels in the rat hippocampus after prenatal cocaine exposure (Yan *et al.*, 2004)
Heroin	Reduced BDNF serum levels (Angelucci *et al.*, 2007b) Association between Val66Met BDNF polymorphism and later onset of heroin abuse (Cheng *et al.*, 2005)	
Cannabis	Increased BDNF serum concentrations in drug-naive first-episode schizophrenic patients (Jockers-Scherübl *et al.*, 2004)	

DA = dopamine. Further description of experimental data is given in the text.

Finally, in a population of drug-naive first-episode schizophrenic patients it was found significantly elevated BDNF serum concentrations (by up to 34%) in patients with chronic cannabis abuse prior to disease onset while schizophrenic patients without cannabis consumption showed similar results to normal controls (Jockers-Scherübl *et al.*, 2004).

V. Psychostimulants, Neurotoxicity, and Neurotrophins

A large body of evidence indicates that psychostimulants alter neuronal function and neurotransmission by competing with monoamine transporters, such as DAT and VMAT-2. Since DAT and VMAT-2 are critical regulators of DA disposition within the brain, chronic treatment with psychostimulants leads to the enhancement of the mesocorticolimbic dopaminergic neurons from the VTA to the NAc and the medial prefrontal cortex (mPFC) and leads to abnormal glutamatergic function from the mPFC to the NAc and VTA. Heroin and opioids also inhibit the inhibitory GABAergic interneurons in the VTA through MOR activation and subsequently activate the mesocorticolimbic dopaminergic neurons.

The neural adaptation of DAergic-glutamatergic system following psychostimulant exposure is critically implicated in neuropsychotoxic effects of these drugs. An excess of DA in these neuronal circuits may damage the neurons and lead to neuronal death, as demonstrated by numerous *in vitro* and *in vivo* studies concerning DA- or l-DOPA-induced neurotoxicity (see for review Asanuma *et al.*, 2003). It is generally believed that the reactive oxygen or nitrogen species generated in the enzymatical oxidation or auto-oxidation of an excess amount of DA induce neuronal damage and/or apoptotic or nonapoptotic cell death (Asanuma *et al.*, 2003). However, it has become evident that other mechanisms of action of psychostimulants are responsible for their neurotoxic effects. For example, recent studies suggest that the serotonergic neurons from the raphe nuclei to limbic areas modulate the mesocorticolimbic DAergic-glutamatergic system and participate in the neuropsychotoxicity induced by psychostimulants.

In addition to these findings, in the recent years, it has been demonstrated that almost all psychostimulants are able to affect neurotrophins in the PNS and CNS. Neurotrophins, such as NGF and BDNF, have relevant action on neurons involved in psychostimulant action such as DA and serotonergic neurons, and can play dual roles: first, in neuronal survival and death, and, second, in activity-dependent plasticity.

Nigral dopaminergic neurons and striatal neurons express TrkB mRNA and protein (Altar *et al.*, 1994; Merlio *et al.*, 1993; Numan and Seroogy, 1999) in both rodents and humans (Alien *et al.*, 1994; Benisty *et al.*, 1998; Nishio *et al.*, 1998) and BDNF has been shown to provide neuroprotection to these neurons. In fact BDNF prevents the spontaneous death of dopaminergic neurons in rat primary mesencephalic cultures (Hyman *et al.*, 1994; Kniisel *et al.*, 1997) and protects TH-immunoreactive neurons from the selective toxin MPP^+(l-methyl-4-phenylpyridinium) (Hyman *et al.*, 1991). In summary, there is a body of work suggesting that exogenous BDNF administration can increase survival and/or promote recovery of injured mesencephalic dopaminergic neurons, at least in some experimental conditions. NGF also protects dopaminergic neurons following neurotoxin injury (Garcia *et al.*, 1992; Hirata *et al.*, 2006) and these effects appear to be mediated by dopaminergic neurons expressing NGF-receptors (Melchior *et al.*, 2003; Nishio *et al.*, 1998). Moreover, reductions in NGF level were observed in animal models of Parkinson's disease and in human parkinsonian patients, showing a relationship between this neurotrophic factor and the neurodegenerative changes observed in Parkinson's disease (Lorigados *et al.*, 1996; Müller *et al.*, 2005).

BDNF has also potent neurotrophic effects on serotonin (5-HT) neurons when infused into midbrain (Siuciak *et al.*, 1994), resulting in a dramatic elevation of 5-HT neuronal fiber density and protection of neurons from neurotoxic damage (Mamounas *et al.*, 1995). At the same time, BDNF expression can also be regulated by serotonin. For example, BDNF mRNA is decreased in the rat hippocampus and increased in the frontal cortex by treatment with serotonin agonists suggesting that 5-HT modulates BDNF mRNA levels in the rat brain (Nibuya *et al.*, 1996; Vaidya *et al.*, 1997; Zetterstròm *et al.*, 1999). It has been reported that 45% of the serotonergic raphe neurons also express both TrkA (Gibbs and Pfaff, 1994; Sobreviela *et al.*, 1994) and p75NTR (Koh *et al.*, 1989), although a functional relationship between NGF and serotonergic neurons has not yet been shown.

Interestingly, interaction of neurotrophins with the DAT has also been reported. The presence of BDNF Val66Met polymorphism has been associated with a variable number of tandem repeat polymorphism of the dopamine transporter gene (DAT VNTR) on generating individual differences in anxiety-related traits in a sample of healthy volunteers (Hünnerkopf *et al.*, 2007). In addition, in rat dorsal striatal synaptosomes, it has been demonstrated that inhibition of tyrosine kinases results in a rapid concentration-dependent decrease in [(3)H] DA uptake because of a reduction in maximal [(3)H]DA uptake velocity and DAT cell surface expression, while acute activation of TrkB by BDNF significantly increase [(3)H]DA uptake (Hoover *et al.*, 2007). Taken together, these results suggest that striatal DAT function and cell surface expression is constitutively upregulated by tyrosine kinase activation and that neurotrophins can mediate this type of regulation.

These observations suggest that the neurotrophin family plays key roles in determining the fate of the neuron, and therefore is likely to be involved in the toxicity of psychostimulants during brain development or in the adult life. Although acting in many different ways, psychostimulants may reduce NGF and BDNF in the PNS and CNS, as summarized in Tables I and II. However, in some instances, an acute treatment with psychostimulants may also cause an increase of neurotrophin levels or their mRNAs and even produce neurotrophic-like effects in cell cultures, as shown by the effects of intermittent dose of amphetamine on PC12 cells. These opposite observations are difficult to interpret. One hypothesis is that an acute or intermittent injury to CNS neurons may induce a series of

Putative scheme of psychostimulant-induced neurotoxicity

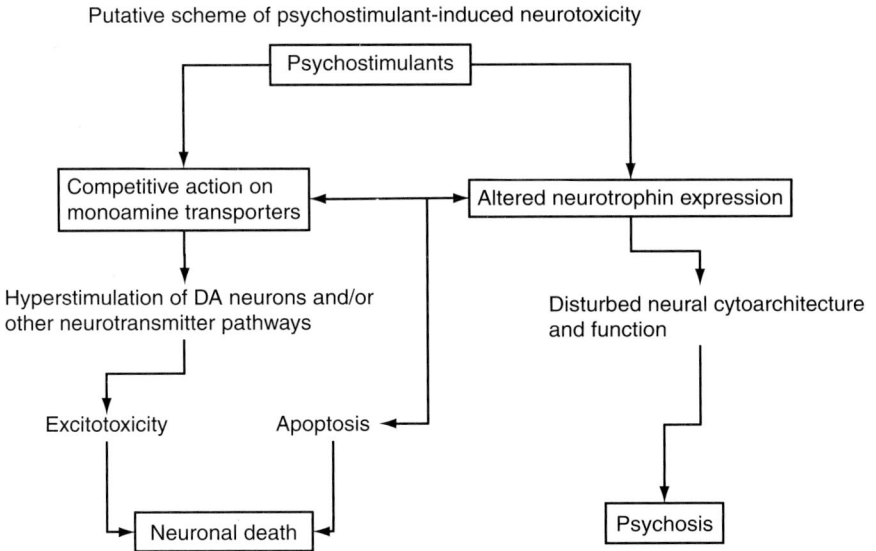

Fig. 1. Putative scheme of psychostimulant-induced neurotoxicity. Psychostimulants enhance DA and/or other monoamine neurotransmission by competing with monoamine transporters. This phenomenon leads to an hyperstimulation of neurons which may result in excitoxic death. Changes in neurotrophin production may result either from a direct action of psychostimulants on neurotrophin-producing neurons or as a consequence of neuronal stimulation. In both cases a reduced long-lasting trophic support to neurons may lead to apoptotic death or cytoarchitectonic rearrangement which predispose the brain to develop psychiatric disturbances.

transient changes leading not only to death of neurons, but also to spontaneous rearrangement of the affected network. One of such proplastic events, detected following injury, is indeed an increased level of neurotrophins.

In any case, these data indicate that psychostimulants can alter neurotrophin synthesis and utilization by neurons through a direct action or as consequence of the induced neuronal stimulation through monoamine transporters. As a consequence, neuronal death may occur because of lack of neurotrophins (apoptotic) or due to mechanisms of neurotransmitter-induced excitotoxicity, as postulated in Fig. 1.

VI. Psychostimulants, Neurotrophins, and Psychosis

Alterations in DA disposition cannot only lead to drug addiction but also give raise to pathological conditions such as Parkinson's disease and schizophrenia, a fact that underscores the importance of DAergic neurotransmission.

Contemporary revisions of the DA hypothesis suggest that schizophrenia symptoms emerge from a functional hyperactivity of DA neurons projecting to the NAc, associated with functional hypoactivity of DA neurons projecting to the frontal cortex (Davis *et al.*, 1991; Goldstein and Deutch, 1992). Other revisions suggest that psychosis and thought disorder may result, in part, from a state of abnormal glutamatergic cortical activity associated with exaggerated DA release or dysregulated DA signaling in the NAc (Csernansky and Bardgett, 1998). The primary addiction hypothesis incorporates evidence of hyperactivity in NAc DA signaling as it relates both to psychotic symptoms and addictive behavior. The ability of typical antipsychotics to attenuate behavioral sensitization and reinforcement produced by psychostimulants agrees with this idea (Richardson *et al.*, 1994; Roberts and Vickers, 1987).

Interestingly, prolonged use of amphetamine, methamphetamine, and cocaine may result in a form of psychosis that is indistinguishable from paranoid type schizophrenia (Pierce and Kalivas, 1997).

It has been also postulated that a lack of neurotrophins is one of the factors contributing to the pathogenesis of schizophrenia (Angelucci *et al.*, 2004; Durany and Thome, 2004). Preclinical and clinical data indicate that dysfunctions of NGF, BDNF, and NT-3 may contribute to impaired brain development, neuroplasticity, and synaptic "dysconnectivity" leading to the schizophrenic syndrome, or at least some of its presentations (Angelucci *et al.*, 2004, 2005; Shoval and Weizman, 2005). Supporting this hypothesis, some studies have demonstrated that schizophrenic patients are characterized by low serum levels of NGF and BDNF (Parikh *et al.*, 2003; Tan *et al.*, 2005; Toyooka *et al.*, 2002) as compared with those of healthy subjects, although other authors report no significant changes (Huang and Lee, 2006; Shimizu *et al.*, 2003). In addition, a consistent number of human (Buckley *et al.*, 2007) and rodent studies (Angelucci *et al.*, 2004) indicate that the beneficial effects of antipsychotic drugs are mediated, at least in part, through BDNF and its receptor, TrkB.

Thus, in view of the agonist action of psychostimulants, either blockers or releasers, on dopaminergic neurons and given the possible role of neurotrophins in the pathogenesis of psychiatric disorders, one plausible hypothesis is that a reduced production of neurotrophins could be a contributing factor for the pathogenesis of psychiatric disorders in chronic drug users.

VII. Conclusions

In summary, a series of experimental evidences indicates that psychostimulants alter dopaminergic and other monoamine (serotonergic) neurotransmission through competition with monoamine transporters. At the same time, it is also clear that neurotrophins may be altered by administration of psychostimulants. Given the role of neurotrophins in neuronal survival and plasticity it is proposed that one of the mechanisms of psychostimulant-induced neurotoxicity is linked to alterations of neurotrophins. Moreover, since altered neurotrophins may be part of the pathogenesis of psychiatric disorders and psychiatric disorders are common in drug users one hypothesis is that psychostimulants can cause psychosis through interfering with neurotrophins synthesis and utilization by CNS neurons. One implication of this hypothesis is that drugs that prevent psychostimulant neurotoxicity have the potential to be useful in the treatment of schizophrenia, as well as other neurodegenerative disorders.

References

Ago, Y., Nakamura, S., Uda, M., Kajii, Y., Abe, M., Baba, A., and Matsuda, T. (2006). Attenuation by the 5-HT1A receptor agonist osemozotan of the behavioral effects of single and repeated methamphetamine in mice. *Neuropharmacology* **51,** 914–922.

Alderson, R. F., Alterman, A. L., Barde, Y. A., and Lindsay, R. M. (1990). Brain-derived neurotrophic factor increases survival and differentiated functions of rat septal cholinergic neurons in culture. *Neuron* **5,** 297–306.

Aldrich, M., and Barker, R. (1976). Cocaine: Chemical, Biological, Social and Treatment Aspects. CRC Press, Cleveland, Oh.

Alien, S. J., Dawbarn, D., Eckford, S. D., Wilcock, G. K., Ashccroft, M., Colebrook, S. M., Feeney, R., and MacGowan, S. H. (1994). Cloning of a non-catalytic forni of human TrkB and distribution of messenger RNA for TrkB in human brain. *Neuroscience* **60,** 825–834.

Altar, C. A., and Di Stefano, P. S. (1998). Neurotrophin trafficking by anterograde transport. *Trends Neurosci.* **21,** 433–437.

Altar, C. A., Siuciak, J. A., Wright, P., Ip, N. Y., Lindsay, R. M., and Wiegand, S. J. (1994). *In situ* hybridization of TrkB and TrkC mRNA in rat forebrain and association with high-affinity binding of [^{125}I]BDNF, [^{125}I]NT-4/5 and [125I]NT-3. *Eur. J. Neurosci.* **6,** 1389–1405.

Amass, L., Nardin, R., Mendelson, J. H., Teoh, S. K., and Woods, B. T. (1992). Quantitative magnetic resonance imaging in heroin- and cocaine-dependent men: A preliminary study. *Psychiatry Res.* **45,** 15–23.

Ameri, A. (1999). The effects of cannabinoids on the brain. *Prog. Neurobiol.* **58,** 315–348.

Angelucci, F., Mathé, A. A., and Aloe, L. (2004). Neurotrophic factors and CNS disorders: Findings in rodent models of depression and schizophrenia. *Prog. Brain Res.* **146,** 151–165.

Angelucci, F., Brenè, S., and Mathé, A. A. (2005). BDNF in schizophrenia, depression and BDNF in schizophrenia, depression and corresponding animal models. *Mol. Psychiatry* **10,** 345–352 (Review).

Angelucci, F., Gruber, S. H., El Khoury, A., Tonali, P. A., and Mathé, A. A. (2007a). Chronic amphetamine treatment reduces NGF and BDNF in the rat brain. *Eur. Neuropsychopharmacol.* **17,** 756–762.

Angelucci, F., Ricci, V., Pomponi, M., Conte, G., Mathé, A. A., Tonali, P. A., and Bria, P. (2007b). Chronic heroin and cocaine abuse is associated with decreased serum concentrations of the nerve growth factor and brain-derived neurotrophic factor. *J. Psychopharmacol.* **21,** 820–825.

Angelucci, F., Ricci, V., Spalletta, G., Pomponi, M., Tonioni, F., Caltagirone, C., and Bria, P. (2008). Reduced serum concentrations of nerve growth factor, but not brain-derived neurotrophic factor, in chronic cannabis abusers. *Eur. Neuropsychopharmacol.* **18,** 882–887.

Asanuma, M., Miyazaki, I., and Ogawa, N. (2003). Dopamine- or L-DOPA-induced neurotoxicity: The role of dopamine quinone formation and tyrosinase in a model of Parkinson's disease. *Neurotox. Res.* **5,** 165–176.

Auclair, N., Otani, S., Soubrie, P., and Crepel, F. (2000). Cannabinoids modulate synaptic strength plasticity at glutamatergic synapses of rat prefrontal cortex pyramidal neurons. *J. Neurophysiol.* **83,** 3287–3293.

Barbacid, M. (1995). Structural and functional properties of the TRK family of neurotrophin receptors. *Ann. NY Acad. Sci.* **766,** 442–458.

Barde, Y. A., Edgar, D., and Thoenen, H. (1982). Purification of a new neurotrophic factor from mammalian brain. *Embo J.* **1,** 549–553.

Beattie, M. S., Harrington, A. W., Lee, R., Kim, J. Y., Boyce, S. L., Longo, F. M., Bresnahan, J. C., Hempstead, B. L., and Yoon, S. O. (2002). ProNGF Induces p75-mediated death of oligodendrocytes following spinal cord injury. *Neuron* **36,** 375–386.

Benisty, S., Boisiere, F., Faucheux, B., Agid, Y., and Hirsch, E. C. (1998). TrkB messenger RNA expression in normal human brain and in the substantia nigra of parkinsonian patients: An *in situ* hybridization study. *Neuroscience* **86,** 813–826.

Berglind, W. J., See, R. E., Fuchs, R. A., Ghee, S. M., Whitfield, T. W., Jr., Miller, S. W., and McGinty, J. F. A. (2007). BDNF infusion into the medial prefrontal cortex suppresses cocaine seeking in rats. *Eur. J. Neurosci.* **26,** 757–766.

Bortolozzi, A., Diaz-Mataix, L., Scorza, M. C., Celada, P., and Artigas, F. (2005). The activation of 5-HT receptors in prefrontal cortex enhances dopaminergic activity. *J. Neurochem.* **95,** 1597–1607.

Bothwell, M. (1996). p75NTR: A receptor after all. *Science.* **272,** 506–507.

Buckley, P. F., Mahadik, S., Pillai, A., and Terry, A., Jr. (2007). Neurotrophins and schizophrenia. *Schizophr Res.* **94,** 1–11.

Buening, M. K., and Gibb, J. W. (1974). Influence of methamphetamine and neuroleptic drugs on tyrosine hydroxylase activity. *Eur. J. Pharmacol.* **26,** 30–34.

Cala, L. A., and Mastaglia, F. L. (1980). Computerized axial tomography in the detection of brain damage: 1. Alcohol, nutritional deficiency and drugs of addiction. *Med. J. Aust.* **2,** 193–198.

Chao, M., Casaccia-Bonnefil, P., Carter, B., Chittka, A., Kong, H., Yoon, S.O. (1998). Neurotrophin receptors: Mediators of life and death. *Brain Res. Brain Res. Rev.* **26,** 295–301.

Chao, M. V., and Hempstead, B. L. (1995). p75 and Trk: A two-receptor system. *Trends in Neuroscience.* **18,** 321–326.

Chaturvedi, R. K., Shukla, S., Seth, K., and Agrawal, A. K. (2006). Nerve growth factor increases survival of dopaminergic graft, rescue nigral dopaminergic neurons and restores functional deficits in rat model of Parkinson's disease. *Neurosci. Lett.* **398,** 44–49.

Cheng, C. Y., Hong, C. J., Yu, Y. W., Chen, T. J., Wu, H. C., and Tsai, S. J. (2005). Brain-derived neurotrophic factor (Val66Met) genetic polymorphism is associated with substance abuse in males. *Brain Res. Mol. Brain Res.* **140,** 86–90.

Chen, J. P., Paredes, W., Li, J., Smith, D., Lowinson, J., and Gardner, E. L. (1990). Delta 9-tetrahydrocannabinol produces naloxoneblockable enhancement of presynaptic basal dopamine efflux in nucleus accumbens of conscious freely-moving rats as measured by intracerebral microdialysis. *Psychopharmacology* **102,** 156–162.

Chen, J. P., Paredes, W., Lowinson, J. H., and Gardner, E. L. (1991). Strainspecific facilitation of dopamine efflux by delta 9-tetrahydrocannabinol in the nucleus accumbens of rat: An *in vivo* microdialysis study. *Neurosci. Lett.* **129,** 136–180.

Clarke, P. B. S., Jakubovic, A., and Fibiger, H. C. (1988). Anatomical analysis of the involvement of mesolimbocortical dopamine in the locomotor stimulant actions of D-amphetamine and apomorphine. *Psychopharmacology* **96,** 511–520.

Csernansky, J. G., and Bardgett, M. E. (1998). Limbic-cortical neuronal damage and the pathophysiology of schizophrenia. *Schizophr. Bull.* **24,** 238–248.

Davis, K. L., Kahn, R. S., Ko, G., and Davidson, M. (1991). Dopamine and schizophrenia: A review and reconceptualization. *Am. J. Psychiatry* **148,** 1474–1486.

Dechant, G. (2001). Molecular interactions between neurotrophin receptors. *Cell Tissue Res.* **305,** 229–238.

Devane, W. A., Dysarz, F. A., III, Johnson, M. R., Melvin, L. S., and Howlett, A. C. (1988). Determination and characterization of a cannabinoid receptor in rat brain. *Mol. Pharmacol.* **34,** 605–613.

Durany, N., and Thome, J. (2004). Neurotrophic factors and the pathophysiological of schizophrenic psychoses. *Eur. Psychiatr.* **19,** 326–337.

Eisch, A. J., Gaffney, M., Weihmuller, F. B., O'Dell, S. J., and Marshall, J. F. (1992). Striatal subregions are differentially vulnerable to the neurotoxic effects of methamphetamine. *Brain Res.* **598,** 321–326.

Filip, M., Faron-Górecka, A., Kuśmider, M., Gołda, A., Frankowska, M., and Dziedzicka-Wasylewska, M. (2006). Alterations in BDNF and trkB mRNAs following acute or sensitizing cocaine treatments and withdrawal. *Brain Res.* **107,** 218–225.

Flanagin, B. A., Cook, E. H., Jr., and de Wit, H. (2006). An association study of the brain-derived neurotrophic factor Val66Met polymorphism and amphetamine response. *Am. J. Med. Genet. B. Neuropsychiatr. Genet.* **141,** 576–583.

French, E. D. (1997). Delta9-Tetrahydrocannabinol excites rat VTA dopamine neurons through activation of cannabinoid CB1 but not opioid receptors. *Neurosci. Lett.* **226,** 159–162.

Fumagalli, F., Di Pasquale, L., Caffino, L., Racagni, G., and Riva, M. A. (2007). Repeated exposure to cocaine differently modulates BDNF mRNA and protein levels in rat striatum and prefrontal cortex. *Eur. J. Neurosci.* **26,** 2756–2763.

Galiegue, S., Mary, S., Marchand, J., Dussossoy, D., Carriere, D., Carayon, P., Bouaboula, M., Shire, D., Le Fur, G., and Casellas, P. (1995). Expression of central and peripheral cannabinoid receptors in human immune tissues and leukocyte subpopulations. *Eur. J. Biochem.* **232,** 54–61.

Gaoni, Y., and Mechoulam, R. (1964). Isolation, structure, and partial synthesis of an active constituent of hashish. *J. Am. Chem. Soc.* **86,** 1646–1647.

Garcia, E., Rios, C., and Stelo, J. (1992). Ventricular injection of nerve growth factor increases dopamine content in the striata of MPTP-treated mice. *Neurochem. Res.* **17,** 979–982.

Ghosh, A., Carnahan, J., and Greenberg, M. E. (1994). Requirement for BDNF in activity-dependent survival of cortical neurons. *Science* **263,** 1618–1623.

Gibb, J. W., and Kogan, F. J. (1979). Influence of dopamine synthesis on methamphetamine-induced changes in striatal and adrenal tyrosine hydroxylase activity. *Naunyn-Schmiedeberg's Arch. Pharmacol.* **310,** 185–187.

Gibbs, R. B., and Pfaff, D. W. (1994). *In situ* hybridization detection of trkA mRNA in brain: Distribution, colocalization with p75NGFR and up-regulation by nerve growth factor. *J. Comp. Neurol.* **341,** 324–339.

Gissen, A. J., Covino, B. J., and Gregus, J. (1980). Differential sensitivities of mammalian nerve fibers to local anesthetic agents. *Anesthesiology* **53,** 467–474.

Gold, M. S. (1993). Opiate addiction and the locus coeruleus. The clinical utility of clonidine, naltrexone, methadone, and buprenorphine. *Psychiatr. Clin. North. Am.* **16,** 61–73.

Goldstein, M., and Deutch, A. Y. (1992). Dopaminergic mechanisms in the pathogenesis schizophrenia. *Faseb. J.* **6,** 2413–2421.

Griesbach, G. S., Hovda, D. A., Gomez-Pinilla, F., and Sutton, R. L. (2008). Voluntary exercise or amphetamine treatment, but not the combination, increases hippocampal brain-derived neuro-trophic factor and synapsin I following cortical contusion injury in rats. *Neuroscience* **154,** 530–540.

Hallbook, F. (1999). Evolution of the vertebrate neurotrophin and Trk receptor gene families. *Curr. Opin. Neurobiol.* **9,** 616–621.

Harrington, A. W., Leiner, B., Blechschmitt, C., Arevalo, J. C., Lee, R., Mörl, K., Meyer, M., Hempstead, B. L., Yoon, S. O., and Giehl, K. M. (2004). Secreted proNGF is a pathophysiological death-inducing ligand after adult CNS injury. *Proc. Natl. Acad. Sci. USA* **101,** 6226–6230.

Heikkila, R. E., Orlansky, H., and Cohen, G. (1975). Studies on the distinction between uptake inhibition and release of w3Hxdopamine in rat brain tissue slices. *Biochem. Pharmacol.* **24,** 847–852.

Hermann, D., Sartorius, A., Welzel, H., Walter, S., Skopp, G., Ende, G., and Mann, K. (2007). Dorsolateral prefrontal cortex N-acetylaspartate/total creatine (NAA/tCr) loss in male recrea-tional cannabis users. *Biol. Psychiatry* **61,** 1281–1289.

Hirata, Y., Meguro, T., and Kiuchi, K. (2006). Differential effect of nerve growth factor on dopami-nergic neurotoxin-induced apoptosis. *J. Neurochem.* **99,** 416–425.

Hoover, B. R., Everett, C. V., Sorkin, A., and Zahniser, N. R. (2007). Rapid regulation of dopamine transporters by tyrosine kinases in rat neuronal preparations. *J. Neurochem.* **101,** 1258–1271.

Hotchkiss, A. J., and Gibb, J. W. (1980). Long-term effects of multiple doses of methamphetamine on tryptophan hydroxylase and tyrosine hydroxylase in rat brain. *J. Pharmacol. Exp. Ther.* **214,** 257–262.

Howlett, A. C., Barth, F., Bonner, T. I., Cabral, G., Casellas, P., Devane, W. A., Felder, C. C., Herkenham, M., Mackie, K., Martin, B. R., Mechoulam, R., Pertwee, R. G., *et al.* (2002). International Union of Pharmacology. XXVII. Classification of cannabinoid receptors. *Pharmacol. Rev.* **54,** 161–202.

Howlett, A. C., Breivogel, C. S., Childers, S. R., Deadwyler, S. A., Hampson, R. E., and Porrino, L. J. (2004). Cannabinoid physiology and pharmacology: 30 years of progress. *Neuropharmacology* **47,** 345–358.

Huang, T. L., and Lee, C. T. (2006). Associations between serum brain-derived neurotrophic factor levels and clinical phenotypes in schizophrenia patients. *J. Psychiatr. Res.* **40,** 664–668.

Hünnerkopf, R., Strobel, A., Gutknecht, L., Brocke, B., and Lesch, K. P. (2007). Interaction between BDNF Val66Met and dopamine transporter gene variation influences anxiety-related traits. *Neuropsychopharmacology* **32,** 2552–2560.

Hu, S., Sheng, W. S., Lokensgard, J. R., and Peterson, P. K. (2002). Morphine induces apoptosis of human microglia and neurons. *Neuropharmacology* **42,** 829–836.

Hutchinson, M. R., and Somogyi, A. A. (2002). Diacetylmorphine degradation to 6-monoacetylmor-phine and morphine in cell culture: Implications for *in vitro* studies. *Eur. J. Pharmacol.* **453,** 27–32.

Hyman, C., Hofer, M., Barde, Y. A., Juhasz, M., Yancopoulos, G. D., Squinto, S. P., and Lindsay, R. M. (1991). BDNF is a neurotrophic factor for dopaminergic neurons of the substantia nigra. *Nature* **350,** 230–232.

Hyman, C., Juhasz, M., Jackson, C., Wright, P., Ip, N. Y., and Lindsay, R. M. (1994). Overlapping and distinct actions of the neurotrophins BDNF, NT-3 and NT-4/5 on cultured dopaminergic and GABAergic neurons of the ventral mesencephalon. *J. Neurosci.* **14,** 335–347.

Ibanez, C. F. (1998). Emerging themes in structural biology of neurotrophic factors. *Trends Neurosci.* **21,** 438–444.

Insel, T. R., Battaglia, G., Johannessen, J. N., Marra, S., and De Souza, E. B. (1989). 3,4-Methyle-nedioxymethamphetamine-AecstasyB. Selectively destroys brain serotonin terminals in rhesus monkeys. *J. Pharmacol. Exp. Ther.* **249,** 713–720.

Itoh, K., Hashimoto, K., Shimizu, E., Sekine, Y., Ozaki, N., Inada, T., Harano, M., Iwata, N., Komiyama, T., Yamada, M., Sora, I., Nakata, K., *et al.* (2005). Association study between brain-derived neurotrophic factor gene polymorphisms and methamphetamine abusers in Japan. *Am. J. Med. Genet. B. Neuropsychiatr. Genet.* **132**, 70–73.

Jockers-Scherübl, M. C., Matthies, U., Danker-Hopfe, H., Lang, U. E., Mahlberg, R., and Hellweg, R. (2003). Chronic cannabis abuse raises nerve growth factor serum concentrations in drug-naive schizophrenic patients. *J. Psychopharmacol.* **17**, 439–445.

Jockers-Scherübl, M. C., Danker-Hopfe, H., Mahlberg, R., Selig, F., Rentzsch, J., Schürer, F., Lang, U. E., and Hellweg, R. (2004). Brain-derived neurotrophic factor serum concentrations are increased in drug-naive schizophrenic patients with chronic cannabis abuse and multiple substance abuse. *Neurosci. Lett.* **371**, 79–83.

Johnson, S. W., and North, R. A. (1992). Opioids excite dopamine neurons by hyperpolarization of local interneurons. *J. Neurosci.* **12**, 483–488.

Johnson, M., Letter, A. A., Merchant, K., Hanson, G. R., and Gibb, J. W. (1988). Effects of 3,4-methylenedioxyamphetamine and 3,4-methylenedioxymethamphetamine isomers on central serotonergic, dopaminergic and nigral neurotensin systems of the rat. *J. Pharmacol. Exp. Ther.* **244**, 977–982.

Kaplan, D. R., and Miller, F. D. (1997). Signal transduction by the neurotrophin receptors. *Curr. Opin. Cell Biol.* **9**, 213–221.

Kelley, P. H., and Iverson, S. D. (1975). Selective 6-OHDA-induced destruction of mesolimbic dopamine neurons: Abolition of psychostimulant-induced locomotor activity in rats. *Eur. J. Pharmacol.* **40**, 45–56.

Kim, D. J., Roh, S., Kim, Y., Yoon, S. J., Lee, H. K., Han, C. S., and Kim, Y. K. (2005). High concentrations of plasma brain-derived neurotrophic factor in methamphetamine users. *Neurosci. Lett.* **388**, 112–115.

Klein, R. (1994). Role of neurotrophins in mouse neuronal development. *Faseb J.* **8**, 738–744.

Knüsel, B., Gao, H., Okazaki, T., Yoshida, T., Mori, N., Hefti, F., and Kaplan, D. R. (1997). Ligand-induced down-regulation of Trk messenger RNA, protein and tyrosine phosphorilation in rat cortical neurons. *Neuroscience* **78**, 851–862.

Koh, S., Oyler, G. A., and Higgins, G. A. (1989). Localization of nerve growth factor receptor messenger RNA and protein in the adult rat brain. *Exp. Neurol.* **106**, 209–221.

Koob, G. F., Ahmed, S. H., Boutrel, B., Chen, S. A., Kenny, P. J., Markou, A., O'Dell, L. E., Parsons, L. H., and Sanna, P. P. (2004). Neurobiological mechanisms in the transition from drug use to drug dependence. *Neurosci. Biobehav. Rev.* **27**, 739–749.

Korsching, S. (1993). The neurotrophic factor concept: A reexamination. *J. Neurosci.* **13**, 2739–2748.

Korsching, S., Auburger, G., Heumann, R., Scott, J., and Thoenen, H. (1985). Levels of nerve growth factor and its mRNA in the central nervous system of the rat correlate with cholinergic innervation. *Embo J.* **74**, 1389–1393.

Kuhar, M. J., and Pilotte, N. S. (1996). Neurochemical changes in cocaine withdrawal. *Trends Pharmacol. Sci.* **17**, 260–264.

LaVoie, M. J., and Hastings, T. G. (1999). Dopamine quinone formation and protein modification associated with the striatal neurotoxicity of methamphetamine: Evidence against a role for extracellular dopamine. *J. Neurosci.* **19**, 1484–1491.

Leibrock, J., Lottspeich, F., Hohn, A., Hofer, M., Hengerer, B., Masiakowski, P., Thoenen, H., and Barde, Y. A. (1989). Molecular cloning and expression of brain-derived neurotrophic factor. *Nature* **341**, 149–152.

Levi-Montalcini, R. (1987). The nerve growth factor 35 years later. *Science* **237**, 1154–1162.

Levi-Montalcini, R. (2004). The nerve growth factor and the neuroscience chess board. *Prog. Brain Res.* **146**, 525–527.

Levi-Montalcini, R., Skaper, S. D., Dal Toso, R., Petrelli, L., and Leon, A. (1996). Nerve growth factor: From neurotrophin to neurokine. *Trends Neurosci.* **19**, 514–520.

Lewis, M. E., Brown, R. M., Brownstein, M. J., Hart, T., and Stein, D. G. (1979). Nerve growth factor: Effects on D-amphetamine-induced activity and brain monoamines. *Brain Res.* **176**, 297–310.

Lindholm, D., Carroll, P., Tzimagiogis, G., and Thoenen, H. (1996). Autocrine-paracrine regulation of hippocampal neuron survival by IGF-1 and the neurotrophins BDNF, NT-3, and NT-4. *Eur. J. Neurosci.* **8**, 1452–1460.

Lorigados, L., Alvarez, P., Pavon, N., Serrano, T., Bianco, L., and Macias, R. (1996). NGF in experimental models of Parkinson disease. *Mol. Chem. Neuropathol.* **28**, 225–228.

Mamounas, L. A., Blue, M., Siuciak, J. A., and Aitar, C. A. (1995). Brain-derived neurotrophic factor promotes the survival and sprouting of serotoninergic axons in rat brain. *J. Neurosci.* **15**, 7929–7939.

Mao, J., Sung, B., Ji, R. R., and Lim, G. (2002). Neuronal apoptosis associated with morphine tolerance: Evidence for an opioid-induced neurotoxic mechanism. *J. Neurosci.* **22**, 7650–7661.

Martin-Iverson, M. T., Todd, K. G., and Altar, C. A. (1994). Brain-derived neurotrophic factor and neurotrophin-3 activate striatal dopamine and serotonin metabolism and related behaviors: Interactions with amphetamine. *J. Neurosci.* **14**, 1262–1270.

Matsuda, L. A., Lolait, S. J., Brownstein, M. J., Young, A. C., and Bonner, T. I. (1990). Structure of a cannabinoid receptor and functional expression of the cloned cDNA. *Nature* **346**, 561–564.

McDonald, N. Q., and Blundell, T. L. (1991). Crystallization and characterization of the high molecular weight forni of nerve growth factor (7 S NGF). *J. Mol. Biol.* **219**, 595–601.

Melchior, B., Nerrière-Daguin, V., Laplaud, D. A., Rémy, S., Wiertlewski, S., Neveu, I., Naveilhan, P., Meakin, S. O., and Brachet, P. (2003). Ectopic expression of the TrkA receptor in adult dopaminergic mesencephalic neurons promotes retrograde axonal NGF transport and NGF-dependent neuroprotection. *Exp. Neurol.* **183**, 367–378.

Melis, M., Gessa, G. L., and Diana, M. (2000). Different mechanisms for dopaminergic excitation induced by opiates cannabinoids in the rat midbrain. *Prog. Neuro-Psychopharmacol. Biol. Psychiatry* **24**, 993–1006.

Meredith, G. E., Callen, S., and Scheuer, D. A. (2002). Brain-derived neurotrophic factor expression is increased in the rat amygdala, piriform cortex and hypothalamus following repeated amphetamine administration. *Brain Res.* **949**, 218–227.

Merlio, J. P., Ernfors, P., Kokaia, Z., Middlemas, D. S., Bengzon, J., Kokaia, M., Smith, M. L., Siesjö, B. K., Hunter, T., and Lindvall, O. (1993). Increased production of the TrkB protein tyrosine kinase receptor after brain insults. *Neuron* **10**, 151–164.

Morgan, M. E., and Gibb, J. W. (1980). Short-term and long-term effects of methamphetamine on biogenic amine metabolism in extra-striatal dopaminergic nuclei. *Neuropharmacology* **10**, 989–995.

Müller, T., Lang, U. E., Muhlack, S., Welnic, J., and Hellweg, R. (2005). Impact of levodopa on reduced nerve growth factor levels in patients with Parkinson disease. *Clin. Neuropharmacol.* **28**, 238–240.

Müller, C. P., Carey, R. J., Huston, J. P., and De Souza Silva, M. A. (2007). Serotonin and psychostimulant addiction: Focus on 5-HT1A-receptors. *Prog. Neurobiol.* **81**, 133–178.

Munro, S., Thomas, K. L., and Abu-Shaar, M. (1993). Molecular characterization of a peripheral receptor for cannabinoids. *Nature* **365**, 61–65.

Narita, M., Aoki, K., Takagi, M., Yajima, Y., and Suzuki, T. (2003). Implication of brain-derived neurotrophic factor in the release of dopamine and dopamine-related behaviors induced by methamphetamine. *Neuroscience* **119**, 767–775.

Nestler, E. J. (1993). Cellular responses to chronic treatment with drugs of abuse. *Crit. Rev. Neurobiol.* **7**, 23–39.

Nibuya, M., Nestler, E. J., and Duman, R. S. (1996). Chronic antidepressant administration increases the expression of cAMP response element binding protein (CREB) in rat hippocampus. *J. Neurosci.* **16**, 2365–2372.

Nicolaysen, L. C., and Justice, J. B. (1988). Effects of cocaine on release and dopamine uptake of dopamine *in vivo*: Differentiation by mathematical modeling. *Pharmacol. Biochem. Behav.* **31,** 327–335.

Nishio, T., Furukawa, S., Akiguchi, I., and Sunohara, N. (1998). Medial nigral dopamine neurons have rich neurotrophin support in humans. *Neuroreport* **9,** 2847–2851.

Numan, S., and Seroogy, K. B. (1999). Expression of TrkB and TrkC mRNAs by adult midbrain dopamine neurons a double *in situ* hybridization study. *J. Comp. Neurol.* **403,** 295–308.

Oliveira, M. T., Rego, A. C., Morgadinho, M. T., Macedo, T. R., and Oliveira, C. R. (2002). Toxic effects of opioid and stimulant drugs on undifferentiated PC12 cells. *Ann. NY Acad. Sci.* **965,** 487–496.

Parikh, V., Evans, D. R., Khan, M. M., and Mahadik, S. P. (2003). Nerve growth factor in never-medicated first-episode psychotic and medicated chronic schizophrenic patients: Possible implications for treatment outcome. *Schizophr. Res.* **60,** 117–123.

Park, Y. H., Kantor, L., Wang, K. K., and Gnegy, M. E. (2002). Repeated, intermittent treatment with amphetamine induces neurite outgrowth in rat pheochromocytoma cells (PC12 cells). *Brain Res.* **951,** 43–52.

Pezawas, L. M., Fischer, G., Diamant, K., Schneider, C., Schindler, S., Thurnher, M., Ploechl, W., Eder, H., and Kasper, S. (1988). Cerebral CT findings in male opioid-dependent patients: Stereological, planimetric, and linear measurements. *Psychiatr. Res.* **83,** 139–147.

Pierce, R. C., and Kalivas, P. W. (1997). A circuitry model of the expression of behavioral sensitization to amphetamine-like psychostimulants. *Brain Res. Rev.* **25,** 192–216.

Pistis, M., Muntoni, A. L., Pillolla, G., and Gessa, G. L. (2002). Cannabinoids inhibit excitatory inputs to neurons in the shell of the nucleus accumbens: An *in vivo* electrophysiological study. *Eur. J. Neurosci.* **15,** 1795–1802.

Pontieri, F. E., Tanda, G., and Di Chiara, G. (1995). Intravenous cocaine, morphine, and amphetamine preferentially increase extracellular dopamine in the "shell" as compared with the "core" of the rat nucleus accumbens. *Proc. Natl. Acad. Sci. USA* **92,** 12304–12308.

Pu, L., Liu, Q. S., and Poo, M. M. (2006). BDNF-dependent synaptic sensitization in midbrain dopamine neurons after cocaine withdrawal. *Nat. Neurosci.* **9,** 605–607.

Ricaurte, G. A., Fuller, R. W., Perry, K. W., Seiden, L. S., and Schuster, C. R. (1983). Fluoxetine increases long-lasting neostriatal dopamine depletion after administration of D-methamphetamine and D-amphetamine. *Neuropharmacology* **2,** 1165–1169.

Richardson, N. R., Smith, A. M., and Roberts, D. C. (1994). A single injection of either flupenthixol decanoate or haloperidol decanoate produces long-term changes in cocaine self-administration in rats. *Drug Alcohol Depend.* **36,** 23–25.

Roberts, D. C., and Vickers, G. (1987). The effects of haloperidol on cocaine self-administration is augmented with repeated administrations. *Psychopharmacology* **93,** 526–528.

Robinson, T., and Becker, J. (1986). Enduring changes in brain and behavior produced by chronic amphetamine administration: A review and evaluation of animal models of amphetamine psychosis. *Brain Res.* **396,** 157–198.

Rocamora, N., Pascual, M., Acsady, L., de Lecea, L., Freund, T. F., and Soriano, E. (1996). Expression of NGF and NT3 mRNAs in hippocampal interneurons innervated by the GABAergic septohippocampal pathway. *J. Neurosci.* **16,** 3991–4004.

Rose, J. S., Branchey, M., Buydens-Branchey, L., Stapleton, J. M., Chasten, K., Werrell, A., and Maayan, M. L. (1996). Cerebral perfusion in early and late opiate withdrawal: A technetium 99m-HMPAO SPECT study. *Psychiatr. Res.* **67,** 39–47.

Rosenthal, A., Goeddel, D. V., Nguyen, T., Martin, E., Burton, L. E., Shih, A., Laramee, G. R., Wurm, F., Mason, A., and Nikolics, K. (1991). Primary structure and biological activity of human brain-derived neurotrophic factor. *Endocrinology* **129,** 1289–1294.

Satel, S., Schwick, S., and Gawin, F. (1991). Clinical features of cocaine-induced paranoia. *Am. J. Psychiatry.* **148,** 121–127.

Schmidt, C. J., Ritter, J. K., Sonsalla, P. K., Hanson, G. R., and Gibb, J. W. (1985). Role of dopamine in the neurotoxic effects of methamphetamine. *J. Pharmacol. Exp. Ther.* **233,** 539–544.

Schoenbaum, G., Stalnaker, T. A., and Shaham, Y. A. (2007). Role for BDNF in cocaine reward and relapse. *Nat. Neurosci.* **10,** 935–936.

Seiden, L. S., Fischman, M. W., and Schuster, C. R. (1976). Long-term methamphetamine induced changes in brain catecholamines in tolerant rhesus monkeys. *Drug Alcohol. Depend.* **1,** 215–219.

Shimizu, E., Hashimoto, K., Watanabe, H., Komatsu, N., Okamura, N., Koike, K., Shinoda, N., Nakazato, M., Kumakiri, C., Okada, S., and Iyo, M. (2003). Serum brain-derived neurotrophic factor (BDNF) levels in schizophrenia are indistinguishable from controls. *Neurosci. Lett.* **351,** 111–114.

Shoval, G., and Weizman, A. (2005). The possible role of neurotrophins in the pathogenesis and therapy of schizophrenia. *Eur. Neuropsychopharmacol.* **15,** 319–329(Review).

Siuciak, J. A., Aitar, C. A., Wiegand, S. J., and Lindsay, R. M. (1994). Antinociceptive effect of brain-derived neurotrophic factor and neurotrophin-3. *Brain Res.* **633,** 326–330.

Sobreviela, T., Clary, D. O., Reichardt, L. F., Brandabur, M. M., Kordower, J. H., and Mufson, E. J. (1994). TrkA-immunoreactive profiles in the central nervous system: Colocalization with neurons containing p75 nerve growth factor receptor, choline acetyltransferase, and serotonin. *J. Comp. Neurol.* **350,** 587–611.

Sonsalla, P. K., Gibb, J. W., and Hanson, G. R. (1986). Roles of D1 and D2 dopamine receptor subtypes in mediating the methamphetamine-induced changes in monoamine systems. *J. Pharmacol. Exp. Ther.* **238,** 932–937.

Spanagel, R., Herz, A., and Shippenberg, T. S. (1990). The effects of opioid peptides on dopamine release in the nucleus accumbens: An *in vivo* microdialysis study. *J. Neurochem.* **55,** 1734–1740.

Tan, Y. L., Zhou, D. F., Cao, L. Y., Zou, Y. Z., and Zhang, X. Y. (2005). Decreased BDNF in serum of patients with chronic schizophrenia on long-term treatment with antipsychotics. *Neurosci. Lett.* **382,** 27–32.

Tanda, G., Pontieri, F. E., and Di Chiara, G. (1997). Cannabinoid heroin activation of mesolimbic dopamine transmission by a common mu1 opioid receptor mechanism (comment). *Science* **276,** 2048–2050.

Tosk, J. M., Farag, M., Ho, J. Y., Lee, C. C., Maximos, B. B., and Yung, H. H. (1996). The effects of nerve growth factor and ganglioside GM1 on the anti-proliferative activity of cocaine in PC12 cells. *Life Sci.* **59**(20), 1731–1737.

Toyooka, K., Asama, K., Watanabe, Y., Muratake, T., Takahashi, M., Someya, T., and Nawa, H. (2002). Decreased levels of brain-derived neurotrophic factor in serum of chronic schizophrenic patients. *Psychiatr. Res.* **110,** 249–257.

Vaidya, V. A., Marek, G. J., Aghajanian, G. K., and Duman, R. S. (1997). 5-HT$_2$a receptor-mediated regulation of brain-derived neurotrophic factor mRNA in the hippocampus and the neocortex. *J. Neurosci.* **17,** 2785–2795.

Volkow, N. D., Valentine, A., and Kulkarni, M. (1998). Modifications' radiologiques et neurologiques chez les toxicomanes: Etudes par resonance magnetique. *J. Neuroradiol.* **15,** 288–293.

Wagner, G. C., Lucot, J. B., Schuster, C. R., and Seiden, L. S. (1983). Alphamethyltyrosine attenuates and reserpine increases methamphetamineinduced neuronal changes. *Brain Res* **270,** 258–285.

White, J. M., and Irvine, R. J. (1999). Mechanisms of fatal opioid overdose. *Addiction* **94,** 961–972.

Wolf, S. L., and Mikhael, M. A. (1979). Computerized transaxial tomographic and neuropsycholic evaluations in chronic alcoholics and heroin abusers. *Am. J. Psychiatry* **136,** 598–602.

Wu, V. W., Mo, Q., Yabe, T., Schwartz, J. P., and Robinson, S. E. (2001). Perinatal opioids reduce striatal nerve growth factor content in rat striatum. *Eur. J. Pharmacol.* **41,** 211–214.

Yan, Q. S., Zheng, S. Z., and Yan, S. E. (2004). Prenatal cocaine exposure decreases brain-derived neurotrophic factor proteins in the rat brain. *Brain Res.* **1009,** 228–233.

Yan, Q. S., Feng, M. J., and Yan, S. E. (2007). RNA interference-mediated inhibition of brain-derived neurotrophic factor expression increases cocaine's cytotoxicity in cultured cells. *Neurosci. Lett.* **414,** 165–169.

Zachor, D. A., Moore, J. F., Brezausek, C. M., Theibert, A. B., and Percy, A. K. (2000a). Cocaine inhibition of neuronal differentiation in NGF-induced PC12 cells is independent of ras signaling. *Int. J. Dev. Neurosci.* **18,** 765–772.

Zachor, D. A., Moore, J. F., Brezausek, C., Theibert, A., and Percy, A. K. (2000a). Cocaine inhibits NGF-induced PC12 cells differentiation through D(1)-type dopamine receptors. *Brain Res* **869,** 85–97.

Zetterström, T. S. C., Pei, Q., Madhav, T. R., Coppell, A. L., Lewis, L., and Grahame-Smith, D. G. (1999). Manipulation of brain 5-HT levels affect genes expression for BDNF in rat brain. *Neuropharmacology* **38,** 1063–1073.

DOSING TIME-DEPENDENT ACTIONS OF PSYCHOSTIMULANTS

H. Manev and T. Uz

Department of Psychiatry, The Psychiatric Institute,
University of Illinois at Chicago, Chicago, Illinois 60612, USA

The concept of the dosing time-dependent (DTD) actions of drugs has been used to describe the effects of diurnal rhythms on pharmacological responsiveness. Notwithstanding the importance of diurnal variability in drug pharmacokinetics and bioavailability, it appears that in the central nervous system (CNS), the DTD actions of psychotropic drugs involve diurnal changes in the CNS-specific expression of genes encoding for psychotropic drug targets and transcription factors known as clock genes. In this review, we focused our discussion on the DTD effects of the psychostimulants cocaine and amphetamines. Both cocaine and amphetamines produce differential lasting behavioral alterations, that is, locomotor sensitization, depending on the time of the day they are administered. This exemplifies a DTD action of these drugs. The DTD effects of these psychostimulants correlate with diurnal changes in the system of transcription factors termed clock genes, for example, *Period 1*, and with changes in the availability of certain subtypes of dopamine receptors, for example, D2 and D3. Diurnal synthesis and release of the pineal hormone melatonin influence the DTD behavioral actions of cocaine and amphetamines. The molecular mechanism of

INTERNATIONAL REVIEW OF
NEUROBIOLOGY, VOL. 88
DOI: 10.1016/S0074-7742(09)88002-1

25

melatonin's effects on the responsiveness of CNS to psychostimulants appears to involve melatonin receptors and clock genes. It is proposed that the DTD characteristics of psychostimulant action and the contributions of the melatonergic system may have clinical implications that include treatments for the attention deficit hyperactivity disorder and possibly neurotoxicity/neuroprotection.

I. Introduction: The Concept of the Dosing Time-Dependent Actions of Drugs

The concept of the dosing time-dependent (DTD) actions of drugs has been used to describe the effects of diurnal rhythms on pharmacological responsiveness. This phenomenon is also a topic of concern for chronopharmacology. It has been known for a long time that biological rhythms may influence drug response and several chronopharmacological studies have underlined the influence of the time of day on drug pharmacodynamics and pharmacokinetics (Bruguerolle *et al.*, 2008). These realizations have led to the recent interest in the development of drug delivery systems that complement these diurnal rhythms (Youan, 2004).

Although chronopharmacology is involved in all biological systems, its relevance is probably most evident in behavioral actions of psychotrophic drugs. For example, the sleep–awake cycle demonstrates a clear diurnal rhythm. The DTD effects of hypnotics can be experimentally demonstrated in mice (Sato *et al.*, 2005). Such preclinical studies suggest that DTD influences the effects of anesthetics and that it would be clinically beneficial to take DTD into account when selecting the time of administration and the doses of these drugs. Similarly, the behavioral effects of antidepressant selective serotonin reuptake inhibitors (SSRIs) were experimentally examined in mice and these behaviors also demonstrated remarkable diurnal variability (Ushijima *et al.*, 2005). Thus, the anti-immobility effect of SSRIs in a forced swimming test was most potent at an early part of the dark phase. These authors demonstrated that the mechanism of this DTD action of SSRIs is not diurnal pharmacokinetics; rather, it is a diurnal rhythm of brain's expression of the serotonin transporter, which is the target for the action of SSRIs (Ushijima *et al.*, 2005).

Notwithstanding the importance of diurnal variability in drug pharmacokinetics and bioavailability, it appears that in the central nervous system (CNS), the DTD actions of psychotropic drugs involve diurnal changes in the CNS-specific expression of genes encoding for psychotropic drug targets.

Currently, transcription factors known as clock genes are considered the likely driving force in setting diurnal CNS susceptibility to responsiveness to psychoactive drugs (Manev and Uz, 2006). In the following chapters, we will focus our discussion on the DTD effects of the psychostimulants cocaine and amphetamine, and on the putative role of dopamine receptors, and clock genes and their hormonal modifiers (e.g., melatonin) in DTD effects.

II. Diurnal Variations in the Behavioral Actions of Psychostimulants

An example of circadian effects of therapeutically used psychostimulants is their military use, especially in units that operate over long hours in which soldiers are deprived of sleep (Emonson and Vanderbeek, 1995). During prolonged military operations, pilots are regularly kept awake for hours and days without fulfilling their biological sleep requirements. This consequently affects their natural circadian rhythm (Eliyahu *et al.*, 2007). On the other hand, the abuse of psychostimulants has been shown to conform to seasonal (Satel and Gawin, 1989) and diurnal (Erickson *et al.*, 1998) rhythms. Not only do drugs such as cocaine and amphetamines affect the brain's circadian molecular machinery (Uz *et al.*, 2005a), but also their long-term molecular and behavioral effects differ depending on the time of the day these drugs are administered.

A. COCAINE

The use of cocaine appears to be diurnally affected; in one study, most abuse occurred during the afternoon and cocaine-caused health problems peaked soon thereafter (Erickson *et al.*, 1998). In another study, it was reported that cases of opiate, ethanol, phencyclidine, benzodiazepine, and cocaine abuse and/or overdose occur with significant ultradian, circadian, and circannual periodicities, and that cocaine overdoses were consistently presented in a bimodal rhythm at 7 AM and 7 PM (Raymond *et al.*, 1992).

Baird and Gauvin (2000) examined the influence of the time of day on the intravenous self-administration of cocaine and its associated pharmacokinetic profile in male Sprague–Dawley rats. They found that groups tested at 1 AM and 1 PM exhibited enhanced sensitivity to the reinforcing properties of low-dose cocaine compared to groups tested at 7 AM and 7 PM. The observed differences in apparent sensitivity of experimental subjects to low-dose cocaine were not related to ongoing patterns of general locomotor activity and were not accompanied by corresponding variances in the pharmacokinetic profiles of cocaine. Another study in rats found that in a model of cocaine self-admistration, these animals generally administered about 80% of the total number of injections during the dark phase (Roberts *et al.*, 2002).

Repeated administration of psychostimulants including cocaine causes sensitization, the increased behavioral response of an animal to repeated drug exposure. It has been proposed that this behavioral response may be considered a preclinical model of human psychostimulant addiction. In humans, the corresponding dependence behaviors include the escalating patterns of cocaine use and the increasing cravings for the drug over time. Furthermore, corresponding to behavioral sensitization in animals are clinical cocaine-induced paranoia and psychosis (Post and Kopanda, 1976). Uz *et al.* (2002) first found that cocaine sensitization in mice demonstrates a remarkable diurnal variability; it can be induced during the day but not at night. Subsequent studies demonstrated that similar diurnal rhythms in cocaine sensitization are common across rodent species (Akhisaroglu *et al.*, 2004). In rats, long-term sensitization to cocaine was expressed only at the onset of darkness (Sleipness *et al.*, 2005).

Conditioned place preference is a commonly used behavioral paradigm for drug seeking behavior in which subjects demonstrate a learned association to specific environmental stimuli (i.e., cues) with the positively rewarding or aversive effects of a drug. When these experiments were performed both during the day and at night, mice demonstrated a significant decrease in cocaine-induced conditioned place preference at night compared to daytime (Kurtuncu *et al.*, 2004).

In addition to DTD responses to cocaine, the effect of cocaine on normal diurnal rhythms of activity also is well documented. For example, the offspring of rats exposed prenatally to cocaine showed altered activity levels, and these effects were influenced by the time of day (Church and Tilak, 1996). More recently, experiments in zebrafish demonstrated that prenatal cocaine exposure had lasting effects on neuronal development and circadian mechanisms. Moreover, the effects of prenatal cocaine depended on the time of treatment, that is, more robust during the day, and these effects were independent of whether the embryos were raised under a light–dark cycle or in constant light (Shang and Zhdanova, 2007).

B. Amphetamines

As early as 1973, Evans *et al.* (1973) reported that in a model of operant behavior (i.e., avoidance), methamphetamine produced the greatest increase in lever pressing when administered during the dark period and this effect was independent of diurnal differences in base-line response rates. Furthermore, methamphetamine increased locomotor activity equally during the light and during the dark. Similar to findings with cocaine (Baird and Gauvin, 2000), no differences were found in the day-versus-night pharmacokinetic profile of methamphetamine (Shappell *et al.*, 1996) that would account for the DTD effects of amphetamine.

Further studies with amphetamine demonstrated that DTD amphetamine responses are evident in stereotypic behavior but not in locomotor activity (Gaytan *et al.*, 1998). Thus, significant diurnal variation was observed in the occurrence and duration of amphetamine-induced head-shaking

(Urba-Holmgren *et al.*, 1977). The lowest values were obtained near noon and the highest around midnight.

In cats, amphetamine reduces caudate activity more markedly at night (Arushanian and Pavlov, 1983). Haloperidol, on the other hand, decreased more significantly the thresholds of arrest reaction induced by stimulation of the caudate nucleus and abolished the amphetamine-induced stereotypy when administered during the day. These authors proposed that in cats, accumulation of nigro-striatal dopamine may be more prominent during the dark.

In rats, tolerance to the locomotor stimulant actions of amphetamine is both dose- and light/dark cycle-dependent (Martin-Iverson and Iversen, 1989). Locomotor stimulation induced by the two highest doses (10 and 20 mg/kg/day) remained high during both the day and the night, whereas tolerance was developed only to the effects of the two lower doses (2 and 6 mg/kg/day) and only during the day.

Similar to repeated cocaine exposure, repeated administration of amphetamine leads to locomotor sensitization and this amphetamine sensitization is also time-dependent. However, in contrast to cocaine, the strongest sensitized response to amphetamine occurred during the middle of the dark cycle (Gaytan *et al.*, 1999). On the other hand, the development of sensitization to methylphenidate, which was also time-dependent was similar to cocaine; the most robust sensitization occurred during the light phase, while no sensitization was observed during the middle of the dark phase (Gaytan *et al.*, 2000).

III. Diurnal Variations in the CNS Systems

It should be stressed at the outset that all biological activities, including CNS functioning, express some type of rhythm and that many are expressed in direct or indirect relationship to day/night or light/dark cycles. It is beyond the scope of this review to discuss all possible diurnal rhythms in the brain. Instead, we will limit our discussion to two molecular systems directly related to the actions of cocaine and amphetamine. These are the receptors and the transcription factors encoded by the so-called "clock genes."

A. CLOCK GENES

Clock genes represent a family of genes encoding for proteins that demonstrate remarkable circadian rhythms in their expression and function. They act primarily as transcription factors capable of modifying their own expression as well as the expression of various other target genes (Reppert and Weaver, 2002). In addition to their expression in the brain area known as the suprachiasmatic nucleus, clock genes are expressed and functional throughout the brain and in many peripheral tissues.

A recent review by McClung (2007) summarized studies pointing to a link between abnormal or disrupted circadian rhythms and the development of addictions including the influence of clock genes on drug-induced behaviors. Studies performed in fruit flies (*Drosophila melanogaster*) discovered that clock genes are required for cocaine sensitizations (Andretic *et al.*, 1999; Dimitrijevic *et al.*, 2004). Subsequent experiments with transgenic mice deficient for various clock genes (e.g., *Per1*, *Per2*, *Clock*, and *NPAS2*) demonstrated a role for these transcription factors in the development of behavioral responses to drugs of abuse (Abarca *et al.*, 2002; McClung *et al.*, 2003, 2005). Thus, Abarca *et al.* (2002) used the *Per1* knockout and found abolished cocaine reward in these mice compared to their wild-type controls. Furthermore, in mice, cocaine-induced drug seeking (reward) presents diurnal rhythms with less reward to cocaine at night when striatal PER1 levels are the lowest (Kurtuncu *et al.*, 2004). In the rat striatum, PER1 protein levels show a diurnal rhythm that is correlated with diurnal cocaine-sensitization differences (Akhisaroglu *et al.*, 2004).

Interestingly, a feedback loop exists in which drugs of abuse such as cocaine influence the expression of clock genes (Uz *et al.*, 2005a), and may thereby alter/entrain circadian/diurnal rhythms. Thus a concept is emerging in which drugs of abuse can influence the circadian program, which in turn influences the effects of drugs of abuse (Manev and Uz, 2006; Perreau-Lenz *et al.*, 2007).

It has been noted that at least two clock-gene products, CLOCK and BMAL1, function as transcription factors by binding to E-box enhancers in the promoter regions of other clock genes, and it was proposed that clock proteins could regulate the expression of drug abuse-relevant nonclock genes [i.e., clock-controlled genes (CCG)] that possess E boxes in their promoter regions (Manev and Uz, 2006). This concept was recently experimentally confirmed by findings that the clock proteins BMAL1, NPAS2, and PER2 acting on a gene promoter mediate the transcriptional regulation of the CCG monoamine oxidase A, an enzyme that catalyzes the oxidation of the monoamines noradrenaline and serotonin (Hampp *et al.*, 2008).

Psychoactive drugs, including drugs of abuse, could interfere with the E-box-dependent actions of clock genes (Fig. 1). Thereby, these drugs are capable of altering the expression of CNS genes encoding for proteins important for brain functioning, including proteins (e.g., receptors and transporters) that serve as effectors (e.g., specific binding sites) for psychoactive drugs.

B. Dopamine Receptors

Animal studies have suggested that the DTD actions of cocaine and amphetamine could in part be attributed to diurnal variations in dopamine receptors. Recently, a positron emission tomography (PET) study with the radioligands

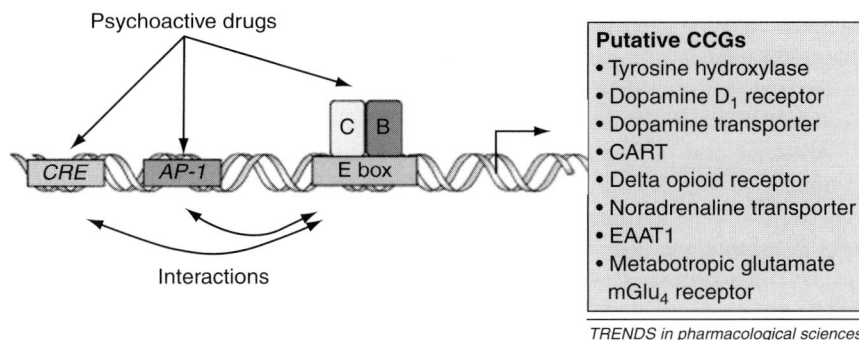

FIG. 1. A proposed role for psychoactive drug-induced regulation of E-box-mediated transcriptional activity of CCGs. Clock genes as transcription factors might regulate their self-expression and that of another set of genes (CCGs) by binding to the E-box sites located at their promoter regions: clock protein–DNA binding has been demonstrated for genes outside the clock family, such as those that encode vasopressin and prolactin. Moreover, not only canonical (CACGTG) but also noncanonical (CANNTG) E-box sites are potential candidates for such binding and functionality. Often, cAMP response element (CRE) and/or activating protein 1 (AP-1) sites in the vicinity of an E-box site might interact with the E box for the *trans*-activation of CCGs. Psychoactive drugs might affect the *trans*-activation of CCGs by regulating: (i) transcription factors for those sites (e.g., CREB and CLOCK); (ii) binding properties of transcription factors with their specific sites; and (iii) interactions between the E box and other sites. Abbreviations: B, BMAL1; CART, cocaine- and amphetamine-regulated transcript; C, CLOCK; EAAT1, excitatory amino acid transporter type 1. The putative CCGs have been selected from published literature because they contain E-box sites in their promoter regions. *Reprinted from reference Manev and Uz, 2006, with permission from Elsevier.

[11C]raclopride and [11C]FLB 457 was conducted to determine dopamine D2-receptor binding in 16 human subjects in the morning and the evening of the same day (Cervenka *et al.*, 2008). When age was taken into account in the analysis, a pattern emerged in which individuals in the lower age range showed reduced evening binding while in older subjects, binding potential increased.

The effect of chronic cocaine administration on the *in vivo* occupation of dopamine receptor subtypes was examined using an irreversible receptor blocker *N*-ethoxycarbonyl-2-ethoxy-1,2-dihydroquinoline (EEDQ) (Burger and Martin-Iverson, 1993). Cocaine increases dopamine D1 receptor occupation by dopamine during the day and decreases it at night. Since no day/night differences were found in D2 receptor density, it was proposed that cocaine treatment selectively alters the level of dopamine at sites containing D1 receptors with differential effects depending on the time of the day/night cycle (Burger and Martin-Iverson, 1993).

Experiments with the dopamine D1 agonist SKF 38393 and the D2 agonist PHNO suggested that the motor stimulant effects of a D2 receptor agonist, behavioral D1/D2 receptor interactions, and the development of sensitization and tolerance to the locomotor effects of D2 receptor agonists are determined by circadian rhythms and lighting conditions (Martin-Iverson and Yamada, 1992). Tolerance to locomotor stimulation by PHNO developed during the day but not at night; this day-time tolerance was reversed to sensitization by cotreatment with a melatonin receptor agonist 2-iodo-melatonin (Munro and Martin-Iverson, 2000).

Andretic and Hirsh (2000) used a dopamine receptor D2/D3 agonist quinpirole in *Drosophila* and found DTD locomotion changes, with the highest responsiveness to quinpirole during the subjective night. Furthermore, these authors established that dopamine receptor responsiveness is under circadian control and depends on the normal function of the clock gene *period*. The DTD actions of quinpirole were also confirmed in mice (Akhisaroglu *et al.*, 2005). These authors demonstrate diurnal rhythm-dependent variations in both quinpirole-induced locomotor behaviors and in striatal D2 and D3 protein levels. The diurnal rhythms of striatal D2 and D3 protein levels are opposing, resulting in a high D2/D3 ratio during the day and a low D2/D3 ratio at night (Fig. 2). The striatal D2 mRNA content in mice showed a rhythm similar to the D2 protein levels; the strongest expression of D2 mRNA was found around mid-light and the minimum around mid-dark (Viyoch *et al.*, 2001). Protracted quinpirole treatment differentially altered striatal D2/D3 rhythms depending on the time of injection (i.e., day or night) (Akhisaroglu *et al.*, 2005).

IV. Diurnal Variations and Melatonin

The biology of the pineal hormone melatonin was recently extensively reviewed (Simonneaux and Ribelayga, 2003). Briefly, melatonin (5-methoxy-*N*-acetyltryptamine) is synthesized in the pineal gland during the dark phase of the light/dark cycle and is rapidly delivered to the body via the bloodstream. The daily rhythm of pineal melatonin is considered to be a circadian mediator used by the body to deliver its circadian message to cells containing melatonin receptors. After the discovery of the neuroprotective effects of supraphysiological melatonin concentrations (Giusti *et al.*, 1995; Uz *et al.*, 1996), this hormone has been considered for pharmacological neuroprotection for a number of neurodegenerative disorders. On the other hand, synthetic melatonin receptor ligands are used to treat insomnia (Pandi-Perumal *et al.*, 2007).

A. MELATONIN RECEPTORS

In mammals, the molecular signaling of endogenously released melatonin is mediated by the G-protein-coupled receptors (GPCRs) MT1 and MT2. In neurons, the signal transduction triggered by these receptors involves multiple

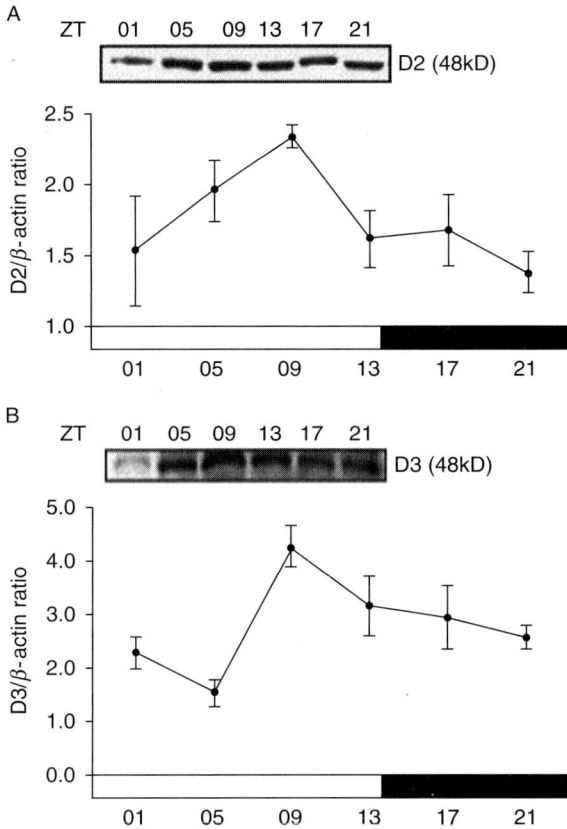

FIG. 2. Diurnal rhythms of D2 and D3 dopamine receptor protein levels in the mouse striatum. Mice (3 per group) were sacrificed every 4 h starting at Zeitgeber time (ZT) 01 (ZT00 is defined as lights on). Striatal samples were processed for Western blotting. Demonstrative Western blot images in panels (A) and (B) show striatal 24-h-rhythms of D2 and D3 signals, respectively. For the quantitative assay, D2 (A) and D3 (B) signals were normalized by their corresponding β-actin signal. The results (mean ± S.E.M.) are shown as ratios of each protein to the corresponding β-actin values. *Reprinted from reference Akhisaroglu et al., 2005, with permission from Elsevier.

signaling pathways (for review, see Jockers et al., 2008), including cAMP, extracellular-signal-regulated kinase (ERK), and serine/threonine kinase Akt (Imbesi et al., 2008b). Recent research points to the mechanisms of oligomerization/ dimerization as general characteristics of GPCRs (Pin et al., 2007). It was proposed that GPCR dimerization has a profound impact on receptor pharmacology, signaling, and regulation. Ample evidence indicates that MT1 and MT2 receptors are capable of dimerizing. The relative propensity for MT1 homodimer and MT1/

MT2 heterodimer formation is similar, whereas the propensity for MT2 homo-dimer formation is three- to fourfold lower (Ayoub *et al.*, 2004). Using radioligand binding assays, these authors demonstrated that heterodimers contain two func-tional ligand binding sites and that occupation of either binding site causes conformational changes in the heterodimer. It was hypothesized that in the absence of MT2, these conformational changes would be confined to the MT1 homodimers and that this receptor configuration may guide melatonin signal transduction toward an increase in brain-derived neurotrophic factor (BDNF) translation (Imbesi *et al.*, 2008a). BDNF has been proposed as an important contributor to addiction-related neuroplasticity (Thomas *et al.*, 2008) and BDNF alterations via MT1-mediated signaling could influence the actions of drugs of abuse. For example, as noted earlier, wild-type rats and mice show prominent diurnal behavioral differences in cocaine sensitization: they show gradually increasing behavioral responses to the same dose of cocaine given repeatedly for several days only if the injections are administered during the day and not at night. This diurnal difference is abolished in MT1 receptor knockout mice, which become sensitized to cocaine both during the day and at night (Uz *et al.*, 2005b).

B. Melatonin and Clock Genes

Clock genes located in the suprachiasmatic nucleus of the hypothalamus, which projects to the pineal gland via a multisynaptic pathway, drive the circadian rhythms of pineal melatonin synthesis and release (Reppert and Weaver, 2002). In addition, clock genes are also expressed and functional in the pineal gland. Simonneaux *et al.* (2004) found that *Per1*, *Per3*, *Cry2*, and *Cry1* clock genes are expressed in the pineal gland and that their transcription is increased at night. These authors showed that the expression of *Per1* and *Cry2* in the rat pineal gland is regulated by changes in norepinephrine, in a manner similar to the melatonin rhythm-generating enzyme arylalkylamine *N*-acetyltransferase (AANAT), whereas the expression of *Per3* and *Cry1* displayed a daily rhythm not regulated by norepi-nephrine, suggesting the involvement of another day/night regulated transmitter.

Generally, the rhythm of expression of clock genes in brain areas other than the suprachiasmatic nucleus appears to be regulated by multiple signaling path-ways including melatonin signaling and melatonin receptors. For example, in the striatum and cells of the hypophyseal pars tuberalis of mice, diurnal changes of *Per1* expression depend on a functioning pineal gland and its product melatonin. Thus, the diurnal rhythms of the striatal *Per1* mRNA and PER1 protein were observed in normal mice but not in pinealectomized mice (Uz *et al.*, 2003). Furthermore, these rhythms were absent in mice unable to synthesize melatonin (Akhisaroglu *et al.*, 2004). In the pars tuberalis, the diurnal rhythms of *Per1*, *Clock*, and *Bmal1* genes were abolished in mice deficient for melatonin MT1 receptors (Jilg *et al.*, 2005; von Gall *et al.*, 2005).

Psychoactive drugs including cocaine are capable of altering the CNS expression of both melatonin receptors (Imbesi *et al.*, 2006) and clock genes (Lynch *et al.*, 2008; Uz *et al.*, 2005a). Considering the remarkable diurnal rhythms of melatonin synthesis and activity, it appears that a complex interplay between melatonin signaling, regulation of transcription factors known as clock genes, and the long-term effects of psychoactive drugs may be involved in determining the type and the extent of the DTD effects of psychostimulants.

C. MELATONIN AND PSYCHOSTIMULANTS

In 1994, Palaoglu *et al.* (1994) showed that pinealectomy affects the behaviors induced by low and high doses of amphetamine. Hence, in sham operated animals, low-dose amphetamine induced significant locomotor stimulation but no stereotyped activity. High-dose amphetamine induced stereotyped activity. After pinealectomy, even low-dose amphetamine produced stereotyped activity. This differential effect of amphetamine seen in pinealectomized rats was completely restored after pineal transplantation.

On the other hand, Sircar (2000) reported on the effects of pharmacological applications of melatonin on cocaine-induced behavioral sensitization. Hence, rats injected daily with melatonin prior to cocaine injections failed to elicit cocaine sensitization. Furthermore, in rats withdrawn from a chronic cocaine treatment, melatonin treatment was able to counteract withdrawal-associated anxiety (Zhdanova and Giorgetti, 2002).

Subsequent findings pointed to a role for endogenous/pineal melatonin in the development of cocaine sensitization. This evidence comes from experiments with mice mutant for the critical melatonin-synthesizing enzyme, AANAT. Although intraperitoneal cocaine treatment increased the locomotor activity of both normal and AANAT mutant mice, cocaine sensitization developed differently in these mice (Uz *et al.*, 2002). Thus, repeated cocaine injections (3 days) during the day/light resulted in behavioral sensitization in both normal and AANAT mutant mice, but treatment at night/dark triggered sensitization in AANAT-deficient mice only. This circadian cocaine sensitization was studied at times when striatal PER1 protein levels in normal mice were high and low, respectively (Uz *et al.*, 2003). Only mice with circadian changes in striatal Per1 gene expression showed the night-time absence of cocaine sensitization, whereas pinealectomized or AANAT mutant mice did not show circadian changes in striatal Per1 and were sensitized to cocaine regardless of diurnal rhythm. Furthermore, it was found that the pineal gland influences cocaine reward in mice (Kurtuncu *et al.*, 2004). Controls with intact pineal glands demonstrated a significant decrease in cocaine-induced conditioned place preference at night compared to the daytime, whereas pinealectomized mice did not show any diurnal differences.

The exact mechanism of melatonin's action on dopamine-mediated behavioral effects of psychostimulants is not completely understood. One possibility is the inhibition of dopamine release by melatonin, which has been demonstrated in specific areas of the mammalian CNS (hypothalamus, hippocampus, medulla-pons, and retina), which occurs along with the antidopaminergic activities of melatonin in the striatum (Zisapel, 2001).

V. Possible Clinical Implications

The concept of the DTD actions of drugs suggests numerous clinical implications. The fact that psychoactive drugs express DTD differential behavioral properties indicates the possibility for a DTD differential molecular neuroplasticity and the DTD amenability of CNS for therapeutic interventions. In this review, we summarize evidence of a possible role of the melatonin receptor-mediated signaling in influencing the actions and/or consequences of psychostimulants such as cocaine and amphetamine. The endogenous melatonergic pathway may be one of the important systems in setting the diurnal CNS susceptibility to psychostimulants. At the same time, this pathway is modified by psychostimulants and may be responsible for the neuroplasticity-dependent mechanisms of addiction. By the same token, pharmacological interventions in the melatonergic system could provide a novel approach to dealing with clinical issues of the actions of psychostimulants.

A. Attention Deficit Hyperactivity Disorder

Three patterns of behavior indicate attention deficit hyperactivity disorder (ADHD); being inattentive, hyperactive, and impulsive far more than others of the same age. It is estimated that between 3% and 5% of children have ADHD, or approximately 2 million children in the United States. The most recent estimated prevalence of adult ADHD is 4.4%. The most common pharmacological treatment for ADHD includes psychostimulants; for example, amphetamine and methylphenidate. Sleep problems are associated with ADHD and typically these problems are not improved and can even deteriorate with stimulant therapy. Hence, considering the DTD action of psychostimulants, it might be possible to adjust both the therapeutic and unwanted effects of ADHD treatments by taking into consideration the time of the day these treatments are administered. Moreover, a couple of clinical trials demonstrated that the addition of melatonin to stimulant therapy improved sleep disorders in ADHD patients (Van der Heijden *et al.*, 2007; Weiss *et al.*, 2006). It should be investigated whether melatonin or the new melatonin receptor agonists (e.g., ramelteon) could improve the therapeutic actions of stimulant therapy.

B. NEUROTOXICITY

It has been noted that the progression/severity of CNS neurodegeneration may be influenced by the time of the day it occurs (Vinall *et al.*, 2000). Based on the neuroprotective action of endogenous melatonin in animal models of stroke (Joo *et al.*, 1998), it was proposed that the outcome of stroke is worse if it occurs during the day than during the night (Manev and Uz, 1998).

In addition to behavioral modifications, drugs such as methamphetamine produce toxic effects in the brain that are thought to be due to oxidative stress. In mice, intraperitoneal administration of melatonin prior to methamphetamine, given at 2-h intervals, significantly attenuated methamphetamine-induced toxic effects on both dopamine and serotonin systems (Hirata *et al.*, 1998). Methamphetamine also increases the striatal content of 3-nitrotyrosine (3-NT), a biomarker of peroxynitrite production. This formation of 3-NT correlated with striatal dopamine depletion. It has been suggested that an increased production of nitric oxide (NO) may be responsible for this type of neurotoxicity. Pretreatment with melatonin was able to completely protect against the formation of 3-NT and depletion of striatal dopamine (Imam *et al.*, 2001).

Recently, Tocharus *et al.* (2008) examined the involvement of NO and altered nitric oxide synthase (NOS) in the neurotoxic effects of amphetamine. In *in vitro* studies, these authors found that amphetamine increases inducible NOS (iNOS) mRNA and that this action of amphetamine was prevented by melatonin. These results suggest that administration of melatonin or melatonin receptor agonists may be neuroprotective against amphetamine toxicity.

VI. Conclusions

The psychostimulants cocaine and amphetamines produce differential lasting behavioral alterations depending on the time of the day they are administered. This exemplifies a DTD action of these drugs.

The DTD effects of these psychostimulants correlate with diurnal changes in the system of transcription factors termed clock genes and with changes in the availability of certain subtypes of dopamine receptors.

Diurnal synthesis and release of the pineal hormone melatonin influence the DTD behavioral actions of cocaine and amphetamines.

The molecular mechanism of melatonin's effects on the responsiveness of CNS to psychostimulants appears to involve melatonin receptors and clock genes.

The DTD characteristics of psychostimulants' action and the contributions of the melatonergic system may have clinical implications that include treatments for ADHD and possibly neurotoxicity/neuroprotection.

References

Abarca, C., Albrecht, U., and Spanagel, R. (2002). Cocaine sensitization and reward are under the influence of circadian genes and rhythm. *Proc. Natl. Acad. Sci. USA* **99,** 9026–9030.

Akhisaroglu, M., Ahmed, R., Kurtuncu, M., Manev, H., and Uz, T. (2004). Diurnal rhythms in cocaine sensitization and in Period1 levels are common across rodent species. *Pharmacol. Biochem. Behav.* **79,** 37–42.

Akhisaroglu, M., Kurtuncu, M., Manev, H., and Uz, T. (2005). Diurnal rhythms in quinpirole-induced locomotor behaviors and striatal D2/D3 receptor levels in mice. *Pharmacol. Biochem. Behav.* **80,** 371–377.

Andretic, R., and Hirsh, J. (2000). Circadian modulation of dopamine receptor responsiveness in *Drosophila melanogaster. Proc. Natl. Acad. Sci. USA* **97,** 1873–1878.

Andretic, R., Chaney, S., and Hirsh, J. (1999). Requirement of circadian genes for cocaine sensitization in *Drosophila. Science* **285,** 1066–1068.

Arushanian, E. B., and Pavlov, V. D. (1983). Diurnal fluctuations in the activity of the caudate nucleus and its sensitivity to psychotropic agents in cats. *Biull. Eksp. Biol. Med.* **96,** 69–72.

Ayoub, M. A., Levoye, A., Delagrange, P., and Jockers, R. (2004). Preferential formation of MT1/MT2 melatonin receptor heterodimers with distinct ligand interaction properties compared with MT2 homodimers. *Mol. Pharmacol.* **66,** 312–321.

Baird, T. J., and Gauvin, D. (2000). Characterization of cocaine self-administration and pharmacokinetics as a function of time of day in the rat. *Pharmacol. Biochem. Behav.* **65,** 289–299.

Bruguerolle, B., Boulamery, A., and Simon, N. (2008). Biological rhythms: A neglected factor of variability in pharmacokinetic studies. *J. Pharm. Sci.* **97,** 1099–1108.

Burger, L. Y., and Martin-Iverson, M. T. (1993). Day/night differences in D1 but not D2 DA receptor protection from EEDQ denaturation in rats treated with continuous cocaine. *Synapse* **13,** 20–29.

Cervenka, S., Halldin, C., and Farde, L. (2008). Age-related diurnal effect on D2 receptor binding: A preliminary PET study. *Int. J. Neuropsychopharmacol.* **10,** 1–8.

Church, M. W., and Tilak, J. P. (1996). Differential effects of prenatal cocaine and retinoic acid on activity level throughout day and night. *Pharmacol. Biochem. Behav.* **55,** 595–605.

Dimitrijevic, N., Dzitoyeva, S., and Manev, H. (2004). An automated assay of the behavioral effects of cocaine injections in adult *Drosophila. J. Neurosci. Methods* **137,** 181–184.

Eliyahu, U., Berlin, S., Hadad, E., Heled, Y., and Moran, D. S. (2007). Psychostimulants and military operations. *Mil. Med.* **172,** 383–387.

Emonson, D. L., and Vanderbeek, R. D. (1995). The use of amphetamines in U.S. Air Force tactical operations during Desert Shield and Storm. *Aviat. Space Environ. Med.* **66,** 260–263.

Erickson, T. B., Lee, J., Zautcke, J. L., and Morris, R. (1998). Analysis of cocaine chronotoxicology in an urban ED. *Am. J. Emerg. Med.* **16,** 568–571.

Evans, H. L., Ghiselli, W. B., and Patton, R. A. (1973). Diurnal rhythm in behavioral effects of methamphetamine, *p*-chloramethamphetamine and scopolamine. *J. Pharmacol. Exp. Ther.* **186,** 10–17.

Gaytan, O., Swann, A., and Dafny, N. (1998). Diurnal differences in rat's motor response to amphetamine. *Eur. J. Pharmacol.* **345,** 119–128.

Gaytan, O., Lewis, C., Swann, A., and Dafny, N. (1999). Diurnal differences in amphetamine sensitization. *Eur. J. Pharmacol.* **374,** 1–9.

Gaytan, O., Yang, P., Swann, A., and Dafny, N. (2000). Diurnal differences in sensitization to methylphenidate. *Brain Res.* **864,** 24–39.

Giusti, P., Gusella, M., Lipartiti, M., Milani, D., Zhu, W., Vicini, S., and Manev, H. (1995). Melatonin protects primary cultures of cerebellar granule neurons from kainate but not from *N*-methyl-D-aspartate excitotoxicity. *Exp. Neurol.* **131,** 39–46.

Hampp, G., Ripperger, J. A., Houben, T., Schmutz, I., Blex, C., Perreau-Lenz, S., Brunk, I., Spanagel, R., Ahnert-Hilger, G., Meijer, J. H., and Albrecht, U. (2008). Regulation of monoamine oxidase a by circadian-clock components implies clock influence on mood. *Curr. Biol.* **18,** 678–683.

Hirata, H., Asanuma, M., and Cadet, J. L. (1998). Melatonin attenuates methamphetamine-induced toxic effects on dopamine and serotonin terminals in mouse brain. *Synapse* **30,** 150–155.

Imam, S. Z., el-Yazal, J., Newport, G. D., Itzhak, Y., Cadet, J. L., Slikker, W. Jr., and Ali, S. F. (2001). Methamphetamine-induced dopaminergic neurotoxicity: Role of peroxynitrite and neuroprotective role of antioxidants and peroxynitrite decomposition catalysts. *Ann. N. Y. Acad. Sci.* **939,** 366–380.

Imbesi, M., Uz, T., Yildiz, S., Arslan, A. D., and Manev, H. (2006). Drug- and region-specific effects of protracted antidepressant and cocaine treatment on the content of melatonin MT1 and MT2 receptor mRNA in the mouse brain. *Int. J. Neuroprot. Neuroregener.* **2,** 185–189.

Imbesi, M., Uz, T., and Manev, H. (2008a). Role of melatonin receptors in the effects of melatonin on BDNF and neuroprotection in mouse cerebellar neurons. *J. Neural. Transm.* **115,** 1495–1499.

Imbesi, M., Uz, T., Dzitoyeva, S., and Manev, H. (2008b). Melatonin signaling in mouse cerebellar granule cells with variable native MT1 and MT2 melatonin receptors. *Brain Res.* **1227,** 19–25.

Jilg, A., Moek, J., Weaver, D. R., Korf, H. W., Stehle, J. H., and von Gall, C. (2005). Rhythms in clock proteins in the mouse pars tuberalis depend on MT1 melatonin receptor signalling. *Eur. J. Neurosci.* **22,** 2845–2854.

Jockers, R., Maurice, P., Boutin, J. A., and Delagrange, P. (2008). Melatonin receptors, heterodimerization, signal transduction and binding sites: what's new? *Br. J. Pharmacol.* **154,** 1182–1195.

Joo, J. Y., Uz, T., and Manev, H. (1998). Opposite effects of pinealectomy and melatonin administration on brain damage following cerebral focal ischemia in rat. *Restor. Neurol. Neurosci.* **13,** 185–191.

Kurtuncu, M., Arslan, A. D., Akhisaroglu, M., Manev, H., and Uz, T. (2004). Involvement of the pineal gland in diurnal cocaine reward in mice. *Eur. J. Pharmacol.* **489,** 203–205.

Lynch, W. J., Girgenti, M. J., Breslin, F. J., Newton, S. S., and Taylor, J. R. (2008). Gene profiling the response to repeated cocaine self-administration in dorsal striatum: A focus on circadian genes. *Brain Res.* **1213,** 166–177.

Manev, H., and Uz, T. (1998). The role of the light–dark cycle and melatonin in stroke outcome. *J. Stroke Cerebrovasc. Dis.* **7,** 165–167.

Manev, H., and Uz, T. (2006). Clock genes: Influencing and being influenced by psychoactive drugs. *Trends Pharmacol. Sci.* **27,** 186–189.

Martin-Iverson, M. T., and Iversen, S. D. (1989). Day and night locomotor activity effects during administration of (+)-amphetamine. *Pharmacol. Biochem. Behav.* **34,** 465–471.

Martin-Iverson, M. T., and Yamada, N. (1992). Synergistic behavioural effects of dopamine D1 and D2 receptor agonists are determined by circadian rhythms. *Eur. J. Pharmacol.* **215,** 119–125.

McClung, C. A. (2007). Circadian rhythms, the mesolimbic dopaminergic circuit, and drug addiction. *ScientificWorldJournal* **7**, 194–202.

McClung, C. A., Cooper, D. C., Sidiropoulou, K., Young, Q. L., Sanchez, N., Vitaterna, M., Garcia, J. A., Takahashi, J. S., White, F. J., and Nestler, E. J. (2003). *Clock and NPAS2 differentially regulate cocaine reward. Abstract Viewer/Itinerary Planner.* Society for Neuroscience, Washington, DC, Online. Program No. 112.17.

McClung, C. A., Sidiropoulou, K., Vitaterna, M., Takahashi, J. S., White, F. J., Cooper, D. C., and Nestler, E. J. (2005). Regulation of dopaminergic transmission and cocaine reward by the Clock gene. *Proc. Natl. Acad. Sci. USA* **102**, 9377–9381.

Munro, J. D., and Martin-Iverson, M. T. (2000). Circadian rhythm-dependent development of melatonin effects and tolerance to PHNO in rats. *Pharmacol. Biochem. Behav.* **65**, 495–501.

Palaoglu, S., Palaoglu, O., Akarsu, E. S., Ayhan, I. H., Ozgen, T., and Erbengi, A. (1994). Behavioural assessment of pinealectomy and foetal pineal gland transplantation in rats: Part II. *Acta Neurochir. (Wien)* **128**, 8–12.

Pandi-Perumal, S. R., Srinivasan, V., Poeggeler, B., Hardeland, R., and Cardinali, D. P. (2007). Drug insight: The use of melatonergic agonists for the treatment of insomnia-focus on ramelteon. *Nat. Clin. Pract. Neurol.* **3**, 221–228.

Perreau-Lenz, S., Zghoul, T., and Spanagel, R. (2007). Clock genes running amok. Clock genes and their role in drug addiction and depression. *EMBO Rep.* **8**, S1S20–S23.

Pin, J. P., Neubig, R., Bouvier, M., Devi, L., Filizola, M., Javitch, J. A., Lohse, M. J., Milligan, G., Palczewski, K., Parmentier, M., and Spedding, M. (2007). International Union of Basic and Clinical Pharmacology. LXVII. Recommendations for the recognition and nomenclature of G protein-coupled receptor heteromultimers. *Pharmacol. Rev.* **59**, 5–13.

Post, R. M., and Kopanda, R. T. (1976). Cocaine, kindling, and psychosis. *Am. J. Psychiatry* **133**, 627–634.

Raymond, R. C., Warren, M., Morris, R. W., and Leikin, J. B. (1992). Periodicity of presentations of drugs of abuse and overdose in an emergency department. *Clin. Toxicol.* **30**, 467–478.

Reppert, S. M., and Weaver, D. R. (2002). Coordination of circadian timing in mammals. *Nature* **418**, 935–941.

Roberts, D. C., Brebner, K., Vincler, M., and Lynch, W. J. (2002). Patterns of cocaine self-administration in rats produced by various access conditions under a discrete trials procedure. *Drug Alcohol Depend.* **67**, 291–299.

Satel, S. L., and Gawin, F. H. (1989). Seasonal cocaine abuse. *Am. J. Psychiatry* **146**, 534–535.

Sato, Y., Seo, N., and Kobahashi, E. (2005). The dosing-time dependent effects of intravenous hypnotics in mice. *Anesth. Analg.* **101**, 1706–1708.

Shang, E. H., and Zhdanova, I. V. (2007). The circadian system is a target and modulator of prenatal cocaine effects. *PLoS ONE* **2**, e587.

Shappell, S. A., Kearns, G. L., Valentine, J. L., Neri, D. F., and DeJohn, C. A. (1996). Chronopharmacokinetics and chronopharmacodynamics of dextromethamphetamine in man. *J. Clin. Pharmacol.* **36**, 1051–1063.

Simonneaux, V., and Ribelayga, C. (2003). Generation of the melatonin endocrine message in mammals: A review of the complex regulation of melatonin synthesis by norepinephrine, peptides, and other pineal transmitters. *Pharmacol. Rev.* **55**, 325–395.

Simonneaux, V., Poirel, V. J., Garidou, M. L., Nguyen, D., Diaz-Rodriguez, E., and Pévet, P. (2004). Daily rhythm and regulation of clock gene expression in the rat pineal gland. *Brain Res. Mol. Brain Res.* **120**, 164–172.

Sircar, R. (2000). Effect of melatonin on cocaine-induced behavioral sensitization. *Brain Res.* **857**, 295–299.

Sleipness, E. P., Sorg, B. A., and Jansen, H. T. (2005). Time of day alters long-term sensitization to cocaine in rats. *Brain Res.* **1065**, 132–137.

Thomas, M. J., Kalivas, P. W., and Shaham, Y. (2008). Neuroplasticity in the mesolimbic dopamine system and cocaine addiction. *Br. J. Pharmacol.* **154,** 327–342.

Tocharus, J., Chongthammakun, S., and Govitrapong, P. (2008). Melatonin inhibits amphetamine-induced nitric oxide synthase mRNA overexpression in microglial cell lines. *Neurosci. Lett.* **439,** 134–137.

Urba-Holmgren, R., Holmgren, B., and Aguiar, M. (1977). Circadian variation in an amphetamine induced motor response. *Pharmacol. Biochem. Behav.* **7,** 571–572.

Ushijima, K., Sakaguchi, H., Sato, Y., To, H., Koyanagi, S., Higuchi, S., and Ohdo, S. (2005). Chronopharmacological study of antidepressants in forced swimming test of mice. *J. Pharmacol. Exp. Ther.* **315,** 764–770.

Uz, T., Giusti, P., Franceschini, D., Kharlamov, A., and Manev, H. (1996). Protective effect of melatonin against hippocampal DNA damage induced by intraperitoneal administration of kainate to rats. *Neuroscience* **73,** 631–636.

Uz, T., Javaid, J. I., and Manev, H. (2002). Circadian differences in behavioral sensitization to cocaine: Putative role of arylalkylamine *N*-acetyltransferase. *Life Sci.* **70,** 3069–3075.

Uz, T., Akhisaroglu, M., Ahmed, R., and Manev, H. (2003). The pineal gland is critical for circadian Period1 expression in the striatum and for circadian cocaine sensitization in mice. *Neuropsychopharmacology* **28,** 2117–2123.

Uz, T., Ahmed, R., Akhisaroglu, M., Kurtuncu, M., Imbesi, M., Dirim Arslan, A., and Manev, H. (2005a). Effect of fluoxetine and cocaine on the expression of clock genes in the mouse hippocampus and striatum. *Neuroscience* **134,** 1309–1316.

Uz, T., Kurtuncu, M., and Manev, H. (2005b). Differential role of melatonin receptors MT1 and MT2 in the development of diurnal locomotor sensitization induced by cocaine. Abstract Viewer/Itinerary Planner Society for Neuroscience, Washington, DCProgram No. 799.11.

Van der Heijden, K. B., Smits, M. G., Van Someren, E. J., Ridderinkhof, K. R., and Gunning, W. B. (2007). Effect of melatonin on sleep, behavior, and cognition in ADHD and chronic sleep-onset insomnia. *J. Am. Acad. Child Adolesc. Psychiatry* **46,** 233–241.

Vinall, P. E., Kramer, M. S., Heinel, L. A., and Rosenwasser, R. H. (2000). Temporal changes in sensitivity of rats to cerebral ischemic insult. *J. Neurosurg.* **93,** 82–89.

Viyoch, J., Ohdo, S., Yukawa, E., and Higuchi, S. (2001). Dosing time-dependent tolerance of catalepsy by repetitive administration of haloperidol in mice. *J. Pharmacol. Exp. Ther.* **298,** 964–969.

von Gall, C., Weaver, D. R., Moek, J., Jilg, A., Stehle, J. H., and Korf, H. W. (2005). Melatonin plays a crucial role in the regulation of rhythmic clock gene expression in the mouse pars tuberalis. *Ann. N. Y. Acad. Sci.* **1040,** 508–511.

Weiss, M. D., Wasdell, M. B., Bomben, M. M., Rea, K. J., and Freeman, R. D. (2006). Sleep hygiene and melatonin treatment for children and adolescents with ADHD and initial insomnia. *J. Am. Acad. Child Adolesc. Psychiatry* **45,** 512–519.

Youan, B. B. (2004). Chronopharmaceutics: Gimmick or clinically relevant approach to drug delivery? *J. Control. Release* **98,** 337–353.

Zhdanova, I. V., and Giorgetti, M. (2002). Melatonin alters behavior and cAMP levels in nucleus accumbens induced by cocaine treatment. *Brain Res.* **956,** 323–331.

Zisapel, N. (2001). Melatonin-dopamine interactions: From basic neurochemistry to a clinical setting. *Cell. Mol. Neurobiol.* **21,** 605–616.

DOPAMINE-INDUCED BEHAVIORAL CHANGES AND OXIDATIVE STRESS IN METHAMPHETAMINE-INDUCED NEUROTOXICITY

Taizo Kita,* Ikuko Miyazaki,[†] Masato Asanuma,[†] Mika Takeshima,*
and George C. Wagner[‡]

*Laboratory of Pharmacology, Kyushu Nutrition Welfare University Graduate School
of Food and Nutrition, Kitakyushu, Fukuoka 803–8511, Japan
[†]Department of Brain Science, Okayama University Graduate School of Medicine, Dentistry
and Pharmaceutical Sciences, Okayama 700–8558, Japan
[‡]Department of Psychology, Rutgers University, New Brunswick, New Jersey 08854, USA

High-dose administration of amphetamine-like compounds is associated with acute behavioral toxicity (including stereotypic and self-injurious behavior and schizophrenic-like psychoses) as well as long-lasting damage to dopaminergic neurons. Several mechanisms are thought to be responsible for methamphetamine-induced neurotoxicity including the formation of reactive oxygen species, dopamine quinones, glutamatergic activity, apoptosis, etc. Recently, new factors regarding glial cell line-derived neurotorophic factor, tumor necrosis factor-α, and interferon-γ have also been associated with methamphetamine-induced neurotoxicity. The objective of this review is to link the behavioral and neurotoxic responses of the amphetamines, emphasizing their common underlying mechanism of monoaminergic release together with inhibition of monoamine oxidase activity. The amphetamine-induced release of dopamine and inhibition of monoamine oxidase increases both cytosolic and synaptic levels of dopamine leading to the acute manifestation of stereotypic and self-injurious behavior. In turn, the enhanced extravesicular levels of dopamine lead to oxidative stress through the generation of reactive oxygen species and dopamine quinones, and cause the long-lasting neuronal damage. Thus, we propose that acute behavioral

43

observation of subjects immediately following methamphetamine administration may provide insight into the long-lasting toxicity to dopaminergic neurons.

Abbreviations

AMPT, α-methyl-p-tyrosine; COX, cyclooxygenase; DOPAC, dihydroxyphenylacetic acid; GDNF, glial cell line-derived neurotrophic factor; 6-OHDA, 6-hydroxydopamine; IFN-γ, interferon-γ; MAO, monoamine oxidase; MAP, mitogen-activated protein; METH, methamphetamine; NQO, quinone oxidoreductase; PARP, poly(ADP-ribose) polymerase; PKC, protein kinase C; ROS, reactive oxygen species; SIB, self-injurious behavior; TNF-α, tumor necrosis factor-α; VMAT, vesicular monoamine transporter.

I. Background: Neuronal Damage Caused by Amphetamines

Amphetamine and methamphetamine (METH) are psychomotor stimulants, causing increased arousal, sympathetic activation, reduced food intake, and hyperactivity following low-dose administration. In addition, these drugs exert a reinforcing effect following their self-administration and are, therefore, prone to repeated use and abuse. It is estimated that 35 million people worldwide abuse amphetamine-like compounds, a number twice that for cocaine abuse and more than three times for opiate abuse (Meredith *et al.*, 2005). Following the repeated administration of low doses of amphetamine, a sensitization develops; this heightened responsiveness is manifest as a shorter latency for the onset of the stimulant effects, increased amplitude of these effects, and their appearance following lower doses of the drug. In humans, who repeatedly self-administer amphetamines, this sensitization may manifest as schizophrenic-like symptoms. Paradoxically, with repeated self-administration of the amphetamines, a tolerance develops to their reinforcing effects leading to the self-administration of ever-higher doses needed to produce the reinforcing effect.

Following the repeated administration of these higher doses of the amphetamines, the sensitization to the sympathetomimetic and schizophrenic-like effects becomes more likely. The hyperactivity turns into stereotypic and self-injurious behavior (SIB), together with hyperthermia, intense salivation, and cardiovascular arousal (Kalant, 1973). The amphetamine-induced psychoses closely resemble paranoid schizophrenia with common symptoms including feelings of persecution, violence, suicide, and hallucinations (Connell, 1958; Kalant, 1973; Young and Scoville, 1938). This schizophrenic-like behavior sometimes appears after a single dose of amphetamine but, in at least 75% of the reported cases, requires regular use of the drug for at least 1 month (i.e., sensitization develops). The dose range across individual cases required to produce the psychoses is enormous and, therefore, one might wonder about its generality. However, on further consideration, the large variation in doses required for the sensitization psychoses is likely accounted for by the equally large variations in the type and purity of the administered amphetamine, by the nature of coadministered compounds (including depressant drugs such as morphine, alcohol, and barbiturates as well as other types of stimulants such as cocaine, caffeine, and phenmetrazine), by the various routes of administration (inhalation, oral, i.v.), by the frequency of administration, and by the age and medical history of the subjects. Thus, when these extraneous factors are considered, it appears reasonably clear that with the repeated administration of ever-higher doses of the amphetamines, a psychoses-like condition will develop. It is under these conditions that the amphetamines become neurotoxic, causing long-lasting damage throughout the brain. The focus of this review will be the dopaminergic damage caused by the amphetamines.

The initial studies showed that the repeated administration of high doses of METH caused long-lasting dopaminergic damage used rhesus monkeys receiving i.v. METH eight times each day for a period of up to 6 months. The total daily dose of METH for these monkeys was as high as 52 mg/kg. The monkeys were sacrificed up to 6 months after the last injection of METH and it was found that there was nearly an 80% loss of striatal dopamine (Schuster and Fischman, 1975; Seiden et al., 1975, 1976). Of interest, these monkeys were trained to lever press for their daily ration of food; while the METH initially disrupted performance and the monkeys became food deprived, with its repeated administration, a tolerance developed to the METH and behavioral performance actually improved (Fischman and Schuster, 1975). That is, the 80% loss of striatal dopamine caused by 6 months of high-dose METH administration resulted in a shift to the right in the dose–response curve reflecting the disruptive effects on lever pressing for food following METH. These monkeys showed a neurochemical tolerance to the drug, requiring higher doses of METH to disrupt the behavior.

The first study to show dopaminergic damage in rats following repeated, high-dose exposure to METH was conducted by Kogan et al. (1976). Rats received 15 mg/kg (s.c.) of METH once every 6 h for 30 h (total dose of 90 mg/kg) and were found to have significant reductions in striatal tyrosine hydroxylase activity for at least 12 days after the last injection. Prior studies had demonstrated that the amphetamines decrease tyrosine hydroxylase activity. However, these studies were based upon sacrifice times only hours after the last injection, too short an interval to determine if the reduced tyrosine hydroxylase activity was consequent to the pharmacological effects of the drug (still present in the subject at the time of sacrifice) or due to METH-induced neuronal damage (Fibiger and McGeer, 1971; Koda and Gibb, 1973). Clearly, the Kogan *et al.* (1976) study with its 12-day waiting period reflected damage to dopaminergic neurons as opposed to the pharmacological action of METH. Collectively, these early studies provided evidence that high-dose administration of the amphetamines to monkeys or rats resulted in long-lasting damage to dopaminergic neurons as evidenced by a reduction in steady-state levels of dopamine as well as a reduction in tyrosine hydroxylase activity.

A third biochemical marker of amphetamine-induced dopaminergic damage was reported by Wagner *et al.* (1980a,b), who demonstrated that the repeated administration of high doses of METH to rats resulted in a loss of dopamine transport pumps in the striatum. As a marker of terminal integrity, the magnitude of the loss of the dopamine transport pumps was remarkably consistent with the long-lasting depletion of dopamine levels. Furthermore, in this study, both biochemical markers (i.e., steady-state levels of dopamine and number of dopamine transport pumps) were shown to be reduced by the METH treatment in a dose-dependent fashion.

Morphological evidence of the neurotoxic effects of the amphetamines soon followed these biochemical studies. Nwanze and Jonsson (1981) first showed that dopamine fluorescence was reduced in the caudate of mice following continuous amphetamine infusion (see also Jonsson and Nwanze, 1982). Ricaurte *et al.* (1982) next used the Fink–Heimer method to show that a high-dose METH treatment caused nerve terminal degeneration in the striatum of rats. Finally, Hess *et al.* (1990) and Sonsalla *et al.* (1996) showed that high-dose treatment with METH in mice resulted in a loss of dopaminergic cell bodies in the substantia nigra. These morphological markers of dopaminergic terminal and cell body damage leave no doubt that the amphetamines are potent dopaminergic toxins. It is of interest to note, however, that when one lowers the dose of the amphetamines some, but not all, of the biochemical and morphological markers are detected (e.g., Harvey *et al.*, 2000). Likewise, following lower dose administration of the amphetamines, it appears that some recovery will occur over time (Harvey *et al.*, 2000). However, when higher doses and/or longer term administration of the amphetamines occur in a manner similar to that of human amphetamine abuse, the dopaminergic damage is severe and permanent.

The dopaminergic toxicity following METH has been shown to occur across species. As already noted, the early studies demonstrated dopaminergic toxicity in monkeys, rats, and mice. Wagner *et al.* (1979) demonstrated that high-dose METH treatments to guinea pigs (30 mg/kg/day) again caused long-lasting depletions of striatal dopamine, apparent 2 weeks after the last injection. Of importance, the total dose used across all of these species was quite comparable to the doses self-administered by humans (who administer up to 2 g/day; Kalant, 1973). This suggests that it would be safe to assume that humans who do self-administer these high doses of amphetamine-like compounds would experience the same type of dopaminergic toxicity observed in the other species.

Wilson *et al.* (1996) first reported that humans who self-administered METH incurred dopaminergic damage akin to that observed in other species. The postmortem striatal tissue of 12 METH users was found to have reduced levels of dopamine as well as tyrosine hydroxylase activity and the dopamine transporter. However, their cautious interpretation concluded against neurotoxicity. McCann *et al.* (1998) next used PET scans and observed a similar loss of the dopaminergic transporter in the caudate nucleus of individuals with a history of METH use but who had been abstinent for approximately 3 years before testing. They argue convincingly that their observations, together with those of Wilson *et al.* (1996), lead to the conclusion that the METH use did, in fact, damage dopaminergic neurons. A similar observation by Volkow *et al.* (2001) added the observation that the reduction in the number of dopamine transporters was accompanied by subtle motoric impairments. Collectively, these observations indicate that the two decades of research using other animal species accurately predicted that humans who self-administer high doses of the amphetamines would experience striatal dopaminergic damage. More impressively, in a retrospective analysis it was found that humans with a diagnosis of Parkinson's disease have an elevated rate of prior use/abuse of amphetamines as compared to a control population (Garwood *et al.*, 2006). Beyond this literal interpretation, the study of the dopaminergic toxicity exerted by the amphetamines has led to the development of important animal models of schizophrenia, Parkinson's disease, and autism. These models are concerned with both the mechanisms through which amphetamines exert their neuronal damage as well as the behavioral toxicity that follows.

Amphetamines have two primary mechanisms of action, the inhibition of monoamine oxidase (MAO) and the release of dopamine (Carlsson, 1970). Collectively, these actions lead to an increase in cytoplasmic and extracellular dopamine. With regard to the mechanism through which amphetamines cause dopaminergic toxicity, it is somewhat surprising to note that dopamine itself is prerequisite. That is, when dopamine synthesis is inhibited by pretreatment with α-methyl-p-tyrosine (AMPT: a tyrosine hydroxylase inhibitor), the dopamine levels decline and the subject is protected against the subsequent METH-induced dopaminergic toxicity (Wagner *et al.*, 1983). To be more precise, despite the fact that two groups of rats received the same dose of METH, the group receiving the AMPT pretreatment was protected against the long-lasting dopamine toxicity because the transitory AMPT-induced reduction in dopamine precluded the opportunity for the METH to release dopamine. It was concluded that METH exerts dopaminergic damage by releasing dopamine and inhibiting MAO-B, thus increasing extravesicular levels of dopamine; in turn, the dopamine reacts with dissolved oxygen to create reactive oxygen species (ROS) (Fig. 1).

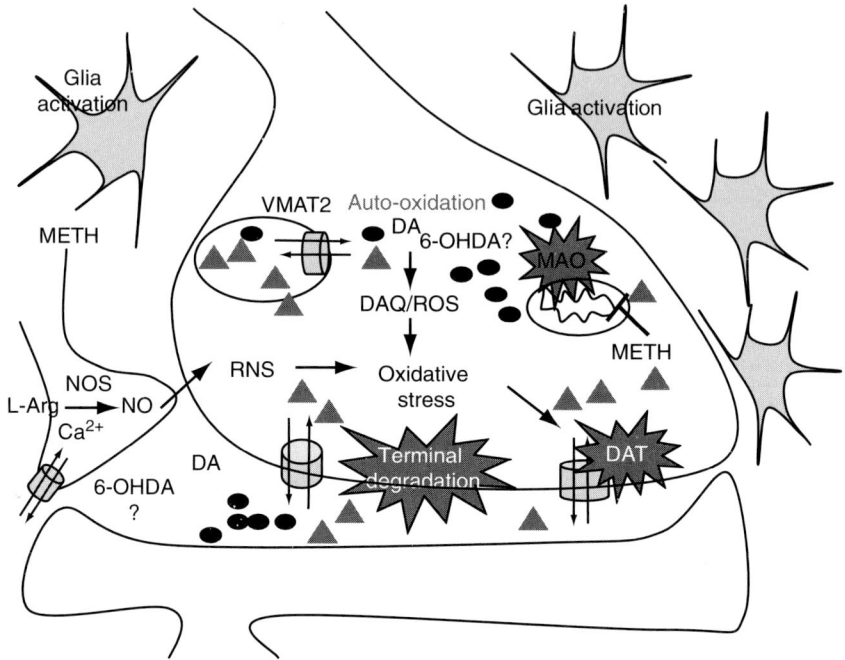

Fig. 1. Schematic model of METH-induced neurotoxicity in striatum. DA, dopamine; DAT, dopamine transporter; DAQ, dopamine quinone; NO, nitric oxide; MAO, monoamine oxidase; METH, methamphetamine; ROS, reactive oxygen species; RNS, reactive nitrogen species; VMAT, vesicular monoamine transporter.

The conclusion that dopamine release is prerequisite for amphetamine-induced dopaminergic neurotoxicity to occur has received substantial support (see Axt *et al.*, 1990; Commins and Seiden, 1986; Schmidt *et al.*, 1985). However, Yuan *et al.* (2001) provided some evidence to the contrary, suggesting that a hyperthermic response is critical. They showed that the ability of AMPT to protect subjects against METH-induced dopaminergic damage was negated when the subjects were placed under a heat lamp, reversing the AMPT-induced hypothermia. This apparent discrepancy may be explained by the fact that the hyperthermic condition caused by the heat lamp likely increases METH-induced dopamine release, thus offsetting the protective effects of the AMPT (Xie *et al.*, 2000). Nonetheless, the Yuan *et al.* (2001) study underscores the obvious conclusion that the mechanism through which the amphetamines cause dopaminergic damage is not fully understood.

Hyperthermia following high-dose administration of the amphetamines has been associated with the long-lasting damage to dopaminergic neurons (Albers and Sonsalla, 1995; Ali *et al.*, 1994; Bowyer *et al.*, 1993, 1994; Wallace *et al.*, 2001).

Indeed, across strains, the intensity of the hyperthermia soon after its administration appears to predict the dopamine depletion observed 3 days later (Halladay *et al.*, 2003). However, it is important to note that the hyperthermia serves as a marker but is not essential for the resulting dopaminergic damage. Amphetamine and METH are still toxic to dopaminergic neurons even if the hyperthermic response is eliminated (Albers and Sonsalla, 1995; Bowyer *et al.*, 1994; Melega *et al.*, 1998; Wagner *et al.*, 1983; Yuan *et al.*, 2001). Likewise, METH-induced dopaminergic toxicity can be eliminated by agents that do not eliminate the hyperthermic response (Albers and Sonsalla, 1995; Cappon *et al.*, 1996; Itzak *et al.*, 2000). Since it is easy to dissociate the hyperthermic response from the dopaminergic toxicity, it should be concluded that it is not the hyperthermic response but, rather, some other mechanism such as dopamine release that is causing the neuronal damage. Recently, Kawasaki *et al.* (2006) revealed that pretreatment with edaravone, a radical scavenger, protected mice against METH-induced dopaminergic neurotoxicity, but not activation of microglia or against METH-induced hyperthermia. Recently, we reported that central interferon-γ (IFN-γ) injections that prevented METH-induced neurotoxicity had no effects on METH-induced hyperthermia, although IFN-γ injected peripherally prevented METH-induced hyperthermia. These observations suggest that the underlying mechanism responsible for hyperthermia is independent of the mechanism responsible for METH-induced neurotoxicity.

The fact that dopamine release is an essential component of the mechanism through which the amphetamines cause dopaminergic damage is also illustrated by the observation that dopamine transport pump blockers do not cause the same damage to dopaminergic neurons, as do the dopaminergic releasers. For example, the repeated administration of high doses of methylphenidate, a dopamine transport blocker, did not result in long-lasting dopaminergic damage despite the fact that care was taken to match doses on a second critical indices of toxicity, lethality (Wagner *et al.*, 1980a,b). In follow-up studies, other dopamine transport blockers such as phencyclidine (Wagner *et al.*, 1984) and cocaine (Ryan *et al.*, 1988) were found not to cause dopaminergic damage while other dopamine releasers such as cathinone (Wagner *et al.*, 1982) were found to cause damage similar to the amphetamines. Of interest, administration of high doses of the dopamine transport pump blockers does result in the hyperthermic response; yet, these agents do not result in sufficient amounts of cytoplasmic or synaptic dopamine to cause the dopaminergic damage.

Collectively, these observations have led to the hypothesis that administration of high doses of the amphetamines leads to dopamine release and inhibition of MAO-B, thereby creating a condition where dopamine reacts with dissolved oxygen to form cytotoxic species. Toward this end, Seiden and Vosmer (1984) were able to detect 6-hydroxydopamine (6-OHDA) in the striatum of METH-treated rats. Likewise, Kita *et al.* (1995) were able to detect 6-OHDA *in vitro* following METH. Finally, pretreatment with antioxidants such as ascorbic acid (Wagner *et al.*, 1986) or vitamin E (DeVito and Wagner, 1989) completely protect subjects against dopaminergic toxicity induced by the amphetamines. These studies revealed that oxidative stress is likely involved in METH-induced dopaminergic toxicity but they do not address the precise mechanism leading to this damage.

Finally, Graham (1978) reported that catecholamines can auto-oxidize to form quinones in the following order: 6-OHDA > dopamine > norepinephrine > epinephrine. This will be the focal point of Section II.

II. Dopamine Quinone Formation as a Dopaminergic Neuron-Specific Neurotoxic Factor in Methamphetamine-Induced Neurotoxicity

Various hypotheses have been proposed regarding the mechanism responsible for METH-induced dopaminergic neurotoxicity (Axt *et al.*, 1990; Cadet and Brannock, 1998; Cubells *et al.*, 1994; Fumagalli *et al.*, 1999; LaVoie and Hastings, 1999; Liu and Edwards, 1997; Marek *et al.*, 1990; Seiden and Vosmer, 1984; Uhl, 1998; Wrona *et al.*, 1997; Yuan *et al.*, 2001). Several studies have demonstrated that endogenous dopamine plays an important role in mediating this damage (Cubells *et al.*, 1994; Liu and Edwards, 1997; Uhl, 1998; Wagner *et al.*, 1983; Wrona *et al.*, 1997). Dopamine release and redistribution from synaptic vesicles to cytoplasmic compartments appears to lead to the auto-oxidation of cytosolic free dopamine (Cadet and Brannock, 1998; Cubells *et al.*, 1994). As noted earlier, initial reports focused on METH-induced dopamine release and consequent increase in extracellular dopamine concentrations was responsible for METH-induced neurotoxicity (Axt *et al.*, 1990; Marek *et al.*, 1990; Seiden and Vosmer, 1984). However, through the use of vesicular monoamine transporter (VMAT)-2 knockout mice, Fumagalli *et al.* (1999) showed that disruption of VMAT potentiated METH-induced neurotoxicity, pointing to a greater contribution of intraneuronal dopamine redistribution as opposed to extraneuronal overflow. Furthermore, LaVoie and Hastings (1999) measured 5-cysteinyl-dopamine levels and concluded that increased intracellular (but not extracellular) dopamine oxidation is better associated with METH neurotoxicity. Thus, although there remains some controversy regarding the mechanism through which METH produces its neurotoxic effects, there is general agreement that dopamine plays a role.

The neurotoxic actions of dopamine and DOPA quinones in the context of dopaminergic-specific oxidative stress has recently been linked to the pathogenesis of Parkinson's disease (Asanuma *et al.*, 2003, 2004; Choi *et al.*, 2003, 2005; LaVoie *et al.*, 2005; Machida *et al.*, 2005; Whitehead *et al.*, 2001). There are two major oxidation pathways to consider for dopamine. First, dopamine as a neurotransmitter is usually stable within the synaptic vesicle under normal physiological conditions. However, when an excess amount of cytosolic dopamine exists, it is easily metabolized via MAO-B or by auto-oxidation to produce cytotoxic ROS that, in turn, lead to the formation of neuromelanin (Sulzer and Zecca, 2000; Sulzer et al., 2000). In the oxidation of dopamine by MAO, dopamine is converted to dihydroxyphenylacetic acid (DOPAC) with hydrogen peroxide as a byproduct. In contrast, the non-enzymatic, spontaneous auto-oxidation of dopamine and L-DOPA produces superoxide and reactive quinones such as dopamine quinones or DOPA quinines, respectively (Graham, 1978; Tse et al., 1976). Dopamine quinones are also generated in the enzymatic oxidation of dopamine by prostaglandin H synthase (cyclooxygenase: COX), lipoxygenase, tyrosinase, and xanthine oxidase (Foppoli, et al., 1997; Graham, 1978; Hastings, 1995; Korytowski et al., 1987; Rosei et al., 1994; Tse et al., 1976). These quinones are easily oxidized to the cyclized aminochromes and are then finally polymerized to form melanin. Since these highly reactive catechol quinones are generated from free cytosolic dopamine outside the synaptic vesicle, they exert cytotoxicity predominantly from within dopaminergic neurons (Sulzer et al., 2000).

The dopamine and DOPA quinones conjugate with sulfhydryl groups of cysteine to form 5-cysteinyl-dopamine and 5-cysteinyl-DOPA, respectively (Fornstedt et al., 1986; Graham, 1978; Ito and Fujita, 1982). Since the cystein residues are often found at the active site of functional proteins, covalent modification by these catechol quinines may irreversibly alter or inhibit protein function leading to the cytotoxicity. In particular, it is of interest that dopamine quinones generated in the brain covalently modify and inactivate tyrosine hydroxylase and the dopamine transporter, subsequently inhibiting both dopamine synthesis and uptake (Kuhn et al., 1999; Whitehead et al., 2001; Xu et al., 1998). More interestingly, LaVoie et al. (2005) showed that dopamine covalently modifies the parkin protein to increase parkin insolubility and inactivate its E3 ubiquitin ligase function. This protein is particularly relevant in the context of dopaminergic toxicity as a mutation of this gene has been associated with familial Parkinson's disease.

To clarify the possible involvement of dopamine quinone formation in METH-induced dopaminergic neurotoxicity, we recently examined changes in cell death and quinoprotein (protein-bound quinone) formation using METH-treated dopaminergic cells and METH-injected mouse brain (Miyazaki et al., 2006). In dopaminergic CATH.a cells, METH treatment dose-dependently increased the levels of quinoprotein and the expression of quinone reductase in parallel with neurotoxicity. Similarly, repeated injections of METH caused an increase in quinoprotein levels coinciding with reduction in the number of dopamine transporters. Furthermore, pretreatment with the quinone reductase inducer, butylated hydroxyanisole, significantly and dose-dependently blocked the METH-induced elevation of quinoprotein, and ameliorated METH-induced cell death in cultured dopaminergic cells.

With respect to the melanin biosynthesis pathway, tyrosinase catalyzes both the hydroxylation of tyrosine to L-DOPA as well as the consequent oxidation of L-DOPA to form melanin (Hearing and Ekel, 1976). In the central nervous system, tyrosinase and its promoter activity are expressed (Tief *et al.*, 1997, 1998) and enzymatically oxidize, not only L-DOPA but also dopamine, to form melanin via the dopamine quinone (Miranda and Botti, 1983). The rapid oxidation of excess cytosolic dopamine to form stable melanin by tyrosinase may prevent the slow progression of cell damage induced by dopamine auto-oxidation and long-term exposure to dopamine quinone (Graham, 1978; Korytowski *et al.*, 1987). Using tyrosinase-null mice, we recently showed that the presence of tyrosinase protects against METH-induced dopaminergic neurotoxicity *in vitro* and *in vivo*. Repeated METH injections caused a reduction of the dopamine transporter in the striatum that was markedly aggravated in the tyrosinase-null mice as compared with that in the METH-injected wild-type mice. Interestingly, the basal quinoprotein level in the striatum of tyrosinase-null mice was higher than that of wild-type mice, suggesting vulnerability in tyrosinase-null mice. These experimental results indicate that dopamine quinone formation is involved in the METH-induced dopaminergic neurotoxicity *in vitro* and *in vivo* as a dopaminergic neuron-specific neurotoxic factor, and that quinone formation-related molecules, such as quinone reductase and tyrosinase (Miyazaki *et al.*, 2006), protect against METH neurotoxicity by reducing intracellular free dopamine and dopamine quinone.

Dopamine quinone formation and consequent dopaminergic cell damage *in vitro* and *in vivo* were successfully prevented by treatment with superoxide dismutase, glutathione, some thiol reagents, and some dopamine agonists through their quinone-quenching activities (Asanuma *et al.*, 2005; Haque *et al.*, 2003a,b; Kuhn *et al.*, 1999; Lai and Yu, 1997; LaVoie and Hastings, 1999; Miyazaki *et al.*, 2005; Offen *et al.*, 1996). Treatment with butylated hydroxyanisole, dimethyl fumarate, or *tert*-butylhydroquinone, which upregulates the activity of NADPH: quinone oxidoreductase (NQO-1:quinone reductase), protects against cell death consequent to quinone formation (Choi *et al.*, 2003; Duffy *et al.*, 1998; Hara *et al.*, 2003; Munday *et al.*, 1998). In particular, dimethyl fumarate increases not only the activity of quinone reductase but also total intracellular glutathione and the activities of glutathione-*S*-transferase and glutathione reductase to reduce the cytotoxicity associated with dopamine quinone formation. Therefore, some reagents that activate NQO quinone reductase may be potential drugs to inhibit dopamine quinone-induced neurotoxicity. Taking these findings into consideration, enhancing activities of quinone formation-inhibiting molecules such as quinone reductase would be a novel approach to prevent METH-induced neurotoxicity. A proposed mechanism of METH-induced dopamine quinone formation and neurotoxicity is depicted in Fig. 2.

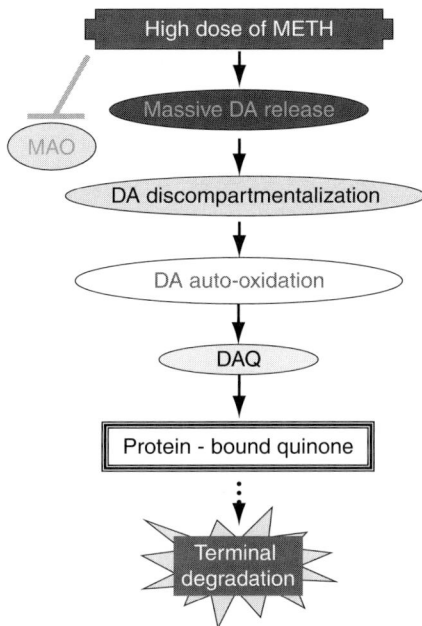

FIG. 2. A proposed mechanism of quinone formation in METH-induced neurotoxicity. DA, dopamine; DAT, dopamine transporter; DAQ, dopamine quinone; NO, nitric oxide; MAO, monoamine oxidase; METH, methamphetamine.

III. Methamphetamine-Induced Behavioral Changes: Relation to Neuronal Damage

A. SENSITIZATION TO METHAMPHETAMINE-INDUCED BEHAVIORAL CHANGES

As noted, the repeated administration of amphetamine-like compounds causes an intense psychological dependence, and continued use or abuse of high doses results in METH-induced psychosis (Ellinwood *et al.*, 1973; Strakowski *et al.*, 1997). In animal models, it is well known that administration of the amphetamines induces hyperlocomotion at low doses and stereotypic behavior at higher doses. It is generally believed that dopaminergic transmission in the nucleus accumbens and striatum mediate amphetamine-induced hyperlo-comotion and stereotypic behavior, respectively (Creese and Iversen, 1974; Kelly and Iversen, 1976; Kelly *et al.*, 1975). The accumbens is innervated by dopamine-containing cells of the ventral tegmental area as well as glutamate-containing cells from the prefrontal cortex (Carlezon and Wise, 1996; Koob and Bloom, 1988; Wise and Bozarth, 1987).

The behavioral sensitization that is observed following the repeated administration of these compounds (Kalivas and Stewart, 1991; Kelly *et al.*, 1975; Robinson and Becker, 1986) is thought to be mediated by a long-lasting neural plasticity manifest as a drug-induced increases in the length of dendrites and increased density and number of spines in the accumbens and frontal cortex (Robinson and Kolb, 1997). In addition, Narita *et al.* (2005) reported that an intra-accumnbens preinjection with 0.025 pmol/rat of rapamycin (a selective inhibitor of p70-S6K, a key factor in the regulation of cell growth and proliferation; Dufner and Thomas, 1999) failed to affect METH-induced conditioned place preference but abolished the development of sensitization of the METH-induced conditioned place preference. In addition, the immunoreactivities in the cytosolic preparation for Western blotting and immunohistochemical density of phosphorylated-p70-S6K were significantly increased in the accumbens of METH-sensitized rats as compared to those treated with chronic saline, an effect not observed in the nonsensitized conditioned place preference rats. The authors speculated that increased translation factors regulated by p70-S6K could lead to a sustained increase in the translational capacity of the synapse. These results may account for the development of behavioral sensitization to METH by both increasing translation of RNAs that are at the synapse and RNAs that are emanating from the nucleus.

It has been shown that amphetamine-induced increases in mitogen-activated protein (MAP) kinese and protein kinase C (PKC) are important for the induction of both the enhancement in transporter-mediated dopamine release and neurite outgrowth (Park *et al.*, 2003). Moreover, MAP kinase and MKPs have been shown to be involved in psychomotor stimulant-induced behavioral sensitization (Pierce and Kalivas, 1997). Narita *et al.* (2003a,b) reported a role for brain-derived neurotrophic factor/tyrosine kinase and Rho/Rho-associated kinase pathways in dopamine release in the accumbens and the expression of dopamine-related behaviors induced by METH. In addition, they found that the upregulation of PKC may be involved in sensitization of synaptic transmission (Narita *et al.*, 2004). These findings indicate that the activated signal transduction by PKC in the accumbens may play an important role in the development of sensitization to the rewarding effects of METH. In addition, they demonstrated that an enhancement of METH-induced dopamine release is completely blocked by chronic pretreatment with a selective PKC inhibitor, chelerythrine chloride, into the accumbens of the METH-induced sensitized rats (Narita *et al.*, 2004).

Recent evidence suggests that tumor necrosis factor-α (TNF-α), a potent proinflammatory cytokine, plays a critical role in the METH-induced dependence and neurotoxicity (Nakajima *et al.*, 2004; Niwa *et al.*, 2007). Nakajima *et al.* (2004) reported that TNF-$\alpha(-/-)$ mice showed enhanced responses to the locomotor-sensitizing, rewarding, and neurotoxic effects of METH compared with wild-type mice. In addition, they also demonstrated that exogenous TNF-α attenuated the METH-induced increase in extracellular striatal dopamine *in vivo* and potentiated striatal dopamine uptake into synaptosomes *in vitro* and *in vivo*, suggesting that TNF-α directly activates vesicular dopamine uptake and diminishes the METH-induced decrease in vesicular dopamine uptake. Therefore, they proposed that TNF-α plays a neuroprotective role in METH-induced drug dependence and neurotoxicity by activating plasmalemmal and vesicular dopamine transporter as well as inhibiting METH-induced increase in extracellular dopamine levels.

Furthermore, IFN-γ, an inflammatory cytokine, may be critically involved in the mechanism of METH-induced neurotoxicity. Yu *et al.* (2002) reported that chronic treatment with METH alters immue function. Recently, Hozumi *et al.* (2008) reported that IFN-γ injected systemically or its related molecule protects against METH-induced neurotoxicity through intracerebral molecular pathways, although it prevents METH-induced hyperthermia through different molecular events. Therefore, these cytokines or their related molecules would be novel therapeutic agents to prevent METH-induced dependence or neurotoxicity.

B. Linking Behavioral and Neurotoxic Actions of the METH

The paradigm most often used to study METH-induced dopamine neurotoxicity involves the repeated administration of high doses of the drug. Immediately following its administration, animals engage in hyper excitability, with much pacing, checking, sniffing, and exploration and almost zero food intake. As the regimen continues, these behaviors become increasingly repetitive often with a heightened amount of SIB to a point of ulceration of the skin (Kita *et al.*, 2003).

The neuronal changes following the high-dose administration of METH are accompanied by functional deficits, most notably, a decrement in locomotor performance as assessed by conditioned avoidance responses paradigm (Walsh and Wagner, 1992). The long-lasting neuronal damage following high-dose administration of METH has been associated with long-lasting behavioral deficits (Wallace *et al.*, 1999) as well as hyperactivity during the dark cycle (Kita *et al.*, 1998a) under baseline conditions as well as following drug challenge (Friedman *et al.*, 1998; Wallace *et al.*, 1999, 2001).

Recently, it has been reported that glia cell line-derived neurotrophic factor (GDNF) is an important regulator of METH-induced dopamine neurotoxicity and motor-system function during aging (Boger *et al.*, 2007). These authors suggest that METH use in young adults, when combined with lower levels of GDNF through life, may precipitate the appearance of Parkinson-like behaviors during aging. Therefore, the dopamine depletion caused by the amphetamines, although never as intense as that found in Parkinson's disease, does represent the excellent model for degenerative disease.

Previously, we found that BALB/c mice exhibited an increased sensitivity to METH-induced striatal dopaminergic toxicity as compared to other strains of mice (Kita *et al.*, 1998b). In addition, we reported that the repeated administration of toxic doses of METH caused intense SIB and that this behavior was related to the both the immediate METH-induced striatal dopamine release as well as the long-lasting METH-induced neurotoxicity (Halladay *et al.*, 2003; Kita *et al.*, 2000a). Furthermore, we demonstrated that both the SIB and the neurotoxicity produced by high doses of METH were significantly reduced by pretreatment with the spin-trap agent, α-phenyl-*N-tert*-butylnitrone. This observation was interpreted as an indication that METH-induced SIB may accurately reflect the formation of ROS (Kita *et al.*, 2000b). Recently, pretreatment with risperidone (a mixed $5\text{-}HT_2/D_2$ antagonist) eliminated the SIB produced by high doses of amphetamine administered to BALB/c mice (Wagner *et al.*, 2004) and pretreatment with haloperidol, SCH23390, 5-HTP, and fluvoxamine also attenuated METH-induced SIB in ddY mice (Mori *et al.*2004). Collectively, these studies indicate that the immediate SIB following high-dose administration of amphetamine-like compounds is the result of both dopaminergic and serotonergic activation.

Recently, Mori *et al.* (2004, 2006, 2007) reported that μ-opioid receptor agonist morphine, but not the δ-opioid receptor agonist SNC80, attenuated METH-induced SIB in mice. In addition, the κ-opioid agonist U50, 488H slightly, but significantly, attenuated METH-induced SIB. Therefore, each μ-, δ-, and κ-opioid receptor agonists modulate METH-induced SIB through different mechanisms.

Finally, it appears that there is also a relationship between excitatory amino acid neurotransmission and incidence of SIB. That is, MK801 significantly attenuated the METH-induced SIB, suggesting that the activation of N-methyl-D-asparatate receptor is involved in the METH-induced SIB (Mori *et al.*, 2007).

Furthermore, they also reported that neuronal nitric oxide synthase inhibitor 7-nitroindazole and poly(ADP-ribose) polymerase (PARP) inhibitor, benzamide attenuate the METH-induced SIB, suggesting that oxidative stress mediated by the activation of neural nitric oxide synthase is associated with METH-induced SIB.

From these evidences, we indicate that METH-induced SIB is closely connect with neuronal damage in brain (mainly in the striatum) and the mechanism of incidence of SIB also is associated with ROS, reactive nitrogen species-induced oxidative stress, monoamine, and opioid neuronal systems (Fig. 3).

IV. Conclusion

The behavioral symptoms of Parkinson's disease do not appear until there is about an 80–90% depletion of striatal dopamine. METH-induced dopamine toxicity seldom reaches this level and, therefore, individuals who have repeatedly administered high doses of METH do not show the tremor, rigidity, or akinesia associated with the disease. In other species, sensitive behavioral assays are required to detect these otherwise preclinical lesions induced by high-dose METH administration. We believe that the acute behavioral hyperactivity and

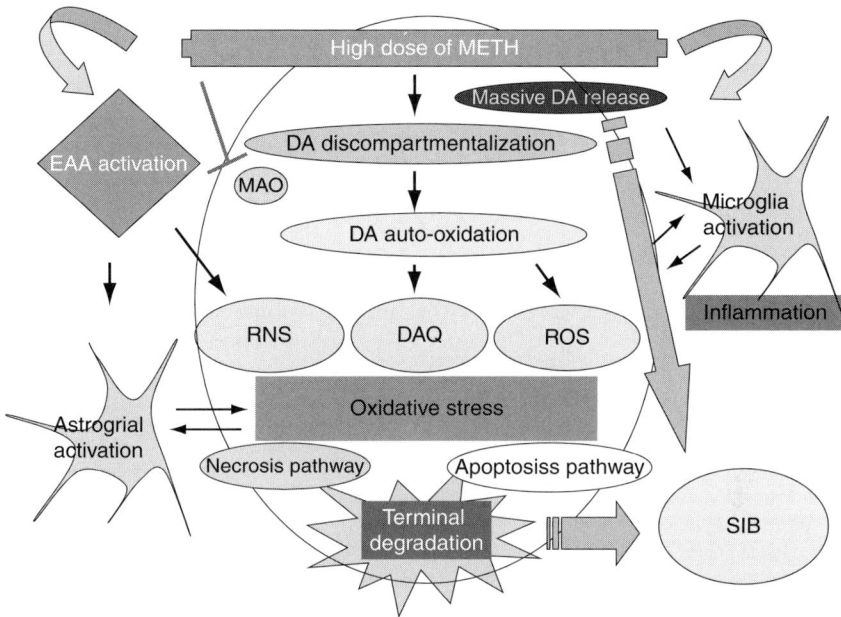

FIG. 3. A proposed mechanism of self-injurious behavior (SIB) in METH-induced neurotoxicity. DA, dopamine; DAT, dopamine transporter; DAQ, dopamine quinone; NO, nitric oxide; MAO, monoamine oxidase; METH, methamphetamine; SIB, self-injurious behavior.

SIB following the high-dose METH administration reflects the redistribution of the dopamine from the vesicular to cytoplasmic stores and, ultimately, to the dopamine release. This action of METH, together with inhibition of MAO, links the acute behavioral effects of METH to its long-lasting neurotoxicity. Endogenous dopamine plays an important role in mediating METH-induced neuronal damage. The dopamine release and redistribution from synaptic vesicles to cytoplasmic compartments is thought to cause dopamine quinone formation as well as ROS generation and to involve METH-induced changes in both the vesicular monoamine transporter-2 and DA transporter function. Therefore, enhancing activities of quinone formation-inhibiting molecules would be a novel approach to prevent METH-induced neurotoxicity.

References

Albers, D. S., and Sonsalla, P. K. (1995). Methamphetamine-induced hyperthermia and dopaminergic neurotoxicity in mice: Pharmacologic profile of protective and nonprotective agents. *J. Pharmacol. Exp. Ther.* **275,** 1104–1114.

Ali, S. F., Newport, G. D., Holson, R. R., Slikker, W., and Bowyer, J. F. (1994). Low environmental temperatures or pharmacologic agents that produce hypothermia decrease methamphetamine neurotoxicity in mice. *Brain Res.* **658,** 33–38.

Asanuma, M., Miyazaki, I., and Ogawa, N. (2003). Dopamine- or L-DOPA-induced neurotoxicity: The role of dopamine quinone formation and tyrosinase in a model of Parkinson's disease. *Neurotox. Res.* **5,** 165–176.

Asanuma, M., Miyazaki, I., Diaz-Corrales, F. J., and Ogawa, N. (2004). Quinone formation as dopaminergic neuron-specific oxidative stress in pathogenesis of sporadic Parkinson's disease and neurotoxin-induced parkinsonism. *Acta Med. Okayama* **58,** 221–233.

Asanuma, M., Miyazaki, I., Diaz-Corrales, F. J., Shimizu, M., Tanaka, K., and Ogawa, N. (2005). Pramipexole has ameliorating effects on levodopa-induced abnormal dopamine turnover in parkinsonian striatum and quenching effects on dopamine-semiquinone generated *in vitro. Neurol. Res.* **27,** 533–539.

Axt, K. J., Commins, D. L., Vosmer, G., and Seiden, L. S. (1990). Alpha-methyl-*p*-tyrosine pretreatment partially prevents methamphetamine-induced endogenous neurotoxin formation. *Brain Res.* **515,** 269–276.

Boger, H. A., Middaugh, L. D., Patrick, K. S., Ramamoorthy, S., Denehy, E. D., Zhu, H., Pacchioni, A. M., Granholm, A. C., and McGinty, J. F. (2007). Long-term consequences of methamphetamine exposure in young adults are exacerbated in glial cell line-derived neurotrophic factor heterozygous mice. *J. Neurosci.* **27,** 8816–8825.

Bowyer, J. F., Gough, B., Slikker, W., Lipe, G. W., Newport, G. D., and Holson, R. R. (1993). Effects of a cold environment or age on methamphetamine-induced dopamine release in the caudate putamen of female rats. *Pharmacol. Biochem. Behav.* **44,** 87–98.

Bowyer, J. F., Davies, D. L., Schmued, L., Broening, H. W., Newport, G. D., Slikker, W., and Holson, R. R. (1994). Further studies of the role of hyperthermia in methamphetamine toxicity. *J. Pharmacol. Exp. Ther.* **268,** 1571–1580.

Cadet, J. L., and Brannock, C. (1998). Free radicals and the pathobiology of brain dopamine systems. *Neurochem. Int.* **32,** 117–131.

Cappon, G. D., Broening, H. W., Pu, C., Morford, L., and Vorhees, C. V. (1996). alpha-phenyl-*N-tert*-Butyl nitrone attenuates methamphetamine-induced depletion of striatal dopamine without altering hyperthermia. *Synapse* **24,** 173–181.

Carlezon, W. A., Jr., and Wise, R. A. (1996). Microinjections of phencyclidine(PCP) and related drugs into nucleus accumbense shell potentiate medial forebrain bundle brain stimulation reward. *Psychopharmacology (Berl.)* **128,** 413–420.

Carlsson, A. (1970). Amphetamine and brain catecholamines. *In* "Amphetamines and Related Compounds", (E. Costa and S. Garattini, Eds.), pp. 298–300. Raven Press, New York.

Choi, H. J., Kim, S. W., Lee, S. Y., and Hwang, O. (2003). Dopamine-dependent cytotoxicity of tetrahydrobiopterin: A possible mechanism for selective neurodegeneration in Parkinson's disease. *J. Neurochem.* **86,** 143–152.

Choi, H. J., Lee, S. Y., Cho, Y., and Hwang, O. (2005). Inhibition of vesicular monoamine transporter enhances vulnerability of dopaminergic cells: Relevance to Parkinson's disease. *Neurochem. Int.* **46,** 329–335.

Commins, D. H., and Seiden, L. S. (1986). alpha-Methyltyrosine blocks methylamphetamine-induced degeneration in the rat somatosensory cortex. *Brain Res.* **365,** 15–20.

Connell, P. H. (1958). *Amphetamine Psychoses.* Chapman and Hall, London.

Creese, I., and Iversen, S. D. (1974). The role of forebrain dopamine systems in amphetamine induced stereotyped behavior in the rat. *Psychopharmaclogia* **39,** 345–357.

Cubells, J. F., Rayport, S., Rajendran, G., and Sulzer, D. (1994). Methamphetamine neurotoxicity involves vacuolation of endocytic organelles and dopamine-dependent intracellular oxidative stress. *J. Neurosci.* **14,** 2260–2271.

De Vito, M. J., and Wagner, G. C. (1989). METH-induced neuronal damage: A possible role for free radicals. *Neuropharmacology* **28**, 1145–1150.

Duffy, S., So, A., and Murphy, T. H. (1998). Activation of endogenous antioxidant defenses in neuronal cells prevents free radical-mediated damage. *J. Neurochem.* **71**, 69–77.

Dufner, A., and Thomas, G. (1999). Ribosomal S6 kinase signaling and the control of translation. *Exp. Cell Res.* **253**, 100–109.

Ellinwood, E. H., Jr., Sudilovsky, A., and Nelson, L. M. (1973). Evolving behavior in the clinical and experimental amphetamine (model) psychosis. *Am. J. Psychiatry* **130**, 1088–1093.

Fibiger, H. C., and McGeer, E. G. (1971). Effect of acute and chronic methamphetamine treatment on tyrosine hydroxylase activity in brain and adrenal medulla. *Eur. J. Pharmacol.* **16**, 176–180.

Fischman, M. W., and Schuster, C. R. (1975). Behavioral, Biochemical and morphological effects of methamphetamine in the rhesus monkey. In "Behavioral Toxicology", (B. Weiss and V. G. Laties, Eds.), pp. 375–399. Plenum Publishing Co., New York.

Foppoli, C., Coccia, R., Cini, C., and Rosei, M. A. (1997). Catecholamines oxidation by xanthine oxidase. *Biochim. Biophys. Acta* **1334**, 200–206.

Fornstedt, B., Rosengren, E., and Carlsson, A. (1986). Occurrence and distribution of 5-S-cysteinyl derivatives of dopamine, dopa and dopac in the brains of eight mammalian species. *Neuropharmacology* **25**, 451–454.

Friedman, S. D., Castaneda, E., and Hodge, G. K. (1998). Long-term monoamine depletion, differential recovery, and subtle behavioral impairment following methamphetamine-induced neurotoxicity. *Pharmacol. Biochem. Behav.* **61**, 35–44.

Fumagalli, F., Gainetdinov, R. R., Wang, Y. M., Valenzano, K. J., Miller, G. W., and Caron, M. G. (1999). Increased methamphetamine neurotoxicity in heterozygous vesicular monoamine transporter 2 knock-out mice. *J. Neurosci.* **19**, 2424–2431.

Garwood, E. R., Bekele, W., McCulloch, C. E., and Christine, C. W. (2006). Amphetamine exposure is elevated in Parkinson's disease. *Neurotoxicology* **27**, 1003–1006.

Graham, D. G. (1978). Oxidative pathways for catecholamines in the genesis of neuromelanin and cytotoxic quinones. *Mol. Pharmacol.* **14**, 633–643.

Halladay, A. K., Kusnecov, A., Michna, L., Kita, T., Hora, C., and Wagner, G. C. (2003). Relationship between methamphetamine-induced dopamine release, hyperthermia, self-injurious behavior and long term dopamine depletion in BALB/c and C57BL/6 mice. *Pharmacol. Toxicol.* **93**, 33–41.

Haque, M. E., Asanuma, M., Higashi, Y., Miyazaki, I., Tanaka, K., and Ogawa, N. (2003a). Apoptosis-inducing neurotoxicity of dopamine and its metabolites via reactive quinone generation in neuroblastoma cells. *Biochim. Biophys. Acta* **1619**, 39–52.

Haque, M. E., Asanuma, M., Higashi, Y., Miyazaki, I., Tanaka, K., and Ogawa, N. (2003b). Overexpression of Cu-Zn superoxide dismutase protects neuroblastoma cells against dopamine cytotoxicity accompanied by increase in their glutathione level. *Neurosci. Res.* **47**, 31–37.

Hara, H., Ohta, M., Ohta, K., Kuno, S., and Adachi, T. (2003). Increase of antioxidative potential by *tert*-butylhydroquinone protects against cell death associated with 6-hydroxydopamine-induced oxidative stress in neuroblastoma SH-SY5Y cells. *Brain Res. Mol. Brain Res.* **119**, 125–131.

Harvey, D. C., Lacan, G., Tanious, S. P., and Melega, W. P. (2000). Recovery from methamphetamine induced long-term nigrostriatal deficits without substantia nigra cell loss. *Brain Res.* **871**, 259–270.

Hastings, T. G. (1995). Enzymatic oxidation of dopamine: The role of prostaglandin H synthase. *J. Neurochem.* **64**, 919–924.

Hearing, V. J., and Ekel, T. M. (1976). Mammalian tyrosinase. A comparison of tyrosine hydroxylation and melanin formation. *Biochem. J.* **157**, 549–557.

Hess, A., Desiderio, C., and McAuliffe, W. G. (1990). Acute neuropathological changes in the caudate nucleus caused by MPTP and methamphetamine: Immunohistochemical studies. *J. Neurocytol.* **19**, 338–342.

Hozumi, H., Asanuma, M., Miyazaki, I., Fukuoka, S., Kikkawa, Y., Kimoto, N., Kitamura, Y., Sendo, T., Kita, T., and Gomita, Y. (2008). Protective effects of interferon-gamma against methamphetamine-induced neurotoxicity. *Toxicol. Lett.* **177,** 123–129.

Ito, S., and Fujita, K. (1982). Conjugation of dopa and 5-S-cysteinyldopa with cysteine mediated by superoxide radical. *Biochem. Pharmacol.* **31,** 2887–2889.

Itzhak, Y., Martin, J. L., and Ali, S. F. (2000). nNOS inhibitors attenuate methamphetamine-induced dopaminergic neurotoxicity but not hyperthermia in mice. *Neuroreport* **11,** 2943–2946.

Jonsson, G., and Nwanze, E. (1982). Selective (+) amphetamine neurotoxicity on striatal dopamine nerve terminals in the mouse. *Br. J. Pharmacol.* **77,** 335–345.

Kalant, O. J. (1973). The Amphetamines Toxicity and Addiction, 2nd edn. Charles C. Thomas Publisher, Springfield, IL.

Kalivas, P. W., and Stewart, J. (1991). Dopamine transmission in the initiation and expression of drug- and stress-induced sensitization of motor activity. *Brain Res. Brain Res. Rev.* **16,** 223–244.

Kawasaki, T., Ishihara, K., Ago, Y., Nakamura, S., Itoh, S., Baba, A., and Matsuda, T. (2006). Protective effect of the radical scavenger edaravone against methamphetamine-induced dopaminergic neurotoxicity in mouse striatum. *Eur. J. Pharmacol.* **542,** 92–99.

Kelly, P. H., and Iversen, S. D. (1976). Selective 6OHDA-induced destruction of mesolimbic dopamine neurons: Abolition of psychostimulant-induced locomotor activity in rats. *Eur. J. Pharmacol.* **40,** 45–56.

Kelly, P. H., Seviour, P. W., and Iversen, S. D. (1975). Amphetamine and apomorphine responses in the rat following 6-hydroxydopamine lesions of the nucleus accumbens septi and corpus striatum. *Brain Res.* **94,** 507–522.

Kita, T., Wagner, G. C., Philbert, M. A., King, L. A., and Lowndes, H. E. (1995). Effects of pargyline and pyrogallol on the methamphetamine-induced dopamine depletion. *Mol. Chem. Neuropathol.* **24,** 31–41.

Kita, T., Takahashi, M., Wagner, G. C., Kubo, K., and Nakashima, T. (1998a). Methamphetamine-induced changes in activity and water intake during light and dark cycles in rats. *Prog. Neuropsychopharmacol. Biol. Psychiatry* **22,** 1185–1196.

Kita, T., Paku, S., Takahashi, M., Kubo, K., Wagner, G. C., and Nakashima, T. (1998b). Methamphetamine-induced neurotoxicity in BALB/c, DBA/2N and C57BL/6N mice. *Neuropharmacology* **37,** 1177–1184.

Kita, T., Matsunari, Y., Saraya, T., Shimada, K., O'Hara, K., Kubo, K., Wagner, G. C., and Nakashima, T. (2000a). Methamphetamine-induced striatal dopamine release, behavior changes and neurotoxicity in BALB/c mice. *Int. J. Dev. Neurosci.* **18,** 521–530.

Kita, T., Matsunari, Y., Saraya, T., Shimada, K., O'Hara, K., Kubo, K., Wagner, G. C., and Nakashima, T. (2000b). Evaluation of the effects of α-phenyl-*N-tert*-butyl nitrone pretreatment on the neurobehavioral effects of methamphetamine. *Life Sci.* **67,** 1559–1571.

Kita, T., Wagner, G. C., and Nakashima, T. (2003). Current research on methamphetamine- induced neurotoxicity: Animal models of monoamine disruption. *J. Pharmacol. Sci.* **92,** 178–195.

Koda, L. Y., and Gibb, J. W. (1973). Adrenal and striatal tyrosine hydroxylase activity after methamphetamine. *J. Pharm. Exp. Ther.* **185,** 42–48.

Kogan, F. J., Nichols, W. K., and Gibb, J. W. (1976). Influence of methamphetamine on nigral and striatal tyrosine hydroxylase activity and on striatal dopamine levels. *Eur. J. Pharmacol.* **36,** 363–371.

Koob, G. F., and Bloom, F. E. (1988). Cellular and molecular mechanisms of drug dependence. *Science* **242,** 715–723.

Korytowski, W., Sarna, T., Kalyanaraman, B., and Sealy, R. C. (1987). Tyrosinase-catalyzed oxidation of dopa and related catechol(amine)s: A kinetic electron spin resonance investigation using spin-stabilization and spin label oximetry. *Biochim. Biophys. Acta* **924,** 383–392.

Kuhn, D. M., Arthur, R. E., Jr., Thomas, D. M., and Elferink, L. A. (1999). Tyrosine hydroxylase is inactivated by catechol-quinones and converted to a redox-cycling quinoprotein: Possible relevance to Parkinson's disease. *J. Neurochem.* **73,** 1309–1317.

Lai, C. T., and Yu, P. H. (1997). Dopamine- and L-β-3,4-dihydroxyphenylalanine hydrochloride (L-Dopa)-induced cytotoxicity towards catecholaminergic neuroblastoma SH-SY5Y cells. Effects of oxidative stress and antioxidative factors. *Biochem. Pharmacol.* **53,** 363–372.

LaVoie, M. J., and Hastings, T. G. (1999). Dopamine quinone formation and protein modification associated with the striatal neurotoxicity of methamphetamine: Evidence against a role for extracellular dopamine. *J. Neurosci.* **19,** 1484–1491.

LaVoie, M. J., Ostaszewski, B. L., Weihofen, A., Schlossmacher, M. G., and Selkoe, D. J. (2005). Dopamine covalently modifies and functionally inactivates parkin. *Nat. Med.* **11,** 1214–1221.

Liu, Y., and Edwards, R. H. (1997). The role of vesicular transport proteins in synaptic transmission and neural degeneration. *Annu. Rev. Neurosci.* **20,** 125–156.

Machida, Y., Chiba, T., Takayanagi, A., Tanaka, Y., Asanuma, M., Ogawa, N., Koyama, A., Iwatsubo, T., Ito, S., Jansen, P. H., Shimizu, N., Tanaka, K., *et al.* (2005). Common anti-apoptotic roles of parkin and alpha-synuclein in human dopaminergic cells. *Biochem. Biophys. Res. Commun.* **332,** 233–240.

Marek, G. J., Vosmer, G., and Seiden, L. S. (1990). Dopamine uptake inhibitors block long-term neurotoxic effects of methamphetamine upon dopaminergic neurons. *Brain Res.* **513,** 274–279.

McCann, U. D., Wong, D. F., Yokoi, F., Villemagne, V., Dannals, R. F., and Ricaurte, G. A. (1998). Reduced striatal dopamine transporter density in abstinent METH and methcathinone users: Evidence from positron emission tomography studies with [^{11}C]WIN-35,428. *J. Neurosci.* **18,** 8417–8422.

Melega, W. P., Lacan, G., Harvey, D. C., Huang, S. C., and Phelps, M. E. (1998). Dizocilpine and reduced body temperature do not prevent methamphetamine-induced neurotoxicity in the vervet monkey: [^{11}C]WIN 35,428 B positron emission tomography studies. *Neurosci. Lett.* **258,** 17–20.

Meredith, C. W., Jaffe, C., Ang-Lee, K., and Saxon, A. J. (2005). Implications of chronic methamphetamine use: A literature review. *Harv. Rev. Psychiatry* **13,** 141–154.

Miranda, M., and Botti, D. (1983). Harding-passey mouse-melanoma tyrosinase inactivation by reaction products and activation by L-epinephrine. *Gen. Pharmacol.* **14,** 231–237.

Miyazaki, I., Asanuma, M., Diaz-Corrales, F. J., Miyoshi, K., and Ogawa, N. (2005). Dopamine agonist pergolide prevents levodopa-induced quinoprotein formation in parkinsonian striatum and shows quenching effects on dopamine-semiquinone generated *in vitro*. *Clin. Neuropharmacol.* **28,** 155–160.

Miyazaki, I., Asanuma, M., Diaz-Corrales, F. J., Fukuda, M., Kitaichi, K., Miyoshi, K., and Ogawa, N. (2006). Methamphetamine-induced dopaminergic neurotoxicity is regulated by quinone formation-related molecules. *FASEB J.* **20,** 571–573.

Mori, T., Ito, S., Kita, T., and Sawaguchi, T. (2004). Effects of dopamine- and serotonin-related compounds on methamphetamine-induced self-injurious behavior in mice. *J. Pharmacol. Sci.* **96,** 459–464.

Mori, T., Ito, S., Kita, T., Narita, M., Suzuki, T., and Sawaguchi, T. (2006). Effects of mu-, delta- and kappa-opioid receptor agonists on methamphetamine-induced self-injurious behavior in mice. *Eur. J. Pharmacol.* **532,** 81–87.

Mori, T., Ito, S., Kita, T., Narita, M., Suzuki, T., Matsubayashi, K., and Sawaguchi, T. (2007). Oxidative stress in methamphetamine-induced self-injurious behavior in mice. *Behav. Pharmacol.* **18,** 239–249.

Munday, R., Smith, B. L., and Munday, C. M. (1998). Effects of butylated hydroxyanisole and dicoumarol on the toxicity of menadione to rats. *Chem. Biol. Interact.* **108,** 155–170.

Nakajima, A., Yamada, K., He, J., Zeng, N., Nitta, A., and Nabeshima, T. (2004). Anatomical substrates for the discriminative stimulus effects of methamphetamine in rats. *J. Neurochem.* **91,** 308–317.

Narita, M., Aoki, K., Takagi, M., Yajima, Y., and Suzuki, T. (2003a). Implication of brain-derived neurotrophic factor in the release of dopamine and dopamine-related behaviors induced by methamphetamine. *Neuroscience* **119,** 767–775.

Narita, M., Takagi, M., Aoki, K., Kuzumaki, K., and Suzuki, T. (2003b). Implication of Rho-associated kinase in the elevation of extracellular dopamine levels and its related behaviors induced by methamphetamine in rats. *J. Neurochem.* **86,** 273–282.

Narita, M., Akai, H., Nagumo, Y., Sunagawa, N., Hasebe, K., Nagase, H., Kita, T., Hara, C., and Suzuki, T. (2004). Implication of protein kinase C in the nucleus accumbens in the development of sensitization to methamohetamine in rats. *Neuroscience* **127,** 941–948.

Narita, M., Akai, H., Kita, T., Nagumo, Y., Narita, M., Sunagawa, N., Hara, C., Hasebe, K., Nagase, H., and Suzuki, T. (2005). Involvement of mitogen-stimulated p70-S6 kinase in the development of sensitization to the methamphetamine-induced rewarding effect in rats. *Neuroscience* **132,** 553–560.

Niwa, M., Nitta, A., Yamada, K., and Nabeshima, T. (2007). The roles of glial cell line-derived neurotrophic factor, tumor necrosis factor-alpha, and an inducer of these factors in drug dependence. *J. Pharmacol. Sci.* **104,** 116–121.

Nwanze, E., and Jonsson, G. (1981). Amphetamine neurotoxicity on dopamine nerve terminals in the caudate nucleus of mice. *Neurosci. Lett.* **26,** 163–168.

Offen, D., Ziv, I., Sternin, H., Melamed, E., and Hochman, A. (1996). Prevention of dopamine-induced cell death by thiol antioxidants: Possible implications for treatment of Parkinson's disease. *Exp. Neurol.* **141,** 32–39.

Park, Y. H., Kantor, L., Guptaroy, B., Zhang, M., Wang, K. K., and Gnegy, M. E. (2003). Repeated amphetamine treatment induces neurite outgrowth and enhanced amphetamine-stimulated dopamine release in rat pheochromocytoma cells (PC12 cells) via a protein kinase C- and mitogen activated protein kinase-dependent mechanism. *J. Neurochem.* **87,** 1546–1557.

Pierce, R. C., and Kalivas, P. W. (1997). A circuitry model of the expression of behavioral sensitization to amphetamine-like psychostimulants. *Brain Res. Brain Res. Rev.* **25,** 1546–1557.

Ricaurte, G. A., Guillery, R. W., Seiden, L. S., Schuster, C. R., and Moore, R. Y. (1982). dopamine nerve terminal degeneration produced by high doses of methylamphetamine in the rat brain. *Brain Res.* **235,** 93–103.

Robinson, T. E., and Becker, J. B. (1986). Enduring changes in brain and behavior produced by chronic amphetamine administration: A review and evaluation of animal models of amphetamine psychosis. *Brain Res.* **396,** 157–198.

Robinson, T. E., and Kolb, B. (1997). Persistent structural modifications in nucleus accumbens and prefrontal cortex neurons produced by previous experience with amphetamine. *J. Neurosci.* **17,** 8491–8497.

Rosei, M. A., Blarzino, C., Foppoli, C., Mosca, L., and Coccia, R. (1994). Lipoxygenase-catalyzed oxidation of catecholamines. *Biochem. Biophys. Res. Commun.* **200,** 344–350.

Ryan, L. J., Martone, M. E., Linder, J. C., and Groves, P. M. (1988). Cocaine, in contrast to d-amphetamine, does not cause axonal terminal degeneration in neostriatum and agranular frontsl cortex of Long-Evans rats. *Life Sci.* **43,** 1403–1409.

Schmidt, C. J., Ritter, J. K., Sonsalla, P. K., Hanson, G. R., and Gibb, J. W. (1985). Role of dopamine in the neurotoxic effects of methamphetamine. *J. Pharmacol. Exp. Ther.* **233,** 539–544.

Schuster, C. R., and Fischman, M. W. (1975). Amphetamine toxicity: Behavioral and neuropathological indexes. *Fed. Proc.* **34,** 1945–1951.

Seiden, L. S., and Vosmer, G. (1984). Formation of 6-hydroxydopamine in caudate nucleus of the rat brain after a single large dose of methylamphetamine. *Pharmacol. Biochem. Behav.* **21,** 29–31.

Seiden, L. S., Fischman, M. W., and Schuster, C. R. (1975). Long-termmethamphetamine-induced changes in brain catecholamines in tolerant rhesus monkeys. *Drug Alcohol Depend.* **1,** 215–219.

Sonsalla, P. K., Jochnowitz, N. D., Zeevalk, G. D., Oostveen, J. A., and Hall, E. D. (1996). Treatment of mice with methamphetamine produces cell loss in the substantia nigra. *Brain Res.* **738,** 172–175.

Strakowski, S. M., Sax, K. W., Setters, M. J., Stanton, S. P., and Keck, P. E., Jr. (1997). Lack of enhanced response to repeated d-amphetamine challenge in first-episode psychosis: Implications for a sensitization model of psychosis in humans. *Biol. Psychiatry* **42,** 749–755.

Sulzer, D., and Zecca, L. (2000). Intraneuronal dopamine-quinone synthesis: A review. *Neurotox. Res.* **1,** 181–195.

Sulzer, D., Bogulavsky, J., Larsen, K. E., Behr, G., Karatekin, E., Kleinman, M. H., Turro, N., Krantz, D., Edwards, R. H., Greene, L. A., and Zecca, L. (2000). Neuromelanin biosynthesis is driven by excess cytosolic catecholamines not accumulated by synaptic vesicles. *Proc. Natl. Acad. Sci. USA* **97,** 11869–11874.

Tief, K., Schmidt, A., and Beermann, F. (1997). Regulation of the tyrosinase promoter in transgenic mice: Expression of a tyrosinase-lacZ fusion gene in embryonic and adult brain. *Pigment Cell Res.* **10,** 153–157.

Tief, K., Schmidt, A., and Beermann, F. (1998). New evidence for presence of tyrosinase in substantia nigra, forebrain and midbrain. *Mol. Brain Res.* **53,** 307–310.

Tse, D. C., McCreery, R. L., and Adams, R. N. (1976). Potential oxidative pathways of brain catecholamines. *J. Med. Chem.* **19,** 37–40.

Uhl, G. R. (1998). Hypothesis: The role of dopaminergic transporters in selective vulnerability of cells in Parkinson's disease. *Ann. Neurol.* **43,** 555–560.

Volkow, N. D., Chang, L., Wang, G. J., Fowler, J. S., Yee, L. M., Franceschi, D., Sedler, M. J., Gatley, S. J., Hitzemann, R., Ding, Y. S., Logan, J., Wong, C., et al. (2001). Association of dopamine transporter reduction with psychomotor impairment in METH abusers. *Am. J. Psychiatry* **158,** 377–382.

Wagner, G. C., Seiden, L. S., and Schuster, C. R. (1979). Methamphetamine-induced changes in brain catecholamines in rats and guinea pigs. *Drug Alcohol Depend.* **4,** 435–438.

Wagner, G. C., Ricaurte, G. A., Seiden, L. S., Schuster, C. R., Miller, R. J., and Westley, J. (1980a). Long-lasting depletions of striatal dopamine and loss of dopamine uptake sites following repeated administration of methamphetamine. *Brain Res.* **181,** 151–160.

Wagner, G. C., Ricaurte, G. A., Johanson, C. E., Schuster, C. R., and Seiden, L. S. (1980b). Amphetamine induces depletion of dopamine and loss of dopamine uptake sites in caudate. *Neurology* **30,** 547–550.

Wagner, G. C., Preston, K. L., Ricaute, G. A., Schuster, C. R., and Seiden, L. S. (1982). Neurochemical similarities between D,L-cathinone and d-amphetamine. *Drug Alchol Depend.* **9,** 279–284.

Wagner, G. C., Lucot, J. B., Schuster, C. R., and Seiden, L. S. (1983). alpha-Methyltyrosine attenuates and reserpine increases methamphetamine-induced neuronal changes. *Brain Res.* **270,** 285–288.

Wagner, G. C., Gardner, J., Tsigas, D. J., and Masters, D. B. (1984). Tolerance following the repeated administration of high doses of phencyclidine: No relation to central catecholamine depletion. *Drug Alcohol Depend.* **13,** 225–234.

Wagner, G. C., Carelli, R. M., and Jarvis, M. F. (1986). Ascorbic acid reduces the dopamine depletion induced by METH and the 1-methyl-4-phenyl pyridinium ion. *Neuropharmacology* **25,** 559–561.

Wagner, G. C., Avena, N., Kita, T., Nakashima, T., Fisher, H., and Halladay, A. K. (2004). Risperidone reduction of amphetamine-induced self-injurious behavior in mice. *Neuropharmacology* **46,** 700–708.

Wallace, T. L., Gudelsky, G. A., and Vorhees, C. V. (1999). Methamphetamine-induced neurotoxicity alters locomotor activity, stereotypic behavior, and stimulated dopamine release in the rat. *J. Neurosci.* **19,** 9141–9148.

Wallace, T. L., Gudelsky, G. A., and Vorhees, C. V. (2001). Neurotoxic regimen of methamphetamine produces evidence of behavioral sensitization in the rat. *Synapse* **39,** 1–7.

Walsh, S. L., and Wagner, G. C. (1992). Motor impairments after methamphetamine-induced neurotoxicity in the rat. *J. Pharmacol. Exp. Ther.* **263,** 617–626.

Whitehead, R. E., Ferrer, J. V., Javitch, J. A., and Justice, J. B. (2001). Reaction of oxidized dopamine with endogenous cysteine residues in the human dopamine transporter. *J. Neurochem.* **76,** 1242–1251.

Wilson, J. M., Kalasinsky, K. S., Levey, A. I., Bergeron, C., Reiber, G., Anthony, R. M., Schmunk, G. A., Shannak, K., Haycock, J. W., and Kish, S. J. (1996). Striatal dopamine nerve terminal markers in human, chronic METH users. *Nat. Med.* **2,** 699–703.

Wise, R. A., and Bozarth, M. A. (1987). A psychomotor stimulant theory of addiction. *Psychol. Rev.* **94,** 469–492.

Wrona, M. Z., Yang, Z., Zhang, F., and Dryhurst, G. (1997). Potential new insights into the molecular mechanisms of methamphetamine-induced neurodegeneration. *NIDA Res. Monogr.* **173,** 146–174.

Xie, T., McCann, U. D., Kim, J., Yuan, J., and Ricaurte, G. A. (2000). Effect of temperature on doaopamine transporter fun ctiuon and intracellular accumulation of methamphetamine: Implications for methasmphetamine-induiced dopaminergic neurotoxicity. *J. Neurosci.* **20,** 7838–7845.

Xu, Y., Stokes, A. H., Roskoski, R. Jr., and Vrana, K. E. (1998). Dopamine, in the presence of tyrosinase, covalently modifies and inactivates tyrosine hydroxylase. *J. Neurosci. Res.* **54,** 691–697.

Young, D., and Scoville, W. B. (1938). Paranoid psychoses in narcolepsy and the possible danger of Benzedrine treatment. *Med. Clin. North Am.* **22,** 637–645.

Yuan, J., Callahan, B. T., McCann, U. D., and Ricaurte, G. A. (2001). Evidence against an essential role of endogenous brain dopamine in methamphetamine-induced dopaminergic neurotoxicity. *J. Neurochem.* **77,** 1338–1347.

Yu, Q., Zhang, D., Walston, M., Zhang, J., Liu, Y., and Watson, R. R. (2002). Chronic methamphetamine exposure alters immune function in normal and retrovirus-infected mice. *Int. Immunopharmacol.* **2,** 951–962.

ACUTE METHAMPHETAMINE INTOXICATION: BRAIN HYPERTHERMIA, BLOOD–BRAIN BARRIER, BRAIN EDEMA, AND MORPHOLOGICAL CELL ABNORMALITIES

Eugene A. Kiyatkin* and Hari S. Sharma*,†

*Behavioral Neuroscience Branch, National Institute on Drug Abuse–Intramural Research Program, National Institutes of Health, Baltimore, Maryland 21224, USA
†Laboratory of Cerebrovascular Research, Department of Surgical Sciences, University Hospital, Uppsala University, SE-75185 Uppsala, Sweden

Methamphetamine (METH) is a powerful and often abused stimulant with potent addictive and neurotoxic properties. While it is generally assumed that multiple chemical substances released in the brain following METH-induced metabolic activation (or oxidative stress) are primary factors underlying damage of neural cells, in this work we present data suggesting a role of brain hyperthermia and associated leakage of the blood–brain barrier (BBB) in acute METH-induced toxicity. First, we show that METH induces a dose-dependent brain and body hyperthermia, which is strongly potentiated by associated physiological activation and in warm environments that prevent proper heat dissipation to the external environment. Second, we demonstrate that acute METH intoxication induces robust, widespread but structure-specific leakage of the BBB, acute glial activation, and increased water content (edema), which are related to drug-induced brain hyperthermia. Third, we document widespread morphological

INTERNATIONAL REVIEW OF
NEUROBIOLOGY, VOL. 88
DOI: 10.1016/S0074-7742(09)88004-5

65

abnormalities of brain cells, including neurons, glia, epithelial, and endothelial cells developing rapidly during acute METH intoxication. These structural abnormalities are tightly related to the extent of brain hyperthermia, leakage of the BBB, and brain edema. While it is unclear whether these rapidly developed morphological abnormalities are reversible, this study demonstrates that METH induces multiple functional and structural perturbations in the brain, determining its acute toxicity and possibly contributing to neurotoxicity.

I. Introduction

Methamphetamine (METH) is a powerful and often abused psychomotor stimulant with potent addictive and neurotoxic properties. In addition to the social harms of addiction, the use of this substance could adversely influence human health, causing acute behavioral and physiological disturbances during intoxication and long-term health complications following chronic use (Kalant, 2001). METH abuse could also be a cofactor enhancing different latent pathological conditions, especially cardiovascular, neurological, and psychiatric, as well as increasing the probability and severity of numerous viral and bacterial infections.

Considering the issue of neurotoxicity, it is usually assumed that METH has direct toxic effects on neural cells, with relative selectivity toward specific cell groups, brain structures, and cellular organelles. In particular, METH preferentially affects midbrain dopamine (DA) cells, damaging fine axonal terminals in the striatum (Ricaurte et al., 1980; Riddle et al., 2006; Woolverton et al., 1989) and resulting in health complications associated with pathologically altered DA transmission. Particularly, perturbations in DA as well as other monoamine systems are an important factor in psychoemotional and movement disorders, including acute METH psychosis and severe depression following long-term METH use and withdrawal (Kalant, 2001).

However, METH also induces metabolic activation (Estler, 1975; Makisumi et al., 1998) and hyperthermia (Alberts and Sonsalla, 1995; Kalant and Kalant, 1975; Sandoval et al., 2000). Enhanced metabolism is tightly related to oxidative stress, which is caused by an imbalance between the production of reactive oxygen and a biological system's ability to detoxify readily the reactive intermediates or easily repair the resulting damage. Disturbances in this normal redox state can cause toxic effects through the production of peroxides and free radicals that damage all components of the cell, including proteins, lipids, and DNA. Oxidative stress as a consequence of brain hypermetabolism is usually viewed as a primary factor in METH-induced neurotoxicity (Cadet et al., 2007; De Vito and Wagner,

1989; Stephans and Yamamoto, 1994; Yamamoto and Zhu, 1998). On the other hand, brain cells are exceptionally temperature-sensitive, with the appearance of irreversible changes in structure and functions at approximately 40 °C, that is, only 3 °C above normal baseline (Chen *et al.*, 2003; Iwagami, 1996; Oifa and Kleshchenov, 1985; Sharma and Hoopes, 2003). Due to the strong temperature dependence of most physicochemical processes governing neural activity (see Kiyatkin, 2005 for review), hyperthermia could also strongly modulate toxic effects of METH on brain cells. It is well known that METH is much more toxic at high environmental temperatures, while toxicity is diminished by low ambient temperatures (Alberts and Sonsalla, 1995; Ali *et al.*, 1994; Bowyer *et al.*, 1993; Farfel and Seiden, 1995; Miller and O'Callaghan, 1994). While it is reasonable to assume that the more harmful effects of METH seen in warm, humid conditions are associated with intrabrain heat accumulation due to enhanced brain metabolism coupled with the diminished ability to dissipate properly metabolic heat, direct data on brain temperature changes induced by METH and its environmental modulation are limited.

In addition to the direct effects of high temperatures on brain cells and potentiation of toxic effects of brain metabolites, brain hyperthermia appears to affect the permeability of the blood–brain barrier (BBB). The BBB is an important border that protects the brain environment, while BBB leakage allows water, ions, and various potentially neurotoxic substances contained in blood plasma (i.e., glutamate) to enter the brain (Rapoport, 1976; Zlokovic, 2008). Although leakage of the BBB has been documented during environmental warming (Cervos-Navarro *et al.*, 1998; Sharma *et al.*, 1992), intense physical exercise (Watson *et al.*, 2005), various types of stress (Esposito *et al.*, 2001; Ovadia *et al.*, 2001; Sharma and Dey, 1986), and morphine withdrawal (Sharma and Ali, 2006), data on METH-induced alterations in the BBB and its relationship to brain temperature are limited.

In this work, we present and discuss several sets of recent data, suggesting a tight link between acute METH toxicity, brain temperature, and alterations of the BBB. Based on the literature and these data, we demonstrate that brain hyperthermia induced by METH plays an important role in the triggering of several pathophysiological mechanisms underlying acute toxicity of this drug and contributing to neurotoxicity.

II. Basic Terminology and Major Topics of This Study

Drug toxicity is generally defined as a degree to which a chemical toxin is able to damage an exposed organism. Toxicity can refer to the effect on a whole organism as well as the effect on a substructure such as a cell (citotoxicity), an organ (organotoxicity), or specific tissue. Therefore, *neurotoxicity* usually refers to adverse effects on the structure or function of the central and/or peripheral nervous system. Clinically, neurotoxicity is measured based on specific symptoms (i.e., muscle weakness, loss of sensation and motor control, tremors, cognitive altera-tions, or autonomic nervous system dysfunctions), but in the experimental work, it is typically defined as neural damage, the analysis of which requires postmortem examination of brain tissue. The term "neural damage," moreover, is not precise, because it could mean morphological or functional abnormalities of specific brain cells or cell groups, which contrast to a certain morphological and functional "norm" and could suggest that these cells will degenerate in the future. This approach allows one to detect early changes that could finally result in cellular death, but it fails to differentiate clearly between reversible and irreversible damage. On the other hand, neural damage could be verified by detection of dead cells, but significant time is always necessary for a normal cell to become damaged, for the damage to become irreversible, and for an irreversibly damaged cell to die and be recognized as dead. Moreover, dead cells are effectively eliminated from brain tissue by phagocytosis and their detection in brain tissue has a relatively short window of time (see Bowyer and Ali, 2006). Irreversible structural abnormalities could be better verified with the use of electronic micro-scopy, but the use of this technique for the study of METH neurotoxicity is limited.

Another important issue of toxicity is the drug's dose dependence. The effects are absent or undetectable at low doses, progress at larger doses, and finally result in lethality at high doses. Toxicity could be measured by the drug's effects on a particular target (organism, organ, tissue, or cell) and, at the level of the whole organism, it could be quantified based on LD50. The LD50 for METH with intraperitoneal (ip) administration is 55 and 57 mg/kgD in rats and mice, respectively (Davis *et al.*, 1987; Yamamoto, 1963). While it is an important measure to grade a degree of toxicity, LD50 could be misleading because the toxic effects of METH in humans differ from those in experimental animals, are strongly modulated by environmental conditions, depend on age and activity state, and are influenced by the use of other substances and individual sensitivity. For example, approximately 80% of rats that received 9 mg/kg of METH (i.e., 1/6 of a traditional LD50) died within 6 h when the injection was done at 29 °C ambient temperature (Brown and Kiyatkin, 2005). This temperature is only 5–6 °C above the laboratory standard, corresponding to normothermia or temper-ature comfort (Romanovsky *et al.*, 2002) and has no evident effects on basal temperature and animal behavior (Kiyatkin and Brown, 2004). Therefore, under specific conditions, the drug could be toxic at the doses incomparably lower than traditional LD50 values. This feature could be especially important for psychomotor stimulant drugs, which are often taken under specific activity states and environmental conditions such as raves, when individuals are highly active, exposed to warm and humid conditions, and their ability to dissipate heat is limited.

Brain temperature and brain thermal homeostasis are other important topics of this work. *Hyperthermia* (or *hypothermia*) in this study defines an increase (or decrease) in temperature. Depending upon location, hyperthermia (or hypothermia) could occur in the brain, muscles, body core, and skin. Traditionally, hyperthermia is contrasted to fever (pyrexia), an increase in internal temperature, which occurs due to temporary elevation of the body's thermoregulatory set-point, a conceptual mechanism that was proposed to explain body temperature regulation (see Romanovsky, 2004 for discussion). Fever is usually viewed as an organism's induced temperature increase, when heat is produced and actively retained in the body by decreased heat dissipation. In contrast, hyperthermia defines body temperature increase that occurs without the consent of the heat control system, that is, over the body thermoregulatory set-point; in this case, the body produces (or absorbs from outside) more heat than it can dissipate. While the discussion of these issues is clearly out of the scope of the present work, brain temperature alterations induced by psychomotor stimulant drugs are determined by their specific actions on neural substrates, affecting brain and body metabolism and inducing behavioral and physiological effects (see Kiyatkin, 2005, 2006 for further review). These drugs also have vasoconstrictive effects, limiting heat dissipation to the external environment and potentiating brain and body hyperthermia. Therefore, as we will show below, hyperthermia induced by METH and other related drugs mimic some important aspects of fever, specifically, active temperature increase.

METH-induced change in the permeability of the BBB is another important topic of this work. The BBB is a highly specialized brain-endothelial structure of the fully differentiated neurovascular system. In concert with pericytes, astrocytes, and microglia, the BBB separates components of the circulating blood from neurons and maintains the chemical composition of the intrabrain environment, which is required for proper functioning of neural circuits, synaptic transmission, synaptic remodeling, angiogenesis, and neurogenesis. BBB breakdown, due to disruption of the tight junctions and altered transport of molecules between the blood and brain, may initiate various pathophysiological mechanisms resulting in robust brain abnormalities and serious health complications (see Zlokovic, 2008 for review). While alterations in the permeability of the BBB are implicated in the development and progression of various neurodegenerative disorders (Alzheimer's disease, Parkinson's disease, amyotrophic lateral sclerosis, and multiple sclerosis), we will demonstrate that METH also seriously alters BBB permeability, triggering a link of pathophysiological processes that contribute to both acute drug toxicity and neurotoxicity.

III. Brain Temperature Responses to METH are Dose-Dependent and Modulated by Activity State and Environmental Conditions

The effects of addictive drugs are usually studied in animals under well-controlled experimental conditions. In addition to standard temperature and humidity, drugs are typically administered after animals' habituation to the testing environment and under quiet resting conditions when baselines are stable and low. In contrast, humans often use the same drugs voluntarily, in different doses, under conditions of psychophysiological activation (that usually precedes drug intake) and in specific environmental conditions that often dramatically differ from those in animal experiments. For example, METH and other psychomotor stimulant drugs (i.e., MDMA) are often taken during raves, when the drug intake is associated with a high degree of psychophysiological and behavioral activation and adverse environmental conditions (hot and humid environment) that seriously impact an organism's thermoregulatory mechanisms. To examine how the effects of psychomotor stimulants are modulated by activity states and environmental conditions, we performed a series of studies, focusing on brain temperature as a primary parameter of interest (Brown and Kiyatkin, 2004, 2005; Brown et al., 2003). First, we examined how METH, at different doses, affects brain temperature and what the relationships between these temperatures and those recorded from various body locations are. Second, we examined how thermal effects of this drug are modulated during associated physiological activation and in a moderately hot environment. To model psychophysiological activation, we used the procedure of social interaction, in which the recorded male rat was exposed to a female rat, resulting in behavioral activation and a clear temperature response. The effects of METH were also compared in two ambient temperatures: 23 and 29 °C. Finally, to clarify the mechanisms underlying brain hyperthermic effects of psychomotor stimulants, we examined how these effects are modulated by functionally diminished blood outflow from the brain. To model this condition, both jugular veins, which provide >90% of blood outflow from the brain (Doepp et al., 2004), were sequentially surgically closed. Animals with blocked jugular veins were seemingly normal, gained weight, showed generally normal responses to environmental challenges and had basal brain or body temperature similar to those in control rats.

Our studies revealed that METH induces dose-dependent temperature increases, which were generally correlative in different brain sites (NAcc, hippocampus), temporal muscle, and body core. At the lowest dose (1 mg/kg, sc), the increase had the smallest amplitude and duration ($\sim$1.0 °C for $\sim$160 min) and was progressively larger and more prolonged at moderate (3 mg/kg; 1.4 °C for 280 min) and high (9 mg/kg: $\sim$3.4 °C for 360 min) doses. At the latter dose (see Fig. 1A), brain hyperthermia even in standard ambient temperatures (22–23 °C)

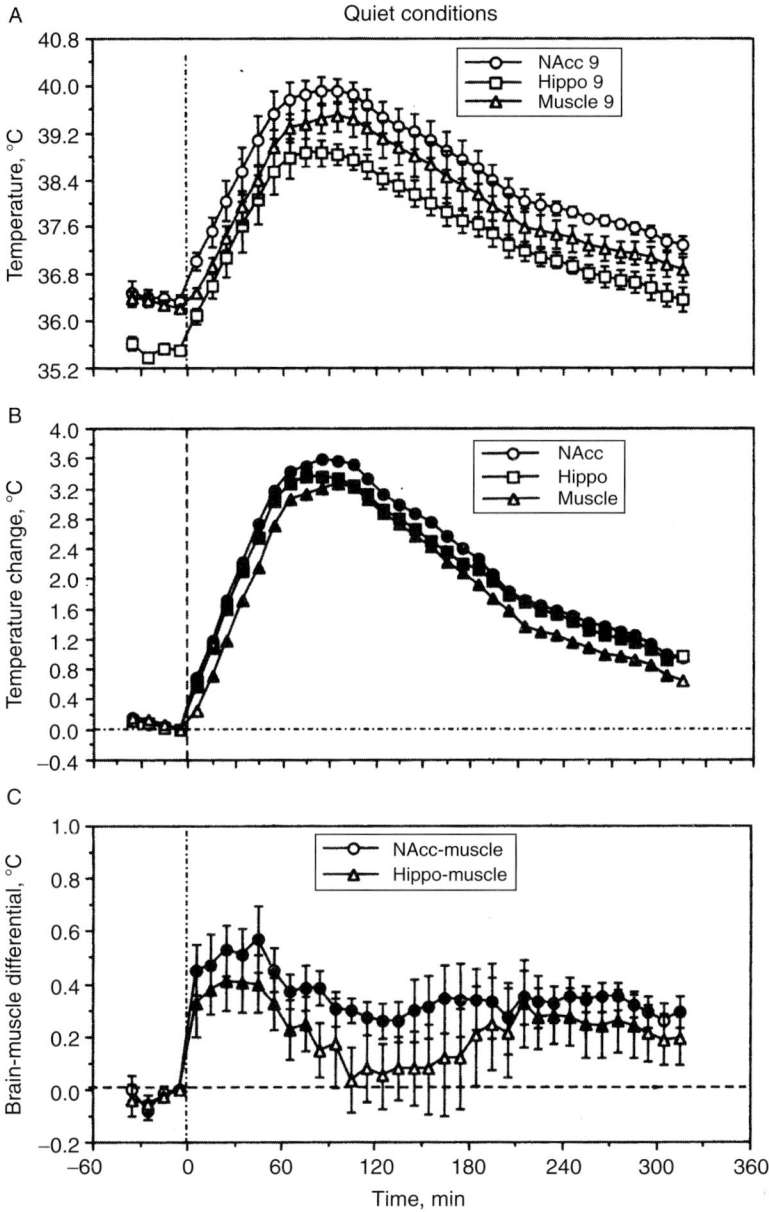

FIG. 1. (Continued)

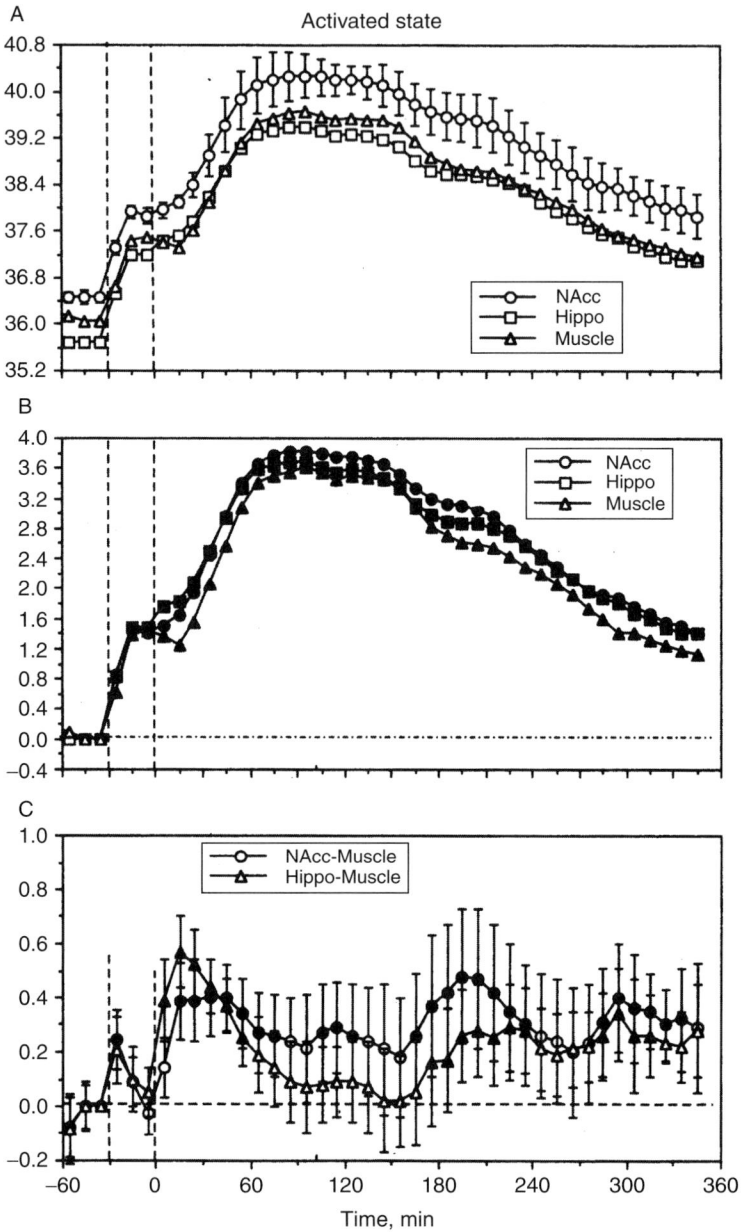

FIG. 1. (Continued)

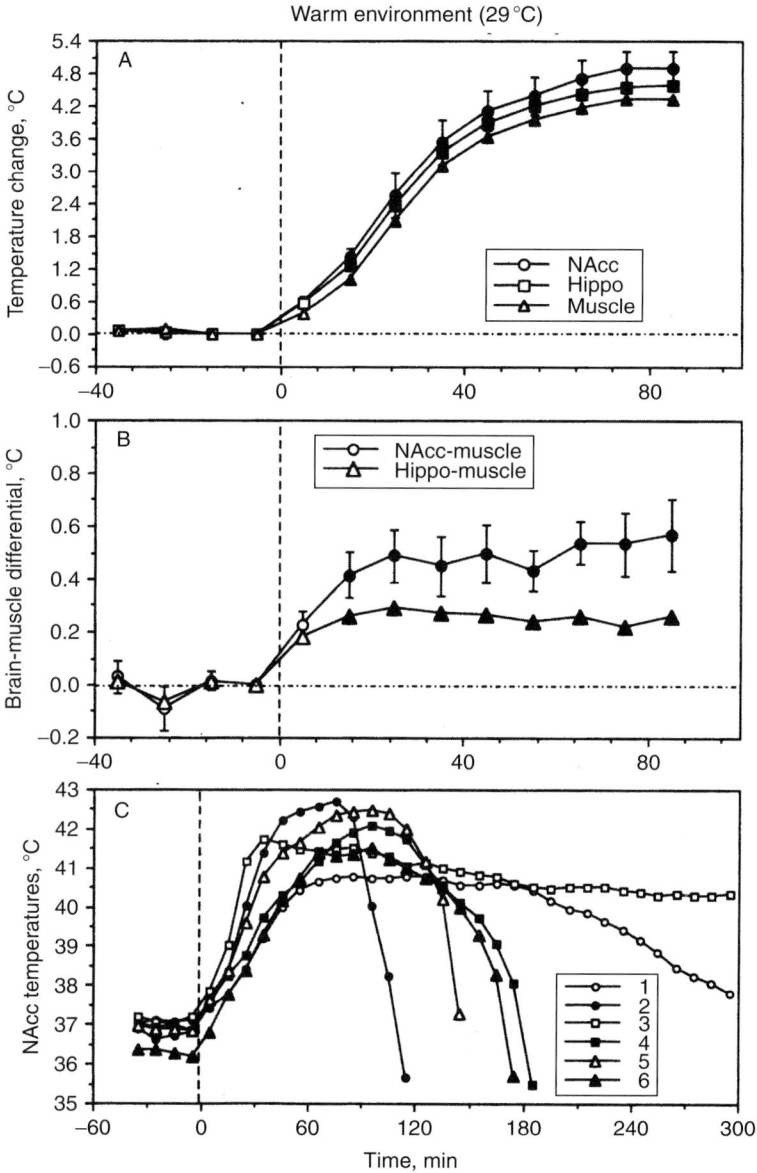

FIG. 1. Effects of methamphetamine (9 mg/kg, sc) on brain (NAcc and hippocampus) and muscle temperatures in rats in three conditions (left: quiet rest at 23 °C; middle: social interaction with female at 23 °C; right: quiet rest at 29 °C). Graphs represent absolute and relative temperature changes as well as brain–muscle differentials. Filled symbols mark values significantly different from baseline.

reached clearly pathological levels ($\sim$4.0 or 3.5 °C above baseline). This temperature increase generally correlated with locomotor hyperactivity, which was evident and strong at 1 mg/kg and greatly progressed (with the addition of strong stereotypy) at higher doses. Although drug-induced temperature changes were generally correlative in brain structures, muscle, and body core, each point has its own temperature. Within the brain, temperatures are distributed according to a dorsoventral gradient (cortex < hippocampus < thalamus, dorsal striatum < ventral striatum, hypothalamus) and for ventrally located structures (i.e., NAcc), they are approximately 1 °C higher than in temporal muscle and about the same as in the body core. Skin has the lowest temperature, about 1 °C less than in the muscle and approximately 2 °C less than in ventral brain structures (ventral striatum, hypothalamus).

The hyperthermic response to METH had two important features (see Fig. 1, left panel). First, although temperatures in the NAcc and hippocampus generally paralleled those in temporal muscle (Fig. 1A), the increases were significantly more rapid and stronger in the brain than temporal muscle, resulting in significant increases in brain–muscle differentials (Fig. 1C). Therefore, it appears that metabolic brain activation associated with intrabrain heat production is the primary cause of brain hyperthermia and a factor behind more delayed and weaker body hyperthermia. At a 9 mg/kg METH dose, the increase in the NAcc–muscle differential was robust and continued for more than 5 h, suggesting a pathological aspect to this brain hyperthermia. Second, temperature increase in brain sites and muscle was consistently associated with a rapid, transient decrease in skin temperature, suggesting acute vasoconstriction. Therefore, decreased heat dissipation to the external environment is another important factor contributing to overall brain and body hyperthermia.

Hyperthermic effects of METH became stronger when the drug was injected during social interaction. As shown in Fig. 1 (middle panel), after presentation of the female, the recorded male showed a strong increase in brain and body temperatures. When METH was injected under these conditions, drug-induced temperature increase became stronger in both brain areas. While the effect was not additive, METH-induced brain hyperthermia during social interaction reached significantly higher values and was maintained for a significantly longer time than in quiet resting conditions.

Hyperthermic effects of METH were also altered when the drug was administered in a warm environment. As shown in Fig. 1 (right panel), mean temperatures after drug administration increased rapidly in all animals, in some of them the increase reached clearly pathological values (>41 °C), and four of six animals died within 3 h. One more animal, which showed a strong and stable hyperthermia ($\sim$40.6 °C) died overnight. Importantly, temperature increases in the brain sites were consistently more rapid and stronger and the increase in NAcc–muscle differential reached pathological levels not seen in any other physiological conditions (see Fig. 1B). However, at the moment of death, brain–muscle differentials rapidly inverted and the brain became cooler than the body (Fig. 1C).

Similar data were also obtained with MDMA, another widely used psychomotor stimulant drug (Brown and Kiyatkin, 2004). Although MDMA induced weaker increases in brain and body temperatures at similar doses (and even induced mild hypothermia at 1 mg/kg dose), these effects were strongly potentiated during social interaction (+89%) and in a warm environment (29 °C; +268%). Similar to METH, three of six tested animals that showed maximal temperature increases (>41 °C) after MDMA at 9 mg/kg died within 4 h and two more rats died at the fifth postdrug hour. Similar to METH, rats also showed dramatic increases in brain–muscle differentials, which peaked close to the moment of death but inverted rapidly when the rat stopped breathing and the brain became relatively cooler than the body. In the case of MDMA, we also evaluated how diminished blood outflow affects drug-induced hyperthermic responses. Compared to control animals, the hyperthermic response to MDMA in rats with blockade of both jugular veins was strongly potentiated (+188%), with a much greater increase in brain–muscle temperature differentials. Therefore, MDMA-induced brain hyperthermia results from intrabrain heat production, while venous outflow is an important means for heat dissipation from the brain to the rest of the body.

Classic features of neurotoxicity induced by amphetamine-like substances (i.e., neuronal necrosis and apoptosis) are usually linked to some toxic products (i.e., nitric oxide, catechol quinones, peroxynitrite) of abnormally increased metabolism of endogenous neurotransmitter substances (Cadet et al., 2007; Kuhn and Geddes, 2000). While these factors may contribute to neural damage following chronic use of these substances, our present data suggest the importance of brain overheating as a factor responsible for fatal decompensation of an organism's vital functions due to an acute drug intoxication. The rise in brain temperature above certain "normal" limits may per se have a directly destructive action on brain cells, which will increase exponentially with slight increases above these limits. The most temperature-sensitive cellular elements are mitochondrial and cellular membranes, in which irreversible transitions in protein structure or arrangements begin to occur at temperatures higher than 40 °C (Iwagami, 1996; Lepock, 2003; Willis et al., 2000). Therefore, 40 °C could be considered the threshold of pathological hyperthermia, which could have long-term negative consequences even if the temperature will later return to its baselines.

Since the rats that died following METH and MDMA intoxication in a moderately warm environment showed some clinical features suggesting brain edema, we hypothesized that the destruction of endothelial cells in the brain and leakage of serum proteins across the BBB induced by high temperature could be an important pathogenic factor responsible for this life-threatening condition.

The next section presents the results of our recent experiments aimed at verification of this hypothesis (Kiyatkin *et al.*, 2007; Sharma and Kiyatkin, 2009).

IV. Acute METH Intoxication Results in BBB Leakage, Acute Glial Activation, and Morphological Alterations of Brain Cells: Role of Brain Temperature

Similar to our previous thermorecording experiments, rats were implanted with three thermocouple electrodes (NAcc, temporal muscle, and skin) and an intravenous (iv) jugular catheter, habituated to the experimental conditions, and divided into three groups. Rats from the first two groups received METH (9 mg/kg, sc) in normal (23 °C) and warm (29 °C) conditions, respectively. When brain temperature peaked or reached clearly pathological values (>41.5 °C), the rats were injected with Evans Blue (EB), rapidly iv anesthetized with pentobarbital, perfused, and their brains were taken for analysis. Control animals received saline and underwent the same procedures as METH-treated rats. The state of BBB permeability and edema were determined by diffusion into brain tissue of Evans Blue dye, an exogenous tracer that is normally retained by the BBB, and measuring brain water and ion (Na^+, K^+, Cl^-) content. Immunohistochemistry was used to evaluate quantitatively brain presence of albumin, a measure of breakdown of the BBB, and glial fibrillary acidic protein (GFAP), an index of astrocytic activation. Albumin is a relatively large plasma protein (molecular weight 59 kDa, molecular diameter 70 Å) that is normally confined to the luminal side of the endothelial cells and is not present in the brain. Similarly, EB, the exogenous protein tracer that binds to plasma albumin, fails to enter the brain from the peripheral circulation (Blezer, 2005; Ehrlich, 1904; Sharma, 2007). Thus, the presence of EB in the brain, the appearance of albumin-positive cells, and albumin immunoreactivity in the neuropil indicate a breakdown of the BBB. GFAP is an intermediate filament protein that is expressed in glial cells (astrocytes). Increased GFAP immunoreactivity (or astrocytic activation) is usually viewed as an index of gliosis or a relatively slow-developing correlate of neural damage (Finch, 2003; Hausmann, 2003). Normal brain tissue has only scattered GFAP-positive cells but rapid GFAP expression has been reported previously during environmental warming and brain trauma (Cervos-Nacarro *et al.*, 1998; Gordth *et al.*, 2006).

To determine morphological abnormalities of brain cells, slices were stained with hematoxylin-eosin or Nissl and analyzed with light microscopy to determine the extent and specifics of structurally abnormal cells. Brain slices were also immunostained for myelin basic protein (MBP) and stained for luxol fast blue to identify myelinated and unmyelinated nerve fibers and study changes in myelin function. For more detailed analyses of cellular and subcellular alterations, we also used transmission electronic microscopy (TEM). To determine the specificity of METH-induced changes in brain parameters, they were determined separately in the cortex, hippocampus, thalamus, and hypothalamus.

A. ALTERATIONS IN THE BBB PERMEABILITY

As shown in Fig. 2, METH induced significant and widespread BBB leakage. Compared to saline-treated controls, both EB (A) and albumin immunoreactivity (B) increased strongly in both METH groups and the changes were significantly larger when the drug was administered at 29 °C than 23 °C. At the time when brains were taken, these two subgroups of animals had significantly higher NAcc (23 °C: 38.92 ± 0.34 °C and 29 °C: 41.37 ± 0.22 °C) and muscle temperatures (23 °C: 37.92 ± 0.32 °C and 29 °C: 40.44 ± 0.19 °C) compared to controls (36.67 ± 0.14 °C and 35.82 ± 0.08 °C in the NAcc and muscle, respectively).

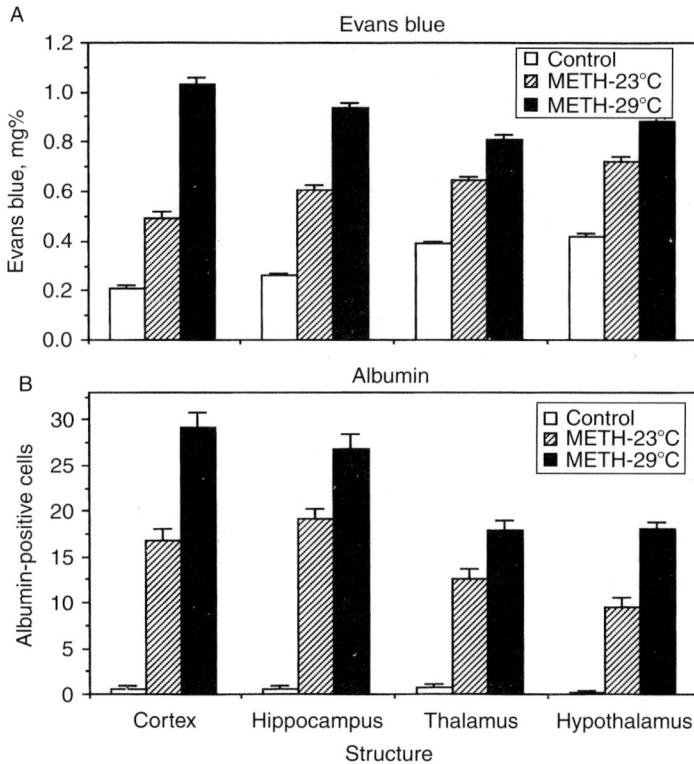

FIG. 2. Continued

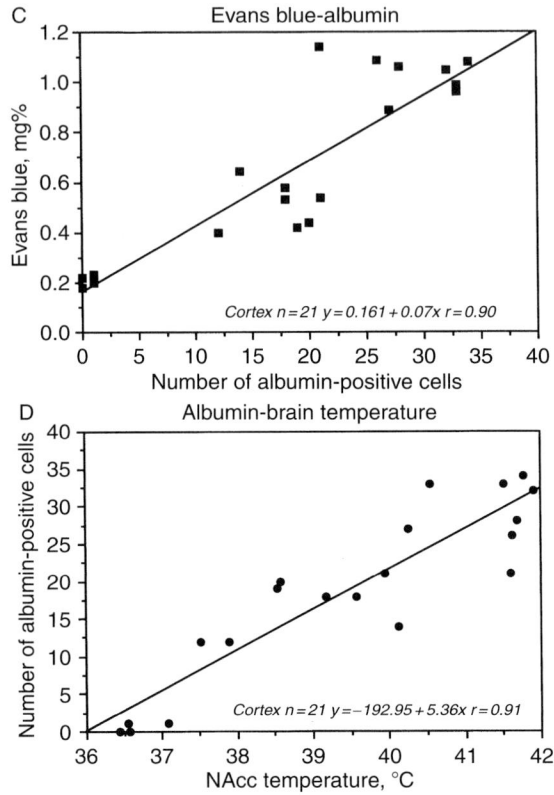

FIG. 2. Mean (±SEM) values of Evans Blue concentrations (A) and numbers of albumin-positive cells (B) in individual brain structures after saline (control) and methamphetamine administration at 23 and 29 °C. (C) Correlative relationships between concentrations of Evans Blue and numbers of albumin-positive cells in the cortex. (D) Correlative relationships between the numbers of albumin-positive cells in the cortex and NAcc temperatures. Each graph shows regression equation, regression line, and correlation coefficient (r).

Moreover, the amplitude of temperature response in both locations was significantly stronger at 29 °C than 23 °C. While the changes in EB occurred in each brain structure, there were some between-structural differences. The cortex showed minimal EB content, followed by the hippocampus, thalamus, and hypothalamus, which all had a higher EB content than cortical tissue. However, the increase with METH-23 °C was stronger in the cortex (+133%) and hippocampus (+134%) and weaker in hypothalamus (+71%) and thalamus (+66%). Similar differences were seen with METH-29 °C; the increase was greatest in the cortex (+390%), followed by hippocampus (+261%), hypothalamus (+109%), and thalamus (+107%).

Albumin immunoreactivity showed similar but stronger changes. As shown in a representative example of immunostained cortical tissue (see Fig. 3A, D, and G), the leakage of albumin in the neuropil was evident in animals that received METH and was more prominent in animals that were exposed to the drug at 29 °C (G) than 23 °C (D). Quantitative analysis of albumin-positive cells (Fig. 2B) showed a significant increase in each brain structure, which was again maximal in the cortex and hippocampus and lower in the thalamus and hypothalamus. The number of albumin-positive cells was significantly higher when METH was administered at 29 °C than 23 °C. Again, the cerebral cortex and hippocampus showed the strongest increase (29.25 ± 1.60 immunopositive cells/section or 49-fold vs control and 26.88 ± 1.13 immunopositive cells/section or 24-fold vs control), followed by the thalamus (17.88 ± 1.13), and hypothalamus (15.00 ± 0.78).

As shown in Fig. 2C, there was a strong liner correlation between concentrations of EB and numbers of albumin-positive cells ($r = 0.90$ for the cortex), suggesting tight relationships between these two measures of BBB permeability. Albumin immunoreactivity was also strongly dependent on brain temperature ($r = 0.91$ for the cortex), with virtually no positive cells at low basal temperatures and a progressive increase at high temperatures (Fig. 2D). This correlation was evident in each brain structure and equally strong with respect to brain and muscle hyperthermia.

B. GLIAL CHANGES

As shown in Fig. 4, only a few GFAP-positive astrocytes were scattered in the normal brains; the hippocampus showed more stained cells than thalamus, cortex, and hypothalamus. In the METH-23 °C group, the number of GFAP-positive cells was significantly larger, with the greatest increase in the thalamus followed by the hippocampus, cortex, and hypothalamus (A). In the METH-29 °C group, the number of GFAP-positive astrocytes was almost doubled in the cortex and hippocampus and 1.5 times higher in the hypothalamus and thalamus compared to the METH-23 °C group. A representative example of GFAP-positive cells during METH treatment is shown in Fig. 3C, F, and I. While a few GFAP-positive cells (arrowhead) are seen in the control thalamus (C), their numbers were larger in the METH-23 °C (F) and especially in the METH-29 °C group (I). In the backgrounds, damaged nerve cells and perineuronal edema were often seen around areas of GFAP expression in these animals.

Changes in GFAP immunoreactivity were tightly and linearly correlated with albumin immunoreactivity (Fig. 4B), suggesting close relationships between

Control   Meth 23°C   Meth 29°C

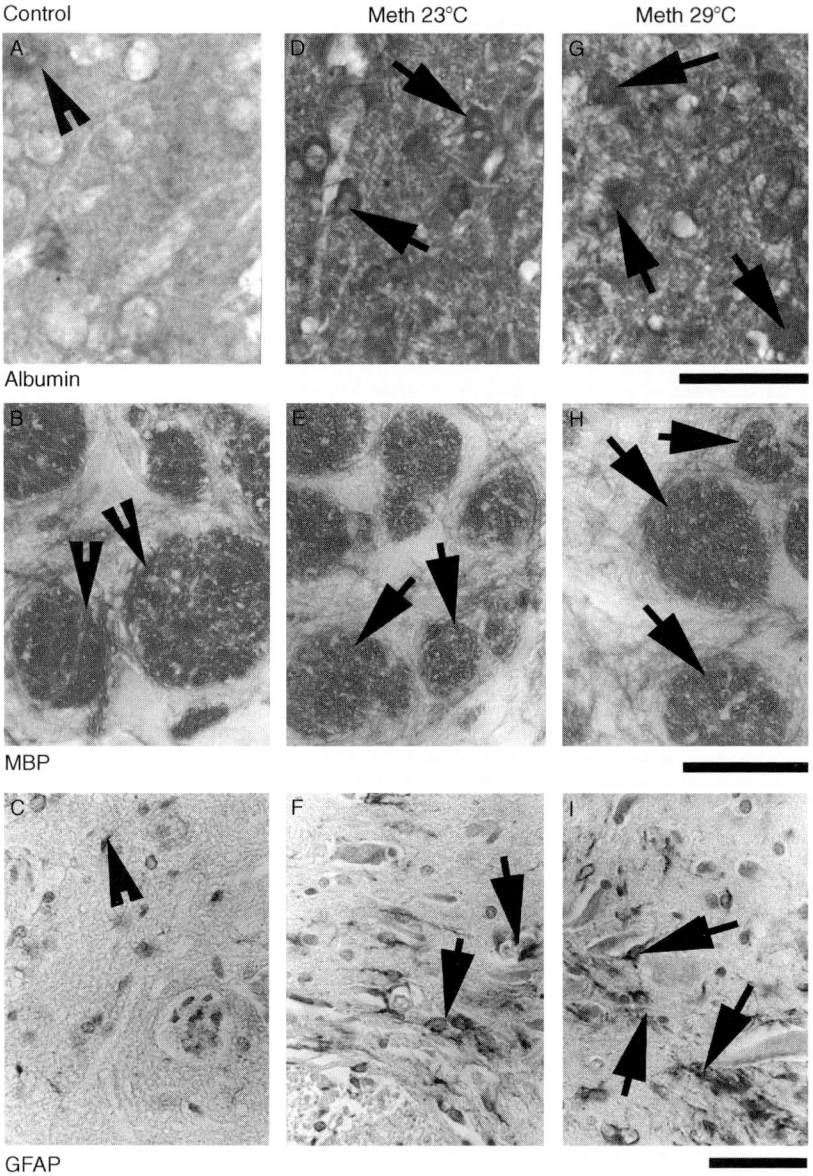

FIG. 3. Immunohistochemical changes in albumin (A, D, G), myelin basic protein (MBP; B, E, H), and glial fibrillary acidic protein (GFAP; C, F, I) in control (left vertical panel) and METH-treated rats (23 °C: middle vertical panel and 29 °C: right vertical panel). Compared to weak albumin immuno-reactivity in control (A, arrowhead), METH-treated rats had stronger immunoreactivity (D and G, arrows). Expansion of neuropil and sponginess is also evident in the surrounding background. Bar

leakage of the BBB and acute glial reaction. GFAP counts were also tightly correlated with brain temperatures (Fig. 4C), suggesting that acute glial reaction is progressively larger depending on the extent of brain temperature elevation.

C. Alterations in Water and Ion Content

In agreement with the previous findings on regional differences in brain water content (Rapoport, 1976), we found that the hippocampus has a higher water content (78.4%) than the cortex (74.5%) and the underlying thalamus (75.5%) and hypothalamus (75.5%) in control conditions (Fig. 5A). METH administered at 23 °C induced a significant and strong increase in water content in each structure, but as a relative change, the increase was similar in the cortex, hippocampus, and thalamus (1.3–1.4%) but weaker in the hypothalamus (0.9%). The increase was dramatically larger in the METH-29 °C group but the pattern did not differ significantly in various structures. Thus, the cortex and hypothalamus exhibited about 2.9% and 3.1% increases, respectively, over control values, while the thalamus and hypothalamus showed 2.8% and 2.6% elevations.

In each brain structure, tissue water content was directly related to albumin immunoreactivity (B) and brain temperatures (C), suggesting tight relationships between brain temperature, changes in BBB permeability, and edema.

METH treatment also resulted in a general increase in Na^+, K^+, and Cl^- contents in various brain areas; this increase was stronger when METH was used at 29 °C than 23 °C (Fig. 6A–C). The regional contents of each ion varied, with the highest absolute levels for Na^+, lower for K^+, and lowest for Cl^-. Na^+, K^+, and Cl^- contents in control were highest in the hippocampus, followed by the cortex, and lowest in the hypothalamus and thalamus. With respect to Na^+, the increase seen in the METH-23 °C group versus control was significant and almost identical in each structure (+30 to 36 mM/kg). With respect to K^+ and Cl^-, the increase was structure-selective, with no changes of K^+ in the cortex and hippocampus and no changes in Cl^- in hippocampus and hypothalamus in the METH-23 °C group. In fact, K^+ levels in the cortex were even lower than in control. However, the levels of all ions were significantly higher in the METH-29 °C

(A, D, G) = 40 μm. In contrast to intense red myelin bundles and dense red fibers in control (B, arrowheads), diminution of red staining in the bundles and fibers (E, arrows) was seen in METH-treated rats (E). This degradation of MBP was most prominent in the rats treated with METH at 29 °C (H, arrows). Bar (B, E, H) = 30 μm. GFAP immunostaining was prominent in METH-treated rat at 29 °C (I, arrows) compared to 23 °C (F, arrows). Control rats (C) occasionally show few GFAP-positive astrocytes (arrowhead). Reactive astrocytes were located around the nerve cells and microvessels, and distributed in wide regions in the neuropil. Damaged neurons can also be seen in the background. Bar (C, F, I) = 40 μm. (See Color Insert.)

KIYATKIN AND SHARMA

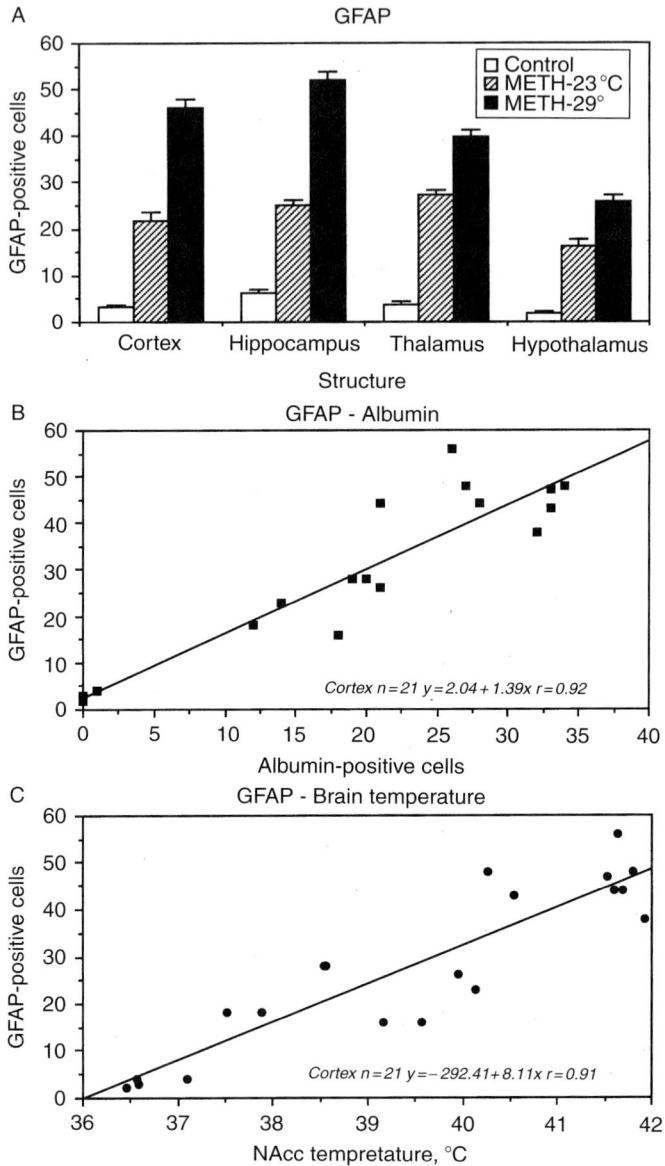

Fig. 4. (A) Mean (±SEM) numbers of GFAP-positive cells in individual brain structures of rat brains taken after saline (control) and methamphetamine administration at 23 and 29 °C. (B) Correlative relationships between the numbers of albumin- and GFAP-positive cells in the cortex. (C) Correlative relationships between the numbers of GFAP-positive cells in the cortex and NAcc temperatures. Each graph shows regression equation, regression line, and correlation coefficient (r).

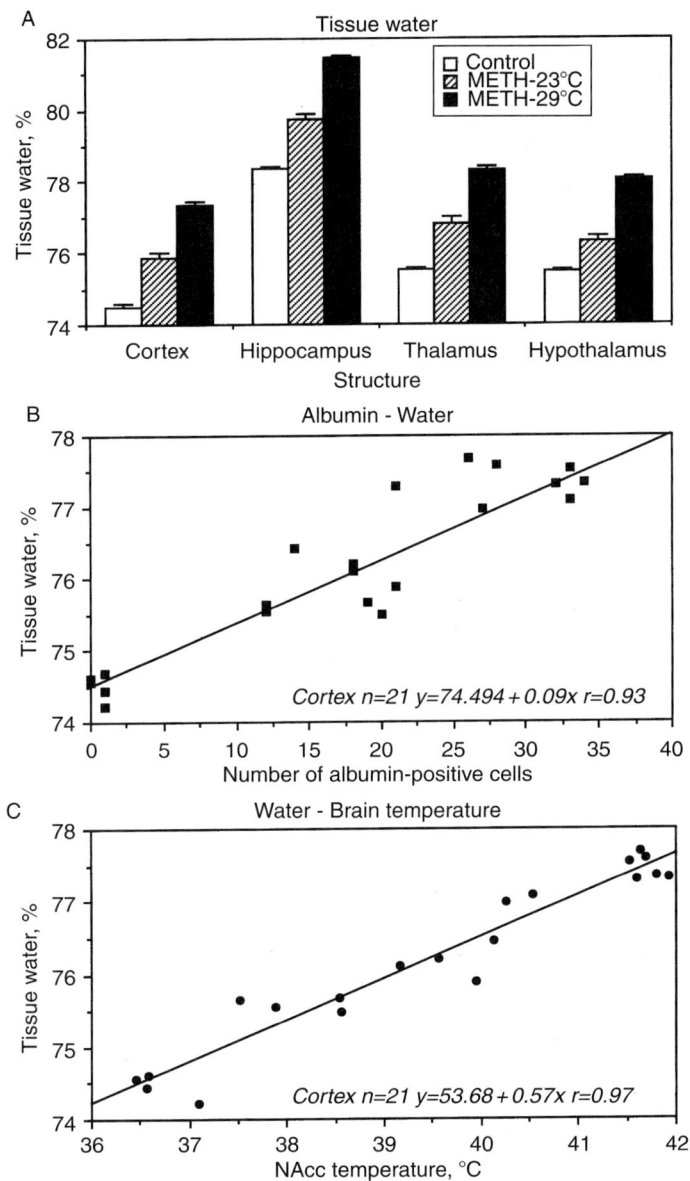

FIG. 5. (A) Mean (±SEM) concentrations of tissue water in individual brain structures after saline (control) and methamphetamine administration at 23 and 29 °C. (B) Correlative relationships between the numbers of albumin-positive cells and water in the cortex. (C) Correlative relationships between tissue water in the cortex and NAcc temperatures. Each graph shows regression equation, regression line, and correlation coefficient (r).

versus control groups. With respect to Na^+, the increase was the most pronounced, with maximal change in hypothalamus ($+72$ mM/kg) and thalamus
($+62$ mM/kg). These two structures also showed maximal increases in Cl^- (32
and 33 nM, respectively), although the increases in this ion were qualitatively
lower than those for Na^+. The increases in K^+ were maximal in the thalamus
($+67$ mM), followed by the hippocampus and hypothalamus (44 and 42 mM), and
minimal in the cortex (29 mM).

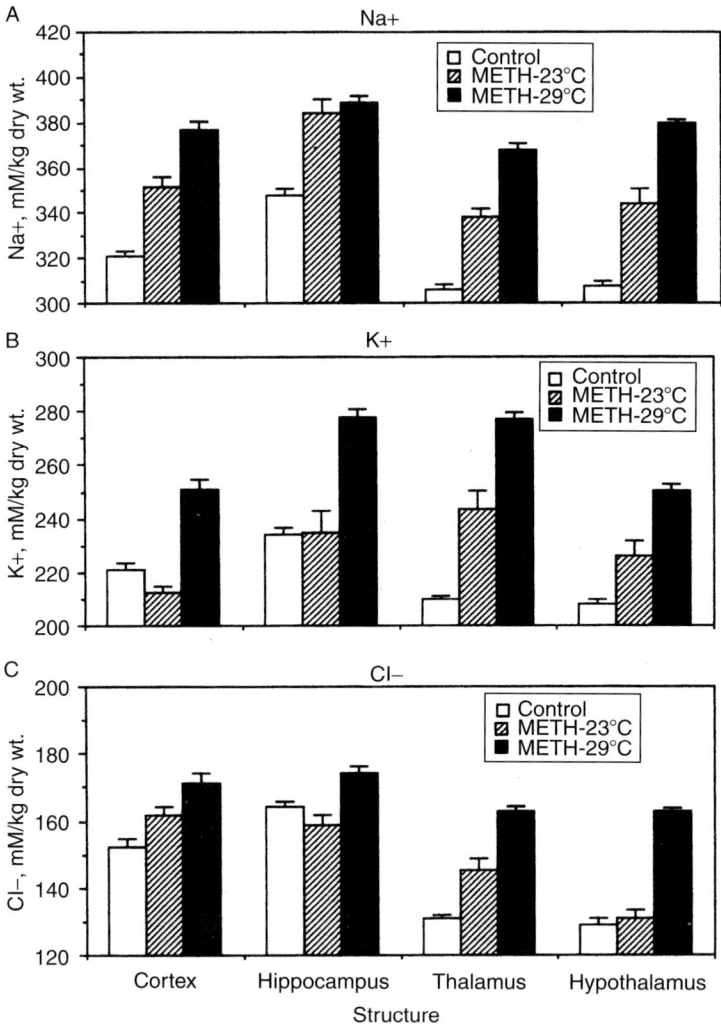

FIG. 6. Continued

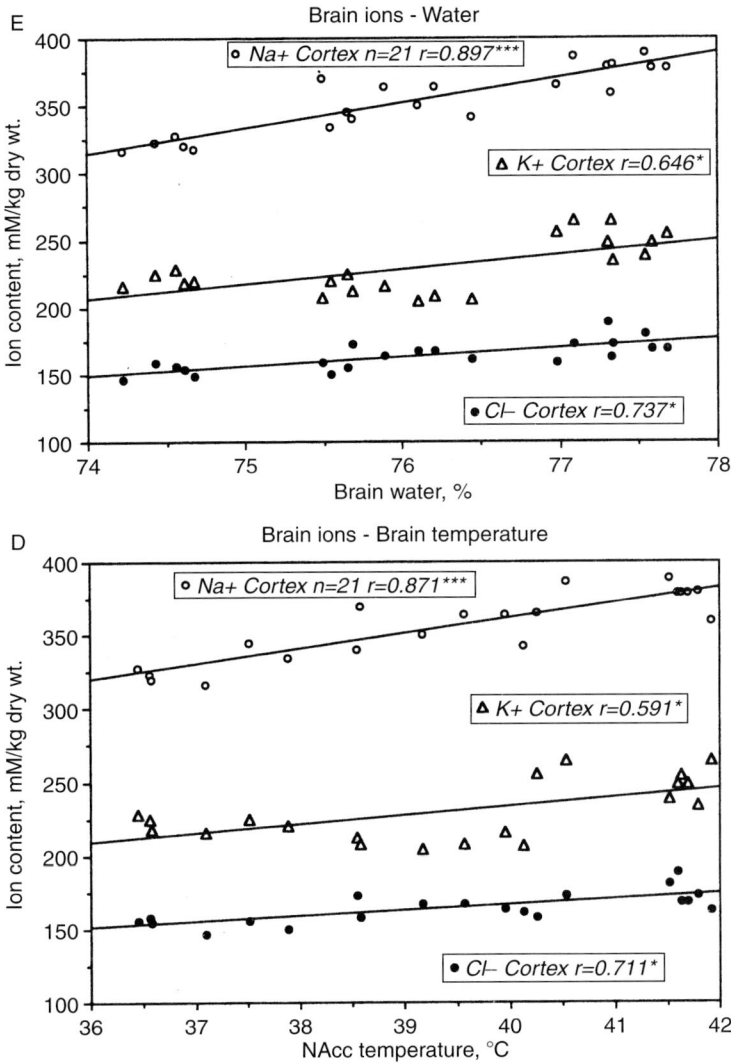

FIG. 6. Mean (±SEM) concentrations of Na$^+$ (A), K$^+$ (B), and Cl$^-$ (C) in individual brain structures of rat brains taken after saline (control) and methamphetamine administration at 23 and 29 °C (A). (E) Correlative relationships between ion content and tissue water in the cortex. (D) Correlative relationships between ion content in the cortex and NAcc temperatures. Each graph shows regression line and correlation coefficient (r).

Although concentrations of each ion significantly correlated with both water content (Fig. 6E) and brain temperatures (D), the correlation was strongest for Na$^+$, weaker for Cl$^-$, and minimal for K$^+$.

D. Morphological Changes

1. *Neuronal Changes*

As shown in Fig. 7, METH treatment resulted in a profound increase in the amount of abnormal neural cells in each studied brain structure. The increase was significant in each region but the highest numbers were seen in the hippocampus followed by the thalamus, hypothalamus, and cortex. In terms of relative change, the hippocampus also showed the maximal effect (×52), followed by

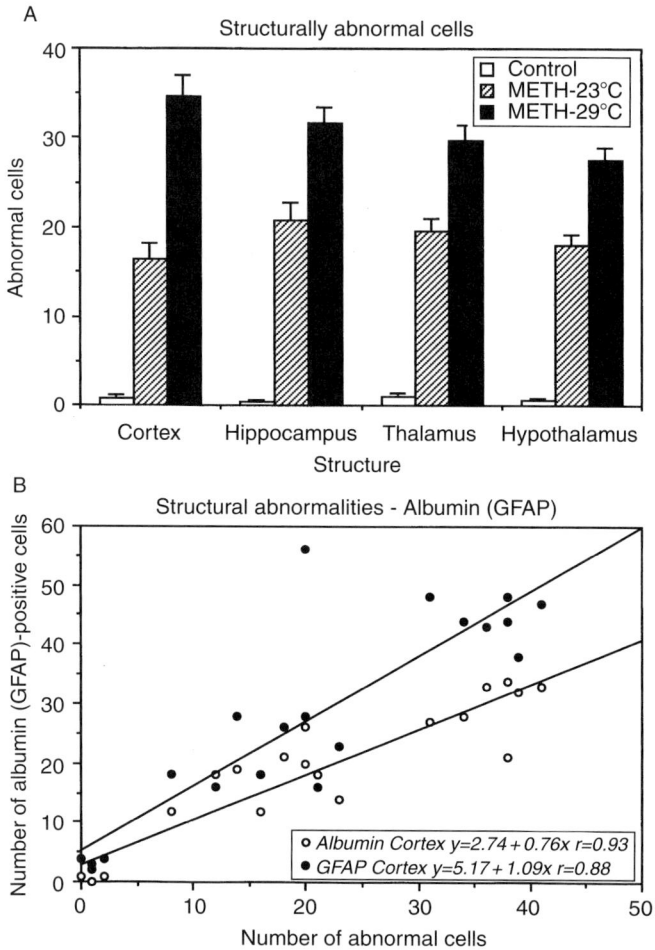

Fig. 7. Continued

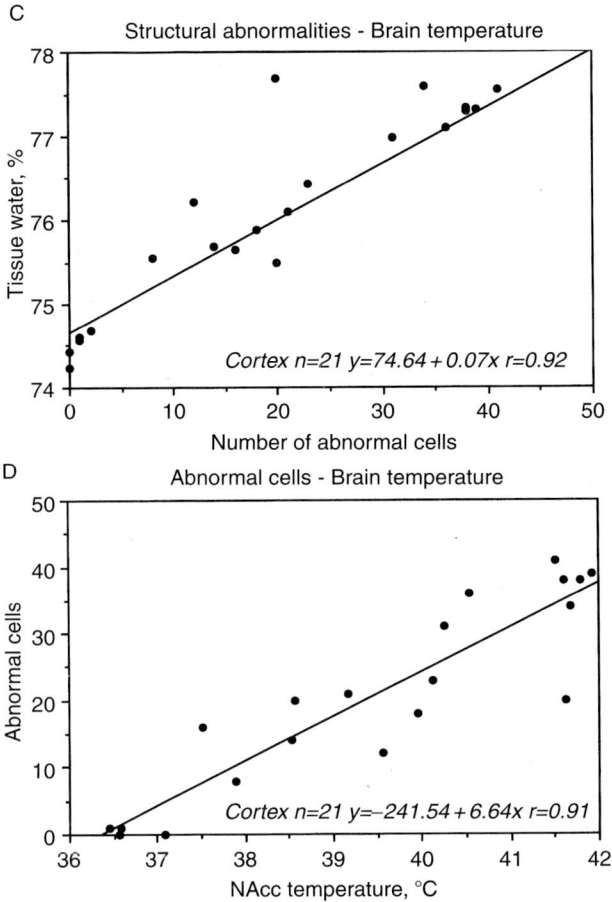

FIG. 7. (A) Mean (±SEM) numbers of structurally abnormal cells in individual brain structures after saline (control) and methamphetamine administration at 23 and 29 °C. (B) Correlative relationships between structural abnormalities and numbers of albumin- and GFAP-positive cells in the cortex. (C) Correlative relationships between the numbers of abnormal cells and tissue water. (D) Correlative relationships between the numbers of abnormal cells in the cortex and NAcc temperatures. Each graph shows regression equation, regression line, and correlation coefficient (r).

hypothalamus ($\times 30$), cortex ($\times 21$), and thalamus ($\times 20$). The incidence of neuronal damage was almost 1.5 times greater in animals that received METH at 29 °C. In this case, the greatest changes were seen in the cortex, followed by the hippocampus, thalamus, and hypothalamus.

Representative examples of Nissl-stained sections showing nerve cell damage in the cerebral cortex are depicted in Fig. 8. Normal animals exhibit healthy

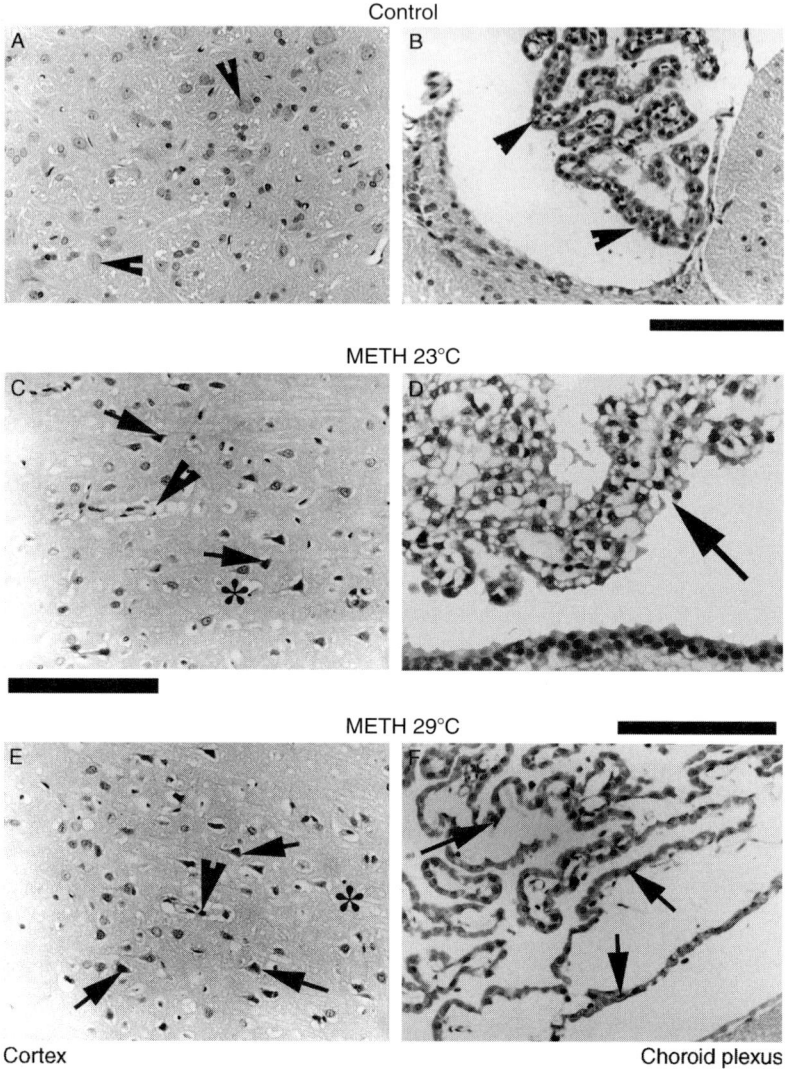

FIG. 8. Nissl-stained sections from the cerebral cortex (left panel) and choroid plexus (right panel) from control (A, B), METH-23 °C (C, D), and METH-29 °C (E, F) treated rats. Most of the nerve cells in the control cortex are healthy with a distinct nucleus in the center. Only a few nerve cells show condense cytoplasm (A, arrowheads). On the other hand, dark and distorted neurons were frequent in METH-treated rats (C and E, arrows). Sponginess (*) and perivascular edema (C and E, arrowheads) are frequent in METH-treated rats; the changes are more pronounced at 29 °C (E) compared to 23 °C (C). Bar (A, C, E) = 50 μm. Nissl-stained choroid epithelial cells in control (B) show compact and densely packer epithelial cells with distinct cell nucleus (B, arrow). METH treatment at 23 °C resulted in mild degeneration of choroid epithelium (D, arrows). The epithelial cell nucleus also appears to be disintegrated (D). These degenerating changes in the choroidal epithelium are most pronounced in the rat that received METH at 29 °C (F, arrows). Bar (B = 50 μm, D = 40 μm, F = 60 μm). (See Color Insert.)

neurons with a distinct nucleus and clear cytoplasm (A). Occasionally, a few neurons show a dense cytoplasm with a slightly excentric nucleus (arrowhead). On the other hand, several pyknotic neurons (C and E, arrows), perineuronal edema, sponginess, and expansion of the cortex were seen in METH-treated rats.

2. Damage to Choroid Epithelial Cells

METH also induced marked damage to the choroid plexus, a substrate of the blood–cerebrospinal fluid (CSF) barrier. While choroid epithelium in control brains consists of compact and dense epithelial cells with a distinct nucleus (Fig. 8B, arrowheads), marked degeneration of epithelial cells with occasional staining of cell nuclei was found in METH-23 °C rats (D). These changes were most prominent in the METH-29 °C group, where massive degeneration of epithelia with no distinction between individual cell nuclei and epithelial cells was apparent (F).

3. Myelin Changes

Since damage of myelin is quite common following BBB breakdown in various experimental and clinical conditions (see Sharma and Westman, 2004 for review), we examined myelin damage in METH-treated rats using MBP immunoreactivity (see Fig. 3B, E, and H). In normal animals, myelinated fibers and bundles are very dense red with MBP immunostaining (B), corresponding to intact myelin sheaths and myelinated axons. In contrast, degradation of MBP (E and H) seen in METH-treated animals reflects markedly degenerated myelinated fibers and axons; this decrease in myelin was greater when METH was administered to rats at 29 °C (H). Degeneration of myelin was commonly seen in the areas showing edematous expansion of the neuropil (H).

4. Relationships Between Brain Damage and Other Parameters

Changes in brain morphology were associated with profound changes in all other analyzed parameters (Fig. 7). As can be seen in Fig. 7B, cortical neural damage during METH intoxication was tightly related to *albumin leakage* ($r = 0.93$, 0.90, 0.89, and 0.90 for the cortex, hippocampus, thalamus, and hypothalamus, respectively). In the control cortex, when albumin leakage was virtually absent, there were no abnormal cells but their number grew progressively in METH-treated brains, depending on the extent of brain hyperthermia. The number of abnormal cells also tightly correlated with extent of *GFAP immunostaining* (Fig. 7B). This correlation was strong, linear, and virtually equal in all other structures ($r = 0.88$, 0.90, 0.90, and 0.90 for the cortex, hippocampus, thalamus, and hypothalamus, respectively), indicating its generality for the brain as a whole.

The number of damaged cortical cells during METH intoxication correlates linearly with *water accumulation* in all brain areas (see Fig. 7C for the cortex). This relationship appears to be a reflection of generalized edema, which progresses when METH is used at 29 °C, resulting in more profound cellular damage. Finally, structural damage was tightly related to *brain temperature* (Fig. 7D). While damaged cell count increased proportionally with temperature, the intensity of damage in each structure was independent of absolute temperatures. While brain temperature was recorded only from the NAcc in this study, it should follow a dorsoventral temperature gradient and be minimal in the cortex, lower in hippocampus and thalamus, and maximal in the hypothalamus (see Kiyatkin, 2005). However, the damage in "warm" hypothalamus during METH intoxication was minimal but was maximal in the "cool" cortex.

More important, METH-induced changes in brain morphology occurred rapidly (range 66–94 and 26–79 min in METH-23 °C and METH-29 °C, respectively) and were independent of *time of METH exposure*. Despite much greater damage in METH-29 °C than in METH-23 °C, the mean time of exposure was shorter (58.13 ± 5.82 vs 81.88 ± 3.96 min). No correlation with exposure time was found in each of the two groups analyzed separately ($r = 0.16$ for 23 °C and $r = 0.06$ for 29 °C) or as a whole ($r = 0.14$).

5. *Ultrastructural Changes Following METH Administration*

Rapid cellular changes seen at light microscopy following METH administration were further confirmed at the ultrastructural level using TEM (Figs. 9 and 10). At this level, several neuronal profiles located in the cortex were examined with special emphasis on the neuronal nucleolus and karyoplasm (Fig. 9, upper panel). As can be seen, a perfectly smooth rounded nuclear membrane containing a centralized nucleolus (arrow) was typical of normal brain tissue (A). The surrounding cytoplasm normally does not contain any vacuoles and the adjacent neuropil is quite compact. In contrast, degeneration of the nucleolus that became slightly excentric was seen in METH-23 °C rats (C). Moreover, the nuclear membrane shows many foldings and vacuolation in the neuronal cytoplasm (*). Perinuclear membranes also often show degenerative changes (C). These nuclear changes were more profound in animals that received METH at 29 °C (E). The nucleolus reached to one end of the karyoplasm and the nuclear envelope showed many irregular foldings (E). Cytoplasmic vacuolation (*) and degeneration of Nissl substance are clearly seen in this group of animals.

Degeneration of myelin and axons found using light microscopy with MBP staining in METH-treated animals was further confirmed at the ultrastructural level (see Fig. 9, lower panel). The most marked changes in myelin vesiculation were seen in rats treated with METH at 29 °C compared to those that received this drug at 23 °C (D and F). Unlike control animals (B), METH-treated rats showed clear signs of damage to myelinated axons (arrows in B and D). Vesiculation of myelin (arrows) and axonal damage (*) are clearly evident in METH-treated rats at 23 °C and are more pronounced in animals that received METH at 29 °C (F). Thus, degeneration of myelinated axons, vesiculation of myelin

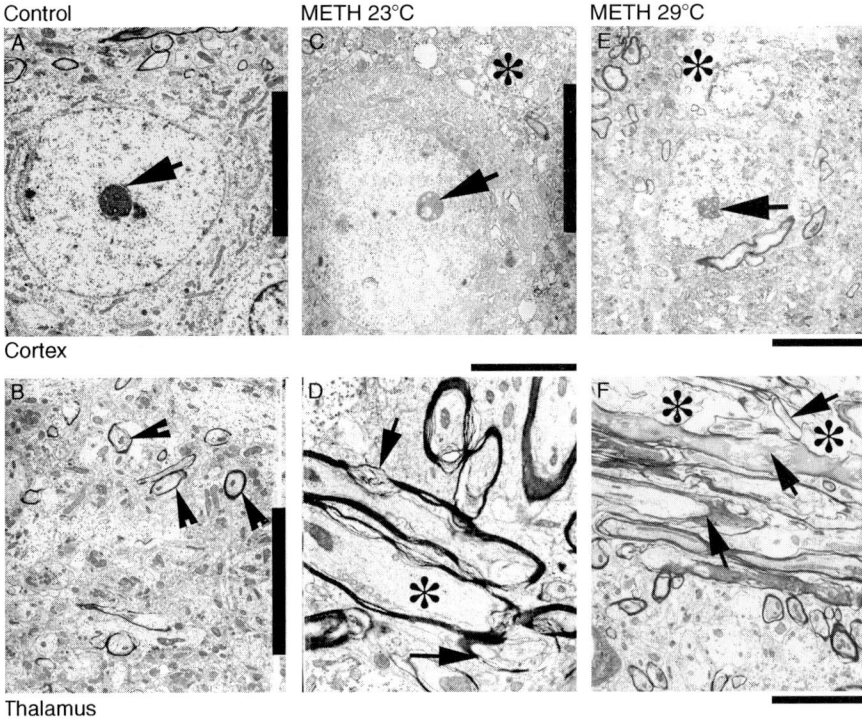

Fig. 9. Low-power transmission electron micrograph from the cortex (upper panel) and thalamus (lower panel) showing neuronal nuclear (A, C, E) and axonal changes (B, D, F) in control (A, B), METH-23 °C (C, D), and METH-29 °C (E, F) groups. The neuronal nucleus in control rat shows a smooth nuclear envelop with dark granular karyoplasm containing a central nucleolus (A, arrow). The nerve cell cytoplasm is compact and condensed without any vacuoles. On the other hand, less electron-dense karyoplasm with an eccentric nucleolus showing degenerative changes is seen in the METH-23 °C group (C, arrow). The nuclear membrane showed irregular foldings and vacuolation (*) in the neuropil including cytoplasm. These changes in the cell nucleus were much more aggravated in rat treated with METH at 29 °C (E); degeneration of nuclear membrane and surrounding neuronal cytoplasm is clearly evident in this slice. The nucleolus is further degenerated (arrow) and became more eccentric (E). Bar (A, C, E) = 1 μm. Axonal changes in the thalamus of METH-treated rats at 23 °C (D) show profound myelin vesiculation (arrow) and edematous swelling (D, *). These changes were stronger in rats treated with METH at 29 °C (F). In this group, the myelin vesiculation (arrows) and degeneration of axons were clearly evident (F, *). On the other hand, normal rat exhibited a compact neuropil with normal myelinated axons (B, arrowheads). Signs of vacuolation and edema are largely absent in control group (A, B). Bar (B = 1500 nm, D = 800 nm, F = 600 nm).

(F, arrows), and edematous swelling of axonal membranes (F, *) were quite frequent in this group. Normal animals exhibit a compact neuropil (B) and a nonvesiculated myelin sheath around the axons (A, arrowheads).

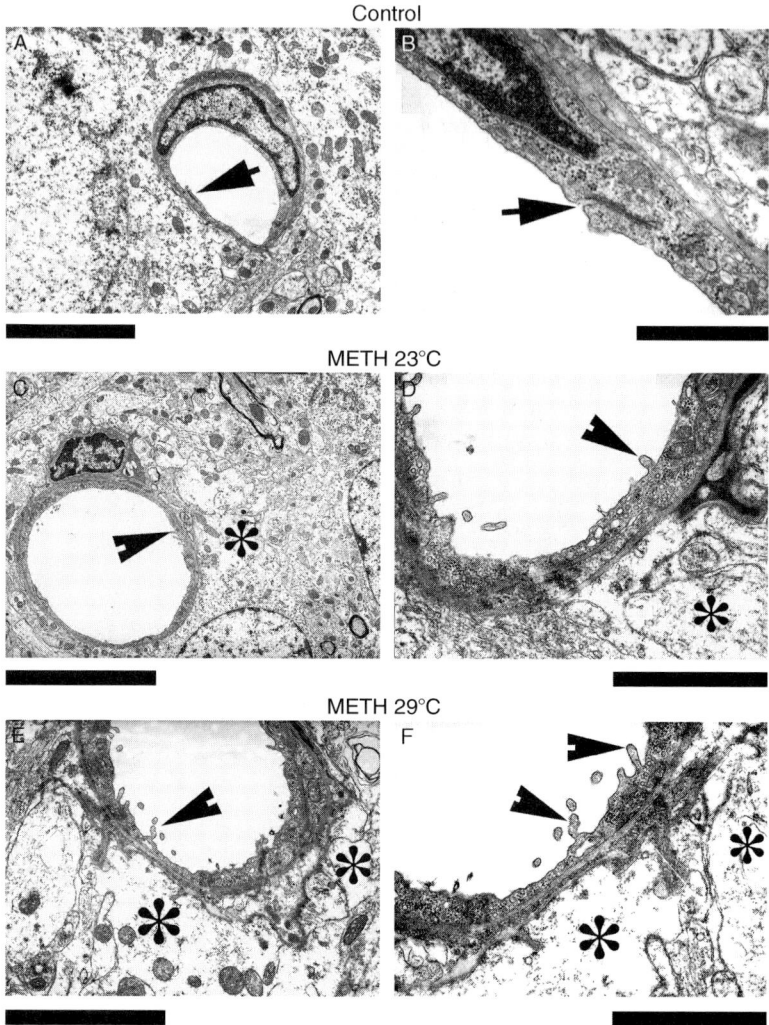

FIG. 10. Low-power (left panel) and high-power (right panel) transmission electron micrographs showing a cerebral capillary and the surrounding neuropil in control (A, B) and METH-treated brain (C, D: 23 °C; E, F: 29 °C). Normal cerebral capillary has smooth luminal surface and a compact, dense neuropil surrounding it and normal tight junctions (A, arrow). The normal capillary also has distinct tight junctions (B, arrow) and the underlying glial cells (astrocyte) do not exhibit any apparent signs of perivascular edema (B). METH treatment at 23 °C resulted in endothelial cell reaction and swelling of the perivascular astrocyte (C, *). The endothelial luminal surface exhibited few distinct bleb formations (arrowhead), indicating the process of enhanced vesicular transport or alterations in membrane transport properties (D, for details see text). Swollen perivascular astrocytes and its processes (*) are evident in this METH-treated rat (D). These ultrastructural changes, for example, bleb formation and perivascular edema, were much more aggravated in the rat after METH treatment at 29 °C (E, F).

Since leakage of EB and albumin immunostaining showed a compromised barrier in animals treated with METH, the endothelial and surrounding neuropil were examined at the ultrastructural level in the cortex and hippocampus (Fig. 10). The cerebral capillary from the normal rat showed a compact neuropil and perivascular structures, endothelial cell surfaces of the lumen were very smooth, and tight junctions (arrow) were clearly distinct (A and B). METH treatment markedly altered the endothelial cell surface in the lumen (C and D) and thus many endothelial cell bleb formations can be seen (D, arrowhead). The perivascular edema (*) and membrane vacuolation in the neuropil were apparent. These perivascular changes and endothelial cell membrane reactions were much more prominent in animals that received METH at 29 °C (E and F, arrowheads). In these rats, swollen perivascular glial cells (*) and degeneration of pericapillary neuropil (E and F) was clearly visible. The endothelial cell membrane showed prominent bleb formation in this group and the endothelial cell cytoplasm was more electron dense (E and F) than the normal endothelium (A and B). Degeneration of perivascular glial cells was also more prominent in this group compared to rats treated with METH at 23 °C (D and F). However, the tight junctions appeared to be intact in both METH-treated groups.

V. Conclusions

While slowly developing, selective, and irreversible damage of specific central neurons is a traditional focus of neurotoxic studies of METH, this work demonstrates that robust morphological abnormalities of neural and non-neural brain cells (i.e., glia, vascular endothelium, and epithelium) occur rapidly (within 30–80 min) in various brain structures during acute METH intoxication. These abnormalities manifest as the distortion of neurons, overexpression of glial cells, vesiculation of myelin, and alterations in vascular endothelium and epithelium of the choroid plexus. While having some structural specificity, these acute morphological abnormalities appear to be widespread, tightly correlating with drug-induced brain and body hyperthermia and alterations in several basic homeostatic brain parameters including permeability of the BBB, tissue water, and ion contents.

Thus, spreading out of small membrane vesicles and elongated bleb formation could be seen in this group (arrowheads). Swelling of astrocytes (*) and disintegration of astrocytic cytoplasm indicating water-filled cells are clearly visible (F, *). The endothelial cell cytoplasm in METH-treated rats shows much condense cytoplasm (D, F) compared to control (A, B). Bar (A, C = 1 μm, E = 2 μm, B = 500 nm, D = 800 nm, F = 600 nm).

A. Mechanisms Underlying METH-Induced Morphological Brain
   Abnormalities

Our study revealed that acute morphological alterations of brain cells during METH intoxication are tightly related to drug-induced increases in brain and muscle temperatures. The tight link between hyperthermia and structural brain damage was evident for the different brain structures and for neuronal (Nissl staining), glial (GFAP immunoreactivity), endothelial (TEM), and epithelial (Nissl staining) cells. Although METH was used in all animals at the same dose, the counts of damaged neurons and abnormal GFAP-positive glial cells were drastically larger at 29 °C than 23 °C, correlating with stronger brain and body hyperthermia. Linear relationships between these parameters and brain temperature were also evident in individual animals, which showed greater morphological abnormalities at higher temperatures, irrespective of the ambient temperature during drug administration.

Since brain cells of various subtypes are exceptionally sensitive to high temperature (Bechtold and Brown, 2003; Chen et al., 2003; Lee et al., 2000; Lin, 1997; Lin et al., 1991; Oifa and Kleshchenov, 1985; Sharma and Hoopes, 2003), hyperthermia could be viewed as an important contributor to morphological abnormalities induced by METH. However, this does not mean that high temperature per se is the cause of these changes. Brain hyperthermia is not only a physical factor that could harm cells, but also an integral physiological index of METH-induced metabolic activation (see Kiyatkin, 2005 for review) that also manifests as an enhanced release of multiple neuroactive substances, lipid peroxidation, and the generation of free radicals—numerous changes combined as oxidative stress (Cadet et al., 2007; Seiden and Sabol, 1996) as well as behavioral and autonomic activation. Although all these factors may contribute to structural brain abnormalities, it is quite difficult to separate them from each other because they are interdependent, representing different manifestations of METH-induced metabolic activation.

METH intoxication also results in a robust increase in BBB permeability, intrabrain water accumulation (edema), and serious alterations in brain ionic homeostasis. These changes, moreover, are tightly related to both the degree of hyperthermia and the intensity of structural brain damage. Therefore, breakdown of the BBB that allows the diffusion of endogenous albumin, water, several ions, and other neuroactive and potentially neurotoxic substances is another important contributor to brain pathology and the primary mechanism underlying decompensation of vital functions and lethality. While different chemical factors activated by METH could be involved in increased BBB permeability and edema formation, brain hyperthermia appears to play a crucial role since both these parameters strongly correlated with brain temperature. Such a role for hyperthermia in the leakage of the BBB and edema has been first suggested for environmental warming (Cervos-Navarro *et al.*, 1998; Sharma *et al.*, 1992); the intensity of these changes was strongly dependent upon the strength of body hyperthermia as evaluated by rectal temperature measurements.

B. COMMON FEATURES AND STRUCTURAL SELECTIVITY OF METH-INDUCED CELLULAR ALTERATIONS

To define the common features and structural specificity of METH-induced morphological and functional perturbations, our examinations were conducted in four brain structures: the cortex, hippocampus, thalamus, and hypothalamus. As shown above, these structures had some differences in tested parameters in normal conditions as well as some degree of specificity following METH impact. Despite evident between-structure differences in tissue water content in control brains, the increase was relatively similar in different brain structures at 23 °C (~1.3%; see Fig. 5) and dramatically amplified at 29 °C (~2.6–3.1%), suggesting severe edema within the brain as a whole. Robust water accumulation in brain tissue was associated with a clearly pathological brain hyperthermia (>41 °C), strong increases in Na^+, K^+, Cl^- in all brain structures, and maximal increases in albumin- (×22–75) and GFAP-positive (×8–13) cells. These animals also showed especially strong brain cell abnormalities that were evident in each structure and with each morphological and histochemical test. Degeneration of cellular elements in the nucleus, condensed cytoplasm, and irregular folding of the nuclear membrane found with electron microscopy denotes clear cell damage, which would possibly lead to cellular disintegration. The extreme character of these changes is related to the fact that if our experiment was not terminated and brains were not taken for examination, some of these animals (especially in the METH-29 °C group) would have naturally died within the next 10–30 min. Therefore, these changes appear to reflect acute pathological brain perturbations that could result in profound functional disturbances and lethality. While edema and ionic misbalance appears to be widespread within the brain, individual brain structures differed in the extent of morphological abnormalities. The numbers of damaged neuronal and glial cells were both greater in the cortex and hippocampus but smaller in the thalamus and hypothalamus. While the reasons of this structural specificity to damage remains unknown, the cortex and hippocampus also showed much stronger increases in EB penetration and albumin immunostaining than the thalamus and hypothalamus, suggesting a tight link between BBB leakage and structural damage. Although brain damage was evident in each cellular subtype (neurons, glial cells, vascular endothelium, and axons), especially profound abnormalities were found in epithelial cells of the choroid plexus—a critical substrate of the CSF barrier (Keep *et al.*, 2005). Because of the role of this organ in the production of the CSF, the damage of epithelial cells and destruction of gap junctions could strengthen edema and ionic disbalance, thus amplifying cellular damage in other brain structures. While it is unclear why this structure is so sensitive to damage during METH intoxication, it has an exceptionally high blood flow (Faraci et al., 1988) and mitochondrial density (Keep and Jones, 1990) compared to other structures, supporting a presumed link between metabolic activity and cellular damage following thermal impact (see Kiyatkin, 2005 for review).

C. Reversibility/Irreversibility of Structural Alterations

Although our data indicate that acute METH intoxication results in rapidly developing morphological abnormalities in neural and non-neural brain cells, it remains unclear whether these abnormalities are reversible or irreversible in nature. Dramatic changes in cellular elements (e.g., degeneration of some neuronal nucleus, endothelial cell membrane bleb formation, vesiculation of myelin, and vacuolation in the neuropil with clear signs of synaptic damage and swelling), especially evident in the METH-29 °C group, appear to be inconsistent with normal cell functions, pointing towards irreversible damage. However, some of these changes appear to be transient and reversible and they could dissipate after basic homeostatic parameters are restored to baseline. In contrast to the unusually large numbers of GFAP-positive glial cells found in this study during acute METH intoxication ($\times 3$–7 and $\times 7$–12 vs control in 23 and 29 °C groups, respectively), much weaker increases in GFAP-positive cells were found in rats in days after acute METH impact (Miller and O'allaghan, 1994; O'Callaghan and Miller, 1994). These changes, moreover, appeared at 12–24 h, peaked at the second day, and remained elevated for 7 days after a single METH exposure. In contrast to the widespread alterations in our study, these changes were evident only in the striatum and to a lesser degree in the cortex, correlating with decreased dopamine levels in these structures. While in light of the profound morphological abnormalities found during acute METH intoxication we can speculate that they could result in irreversible cellular damage, this issue needs to be examined further. While rapid and strong increases in GFAP immunostaining have been reported previously during environmental warming ($\sim$2–3 h; Cervos-Navarro et al., 1998; Sharma et al., 1992) and following acute traumatic injury to the brain and spinal cord (30–40 min; Cervos-Navarro et al., 1998; Gordth et al., 2006), different mechanisms appear to mediate rapid and slow glial reactions. GFAP expression is usually thought of as a late outcome of traumatic, ischemic, or hypoxic insults or a correlate of various neurodegenerative diseases (Finch, 2003; Gordth et al., 2006; Hausmann, 2003), representing astrogliosis (Norton et al., 1992; O'Callaghan, 1993). In contrast, rapid GFAP expression seen in association with strong edema (environmental warming, acute trauma, and METH intoxication) could reflect the interaction of antibodies with GFAP somehow released or made available during membrane damage. Thus, binding sites to GFAP could be increased due to acute breakdown of the BBB and associated edema rather than proliferation of astrocytes or elevated levels of GFAP proteins that require more time. Since damage of astrocytes and swelling of the astrocytic end foot results in increased binding of GFAP antibodies (Bekay *et al.*, 1977; Bondarenko and Chesler, 2001; Gordth *et al.*, 2006), this reaction could reflect acute, possibly reversible, damage of glial cells. Relatively smaller numbers of damaged neural cells in the postintoxication period compared to acute METH intoxication (see Bowyer and Ali, 2006) could also be related to their rapid scavenging, making it difficult to detect them using traditional techniques. Although the issue of the extent of damage and its reversibility remains unanswered and requires additional studies, it is likely that rapid cell abnormalities may initiate cascades that could precipitate cellular and molecular dysfunctions, leading to neurodegeneration—the most dangerous outcome of chronic abuse with amphetamine-like drugs.

Acknowledgments

This research was supported by the Intramural Research Program of the NIH, NIDA and Laerdal Foundation for Acute Medicine, Norway.

References

Alberts, D. S., and Sonsalla, P. K. (1995). Methamphetamine-induced hyperthermia and dopaminergic neurotoxicity in mice: Pharmacological profile of protective and nonprotective agents. *J. Pharmacol. Exp. Ther.* **275**, 1104–1114.

Ali, S. F., Newport, G. D., Holson, R. R., Slikker, W., and Bowyer, J. F. (1994). Low environmental temperatures or pharmacological agents that produce hypothermia decrease methamphetamine neurotoxicity in mice. *Brain Res.* **658**, 33–38.

Bechtold, D. A., and Brown, I. R. (2003). Induction of Hsp27 and Hsp32 stress proteins and vimentin in glial cells of the rat hippocampus following hyperthermia. *Neurochem. Res.* **28**, 1163–1173.

Bekay, L., Lee, J. C., Lee, G. C., and Peng, G. R. (1977). Experimental cerebral concussion: An electron microscopic study. *J. Neurosurg.* **47**, 525–531.

Blezer, E. (2005). Techniques for measuring the blood–brain barrier integrity. *In* "The Blood–Brain Barrier and Its Microenvironment" (E. de Vries and A. Prat, Eds.), pp. 441–456. Taylor & Francis, New York.

Bondarenko, A., and Chesler, M. (2001). Rapid astrocyte death induced by transient hypoxia, acidosis, and extracellular ion shifts. *Glia* **34**, 134–142.

Bowyer, J. F., and Ali, S. (2006). High doses of methamphetamine that cause disruption of the blood–brain barrier in limbic areas produce extensive neuronal degeneration in mouse hippocampus. *Synapse* **60,** 521–532.

Bowyer, J. F., Gough, B., Slikker, W., Lipe, G. W., Wewport, G. D., and Holson, R. R. (1993). Effects of a cold environment or age on methamphetamine-induced dopamine release in the caudate putamen of female rats. *Pharmacol. Biochem. Behav.* **44,** 87–98.

Brown, P. L., and Kiyatkin, E. A. (2004). Brain hyperthermia induced by MDMA ("ecstasy"): Modulation by environmental conditions. *Eur. J. Neurosci.* **20,** 51–58.

Brown, P. L., and Kiyatkin, E. A. (2005). Fata intra-brain heat accumulation induced by methamphetamine at normothermic conditions in rats. *Int. J. Neuroprotect. Neuroregen.* **1,** 86–90.

Brown, P. L., Wise, R. A., and Kiyatkin, E. A. (2003). Brain hyperthermia is induced by methamphetamine and exacerbated by social interaction. *J. Neurosci.* **23,** 3924–3929.

Cadet, J. L., Krasnova, I. N., Jayanthi, S., and Lyles, J. (2007). Neurotoxicity of substituted amphetamines: Molecular and cellular mechanisms. *Neurotox. Res.* **11,** 183–202.

Cervos-Navarro, J., Sharma, H. S., Westman, J., and Bongcum-Rudloff, E. (1998). Glial cell reactions in the central nervous system following heat stress. *Prog. Brain Res.* **115,** 241–274.

Chen, Y. Z., Xu, R. X., Huang, Q. J., Xu, Z. J., Jiang, X. D., and Cai, Y. O. (2003). Effect of hyperthermia on tight junctions between endothelial cells of the blood–brain barrier model in vitro. *Di Yi Jun Yi Da Xue Xue Bao* **23,** 21–24.

De Vito, M. J., and Wagner, G. C. (1989). Methamphetamine-induced neuronal damage: A possible role for free radicals. *Neuropharmacology* **28,** 1145–1150.

Devis, W. M., Hatoum, H. T., and Walters, I. W. (1987). Toxicity of MDA (2.4-methylenedioxyamphetamine) considered for relevance to hazards of MD<A (Ecstasy) abuse. *Alcohol Drug Res.* **7,** 123–134.

Doepp, F., Schreiber, S. J., Munster, T., Rademacher, J., Klingbiel, R., and Valdueza, J. M. (2004). How does the bloods leave the brain? A systematic ultrasound analysis of central venous drainage pattern. *Neuroradiology* **46,** 565–570.

Ehrlich, P. (1904). Ueber die beziehungen von chemischer constitution, verteilung und pharmakologischer wirkung. *In:* Gesammelte Arbeiten zur Immunitätsforschung. Hirschwald, Berlin.

Esposito, P., Cheorghe, D., Kendere, K., Pang, X., Connoly, R., Jaconson, S., and Theodorides, T. C. (2001). Acute stress increases permeability of the blood–brain barrier through activation of brain must cells. *Brain Res.* **888,** 117–127.

Estler, C. J. (1975). Dependence on age of metamphetamine-produced changes in thermoregulation and metabolism. *Experientia* **31,** 1436–1437.

Faraci, F. M., Mayham, W. G., Williams, J. K., and Heistad, D. D. (1988). Effects of vasoactive stimuli on blood flow to choroid plexus. *Am. J. Physiol.* **254,** H286–H291.

Farfel, G. M., and Seiden, L. S. (1995). Role of hyperthermia in the mechanism of protection against serotoninergic toxicity. II. Experiments with methamphetamine, p-chloroamphetamine, fenfluramine, dizocilpine and dextromethorphan. *J. Pharmacol. Exp. Ther.* **272,** 868–875.

Finch, C. E. (2003). Neurons, glia, and plasticity in normal brain aging. *Neurobiol. Aging* **24**(Suppl. 1), S123–S127.

Gordh, T., Chu, H., and Sharma, H. S. (2006). Spinal nerve lesion alters blood–spinal cord barrier function and activates astrocytes in the rat. *Pain* **124,** 211–221.

Hausmann, O. N. (2003). Post-traumatic inflammation following spinal cord injury. *Spinal Cord* **41,** 369–378.

Iwagami, Y. (1996). Changes in the ultrastructure of human cell related to certain biological responses under hyperthermic culture conditions. *Hum. Cell* **9,** 353–366.

Kalant, H. (2001). The pharmacology and toxicology of "ecstasy" (MDMA) and related drugs. *Can. Med. Assoc. J.* **165,** 917–928.

Kalant, H., and Kalant, O. J. (1975). Death in amphetamine users: Causes and rates. *Can. Med. Assoc. J.* **112,** 299–304.

Keep, R. F., and Jones, H. C. (1990). A morphometric study on the development of the lateral choroids plexus, choroids plexus capillaries and ventricular ependyma in the rat. *Dev. Brain Res.* **56,** 47–53.

Keep, R. F., Ennis, S. R., and Xiang, J. (2005). The blood–CSF barrier and cerebral ischemia. *In* "The Blood–Cerebrospinal Fluid Barrier", (W. Zheng and A. Chodobski, Eds.), pp. 345–360. CRC Press, Boca Raton, FL.

Kiyatkin, E. A. (2005). Brain hyperthermia as physiological and pathological phenomena. *Brain Res. Rev.* **50,** 27–56.

Kiyatkin, E. A. (2006). Drug-induced brain hyperthermia: Mechanisms and functional implications. *Int. J. Neuroprotect. Neuroregen.* **2,** 168–174.

Kiyatkin, E. A., and Brown, P. L. (2004). Modulation of physiological brain hyperthermia by environmental temperature and impaired blood outflow in rats. *Physiol. Behav.* **83,** 467–474.

Kiyatkin, E. A., Brown, P. L., and Sharma, H. S. (2007). Brain edema and breakdown of blood–brain barrier during methamphetamine intoxication: Critical role of brain temperature. *Eur. J. Neurosci.* **26,** 1242–1253.

Kuhn, D. M., and Geddes, T. J. (2000). Molecular footprints of neurotoxic amphetamine action. *Ann. N.Y. Acad. Sci.* **914,** 92–103.

Lee, S. Y., Lee, S. H., Akuta, K., Uda, M., and Song, C. W. (2000). Acute histological effects of interstitial hyperthermia on normal rat brain. *Int. J. Hyperthermia* **16,** 73–83.

Lepock, J. R. (2003). Cellular effects of hyperthermia: Relevance to the minimum dose for thermal damage. *Int. J. Hyperthermia* **19,** 252–266.

Lin, M.-T. (1997). Heatstroke-induced cerebral ischemia and neuronal damage. Involvement of cytokines and monoamines. *Ann. N. Y. Acad. Sci.* **813,** 572–580.

Lin, P. S., Quamo, S., Ho, K. C., and Gladding, J. (1991). Hyperthermia enhances the cytotoxic effects of reactive oxygen species to Chinese hamster cells and bovine endothelial cells *in vitro*. *Radiat. Res.* **126,** 43–51.

Makisumi, T., Yoshida, K., Watanabe, T., Tan, N., Murakami, N., and Morimoto, A. (1998). Sympatho-adrenal involvement in methamphetamine-induced hyperthermia through skeletal muscle hypermethanolism. *Eur. J. Pharmacol.* **363,** 107–112.

Miller, D. B., and O'Callaghan, J. P. (1994). Environment-, drug- and stress-induced alterations in body temperature affect the neurotoxicity of substituted amphetamines in the C57BL/6J mouse. *J. Pharmacol. Exp. Ther.* **270,** 752–760.

Norton, W. T., Aquino, D. A., Hozumi, I., Chiu, F.-C., and Brosnan, C. F. (1992). Quantitative aspects of reactive gliolis: A review. *Neurochem. Res.* **17,** 877–885.

O'Callaghan, J. P. (1993). Quantitative features of reactive gliolis following toxicant-induced damage of the CNS. *Ann. N. Y. Acad. Sci.* **679,** 195–210.

O'Callaghan, J. P., and Miller, D. B. (1994). Neurotoxicity profiles of substituted amphetamines in the C57BL/6J mouse. *J. Pharmacol. Exp. Ther.* **270,** 741–751.

Oifa, A. I., and Kleshchnov, V. N. (1985). Ultrastructural analysis of the phenomenon of acute neuronal swelling. *Zh. Nevropatol. Psikhiatr. Im S.S. Korsakova* **85,** 1016–1020.

Ovadia, H., Abramsky, O., Feldman, S., and Weidenfeld, J. (2001). Evaluation of the effects of stress on the blood–brain barrier: Critical role of the brain perfusion time. *Brain Res.* **905,** 21–25.

Papoport, S. I. (1976). Blood–Brain Barrier in Physiology and Medicine. Raven Press, New York, NY.

Ricaurte, G. A., Schuster, C. R., and Seiden, L. S. (1980). Long-term effects of repeated methamphetamine administration on dopamine and serotonin neurons in the rat brain: A regional study. *Brain Res.* **193,** 153–163.

Riddle, E. L., Fleckenstein, A. E., and Hanson, G. R. (2006). Mechanisms of methamphetamine-induced dopaminergic neurotoxicity. *AAPS J.* **8**(2), Article 48(http://www.aapsj.org).

Romanovsky, A. A. (2004). Do fever ana anapyrexia exist? Analysis of set point-based definitions. *Am. J. Physiol.* **287,** R992–R995.

Romanovsky, A. A., Ivanov, A. I., and Shimansky, Y. P. (2002). Ambient temperature for experiments in rats: A new method for determining the zone of thermal neutrality. *J. Appl. Physiol.* **92,** 2667–2679.

Sandoval, V., Hanson, G. R., and Fleckenstein, A. E. (2000). Methamphetamine decreases mouse striatal dopamine transport activity: Roles of hyperthermia and dopamine. *Eur. J. Pharmacol.* **409,** 265–271.

Seiden, L. S., and Sabol, K. E. (1996). Methamphetamine and methylenedioxymethamphetamine neurotoxicity: Possible mechanisms of cell destruction. *NIDA Res. Monogr.* **163,** 251–276.

Sharma, H. S. (2007). Methods to produce hyperthermia-induced brain dysfunction. *Prog. Brain Res.* **162,** 173–199.

Sharma, H. S., and Ali, S. F. (2006). Alterations in blood–brain barrier function by morphine and amphetamine. *Ann. N. Y. Acad. Sci.* **1074,** 198–224.

Sharma, H. S., and Dey, P. K. (1986). Influence of long-term immobilization stress on regional blood–brain permeability, cerebral blood flow and 5-HT levels in conscious normotensive young rats. *J. Neurol. Sci.* **72,** 61–76.

Sharma, H. S., and Hoopes, P. J. (2003). Hyperthermia-induced pathophysiology of the central nervous system. *Int. J. Hyperthermia* **19,** 325–354.

Sharma, H. S., and Kiyatkin, E. A. (2009). Rapid morphological brain abnormalities during acute methamphetamine intoxication in the rat: An experimental study using light and electron microscopy. *J. Chem. Neuroanat.* **37,** 18–32.

Sharma, H. S., and Westman, J. (2004). The blood-spinal cord and brain barriers in health and disease. San Diego, Elsevier/Academic Press, pp. 1–617.

Sharma, H. S., Zimmer, C., Westman, J., and Cervos-Navarro, J. (1992). Acute systemic heat stress increases glial fibrillary acidic protein immunoreactivity in brain: An experimental study in the conscious normotensive young rats. *Neuroscience* **48,** 889–901.

Stephans, S. E., and Yamamoto, B. K. (1994). Methamphetamine-induced neurotoxicity: Role for glutamate and dopamine influx. *Synapse* **17,** 203–209.

Watson, P., Shirreffs, S. M., and Maughan, R. J. (2005). Blood–brain barrier integrity may be threatened by exercise in a warm environment. *Am. J. Physiol.* **288,** R1689–R1694.

Willis, W. T., Jackman, M. R., Bizeau, M. E., Pagliassotti, M. J., and Hazel, J. R. (2000). Hyperthermia impairs liver mitochondrial functions. *Am. J. Physiol.* **278,** R1240–R1246.

Woolverton, W. L., Ricaurte, G. A., Forno, L., and Seiden, L. S. (1989). Long-term effects of chronic methamphetamine administration in rhesus monkeys. *Brain Res.* **486,** 73–78.

Yamamoto, H. (1963). The central effects of xylopinine in mice. *Jpn. J. Pharmacol.* **13,** 230–239.

Yamamoto, B. K., and Zhu, W. (1998). The effect of methamphetamine on the production of free radicals and oxidative stress. *J. Pharmacol. Exp. Ther.* **287,** 107–114.

Zlokovic, B. V. (2008). The blood–brain barrier in health and chronic neurodegenerative disorders. *Neuron* **57,** 178–201.

MOLECULAR BASES OF METHAMPHETAMINE-INDUCED NEURODEGENERATION

Jean Lud Cadet and Irina N. Krasnova

Molecular Neuropsychiatry Branch, NIDA-Intramural Research Program, NIH/DHHS, Baltimore, Maryland 21224, USA

Methamphetamine (METH) is a highly addictive psychostimulant drug, whose abuse has reached epidemic proportions worldwide. The addiction to METH is a major public concern because its chronic abuse is associated with serious health complications including deficits in attention, memory, and executive functions in humans. These neuropsychiatric complications might, in part, be related to drug-induced neurotoxic effects, which include damage to dopaminergic and serotonergic terminals, neuronal apoptosis, as well as activated astroglial and microglial cells in the brain. Thus, the purpose of the present paper is to review cellular and molecular mechanisms that might be responsible for METH neurotoxicity. These include oxidative stress, activation of transcription factors, DNA damage, excitotoxicity, blood–brain barrier breakdown, microglial activation, and various apoptotic pathways. Several approaches that allow protection against METH-induced neurotoxic effects are also discussed. Better understanding of the cellular and molecular mechanisms involved in METH toxicity should help to generate modern therapeutic approaches to prevent or attenuate the long-term consequences of psychostimulant use disorders in humans.

I. Epidemiology of Methamphetamine Abuse

Methamphetamine (METH) is an illegal psychostimulant that is abused by more than 25 million people in the world, which exceeds the amount of people who use heroin and cocaine (Rawson and Condon, 2007; Romanelli and Smith, 2006). METH is the most commonly synthesized illegal drug in the United States and has been cited by law enforcement officials as the leading crime problem in the country (Gettig et al., 2006). The inexpensive production of the drug, its low cost, and its long duration of action make it very desirable to abusers. Recently, METH abuse has reached epidemic proportions in the United States, with populations in the Western, Southern, and Midwest states being the most affected (Gettig et al., 2006; Rawson and Condon, 2007; Topolski, 2007). A 2003 survey found that 5.2% of the American adults have used METH at least once (Roehr, 2005). In addition, 5.3% and 6.2% of high school sophomores and seniors have tried METH (Gettig et al., 2006). METH-related emergency room admissions have also increased from 10 to 52 per 100,000 people between 1992 and 2002 (Roehr, 2005). Although METH abuse has been associated, traditionally, with blue-collar construction workers, truck drivers, and motorcycle gangs, the profile of the typical METH abusing individual has shifted due to the increased popularity among college students, women, and young professionals (Gettig et al., 2006; Romanelli and Smith, 2006). METH use has also highly increased in men who have sex with men resulting in a greater frequency of METH abuse in homosexual and bisexual men than in the general population (Shoptaw, 2006).

II. Clinical Toxicology of METH Abuse

METH abuse causes serious health complications in humans. METH users develop acute clinical symptoms that include agitation, anxiety, aggressive behaviors, paranoia, hypertension, hyperthermia, and psychosis (Albertson et al., 1999; Lynch and House, 1992; Murray, 1998). Ingestions of large doses of the drug can cause life-threatening hyperthermia above 41 °C, cardiac arrythmias, heart attacks, cerebrovascular hemorrhages, strokes, seizures, renal, and liver failure (Albertson et al., 1999; Perez et al., 1999).

Clinical studies have also shown the potential neuropathological consequences of METH abuse, which include deficits in attention, working memory, and executive functions in chronic psychostimulant abusers (Gonzalez et al., 2004; Rippeth et al., 2004; Salo et al., 2002; Sim et al., 2002; Simon et al., 2002, 2004; Woods et al., 2005). The drug can also cause neurodegenerative changes in the brains of human addicts. These include persistent loss of dopamine transporters (DAT) observed in the cortex and caudate-putamen (Sekine *et al.*, 2003; Volkow *et al.*, 2001), loss of serotonin transporters (5-HTT) in the cortex (Sekine *et al.*, 2006), and decrease in the levels of dopamine (DA) and its metabolites in the caudate-putamen of METH abusers (Moszczynska *et al.*, 2004; Wilson *et al.*, 1996).

In agreement with clinical findings, a number of animal studies have shown that METH can cause long-term damage to presynaptic dopaminergic and serotonergic axons in rodents (Ricaurte *et al.*, 1980; Wagner *et al.*, 1980). More recently, it has also been shown that the drug can cause death of the neuronal bodies in the brain through apoptosis (Deng and Cadet, 1999, 2000; Eisch *et al.*, 1998; O'Dell and Marshall, 2000; Zhu *et al.*, 2006). In what follows, we discuss some of the mechanisms that might underlie these METH-induced neurodegenerative effects.

III. Role of Oxidative Stress in METH Toxicity

Toxic effects of METH are thought to depend on the similarity of its chemical structure to DA, which allows the drug to enter DA axons (Iversen, 2006), followed by DA release from synaptic vesicles into cytoplasm and by reverse transport into the synaptic cleft (Sulzer *et al.*, 2005). METH neurotoxicity depends on the formation of DA quinones and superoxide radicals within nerve terminals (LaVoie and Hastings, 1999). DA metabolism by MAO can also increase hydrogen peroxide production, followed by its interactions with metal ions to form very toxic hydroxyl radicals (Cadet and Brannock, 1998). Evidence has accumulated to indicate that METH can also cause oxidative stress by switching the balance between ROS production and the capacity of antioxidant enzyme system to scavenge ROS (Gluck *et al.*, 2001; Harold *et al.*, 2000; Jayanthi *et al.*, 1998). Excessive production of ROS that overwhelms this system can damage cellular components such as lipids, proteins, mitochondrial and nuclear DNA (Potashkin and Meredith, 2006).

A role for oxidative mechanisms in the drug toxicity is consistent with findings that pretreatment with N-acetyl-L-cysteine, ascorbic acid, or vitamin E allows protection against METH-induced depletion of monoaminergic axons (De Vito and Wagner, 1989; Fukami *et al.*, 2004; Wagner *et al.*, 1985). The role for superoxide radicals in the neurotoxic effects of METH on DA axons was tested by injecting METH to transgenic mice that overexpress the human CuZn superoxide dismutase (CuZnSOD) gene (Cadet *et al.*, 1994; Hirata *et al.*, 1996; Jayanthi *et al.*, 1998). These mice have much higher CuZnSOD enzyme activity than control wild-type animals (Jayanthi *et al.*, 1998, 1999) and were protected against METH toxicity. In addition, bromocriptine, which scavengers hydroxyl radicals, also attenuates METH-induced DA depletion in mice (Kondo *et al.*, 1994). Together, these findings support the proposition that DA release caused by METH is accompanied by redox cycling of DA quinones and formation of superoxide radicals. The hypothesis that oxygen-based free radicals are involved in METH toxicity (Cadet and Brannock, 1998) is also supported by reports that the drug can reduce the levels of glutathione (Harold *et al.*, 2000) and antioxidant enzymes (Jayanthi *et al.*, 1998), increase lipid peroxidation (Gluck *et al.*, 2001; Jayanthi *et al.*, 1998), and cause the formation of protein carbonyls (Gluck *et al.*, 2001).

Recent evidence also indicates that changes in nitric oxide (NO) metabolism can contribute to METH-induced oxidative stress and neurotoxicity (Itzhak and Ali, 2006). NO can react with superoxide radicals to form peroxynitrite, a strong oxidant and major neurotoxin (Pacher *et al.*, 2007). Indeed, increased levels of a marker for peroxynitrite production 3-NT have been measured *in vivo* and *in vitro* in response to METH treatment (Imam *et al.*, 2001a). Antioxidants selenium and melationin completely block the formation of 3-NT and striatal DA depletion (Imam *et al.*, 2001a). In addition, free radical scavenger, edaravone, blocked METH-related increase in 3-NT immunoreactivity with subsequent attenuation of DA depletion and of reduction in TH immunoreactivity in the striatum (Kawasaki *et al.*, 2006). Furthermore, the selective neuronal nitric oxide synthase (nNOS) inhibitor, 7-nitroindazole, protects against drug-induced formation of 3-NT, as well as DA and 5-HT depletion in the striatum (Ali and Itzhak, 1998; Di Monte *et al.*, 1996; Itzhak and Ali, 1996). The participation of NO in METH toxicity is also proposed by findings that nNOS knockout mice were protected against the formation of 3-NT and DA terminal degeneration in the striatum (Imam *et al.*, 2001b). Together, these data strongly support the hypothesis that NO and peroxynitrite, in particular, are involved in the mechanisms underlying METH-induced monoaminergic neurotoxicity (Imam *et al.*, 2001b; Itzhak and Ali, 2006; Itzhak *et al.*, 1998).

IV. Involvement of AP-1-Related Transcription Factors in METH-Induced Neurotoxicity

The accumulated data suggested that effects of METH might be mediated, in part, by activation of AP-1 transcription factors. These include upregulation of c-jun, c-fos, jun B, and jun D expression within 2 h after METH administration (Cadet *et al.*, 2001). These changes, in turn, might be related to METH-induced generation of ROS. Specifically, hydroxyl and superoxide radicals can induce the expression of many genes via regulation of AP-1 transcription factors (Dalton *et al.*, 1999). The role for c-fos in METH-induced neuropathological changes has been confirmed by using c-fos heterozygote mice that show increased degeneration of DA axons and increased cell death after psychostimulant treatment (Deng *et al.*, 1999). These findings support a protective role for c-fos against METH damage. The factors that could be involved in this protection include cell adhesion receptors integrins because of decreased basal levels of integrin expression in c-fos heterozygote mice and the further reduction of these receptors in response to toxic doses of METH (Betts *et al.*, 2002). This idea is also supported by the observations that integrins promote cell survival after injury and apoptotic insults via PI3K-Akt pathway which leads to phosphorylation of proapoptotic protein BAD, therefore, reducing its ability to block the antiapoptotic effects of Bcl-2 (Gilcrease, 2007). In contrast, inhibition of integrins increases apoptotic cell death (Gilcrease, 2007).

c-jun is another AP-1 transcription factor that might be involved in METH toxicity, because c-jun knockout mice show partial protection against damaging effects of the drug (Deng *et al.*, 2002b). Moreover, because the c-jun knockout mice and their wild-type littermates show similar degree of dopaminergic toxicity after METH treatment, c-jun appears to only play role in the mediation of neuronal apoptosis in cells postsynaptic to DA axons.

V. Role of DNA Damage in METH-Induced Toxicity

As mentioned earlier, METH can cause neuronal apoptosis in several brain regions, including striatum, cortex, hippocampus, and olfactory bulb (Deng *et al.*, 2001). Because apoptosis is associated with DNA damage, it was possible that treatment with METH might induce responses involved in the repair of the drug-related DNA damage. Data obtained using microarray analyses have shown that METH administration caused changes in the expression of a number of genes that participate in DNA repair, including APEX, PolB, and LIG1 (Cadet *et al.*, 2002). These changes are probably related to METH-induced increase in the levels of free radicals because oxidative stress can cause single and double DNA strand breaks (Li and Trush, 1993). Thus, the upregulation of DNA repair genes following METH treatment suggests that these changes might be compensatory to counteract METH-related ROS-induced DNA damage. If the psychostimulant can cause similar DNA damages in humans, this might account for developmental deficits observed in children born of METH abusing mothers (Smith *et al.*, 2006).

VI. METH Toxicity and Excitotoxicity

METH neurotoxicity might also occur via excitotoxic damage following glutamate release and activation of glutamate receptors. Glutamate toxicity is dependent, in part, on the production of NO (Chung *et al.*, 2005). The hypothesis on the involvement of glutamate in METH toxicity is supported by findings that

METH causes glutamate release in the brain (Abekawa *et al.*, 1994; Baldwin *et al.*, 1993; Mark *et al.*, 2004; Marshall *et al.*, 1993; Nash *et al.*, 1988). In addition, some glutamate antagonists can attenuate METH-induced dopaminergic toxicity (Battaglia *et al.*, 2002; Sonsalla *et al.*, 1989). Glutamate-mediated NO formation might also be involved in METH toxicity because knockout mice deficient in either nNOS or iNOS (inducible nitric oxide synthase) are protected against psychostimulant-induced damage to monoaminergic axons (Itzhak *et al.*, 1998). These data have provided strong support for the idea that the glutamate/NO pathway plays major role in METH neurotoxicity (Imam *et al.*, 2001b; Itzhak and Ali, 2006; Itzhak *et al.*, 1998). Finally, various nNOS inhibitors can also protect against depletion of monoaminergic axons caused by METH administration (Itzhak *et al.*, 2000; Sanchez *et al.*, 2003). In addition to their roles in the damage of monoaminergic axons, oxygen-based radicals and NO may be involved in METH-related neuronal death because CuZnSOD transgenic mice show partial protection against drug-induced apoptosis (Deng and Cadet, 2000).

VII. Role of Blood–Brain Barrier Dysfunction in METH Toxicity

Several recent papers have examined the effects of METH on the blood–brain barrier (BBB) and their potential relationships to METH toxicity (Bowyer and Ali, 2006; Bowyer *et al.*, 2008; Kiyatkin *et al.*, 2007; Sharma and Ali, 2006; Sharma *et al.*, 2007). Using protein tracers and albumin immunohistochemistry, METH was shown to cause marked disruption of BBB at the levels of the cortex, hippocampus, thalamus, hypothalamus, cerebellum, and amygdala (Bowyer and Ali, 2006; Kiyatkin *et al.*, 2007; Sharma *et al.*, 2007). METH-induced BBB breakdown was evidenced by leakage of serum albumin into the brain tissue (Sharma *et al.*, 2007). Doses of METH that cause BBB disturbances also induce neuronal damage, myelin degeneration, and reactive astrocytosis in the parietal and occipital cortices (Sharma *et al.*, 2007). These doses also cause extensive degeneration of pyramidal cells and activation of microglia in amygdala and hippocampus of rats (Bowyer and Ali, 2006). These effects appear to depend on hyperthermia because the psychostimulant failed to induce BBB damage and neurodegeneration in the brains of animals that did not show increased temperature (Bowyer and Ali, 2006). Interestingly, mild BBB dysfunction found in the caudate-putamen after METH treatment was exacerbated by hyperthermia (Bowyer *et al.*, 2008). It is interesting to point out that METH-induced leakage of serum albumin into brain tissue was attenuated by pretreatment with antioxidant, H-290/51, suggesting the involvement of free radicals in BBB damage (Sharma *et al.*, 2007) and further supporting a role for oxidative stress in drug toxicity. Also of interest is the fact the antioxidant was able to attenuate

METH-induced hyperthermia, neuronal damage, myelin degradation, and glial response (Sharma *et al.*, 2007).

VIII. Involvement of Mitochondrial Death Pathway in METH-Induced Apoptosis

In addition to glutamate and NO, the Bcl-2 family of proteins may also be involved in the mechanisms underlying METH neurotoxicity (Cadet *et al.*, 2001; Jayanthi *et al.*, 2001; Stumm *et al.*, 1999). In particular, METH induced increases in proapoptotic proteins, BAX and BID, and decreases in antiapoptotic proteins, Bcl-2 and Bcl-X_L. The upregulation of proapoptotic proteins is consistent with findings that METH treatment caused release of mitochondrial proteins cytochrome c and apoptosis inducing factor (AIF) into the cytosol (Deng *et al.*, 2002a; Jayanthi *et al.*, 2004). AIF and Smac/DIABLO released from mitochondria have also been shown to participate in METH-induced apoptosis (Jayanthi *et al.*, 2004). Their release is followed by activation of caspases 9 and 3, and the breakdown of several structural cellular proteins (Jayanthi *et al.*, 2004). Thus, these findings implicate the mitochondrial pathway in METH-related neuronal death in the brain (Cadet *et al.*, 2005, 2007). This idea is consistent with the findings that overexpression of Bcl-2 can protect against drug-induced apoptosis (Cadet *et al.*, 1997).

IX. Involvement of the Endoplasmic Reticulum-Dependent Death Pathway in METH-Induced Apoptosis

Oxidative stress can trigger cellular damage by causing dysfunctions of cellular organelles such as the endoplasmic reticulum (ER) (Gorlach *et al.*, 2006). In addition to regulating synthesis, folding, and transport of proteins, ER also constitutes the main intracellular store for Ca^{2+}, whose excess can contribute to cell death (Gorlach *et al.*, 2006). At physiological levels, Ca^{2+} released from the ER is taken up by mitochondria to enhance metabolite flow on the outer mitochondrial membrane and to increase ATP production (Kroemer *et al.*, 2007). However, sustained release of Ca^{2+} from the ER stores may initiate calcium-dependent apoptosis via the permeabilization of the outer mitochondrial membrane (Kroemer *et al.*, 2007). ER stress and dysregulation of calcium homeostasis appear to participate in METH-induced cell death because the drug can induce activation of calpain (Jayanthi *et al.*, 2004), a calcium-responsive cytosolic protease involved in ER-dependent apoptosis (Nakagawa and Yuan, 2000). METH has been shown to increase calpain-mediated protolysis of cytoskeletal protein spectrin and microtubule protein tau in the rat cortex, hippocampus (Warren *et al.*, 2005), and striatum (Staszewski and Yamamoto, 2006). In contrast, the calpain inhibitors can attenuate psychostimulant-induced spectrin and tau proteolysis (Warren *et al.*, 2007) as well as neuronal death (Samantaray *et al.*, 2006), strongly implicating ER stress and calpain activation in the mechanisms of METH neuronal degeneration. A role for the ER in METH toxicity is further supported by the findings that apoptotic doses of the drug can increase the expression of proteins such as caspase-12, GRP78/BiP, and CHOP/GADD153 (Jayanthi *et al.*, 2004) that participate in ER-induced apoptosis (Marciniak and Ron, 2006). This METH-related ER stress might be secondary to oxidative stress (Cadet and Brannock, 1998; Cadet *et al.*, 1994; Jayanthi *et al.*, 1998) and to increases in BAX/Bcl-2 ratios induced by this illicit drug (Jayanthi *et al.*, 2001).

X. Microglial Reactions and METH Toxicity

Microglia are the resident immune cells within CNS that function to protect the brain against injury and chemical damage (Raivich, 2005). In the healthy mature brain, microglia typically exist in a resting state characterized by ramified morphology, and monitor the neuronal environment (Block *et al.*, 2007; Raivich, 2005). However, in response to brain injury or damage, microglia are readily activated, undergoing a dramatic transformation from a resting ramified state into an amoeboid morphology. Although microglial activation is necessary for host defense and neuron survival, the overactivation of microglia results in deleterious and neurotoxic consequences. Specifically, microglia contributes to the progress of many neurodegenerative diseases, including Parkinson's (Kim and Joh, 2006), Alzheimer's (Xiang *et al.*, 2006), and Huntington (Sapp *et al.*, 2001) diseases, AIDS-related neuropathy (Gonzalez-Scarano and Martin-Garcia, 2005) as well as in the neurotoxic effects of MPTP (Gao *et al.*, 2002) and kainic acid (Chen *et al.*, 2005). Once activated, microglia become big, migrate to the site of the injury, and cause phagocytosis of dying and dead cells. In addition, microglia secrete a variety of cytokines, reactive oxygen and nitrogen species, and prostaglandins that are known to cause neuronal damage (Block *et al.*, 2007; Perry *et al.*, 2007).

Recently, emerging data have implicated microglial activation as an early event in the neurotoxic cascade that is initiated by METH treatment. Specifically, METH causes strong microglial response in the areas of the brain that show DA axonal degeneration (Thomas *et al.*, 2004b). In contrast, attenuation of METH neurotoxicity by MK-801 and dextromethorphan inhibits microglial activation (Thomas and Kuhn, 2005). Moreover, anti-inflammatory drug ketoprofen allows some protection against METH toxicity and also reduces psychostimulant-induced microgliosis (Asanuma *et al.*, 2003). However, attenuation of microglial activation itself is insufficient to protect against METH neurotoxicity (Sriram *et al.*, 2006b). This microglial activation precedes METH-induced DA axonal degeneration in the striatum, suggesting that microglia might contribute to drug toxicity (LaVoie *et al.*, 2004).

While neurotoxic amphetamines METH, MDMA, amphetamine and *p*-chloroamphetamine cause microglial activation (Thomas *et al.*, 2004a) in addition to reactive astrocytosis (Deng and Cadet, 1999; Krasnova *et al.*, 2005; O'Callaghan *et al.*, 1995; Xu *et al.*, 2005), nonneurotoxic drugs such as fenfluramine and DOI fail to activate microglia (Thomas *et al.*, 2004a). These data establish a link between the neurotoxic amphetamines and microglial activation, suggesting that microglia might be a selective marker for neuronal axonal damage (Thomas *et al.*, 2004a) in agreement with the ability for the same drugs to cause reactive astrocytosis, an established hallmark of neurotoxicity (O'Callaghan and Sriram, 2005).

Microglial cells might potentiate METH-related damage by releasing toxic substances such as superoxide radicals and NO which have already been implicated in drug neurotoxicity (see discussion above). In addition, METH causes increase in the levels of TNF-α and IL-1β (Flora *et al.*, 2002; Sriram *et al.*, 2006a), proinflammatory cytokines that can also contribute to toxicity of the drug. Consistent with these findings, METH neurotoxicity and an increase in a marker for microglial activation PK11195 binding were attenuated in IL-6 null mice (Ladenheim *et al.*, 2000). Together, these observations show that inflammatory reactions in microglia might participate in the molecular pathway underlying METH toxicity in the brain.

A schematic diagram showing molecular mechanisms that lead to METH-induced neuronal degeneration is presented in Fig. 1.

XI. Neuroprotective Mechanisms and METH Toxicity

Several attempts have been made to identify ways to protect the brain against METH toxicity. DA uptake inhibitors and DA receptor antagonists have been shown to provide protection against METH-induced degeneration of striatal DA terminals (Angulo *et al.*, 2004; Jayanthi *et al.*, 2005; Marek *et al.*, 1990; O'Dell *et al.*, 1993; Schmidt and Gibb, 1985; Sonsalla *et al.*, 1986). The DA D1 antagonist, SCH23390, also protects against METH-induced cell death in the striatum (Jayanthi *et al.*, 2005). These neuroprotective effects may depend on changes in DA release because DA receptor antagonists, SCH23390 and eticlopride, were reported to partially block METH-related increases in DA release (O'Dell *et al.*, 1993). DA D1-depending mechanisms are also mediated via Fas/FasL-related events since prior treatment with SCH23390 decreased translocation of NFATc3 and NFATc4 from the cytoplasm to the nucleus and reduced increases in

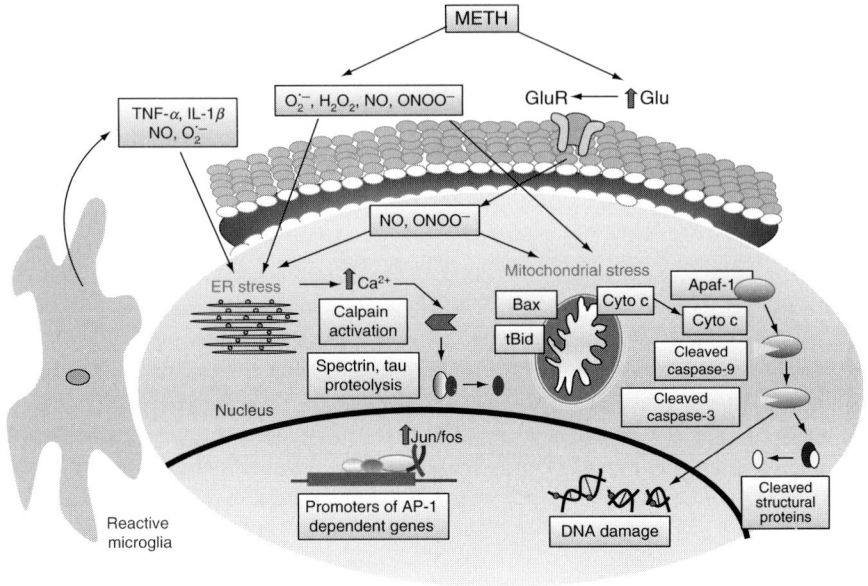

FIG. 1. Mechanisms implicated in METH-induced neurotoxicity. The figure summarizes the pathways that are reviewed in this chapter.

calcineurin expression caused by METH administration (Jayanthi et al., 2005). Pretreatment with SCH23390 also resulted in significant inhibition of the METH-induced increases in the expression of FasL and caspase-3 in rat striatal cells (Jayanthi et al., 2005). Because the dose of SCH23390 used in that study completely blocks METH-induced decreases in DA levels while providing only partial protection against death of striatal neurons (Jayanthi et al., 2005), the possibility exists that METH-induced cell death might involved additional mechanisms independent of stimulation of DA D1 receptors.

In addition to DAT inhibitors and DA receptor antagonists, some trophic factors also provide protection against the toxic effects of METH in vitro (Chou et al., 2008; Mamounas et al., 1995; Matsuzaki et al., 2004; Zhou et al., 2004) and in vivo (Cass, 1996; Chou et al., 2008; Melega et al., 2000). Specifically, brain-derived neurotrophic factor (BDNF) was shown to prevent METH-induced caspase-3 activation and death in primary cultures of cortical neurons (Matsuzaki et al., 2004). The protective effect was blocked by phosphatydilinositol-3-kinase inhibitors and by the expression of kinase-deficient Akt, thus implicating the PI3K/Akt pathway in the effects of BDNF (Matsuzaki et al., 2004). Glial cell line-derived neurotrophic factor (GDNF) administration can also prevent METH-mediated reductions in DA levels in the rat striatum (Cass, 1996; Cass et al., 2006) and hasten the recovery of striatal DA functions in METH-treated rats (Cass *et al.*, 2000). GDNF pretreatment could also partially prevent the loss of DAT binding caused by METH administration in the striatum of vervet monkeys (Melega *et al.*, 2000). Bone morphogenetive protein 7 (BMP7) can also attenuate METH-induced decrease in TH immunoreactivity and cell death in primary DA neurons and in the mouse striatum (Chou *et al.*, 2008) while nerve growth factor was reported to protect R2 cells against drug-related DNA fragmentation and apoptosis (Zhou *et al.*, 2004).

Several studies have demonstrated that METH neurotoxicity can be mediated, in part, via activation of neuroinflammatory responses (Kuhn *et al.*, 2006; Ladenheim *et al.*, 2000; Thomas *et al.*, 2004b). In line with these findings, some cytokines have been shown to induce protection against METH toxicity. For example, METH-induced damage to DA axons was prevented by pretreatment with interferon γ, possibly, by suppression of its neuroinflammatory effects (Hozumi *et al.*, 2008). In addition, interferon γ might cause neuroprotective effects by increasing GDNF production in astrocytes (Appel *et al.*, 1997). Another cytokine, TNF-α, which can enhance autoimmunity and inflammation (Sriram and O'Callaghan, 2007), is protective against METH-induced toxicity (Nakajima *et al.*, 2004). The neuroprotective effects of TNF-α may be mediated by inhibition of METH-related increases in extracellular DA levels in the striatum and by potentiating DA uptake into synaptosomes and synaptic vesicles (Nakajima *et al.*, 2004). The neuroprotection against METH toxicity caused by TNF-α may also be mediated via its effects on temperature regulation, because TNF-α pretreatment prior to drug administration also induced hypothermia, whereas METH-induced hyperthermia was exacerbated in TNF-α knockout mice (Nakajima *et al.*, 2004). In addition, TNF-α has been shown to cause upregulation of MnSOD expression through activation of NFκB in hippocampal neurons (Mattson *et al.*, 1997) because MnSOD transgenic mice are protected against METH toxicity (Maragos *et al.*, 2000).

Estrogen can also protect against METH-induced damage to nigrostriatal DA system (Dluzen and McDermott, 2006). Estrogen treatment caused neuroprotective effects against METH neurotoxicity in female, but not in male mice (D'Astous *et al.*, 2004, 2005; Dluzen *et al.*, 2002; Liu and Dluzen, 2006). Pretreatment with tamoxifen, an estrogen receptor ligand, attenuates METH-induced striatal DA depletion and the decrease in DAT binding via mechanisms that were independent of thermoregulation (Bourque *et al.*, 2007; Dluzen *et al.*, 2001). In addition to its effects on estrogen receptors, tamoxifen can also cause anti-inflammatory responses in microglial cells (Suuronen *et al.*, 2005) and protect glial cells against glutamate toxicity (Shy *et al.*, 2000). Because glutamate release and microgliosis contribute to METH toxicity (see above), the protective effects of tamoxifen probably occur via suppression of both glutamate- and microglia-dependent toxic mechanisms.

Acknowledgments

This work is supported by the Intramural Research Program of the National Institute on Drug Abuse, NIH, DHHS.

References

Abekawa, T., Ohmori, T., and Koyama, T. (1994). Effects of repeated administration of a high dose of methamphetamine on dopamine and glutamate release in rat striatum and nucleus accumbens. *Brain Res.* **643,** 276–281.

Albertson, T. E., Derlet, R. W., and Van Hoozen, B. E. (1999). Methamphetamine and the expanding complications of amphetamines. *West. J. Med.* **170,** 214–219.

Ali, S. F., and Itzhak, Y. (1998). Effects of 7-nitroindazole, an NOS inhibitor on methamphetamine-induced dopaminergic and serotonergic neurotoxicity in mice. *Ann. NY Acad. Sci.* **844,** 122–130.

Angulo, J. A., Angulo, N., and Yu, J. (2004). Antagonists of the neurokinin-1 or dopamine D1 receptors confer protection from methamphetamine on dopamine terminals of the mouse striatum. *Ann. NY Acad. Sci.* **1025,** 171–180.

Appel, E., Kolman, O., Kazimirsky, G., Blumberg, P. M., and Brodie, C. (1997). Regulation of GDNF expression in cultured astrocytes by inflammatory stimuli. *Neuroreport* **8,** 3309–3312.

Asanuma, M., Tsuji, T., Miyazaki, I., Miyoshi, K., and Ogawa, N. (2003). Methamphetamine-induced neurotoxicity in mouse brain is attenuated by ketoprofen, a non-steroidal anti-inflammatory drug. *Neurosci. Lett.* **352,** 13–16.

Baldwin, H. A., Colado, M. I., Murray, T. K., De Souza, R. J., and Green, A. R. (1993). Striatal dopamine release *in vivo* following neurotoxic doses of methamphetamine and effect of the neuroprotective drugs, chlormethiazole and dizocilpine. *Br. J. Pharmacol.* **108,** 590–596.

Battaglia, G., Fornai, F., Busceti, C. L., Aloisi, G., Cerrito, F., De Blasi, A., Melchiorri, D., and Nicoletti, F. (2002). Selective blockade of mGlu5 metabotropic glutamate receptors is protective against methamphetamine neurotoxicity. *J. Neurosci.* **22,** 2135–2141.

Betts, E. S., Krasnova, I. N., McCoy, M. T., Ladenheim, B., and Cadet, J. L. (2002). Analysis of methamphetamine-induced changes in the expression of integrin family members in the cortex of wild-type and c-fos knockout mice. *Neurotox. Res.* **4,** 617–623.

Block, M. L., Zecca, L., and Hong, J. S. (2007). Microglia-mediated neurotoxicity: Uncovering the molecular mechanisms. *Nat. Rev. Neurosci.* **8,** 57–69.

Bourque, M., Liu, B., Dluzen, D. E., and Di Paolo, T. (2007). Tamoxifen protects male mice nigrostriatal dopamine against methamphetamine-induced toxicity. *Biochem. Pharmacol.* **74,** 1413–1423.

Bowyer, J. F., and Ali, S. (2006). High doses of methamphetamine that cause disruption of the blood-brain barrier in limbic regions produce extensive neuronal degeneration in mouse hippocampus. *Synapse* **60,** 521–532.

Bowyer, J. F., Robinson, B., Ali, S., and Schmued, L. C. (2008). Neurotoxic-related changes in tyrosine hydroxylase, microglia, myelin, and the blood-brain barrier in the caudate-putamen from acute methamphetamine exposure. *Synapse* **62,** 193–204.

Cadet, J. L., and Brannock, C. (1998). Free radicals and the pathobiology of brain dopamine systems. *Neurochem. Int.* **32,** 117–131.

Cadet, J. L., Ali, S., and Epstein, C. (1994). Involvement of oxygen-based radicals in methamphet-amine-induced neurotoxicity: Evidence from the use of CuZnSOD transgenic mice. *Ann. NY Acad. Sci.* **738,** 388–391.

Cadet, J. L., Ordonez, S. V., and Ordonez, J. V. (1997). Methamphetamine induces apoptosis in immortalized neural cells: Protection by the proto-oncogene, bcl-2. *Synapse* **25,** 176–184.

Cadet, J. L., Jayanthi, S., McCoy, M. T., Vawter, M., and Ladenheim, B. (2001). Temporal profiling of methamphetamine-induced changes in gene expression in the mouse brain: Evidence from cDNA array. *Synapse* **41,** 40–48.

Cadet, J. L., McCoy, M. T., and Ladenheim, B. (2002). Distinct gene expression signatures in the striata of wild-type and heterozygous c-fos knockout mice following methamphetamine adminis-tration: Evidence from cDNA array analyses. *Synapse* **44,** 211–226.

Cadet, J. L., Jayanthi, S., and Deng, X. (2005). Methamphetamine-induced neuronal apoptosis involves the activation of multiple death pathways. *Rev. Neurotox. Res.* **8,** 199–206.

Cadet, J. L., Krasnova, I. N., Jayanthi, S., and Lyles, J. (2007). Neurotoxicity of substituted amphe-tamines: Molecular and cellular mechanisms. *Neurotox. Res.* **11,** 183–202.

Cass, W. A. (1996). GDNF selectively protects dopamine neurons over serotonin neurons against the neurotoxic effects of methamphetamine. *J. Neurosci.* **16,** 8132–8139.

Cass, W. A., Manning, M. W., and Bailey, S. L. (2000). Restorative effects of GDNF on striatal dopamine release in rats treated with neurotoxic doses of methamphetamine. *Ann. NY Acad. Sci.* **914,** 127–136.

Cass, W. A., Peters, L. E., Harned, M. E., and Seroogy, K. B. (2006). Protection by GDNF and other trophic factors against the dopamine-depleting effects of neurotoxic doses of methamphetamine. *Ann. NY Acad. Sci.* **1074,** 272–281.

Chen, Z., Duan, R. S., Quezada, H. C., Mix, E., Nennesmo, I., Adem, A., Winblad, B., and Zhu, J. (2005). Increased microglial activation and astrogliosis after intranasal administration of kainic acid in C57BL/6 mice. *J. Neurobiol.* **62,** 207–218.

Chou, J., Luo, Y., Kuo, C. C., Powers, K., Shen, H., Harvey, B. K., Hoffer, B. J., and Wang, Y. (2008). Bone morphogenetic protein-7 reduces toxicity induced by high doses of methamphetamine in rodents. *Neuroscience* **151,** 92–103.

Chung, K. K., Dawson, T. M., and Dawson, V. L. (2005). Nitric oxide, S-nitrosylation and neurode-generation. *Cell. Mol. Biol. (Noisy-le-grand).* **51,** 247–254.

D'Astous, M., Gajjar, T. M., Dluzen, D. E., and Di Paolo, T. (2004). Dopamine transporter as a marker of neuroprotection in methamphetamine-lesioned mice treated acutely with estradiol. *Neuroendocrinology* **79,** 296–304.

D'Astous, M., Mickley, K. R., Dluzen, D. E., and Di Paolo, T. (2005). Differential protective properties of estradiol and tamoxifen against methamphetamine-induced nigrostriatal dopaminergic toxicity in mice. *Neuroendocrinology* **82,** 111–120.

Dalton, T. P., Shertzer, H. G., and Puga, A. (1999). Regulation of gene expression by reactive oxygen. *Annu. Rev. Pharmacol. Toxicol.* **39,** 67–101.

Deng, X., and Cadet, J. L. (1999). Methamphetamine administration causes overexpression of nNOS in the mouse striatum. *Brain Res.* **851,** 254–257.

Deng, X., and Cadet, J. L. (2000). Methamphetamine-induced apoptosis is attenuated in the striata of copper-zinc superoxide dismutase transgenic mice. *Brain Res. Mol. Brain Res.* **83,** 121–124.

Deng, X., Ladenheim, B., Tsao, L. I., and Cadet, J. L. (1999). Null mutation of c-fos causes exacerbation of methamphetamine-induced neurotoxicity. *J. Neurosci.* **19,** 10107–10115.

Deng, X., Wang, Y., Chou, J., and Cadet, J. L. (2001). Methamphetamine causes widespread apoptosis in the mouse brain: Evidence from using an improved TUNEL histochemical method. *Brain Res. Mol. Brain Res.* **93,** 64–69.

Deng, X., Cai, N. S., McCoy, M. T., Chen, W., Trush, M. A., and Cadet, J. L. (2002a). Methamphet-amine induces apoptosis in an immortalized rat striatal cell line by activating the mitochondrial cell death pathway. *Neuropharmacology* **42,** 837–845.

Deng, X., Jayanthi, S., Ladenheim, B., Krasnova, I. N., and Cadet, J. L. (2002b). Mice with partial deficiency of c-Jun show attenuation of methamphetamine-induced neuronal apoptosis. *Mol. Pharmacol.* **62,** 993–1000.

De Vito, M. J., and Wagner, G. C. (1989). Methamphetamine-induced neuronal damage: A possible role for free radicals. *Neuropharmacology* **28,** 1145–1150.

Di Monte, D. A., Royland, J. E., Jakowec, M. W., and Langston, J. W. (1996). Role of nitric oxide in methamphetamine neurotoxicity: Protection by 7-nitroindazole, an inhibitor of neuronal nitric oxide synthase. *J. Neurochem.* **67,** 2443–2450.

Dluzen, D. E., and McDermott, J. L. (2006). Estrogen, testosterone, and methamphetamine toxicity. *Ann. NY Acad. Sci.* **1074,** 282–294.

Dluzen, D. E., McDermott, J. L., and Anderson, L. I. (2001). Tamoxifen diminishes methamphetamine-induced striatal dopamine depletion in intact female and male mice. *J. Neuroendocrinol.* **13,** 618–624.

Dluzen, D. E., Anderson, L. I., and Pilati, C. F. (2002). Methamphetamine-gonadal steroid hormonal interactions: Effects upon acute toxicity and striatal dopamine concentrations. *Neurotoxicol. Teratol.* **24,** 267–273.

Eisch, A. J., Schmued, L. C., and Marshall, J. F. (1998). Characterizing cortical neuron injury with Fluoro-Jade labeling after a neurotoxic regimen of methamphetamine. *Synapse* **30,** 329–333.

Flora, G., Lee, Y. W., Nath, A., Maragos, W., Hennig, B., and Toborek, M. (2002). Methamphetamine-induced TNF-alpha gene expression and activation of AP-1 in discrete regions of mouse brain: Potential role of reactive oxygen intermediates and lipid peroxidation. *Neuromolecular Med.* **2,** 71–85.

Fukami, G., Hashimoto, K., Koike, K., Okamura, N., Shimizu, E., and Iyo, M. (2004). Effect of antioxidant *N*-acetyl-L-cysteine on behavioral changes and neurotoxicity in rats after administration of methamphetamine. *Brain Res.* **1016,** 90–95.

Gao, H. M., Jiang, J., Wilson, B., Zhang, W., Hong, J. S., and Liu, B. (2002). Microglial activation-mediated delayed and progressive degeneration of rat nigral dopaminergic neurons: Relevance to Parkinson's disease. *J. Neurochem.* **81,** 1285–1297.

Gettig, J. P., Grady, S. E., and Nowosadzka, I. (2006). Methamphetamine: Putting the brakes on speed. *J. Sch. Nurs.* **22,** 66–73.

Gilcrease, M. Z. (2007). Integrin signaling in epithelial cells. *Cancer Lett.* **247,** 1–25.

Gluck, M. R., Moy, L. Y., Jayatilleke, E., Hogan, K. A., Manzino, L., and Sonsalla, P. K. (2001). Parallel increases in lipid and protein oxidative markers in several mouse brain regions after methamphetamine treatment. *J. Neurochem.* **79,** 152–160.

Gonzalez, R., Rippeth, J. D., Carey, C. L., Heaton, R. K., Moore, D. J., Schweinsburg, B. C., Cherner, M., and Grant, I. (2004). Neurocognitive performance of methamphetamine users discordant for history of marijuana exposure. *Drug Alcohol Depend.* **76,** 181–190.

Gonzalez-Scarano, F., and Martin-Garcia, J. (2005). The neuropathogenesis of AIDS. *Nat. Rev. Immunol.* **5,** 69–81.

Gorlach, A., Klappa, P., and Kietzmann, T. (2006). The endoplasmic reticulum: Folding, calcium homeostasis, signaling, and redox control. *Antioxid. Redox Signal* **8,** 1391–1418.

Harold, C., Wallace, T., Friedman, R., Gudelsky, G., and Yamamoto, B. (2000). Methamphetamine selectively alters brain glutathione. *Eur. J. Pharmacol.* **400,** 99–102.

Hirata, H., Ladenheim, B., Carlson, E., Epstein, C., and Cadet, J. L. (1996). Autoradiographic evidence for methamphetamine-induced striatal dopaminergic loss in mouse brain: Attenuation in CuZn-superoxide dismutase transgenic mice. *Brain Res.* **714,** 95–103.

Hozumi, H., Asanuma, M., Miyazaki, I., Fukuoka, S., Kikkawa, Y., Kimoto, N., Kitamura, Y., Sendo, T., Kita, T., and Gomita, Y. (2008). Protective effects of interferon-gamma against methamphetamine-induced neurotoxicity. *Toxicol. Lett.* **177,** 123–129.

Imam, S. Z., el-Yazal, J., Newport, G. D., Itzhak, Y., Cadet, J. L., Slikker, W. Jr., and Ali, S. F. (2001a). Methamphetamine-induced dopaminergic neurotoxicity: Role of peroxynitrite and neuroprotective role of antioxidants and peroxynitrite decomposition catalysts. *Ann. NY Acad. Sci.* **939,** 366–380.

Imam, S. Z., Newport, G. D., Itzhak, Y., Cadet, J. L., Islam, F., Slikker, W. Jr., and Ali, S. F. (2001b). Peroxynitrite plays a role in methamphetamine-induced dopaminergic neurotoxicity: Evidence from mice lacking neuronal nitric oxide synthase gene or overexpressing copper-zinc superoxide dismutase. *J. Neurochem.* **76,** 745–749.

Itzhak, Y., and Ali, S. F. (1996). The neuronal nitric oxide synthase inhibitor, 7-nitroindazole, protects against methamphetamine-induced neurotoxicity in vivo. *J. Neurochem.* **67,** 1770–1773.

Itzhak, Y., and Ali, S. F. (2006). Role of nitrergic system in behavioral and neurotoxic effects of amphetamine analogs. *Pharmacol. Ther.* **109,** 246–262.

Itzhak, Y., Gandia, C., Huang, P. L., and Ali, S. F. (1998). Resistance of neuronal nitric oxide synthase-deficient mice to methamphetamine-induced dopaminergic neurotoxicity. *J. Pharmacol. Exp. Ther.* **284,** 1040–1047.

Itzhak, Y., Martin, J. L., and Ail, S. F. (2000). nNOS inhibitors attenuate methamphetamine-induced dopaminergic neurotoxicity but not hyperthermia in mice. *Neuroreport* **11,** 2943–2946.

Iversen, L. (2006). Neurotransmitter transporters and their impact on the development of psychopharmacology. *Br. J. Pharmacol.* **147**(Suppl. 1), S82–S88.

Jayanthi, S., Ladenheim, B., and Cadet, J. L. (1998). Methamphetamine-induced changes in antioxidant enzymes and lipid peroxidation in copper/zinc-superoxide dismutase transgenic mice. *Ann. NY Acad. Sci.* **844,** 92–102.

Jayanthi, S., Ladenheim, B., Andrews, A. M., and Cadet, J. L. (1999). Overexpression of human copper/zinc superoxide dismutase in transgenic mice attenuates oxidative stress caused by methylenedioxymethamphetamine (Ecstasy). *Neuroscience* **91,** 1379–1387.

Jayanthi, S., Deng, X., Bordelon, M., McCoy, M. T., and Cadet, J. L. (2001). Methamphetamine causes differential regulation of pro-death and anti-death Bcl-2 genes in the mouse neocortex. *FASEB J.* **15,** 1745–1752.

Jayanthi, S., Deng, X., Noailles, P. A., Ladenheim, B., and Cadet, J. L. (2004). Methamphetamine induces neuronal apoptosis via cross-talks between endoplasmic reticulum and mitochondria-dependent death cascades. *FASEB J.* **18,** 238–251.

Jayanthi, S., Deng, X., Ladenheim, B., McCoy, M. T., Cluster, A., Cai, N. S., and Cadet, J. L. (2005). Calcineurin/NFAT-induced up-regulation of the Fas ligand/Fas death pathway is involved in methamphetamine-induced neuronal apoptosis. *Proc. Natl. Acad. Sci. USA* **102,** 868–873.

Kawasaki, T., Ishihara, K., Ago, Y., Nakamura, S., Itoh, S., Baba, A., and Matsuda, T. (2006). Protective effect of the radical scavenger edaravone against methamphetamine-induced dopaminergic neurotoxicity in mouse striatum. *Eur. J. Pharmacol.* **542,** 92–99.

Kim, Y. S., and Joh, T. H. (2006). Microglia, major player in the brain inflammation: Their roles in the pathogenesis of Parkinson's disease. *Exp. Mol. Med.* **38,** 333–347.

Kiyatkin, E. A., Brown, P. L., and Sharma, H. S. (2007). Brain edema and breakdown of the blood-brain barrier during methamphetamine intoxication: Critical role of brain hyperthermia. *Eur. J. Neurosci.* **26,** 1242–1253.

Kondo, T., Ito, T., and Sugita, Y. (1994). Bromocriptine scavenges methamphetamine-induced hydroxyl radicals and attenuates dopamine depletion in mouse striatum. *Ann. NY Acad. Sci.* **738,** 222–229.

Krasnova, I. N., Ladenheim, B., and Cadet, J. L. (2005). Amphetamine induces apoptosis of medium spiny striatal projection neurons via the mitochondria-dependent pathway. *FASEB J.* **19,** 851–853.

Kroemer, G., Galluzzi, L., and Brenner, C. (2007). Mitochondrial membrane permeabilization in cell death. *Physiol. Rev.* **87,** 99–163.

Kuhn, D. M., Francescutti-Verbeem, D. M., and Thomas, D. M. (2006). Dopamine quinones activate microglia and induce a neurotoxic gene expression profile: Relationship to methamphetamine-induced nerve ending damage. *Ann. NY Acad. Sci.* **1074,** 31–41.

Ladenheim, B., Krasnova, I. N., Deng, X., Oyler, J. M., Polettini, A., Moran, T. H., Huestis, M. A., and Cadet, J. L. (2000). Methamphetamine-induced neurotoxicity is attenuated in transgenic mice with a null mutation for interleukin-6. *Mol. Pharmacol.* **58,** 1247–1256.

LaVoie, M. J., and Hastings, T. G. (1999). Dopamine quinone formation and protein modification associated with the striatal neurotoxicity of methamphetamine: Evidence against a role for extracellular dopamine. *J. Neurosci.* **19,** 1484–1491.

LaVoie, M. J., Card, J. P., and Hastings, T. G. (2004). Microglial activation precedes dopamine terminal pathology in methamphetamine-induced neurotoxicity. *Exp. Neurol.* **187,** 47–57.

Liu, B., and Dluzen, D. E. (2006). Effect of estrogen upon methamphetamine-induced neurotoxicity within the impaired nigrostriatal dopaminergic system. *Synapse* **60,** 354–361.

Li, Y., and Trush, M. A. (1993). DNA damage resulting from the oxidation of hydroquinone by copper: Role for a Cu(II)/Cu(I) redox cycle and reactive oxygen generation. *Carcinogenesis* **14,** 1303–1311.

Lynch, J., and House, M. A. (1992). Cardiovascular effects of methamphetamine. *J. Cardiovasc. Nurs.* **6,** 12–18.

Mamounas, L. A., Blue, M. E., Siuciak, J. A., and Altar, C. A. (1995). Brain-derived neurotrophic factor promotes the survival and sprouting of serotonergic axons in rat brain. *J. Neurosci.* **15,** 7929–7939.

Maragos, W. F., Jakel, R., Chesnut, D., Pocernich, C. B., Butterfield, D. A., St Clair, D., and Cass, W. A. (2000). Methamphetamine toxicity is attenuated in mice that overexpress human manganese superoxide dismutase. *Brain Res.* **878,** 218–222.

Marciniak, S. J., and Ron, D. (2006). Endoplasmic reticulum stress signaling in disease. *Physiol. Rev.* **86,** 1133–1149.

Marek, G. J., Vosmer, G., and Seiden, L. S. (1990). Dopamine uptake inhibitors block long-term neurotoxic effects of methamphetamine upon dopaminergic neurons. *Brain Res.* **513,** 274–279.

Mark, K. A., Soghomonian, J. J., and Yamamoto, B. K. (2004). High-dose methamphetamine acutely activates the striatonigral pathway to increase striatal glutamate and mediate long-term dopamine toxicity. *J. Neurosci.* **24,** 11449–11456.

Marshall, J. F., O'Dell, S. J., and Weihmuller, F. B. (1993). Dopamine-glutamate interactions in methamphetamine-induced neurotoxicity. *J. Neural. Transm. Gen. Sect.* **91,** 241–254.

Matsuzaki, H., Namikawa, K., Kiyama, H., Mori, N., and Sato, K. (2004). Brain-derived neurotrophic factor rescues neuronal death induced by methamphetamine. *Biol. Psychiatry* **55,** 52–60.

Mattson, M. P., Goodman, Y., Luo, H., Fu, W., and Furukawa, K. (1997). Activation of NF-kappaB protects hippocampal neurons against oxidative stress-induced apoptosis: Evidence for induction of manganese superoxide dismutase and suppression of peroxynitrite production and protein tyrosine nitration. *J. Neurosci. Res.* **49,** 681–697.

Melega, W. P., Lacan, G., Desalles, A. A., and Phelps, M. E. (2000). Long-term methamphetamine-induced decreases of [(11)C]WIN 35,428 binding in striatum are reduced by GDNF: PET studies in the vervet monkey. *Synapse* **35,** 243–249.

Moszczynska, A., Fitzmaurice, P., Ang, L., Kalasinsky, K. S., Schmunk, G. A., Peretti, F. J., Aiken, S. S., Wickham, D. J., and Kish, S. J. (2004). Why is parkinsonism not a feature of human methamphetamine users? *Brain* **127,** 363–370.

Murray, J. B. (1998). Psychophysiological aspects of amphetamine-methamphetamine abuse. *J. Psychol.* **132,** 227–237.

Nakagawa, T., and Yuan, J. (2000). Cross-talk between two cysteine protease families. Activation of caspase-12 by calpain in apoptosis. *J. Cell. Biol.* **150,** 887–894.

Nakajima, A., Yamada, K., Nagai, T., Uchiyama, T., Miyamoto, Y., Mamiya, T., He, J., Nitta, A., Mizuno, M., Tran, M. H., Seto, A., Yoshimura, M., *et al.* (2004). Role of tumor necrosis factor-alpha in methamphetamine-induced drug dependence and neurotoxicity. *J. Neurosci.* **24,** 2212–2225.

Nash, J. F. Jr., Meltzer, H. Y., and Gudelsky, G. A. (1988). Elevation of serum prolactin and corticosterone concentrations in the rat after the administration of 3,4-methylenedioxymethamphetamine. *J. Pharmacol. Exp. Ther.* **245,** 873–879.

O'Callaghan, J. P., and Sriram, K. (2005). Glial fibrillary acidic protein and related glial proteins as biomarkers of neurotoxicity. *Expert Opin. Drug Saf.* **4,** 433–442.

O'Callaghan, J. P., Jensen, K. F., and Miller, D. B. (1995). Quantitative aspects of drug and toxicant-induced astrogliosis. *Neurochem. Int.* **26,** 115–124.

O'Dell, S. J., and Marshall, J. F. (2000). Repeated administration of methamphetamine damages cells in the somatosensory cortex: Overlap with cytochrome oxidase-rich barrels. *Synapse* **37,** 32–37.

O'Dell, S. J., Weihmuller, F. B., and Marshall, J. F. (1993). Methamphetamine-induced dopamine overflow and injury to striatal dopamine terminals: Attenuation by dopamine D1 or D2 antagonists. *J. Neurochem.* **60,** 1792–1799.

Pacher, P., Beckman, J. S., and Liaudet, L. (2007). Nitric oxide and peroxynitrite in health and disease. *Physiol. Rev.* **87,** 315–424.

Perez, J. A. Jr., Arsura, E. L., and Strategos, S. (1999). Methamphetamine-related stroke: Four cases. *J. Emerg. Med.* **17,** 469–471.

Perry, V. H., Cunningham, C., and Holmes, C. (2007). Systemic infections and inflammation affect chronic neurodegeneration. *Nat. Rev. Immunol.* **7,** 161–167.

Potashkin, J. A., and Meredith, G. E. (2006). The role of oxidative stress in the dysregulation of gene expression and protein metabolism in neurodegenerative disease. *Antioxid. Redox Signal* **8,** 144–151.

Raivich, G. (2005). Like cops on the beat: The active role of resting microglia. *Trends Neurosci.* **28,** 571–573.

Rawson, R. A., and Condon, T. P. (2007). Why do we need an Addiction supplement focused on methamphetamine? *Addiction* **102**(Suppl. 1), 1–4.

Ricaurte, G. A., Schuster, C. R., and Seiden, L. S. (1980). Long-term effects of repeated methylamphetamine administration on dopamine and serotonin neurons in the rat brain: A regional study. *Brain Res.* **193,** 153–163.

Rippeth, J. D., Heaton, R. K., Carey, C. L., Marcotte, T. D., Moore, D. J., Gonzalez, R., Wolfson, T., and Grant, I. (2004). Methamphetamine dependence increases risk of neuropsychological impairment in HIV infected persons. *J. Int. Neuropsychol. Soc.* **10,** 1–14.

Roehr, B. (2005). Half a million Americans use methamphetamine every week. *BMJ* **331,** 476.

Romanelli, F., and Smith, K. M. (2006). Clinical effects and management of methamphetamine abuse. *Pharmacotherapy* **26,** 1148–1156.

Salo, R., Nordahl, T. E., Possin, K., Leamon, M., Gibson, D. R., Galloway, G. P., Flynn, N. M., Henik, A., Pfefferbaum, A., and Sullivan, E. V. (2002). Preliminary evidence of reduced cognitive inhibition in methamphetamine-dependent individuals. *Psychiatry Res.* **111,** 65–74.

Samantaray, S., Ray, S. K., Ali, S. F., and Banik, N. L. (2006). Calpain activation in apoptosis of motoneurons in cell culture models of experimental parkinsonism. *Ann. NY Acad. Sci.* **1074,** 349–356.

Sanchez, V., Zeini, M., Camarero, J., O'Shea, E., Bosca, L., Green, A. R., and Colado, M. I. (2003). The nNOS inhibitor, AR-R17477AR, prevents the loss of NF68 immunoreactivity induced by methamphetamine in the mouse striatum. *J. Neurochem.* **85,** 515–524.

Sapp, E., Kegel, K. B., Aronin, N., Hashikawa, T., Uchiyama, Y., Tohyama, K., Bhide, P. G., Vonsattel, J. P., and DiFiglia, M. (2001). Early and progressive accumulation of reactive microglia in the Huntington disease brain. *J. Neuropathol. Exp. Neurol.* **60,** 161–172.

Schmidt, C. J., and Gibb, J. W. (1985). Role of the dopamine uptake carrier in the neurochemical response to methamphetamine: Effects of amfonelic acid. *Eur. J. Pharmacol.* **109,** 73–80.

Sekine, Y., Minabe, Y., Ouchi, Y., Takei, N., Iyo, M., Nakamura, K., Suzuki, K., Tsukada, H., Okada, H., Yoshikawa, E., Futatsubashi, M., and Mori, N. (2003). Association of dopamine transporter loss in the orbitofrontal and dorsolateral prefrontal cortices with methamphetamine-related psychiatric symptoms. *Am. J. Psychiatry* **160**, 1699–1701.

Sekine, Y., Ouchi, Y., Takei, N., Yoshikawa, E., Nakamura, K., Futatsubashi, M., Okada, H., Minabe, Y., Suzuki, K., Iwata, Y., Tsuchiya, K. J., Tsukada, H., *et al.* (2006). Brain serotonin transporter density and aggression in abstinent methamphetamine abusers. *Arch. Gen. Psychiatry* **63**, 90–100.

Sharma, H. S., and Ali, S. F. (2006). Alterations in blood-brain barrier function by morphine and methamphetamine. *Ann. NY Acad. Sci.* **1074**, 198–224.

Sharma, H. S., Sjoquist, P. O., and Ali, S. F. (2007). Drugs of abuse-induced hyperthermia, blood-brain barrier dysfunction and neurotoxicity: Neuroprotective effects of a new antioxidant compound H-290/51. *Curr. Pharm. Des.* **13**, 1903–1923.

Shoptaw, S. (2006). Methamphetamine use in urban gay and bisexual populations. *Top. HIV Med.* **14**, 84–87.

Shy, H., Malaiyandi, L., and Timiras, P. S. (2000). Protective action of 17beta-estradiol and tamoxifen on glutamate toxicity in glial cells. *Int. J. Dev. Neurosci.* **18**, 289–297.

Simon, S. L., Domier, C. P., Sim, T., Richardson, K., Rawson, R. A., and Ling, W. (2002). Cognitive performance of current methamphetamine and cocaine abusers. *J. Addict. Dis.* **21**, 61–74.

Simon, S. L., Dacey, J., Glynn, S., Rawson, R., and Ling, W. (2004). The effect of relapse on cognition in abstinent methamphetamine abusers. *J. Subst. Abuse. Treat.* **27**, 59–66.

Sim, T., Simon, S. L., Domier, C. P., Richardson, K., Rawson, R. A., and Ling, W. (2002). Cognitive deficits among methamphetamine users with attention deficit hyperactivity disorder symptomatology. *J. Addict. Dis.* **21**, 75–89.

Smith, L. M., LaGasse, L. L., Derauf, C., Grant, P., Shah, R., Arria, A., Huestis, M., Haning, W., Strauss, A., Grotta, S. D., Liu, J., Lester, B. M., *et al.* (2006). The infant development, environment, and lifestyle study: Effects of prenatal methamphetamine exposure, polydrug exposure, and poverty on intrauterine growth. *Pediatrics* **118**, 1149–1156.

Sonsalla, P. K., Gibb, J. W., and Hanson, G. R. (1986). Roles of D1 and D2 dopamine receptor subtypes in mediating the methamphetamine-induced changes in monoamine systems. *J. Pharmacol. Exp. Ther.* **238**, 932–937.

Sonsalla, P. K., Nicklas, W. J., and Heikkila, R. E. (1989). Role for excitatory amino acids in methamphetamine-induced nigrostriatal dopaminergic toxicity. *Science* **243**, 398–400.

Sriram, K., and O'Callaghan, J. P. (2007). Divergent roles for tumor necrosis factor-alpha in the brain. *J. Neuroimmune Pharmacol.* **2**, 140–153.

Sriram, K., Matheson, J. M., Benkovic, S. A., Miller, D. B., Luster, M. I., and O'Callaghan, J. P. (2006a). Deficiency of TNF receptors suppresses microglial activation and alters the susceptibility of brain regions to MPTP-induced neurotoxicity: Role of TNF-alpha. *FASEB J.* **20**, 670–682.

Sriram, K., Miller, D. B., and O'Callaghan, J. P. (2006b). Minocycline attenuates microglial activation but fails to mitigate striatal dopaminergic neurotoxicity: Role of tumor necrosis factor-alpha. *J. Neurochem.* **96**, 706–718.

Staszewski, R. D., and Yamamoto, B. K. (2006). Methamphetamine-induced spectrin proteolysis in the rat striatum. *J. Neurochem.* **96**, 1267–1276.

Stumm, G., Schlegel, J., Schafer, T., Wurz, C., Mennel, H. D., Krieg, J. C., and Vedder, H. (1999). Amphetamines induce apoptosis and regulation of bcl-x splice variants in neocortical neurons. *FASEB J.* **13**, 1065–1072.

Sulzer, D., Sonders, M. S., Poulsen, N. W., and Galli, A. (2005). Mechanisms of neurotransmitter release by amphetamines: A review. *Prog. Neurobiol.* **75**, 406–433.

Suuronen, T., Nuutinen, T., Huuskonen, J., Ojala, J., Thornell, A., and Salminen, A. (2005). Anti-inflammatory effect of selective estrogen receptor modulators (SERMs) in microglial cells. *Inflamm. Res.* **54**, 194–203.

Thomas, D. M., and Kuhn, D. M. (2005). Attenuated microglial activation mediates tolerance to the neurotoxic effects of methamphetamine. *J. Neurochem.* **92,** 790–797.

Thomas, D. M., Dowgiert, J., Geddes, T. J., Francescutti-Verbeem, D., Liu, X., and Kuhn, D. M. (2004a). Microglial activation is a pharmacologically specific marker for the neurotoxic amphetamines. *Neurosci. Lett.* **367,** 349–354.

Thomas, D. M., Walker, P. D., Benjamins, J. A., Geddes, T. J., and Kuhn, D. M. (2004b). Methamphetamine neurotoxicity in dopamine nerve endings of the striatum is associated with microglial activation. *J. Pharmacol. Exp. Ther.* **311,** 1–7.

Topolski, J. M. (2007). Epidemiology of methamphetamine abuse in Missouri. *Mol. Med.* **104,** 82–88.

Volkow, N. D., Chang, L., Wang, G. J., Fowler, J. S., Leonido-Yee, M., Franceschi, D., Sedler, M. J., Gatley, S. J., Hitzemann, R., Ding, Y. S., Logan, J., Wong, C., *et al.* (2001). Association of dopamine transporter reduction with psychomotor impairment in methamphetamine abusers. *Am. J. Psychiatry* **158,** 377–382.

Wagner, G. C., Ricaurte, G. A., Seiden, L. S., Schuster, C. R., Miller, R. J., and Westley, J. (1980). Long-lasting depletions of striatal dopamine and loss of dopamine uptake sites following repeated administration of methamphetamine. *Brain Res.* **181,** 151–160.

Wagner, G. C., Carelli, R. M., and Jarvis, M. F. (1985). Pretreatment with ascorbic acid attenuates the neurotoxic effects of methamphetamine in rats. *Res. Commun. Chem. Pathol. Pharmacol.* **47,** 221–228.

Warren, M. W., Kobeissy, F. H., Liu, M. C., Hayes, R. L., Gold, M. S., and Wang, K. K. (2005). Concurrent calpain and caspase-3 mediated proteolysis of alpha II-spectrin and tau in rat brain after methamphetamine exposure: A similar profile to traumatic brain injury. *Life Sci.* **78,** 301–309.

Warren, M. W., Zheng, W., Kobeissy, F. H., Cheng Liu, M., Hayes, R. L., Gold, M. S., Larner, S. F., and Wang, K. K. (2007). Calpain- and caspase-mediated alphaII-spectrin and tau proteolysis in rat cerebrocortical neuronal cultures after ecstasy or methamphetamine exposure. *Int. J. Neuropsychopharmacol.* **10,** 479–489.

Wilson, J. M., Kalasinsky, K. S., Levey, A. I., Bergeron, C., Reiber, G., Anthony, R. M., Schmunk, G. A., Shannak, K., Haycock, J. W., and Kish, S. J. (1996). Striatal dopamine nerve terminal markers in human, chronic methamphetamine users. *Nat. Med.* **2,** 699–703.

Woods, S. P., Rippeth, J. D., Conover, E., Gongvatana, A., Gonzalez, R., Carey, C. L., Cherner, M., Heaton, R. K., and Grant, I. (2005). Deficient strategic control of verbal encoding and retrieval in individuals with methamphetamine dependence. *Neuropsychology* **19,** 35–43.

Xiang, Z., Haroutunian, V., Ho, L., Purohit, D., and Pasinetti, G. M. (2006). Microglia activation in the brain as inflammatory biomarker of Alzheimer's disease neuropathology and clinical dementia. *Dis. Markers* **22,** 95–102.

Xu, W., Zhu, J. P., and Angulo, J. A. (2005). Induction of striatal pre- and postsynaptic damage by methamphetamine requires the dopamine receptors. *Synapse* **58,** 110–121.

Zhou, J. L., Liang, J. H., Zheng, J. W., and Li, C. L. (2004). Nerve growth factor protects R2 cells against neurotoxicity induced by methamphetamine. *Toxicol. Lett.* **150,** 221–227.

Zhu, J. P., Xu, W., and Angulo, J. A. (2006). Methamphetamine-induced cell death: Selective vulnerability in neuronal subpopulations of the striatum in mice. *Neuroscience* **140,** 607–622.

INVOLVEMENT OF NICOTINIC RECEPTORS IN METHAMPHETAMINE- AND MDMA-INDUCED NEUROTOXICITY: PHARMACOLOGICAL IMPLICATIONS

E. Escubedo, J. Camarasa, C. Chipana, S. García-Ratés, and D. Pubill

Unitat de Farmacologia i Farmacognòsia, Facultat de Farmàcia,
Universitat de Barcelona, Barcelona 08028, Spain

During the last years, we have focused on the study of the neurotoxic effects of 3,4-methylenedioxymethamphetamine (MDMA) and methamphetamine (METH) on the central nervous system (CNS) and their pharmacological prevention methods. In the process of this research, we have used a semipurified synaptosomal preparation from striatum of mice or rats as a reliable *in vitro* model to study reactive oxygen species (ROS) production by these amphetamine derivatives, which is well-correlated with their dopaminergic injury in *in vivo* models. Using this preparation, we have demonstrated that blockade of $\alpha7$ nicotinic receptors with methyllycaconitine (MLA) prevents ROS production induced by MDMA and METH. Consequently, *in vivo*, MLA significantly prevents MDMA- and METH-induced neurotoxicity at dopaminergic level (mouse striatum), without affecting hyperthermia induced by these amphetamines.

Additionally, when neuroprotection was assayed with memantine (MEM), a dual antagonist of NMDA and $\alpha 7$ receptors, an effective neuroprotection was obtained also ahead of serotonergic injury induced by MDMA in rats. MEM also prevents MDMA effect on serotonin transporter functionality and METH effect on dopamine transporter (DAT), suggesting that behavioral effects of these psychostimulants can also be modulated by MEM.

Finally, we have demonstrated that MEM prevents the impaired memory function induced by MDMA, and also, using binding studies with radioligands, we have characterized the interaction of these substances with nicotinic receptors.

Studies at molecular level showed that both MDMA and METH displaced competitively the binding of radioligands with homomeric $\alpha 7$ and heteromeric nicotinic acetylcholine receptors (nAChRs), indicating that they can directly interact with them. In all the cases, MDMA displayed higher affinity than METH and it was higher for heteromeric than for $\alpha 7$ subtype. Pre-incubation of differentiated PC12 cells with MDMA or METH induces nAChR upregulation in a concentration- and time-dependent manner, as many nicotinic ligands do, supporting their functional interaction with nAChRs. Such interaction expands the pharmacological profile of amphetamines and can account for some of their effects.

Abbreviations

5-HT, Serotonin; 7-NI, 7-Nitroindazole; AMPH, D-Amphetamine; CAT, Catalase; CTRL, Control; DA, Dopamine; DAT, Dopamine transporter; DBE, Dihydro-β-erythroidine; DCF, $2',7'$-Dichlorofluorescein; DCFH-DA, $2',7'$-Dichlorofluorescin diacetate; GLU, Glutamic acid; MAO, Monoamine oxidase; MDMA, 3,4-Methylenedioxymethamphetamine; MEM, Memantine; METH, Methamphetamine; MLA, Methyllycaconitine; nAChR, Nicotinic acetylcholine receptor; nNOS, Neuronal NO synthase; NO, Nitric oxide; ONOO$^-$, Peroxynitrite; PCP, Phencyclidine; PKC, Protein kinase C; ROS, Reactive oxygen species; SERT, Serotonin transporter; SOD, Superoxide dismutase; THC, Tetrahydrocannabinol; VMAT, Vesicular monoamine transporter.

I. Long-Term Effects of Amphetamine Derivatives on CNS

Amphetamine derivatives such as methamphetamine (METH) and 3,4-methylenedioxymethamphetamine (MDMA, ecstasy) are drugs that are widely abused in the United States and Europe, where they are taken in a recreational context. The growing incidence of the use of these drugs has led to concern because of the extensive evidence that they are neurotoxic in animal models (for reviews, see Davidson et al., 2001 and Lyles and Cadet, 2003). It is still uncertain what neurotoxic consequences the acute or long-term use of these substances in humans could have, and a great deal of research is being done on the subject. In fact, cognitive impairment and dopaminergic/serotonergic deficits have been described in chronic abusers of these drugs (McCann et al., 1999; Parrott et al., 1998; Reneman et al., 2002; Volkow et al., 2001).

Preclinical studies have revealed that both METH and MDMA chronic abuses produce long-term damage to dopaminergic and serotonergic nerve terminals in multiple brain areas. Specifically, METH administration has been shown to produce long-lasting depletion in the striatal content of dopamine (DA) and its metabolites (Ricaurte et al., 1982), decrease in tyrosine hydroxylase activity (Ellison et al., 1978; Hotchkiss and Gibb, 1980), and loss of dopamine transporters (DAT) (Escubedo et al., 1998; Ricaurte et al., 1984); while, MDMA-induced depletions have typically been more specific to serotonergic terminal markers in rats and primates, with little effect on dopaminergic terminals (Pubill et al., 2003; Quinton and Yamamoto, 2006), although in mice it behaves as a dopaminergic neurotoxin in addition to serotonergic depletions (Stone et al., 1987).

The mechanisms through which amphetamine derivatives induce neurotoxicity are complex and still being investigated, but there is a clear evidence of some key phenomena that contribute (probably acting coordinately) to such effects, namely vesicular (vesicular monoamine transporter, VMAT-2) and plasmalemmal (DAT) DA transporters function, the mitochondria and energy balance, glutamate, DA receptors, hyperthermia, and reactive species (reviewed by Riddle et al., 2006). Among these, the latter seem to be the final executors of neuronal damage, reacting with functional and structural molecules and inducing degenerative changes. The reactive species hypothesis about substituted amphetamine-induced toxicity was proposed as early as in 1989 by Gibb and co-workers (Gibb et al., 1989). The sources of reactive species can be various, including the auto-oxidation of cytosolic DA (Chiueh et al., 1993); increased release of glutamate leading to mitochondrial dysfunction and nitric oxide (NO) production (Garthwaite and Boulton, 1995; Nash and Yamamoto, 1992); activation of D1 receptors within the striatum leading to increases in neuronal NO synthase (nNOS) mRNA expression (Wang and Lau, 2001) and presumably increased NO and reactive species production; and inhibition of mitochondrial function that increases mitochondrial-mediated reactive species generation. Also, activation of microglia (source of reactive species) has been reported after METH treatment (Escubedo *et al.*, 1998; Guilarte *et al.*, 2003; Pubill *et al.*, 2003; Thomas *et al.*, 2004).

Although in our experiments MDMA (20 mg/kg s.c. b.i.d. for 4 days) induced an important neurotoxic effect on serotonergic terminals in the cortex, striatum, and hippocampus 3–7 days later, no microglial activation was noticeable at either 3 or 7 days after MDMA treatment. This lack of effect on microglial cells was assessed by [^{3}H]PK 11195 binding and OX-6 immunostaining, which were unchanged in the striatum and cortex after MDMA treatment. Furthermore, in MDMA-treated rats, neither HSP27 expression nor an increase in HSP27 immunoreactivity was detected. This result, together with the lack of an increase in glial fibrilliary acidic protein immunoreactivity, indicates no astroglial activation at either 3 or 7 days post treatment. Without microglial activation, an inflammatory process would not accompany the lesion induced by MDMA. The differences in glial activation between METH and MDMA observed in this study (Pubill *et al.*, 2003) could have implications for the prognosis of the injury induced by these drugs. Additionally, in the case of MDMA, a role of a metabolic reactive derivative in the neurotoxic process has also been proposed (Monks *et al.*, 2004).

II. The Role of Reactive Species in Amphetamine Derivatives-Induced Neurotoxicity

The major role of reactive species in amphetamine derivatives-induced neurotoxicity is demonstrated by the fact that inhibition of their formation or their action affords neuroprotection against these substances. Among reactive oxygen and nitrogen species, NO and peroxynitrite ($ONOO^-$) play a key role in the neuronal damage induced by amphetamine derivatives. NO is an inhibitor of mitochondrial respiratory chain complexes II and IV and rapidly reacts with superoxide to yield $ONOO^-$, which is a powerful oxidant and a potentially irreversible inhibitor of complexes II and III. The $ONOO^-$ formed inside mitochondria impairs mitochondrial functions and integrity. Also $ONOO^-$ oxidizes glutathione (GSH), α-tocopherol, and ascorbate, thereby compromising essential antioxidant pools within mitochondria (Vatassery, 1996; Vatassery *et al.*, 2004). NO and Ca^{2+} synergistically inactivate mitochondrial complex I and cause a loss of cytochrome c, probably via formation of $ONOO^-$ (Jekabsone *et al.*, 2003). In fact, NO synthase is activated by Ca^{2+}-calmodulin. Ca^{2+} could enter through NMDA receptors and/or any other way of calcium influx, as that of the nicotinic receptor-mediated pathway, which will be discussed later.

Enhancement of the antioxidant resources of the cells such as glutathione peroxidase (Hom *et al.*, 1997) or isoforms of superoxide dismutase (SOD)

(Jayanthi *et al.*, 1999; Maragos *et al.*, 2000) is neuroprotective against amphetamine derivatives. Also, antioxidants such as selenium and melatonin, (Imam and Ali, 2000; Imam *et al.*, 2001), L-carnitine (Virmani *et al.*, 2002), *N*-acetyl cysteine (Fukami *et al.*, 2004), or the endogenous antioxidant carnosine (Pubill *et al.*, 2002) are also neuroprotective against this damage.

In the present article we will review the recent research that we have carried out on the mechanisms involved in the production of reactive oxygen species (ROS) induced by amphetamine derivatives (mainly METH and MDMA). The use of a synaptosomal preparation has allowed us to study such mechanisms in a model of isolated synaptic terminal, without the interference of other systems or hypertermia. These investigations led us to identify a new target for METH and MDMA: neuronal nicotinic acetylcholine receptors (nAChRs), specially the homomeric α7 subtype, whose blockade inhibits the oxidative effects of METH and MDMA as well as the impairment of DAT function. Moreover, we will describe how blockade of α7 nAChR protects against METH- and MDMA-induced neurotoxicitiy.

A deficit in learning and memory is a highly consistent finding of cross-sectional research with experienced adult ecstasy users (Morgan *et al.*, 2002; Rodgers *et al.*, 2006). In a recent paper (Camarasa *et al.*, 2008), the preventive effect of an α7 antagonist on MDMA-induced memory impairment suggests a new therapeutic approach to the treatment of long-term cognitive effects of amphetamine derivatives, also showing promising results in the treatment of amphetamine addiction (Levi and Borne, 2002).

MDMA has been proposed as a tool in psychotherapy, but neurotoxicity is a drawback for its clinical use. The availability of a neuroprotective treatment could lead to a reconsideration of such an application.

At molecular level, *in vitro* experiments carried out (Garcia-Rates *et al.*, 2007) demonstrate that METH and MDMA interact with α7 and α4,β2 nAChRs and behave as nicotine, inducing nAChR upregulation, an effect that would have important implications in drug addiction and vulnerability.

III. Model of Semipurified Synaptosomes

An extensive series of pharmacological and toxicological studies has strongly implicated ROS in METH-induced dopaminergic neurotoxicity. However, much of the data supporting this hypothesis is confounded by drug effects on core temperature, which strongly influences METH effects (Ali *et al.*, 1996; O'Callaghan and Miller, 1994).

Repeated administration of moderate, high doses of METH to rats has long been associated with lethality that is dependent on environmental temperature and animal housing conditions during METH exposure. METH induces a significant increase in body temperature and this hyperthermia is potentiated by a high ambient temperature (Pubill *et al.*, 2002) as a consequence of a thermo-deregulatory effect of METH.

The duration and the magnitude of hyperthermia produced during this exposure correlated well with subsequent depletions in striatal DA. It is thought that hyperthermia might also potentiate the production of 6-hydroxy-dopamine or related ROS reported to be formed during METH exposure (Seiden *et al.*, 2001) or potentiate the propagation of ROS damage to lipids. Consecutively to the fact that METH-induced hyperthermia potentiates its neurotoxicity (Tata *et al.*, 2007), it has been established that prevention of METH-associated hyper-thermia attenuates long-term decreases in DA and serotonin (5-HT) content (Bowyer *et al.*, 1994).

For the elucidation of the mechanisms underlying METH-induced dopami-nergic neurotoxicity, an *in vitro* model is necessary to avoid this hyperthermic influence. The knowledge of this mechanism is important because it could provide clues regarding the mechanism of cell death in some neurodegenerative diseases where DA-derived ROSs are supposed to play a role, and also to provide neuroprotection. The difficulty in identifying the mechanisms of METH neuro-toxicity can, to some extent, be attributed to the unavailability of a fully validated *in vitro* model. Although dopaminergic cells in culture have been available for a number of years and, indeed, have been fruitfully used to study some aspects of METH neurotoxicity (Bennett *et al.*, 1993; Jimenez *et al.*, 2004; Miyazaki *et al.*, 2006; Warren *et al.*, 2007), there are some core features of METH-induced DA neurotoxicity that have not been established in the cell culture system.

Our study was undertaken with the goal to develop an alternative *in vitro* model that might be useful for studying the molecular mechanisms of METH- induced DA neurotoxicity. Consequently, it had to contain dopaminergic terminals and it also had to fulfill an important requirement: no free mitochondria in the medium, because they could generate inconsistent results about ROS generation. With this purpose, we used the fast and simple method for isolating synaptosomes described by Myhre and Fonnum (2001) with minor modifications (for the detailed method, see Pubill *et al.*, 2005). Synaptosomes' integrity was assessed by electron microscopy, which confirmed that the purity of the synaptosome fraction obtained was relatively high, with little contamination of free mitochondria (Fig. 1).

IV. Methamphetamine-Induced ROS Production in Semipurified Synaptosomes

Using this model, the formation of intrasynaptosomal ROS was measured using the fluorochrome 2′,7′-dichlorofluorescin diacetate (DCFH-DA). Fluores-cence measurements were performed on a flow cytometer equipped with an

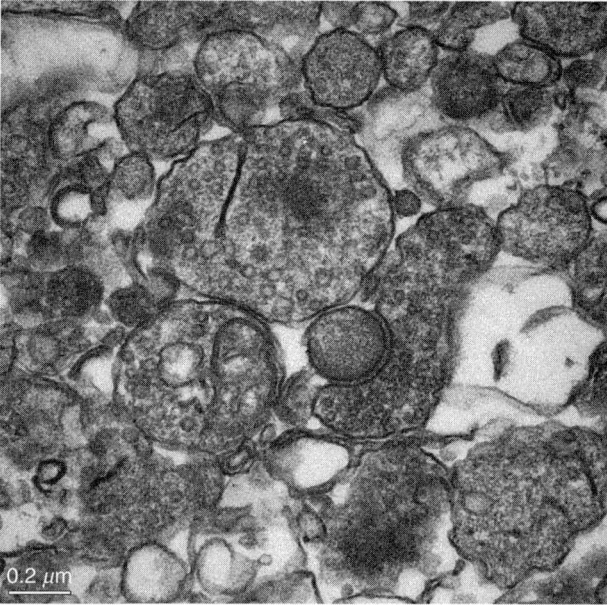

FIG. 1. Representative electron microscopy picture of synaptosomal fraction obtained by the method described in Pubill *et al.*, 2005 (X 65,000).

argon laser. The sample was diluted to obtain a flow rate of 500–900 synaptosomes/s and each sample was measured for 1 min.

METH increases 2′,7′-dichlorofluorescein (DCF) fluorescence when added to our preparation, which indicates that it induces ROS production. This increase is observed inside the synaptosomes (Pubill *et al.*, 2005). The fluorescence histogram shifts to the right (Fig. 2A) and individual synaptosomes show increased fluorescence (Fig. 2C) compared with untreated synaptosomes (Fig. 2B). ROS production reaches a plateau after 30 min of incubation with METH and is maintained for at least 2 h (Pubill *et al.*, 2005).

Incubation of synaptosomes with METH for a few minutes causes release of DA from presynaptic nerve terminals and inhibits DA uptake, probably by reversion of DAT functionality. Released DA can undergo oxidation and generate ROS but, as we measured fluorescence associated with synaptosomes, this putative source of ROS was not taken into account. ROS formation in the extracellular medium does not explain the specificity of METH degeneration for DA terminals, in that the oxidation of extraneuronal DA would be expected to nonspecifically damage all neighboring neurons, not just dopaminergic ones. With the present model, we have described an intracellular oxidative effect of METH that is more likely to induce damage in neuronal elements.

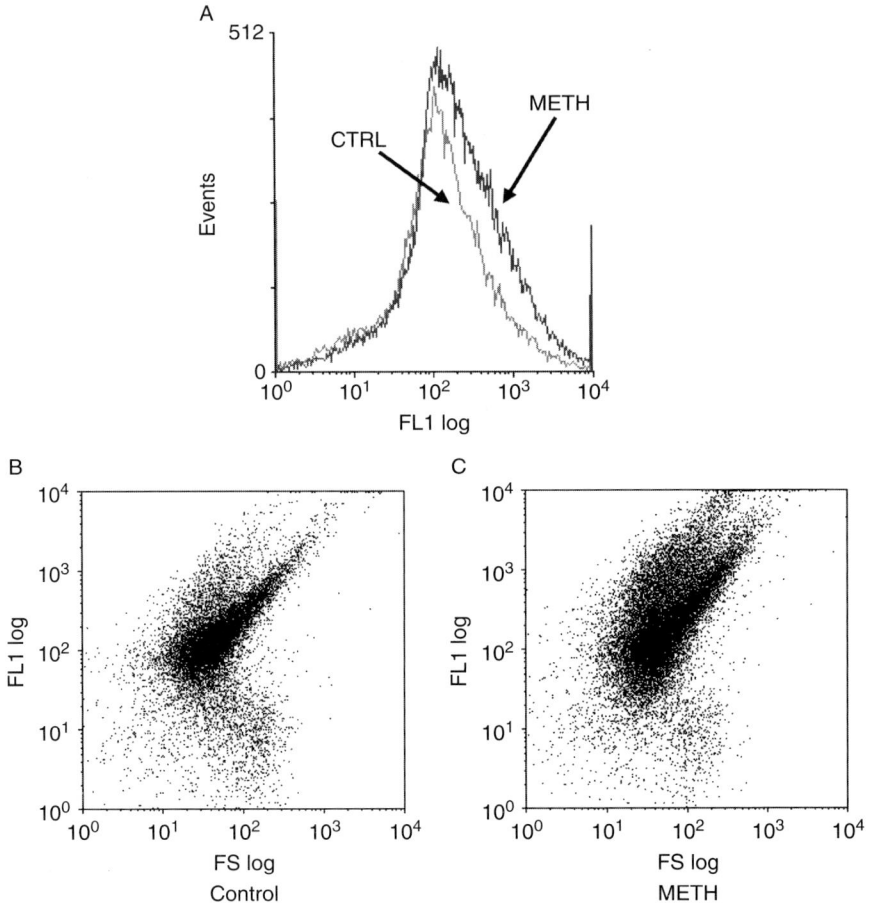

FIG. 2. Representative flow cytometry histograms (A) and dot plots (B and C) showing the change in DCF fluorescence of rat striatal synaptosomes after 2 h incubation at 37 °C alone (Control, CTRL) or with 2 mM METH.

A. DA INVOLVEMENT IN METH-INDUCED ROS PRODUCTION

Several authors point to DA as one of the main sources of ROS induced by amphetamines (Hastings *et al.*, 1996; Stephans and Yamamoto, 1994). When we used synaptosomes from previously DA-depleted rats (with reserpine or reserpine plus α-methyl-p-tyrosine) the METH-induced ROS production was inhibited (Fig. 3). These results corroborate DA as the main source of ROS detected. METH, by altering the intracellular pH gradient, prevents VMAT function and promotes DA release from vesicles to cytosol (Sulzer and Rayport, 1990)

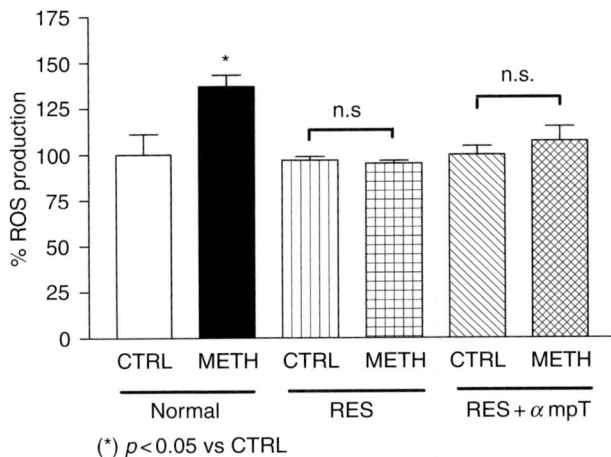

Fig. 3. Effect of catecholamine depletion on METH-induced ROS in rat striatal synaptosomes. Rats were pretreated with saline (normal), reserpine (RES), or with reserpine plus α-methyl-p-tyrosine (RES + AMPT). Synaptosomes were obtained from pretreated rats, and incubated alone (CTRL) or with 2 mM METH. *$p < 0.05$ versus normal control group.

where it can be oxidized. By this way, *in vitro* incubation of synaptosomes with substances that block VMAT (reserpine) prevents METH oxidative effect.

B. nNOS and PKC in METH-Induced ROS Production

A number of *in vivo* studies demonstrate the involvement of nNOS in METH neurotoxicity. Thus, METH administration causes overexpression of nNOS in mouse striatum (Deng and Cadet, 1999). In our model, the inhibitor of nNOS, 7-nitroindazole (7-NI), completely abolished METH-induced ROS production, demonstrating a role of this enzyme in METH oxidative effects. Moreover, the activation of nNOS would be needed, in addition to an increase in cytosolic DA to generate additional DCFH oxidation, as can be deduced from the experiments in which only free cytolosic DA was increased by preincubating the synaptosomes with DA plus reserpine (to prevent its vesicular uptake). In such experiments, no increased ROS were found in the absence of METH.

Activation of nNOS produces NO, which reacts with the peroxide radicals that would originate from DA autooxidation, producing the more toxic radical $ONOO^-$. This oxidant has been found to inhibit DAT functionality (Demiryurek *et al.*, 1998; Park *et al.*, 2002), avoiding DA release. Such an inhibition would favor cytosolic DA accumulation, which would increase oxidative species inside the synaptosomes. This radical has been postulated as the main agent for the damage in cell structures (Demiryurek *et al.*, 1998). Accordingly, inhibition of NO formation through a variety of methodological approaches confers neuroprotection against METH or MDMA: nNOS-deficient mice are resistant to METH-induced dopaminergic neurotoxicity (Itzhak *et al.*, 1998); administration of NO synthase inhibitors such as S-methylcitrulline or AR-R17477 attenuate the dopaminergic and serotonergic neurotoxicity of MDMA and METH (Ali and Itzhak, 1998; Colado *et al.*, 2001; Darvesh *et al.*, 2005; Itzhak *et al.*, 2000; Sanchez *et al.*, 2003). Finally, selenium (more effective as a scavenger of two-electron oxidants, such as $ONOO^-$, and not particularly reactive toward single-electron oxidants, such as NO and superoxide) shows a high neuroprotective effect in METH-induced neurotoxicity (Imam *et al.*, 1999).

Protein kinase C (PKC) has been implicated in various aspects of DAT function and direct phosphorylation (Foster *et al.*, 2002). In 2005, it has been described (Narita *et al.*, 2005) that treatment of cortical neuron/glia co-cultures with METH causes the activation of astrocytes via PKC. In our model, inhibition of PKC (by NPC 15437) fully prevented METH-induced ROS, corroborating a key role of PKC in this process. If PKC contributes to DAT phosphorylation, which produces a change in this transporter (Giambalvo, 2003; Sandoval *et al.*, 2001; Zhang *et al.*, 1997), the prevention by NPC 15437 is explained by maintenance of DAT function and probably the efflux of excessive cytoplasmatic DA.

Summarizing, in addition to an increase in cytosolic DA, activation of nNOS and PKC (blocking DA release through DAT) are needed, to generate ROS inside the dopaminergic terminal. Both, PKC and nNOS are enzymes that require calcium to be activated. Consequently, when calcium of the medium is chelated with ethylene glycol tetraacetic acid (EGTA), the oxidative effect of METH is prevented.

Finally, we postulated an integrative mechanism by which METH induces ROS production in striatal synaptosomes: "At high concentrations, METH enters the synaptosome, mainly by passive diffusion (as cocaine did not prevent METH-oxidative effect) and promotes DA release from synaptic vesicles to the cytosol and from cytosol to synaptic cleft via DAT. Increased cytosolic DA can suffer from autoxidation and generate initial ROS which can modify DAT function. Additionally, METH induces an increase in intrasynaptosomal Ca^{2+}, which would activate nNOS and PKC. PKC activation would lead to phosphorylation of proteins such as DAT promoting, together with $ONOO^-$, a reduction of DAT activity and accumulation of cytosolic DA that would impair the initial oxidative stress." However, a key point remains to be elucidated. What is the mechanism by which METH increases cytosolic calcium?

V. Different Oxidative Profile and Nitric Oxide Dependence of Amphetamine and MDMA

On the basis of these findings, the new goal was to study and compare the mechanisms by which D-amphetamine (AMPH) and MDMA also cause oxidative stress, as a triggering factor for their neurotoxicity (Chipana *et al.*, 2008b).

Again, the use of an *in vitro* model made it possible to obviate the influence of hyperthermia in this phenomenon.

The intracellular production of ROS is indicated by the fact that AMPH, like MDMA, increases DCF fluorescence (measured inside synaptosomes). The concentration–response curve varies depending on the compound tested (Fig. 4A). For AMPH, a double maximum was determined (a significant oxidative effect was detected between 0.1 and 1 μM that did not appear between 10 and 100 μM and

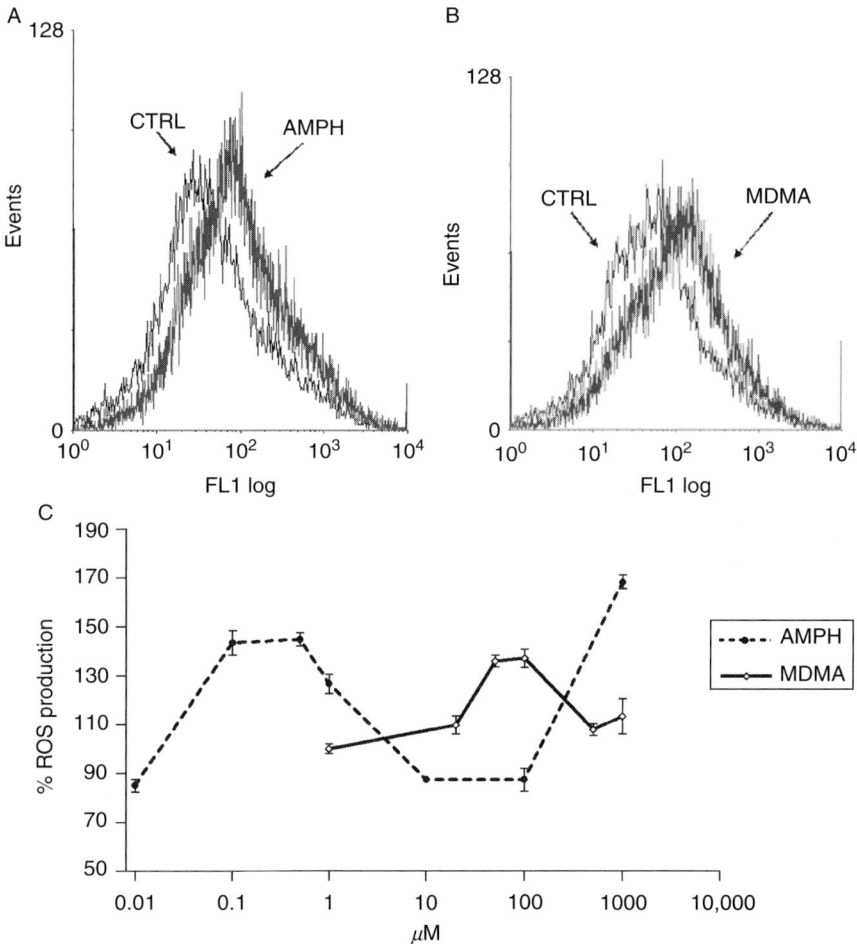

FIG. 4. Representative flow cytometry histograms showing the change in dichlorofluorescein fluorescence (curve shifted to the right) of mouse striatal synaptosomes after 1.5 h of incubation at 37 °C alone (Ctrl) or with AMPH (1 mM, panel A) or MDMA (50 μM, panel B). Panel C: Concentration-response curve of AMPH and MDMA-induced ROS production in mouse striatal synaptosomes. Data represent the means $\pm$ S.E.M. of three experiments carried out in triplicate.

reappeared at 1 mM). The MDMA dose-response curve showed an inverted "U" shape and the maximal oxidative effect appeared at 50–100 μM (Fig. 4B).

The inhibition of AMPH (0.1 μM) ROS production by cocaine, but not that of AMPH at high concentrations, points to a different mechanism of amphetamine incorporation into the synaptosome. At low concentrations, AMPH is exchanged for DA via DAT, but at high concentrations, this lipophilic compound can diffuse into nerve terminals through the plasmalemmal membrane. Then, AMPH incorporation into dopaminergic terminals is concentration dependent. For MDMA, only a DAT-dependent oxidation is observed, pointing to a concentration-dependent mechanism of incorporation into the nerve terminal.

From experiments performed on reserpinized animals, it can be deduced that vesicular DA is the main factor responsible for the oxidative effect of both compounds. In the low micromolar range, amphetamines block monoamine oxidase A (MAO-A), which brings about the enzymatic degradation of DA. But DA can also be metabolized by MAO-B, yielding reactive DA species as well as hydrogen peroxide. Then, oxidative effect showed at low concentrations of amphetamine derivatives can be due to DA oxidation by MAO-B. This hypothesis was confirmed by the prevention of this oxidative effect with L-deprenyl (MAO-B inhibitor) (Fig. 5). Additionally, the decreases shown in the concentration–response curves (AMPH peaked at 0.1–1 μM, then decreased and reappeared at 1 mM; MDMA peaked at 50–100 μM and then decreased) are probably due to inhibition of MAO-B by AMPH and MDMA, respectively.

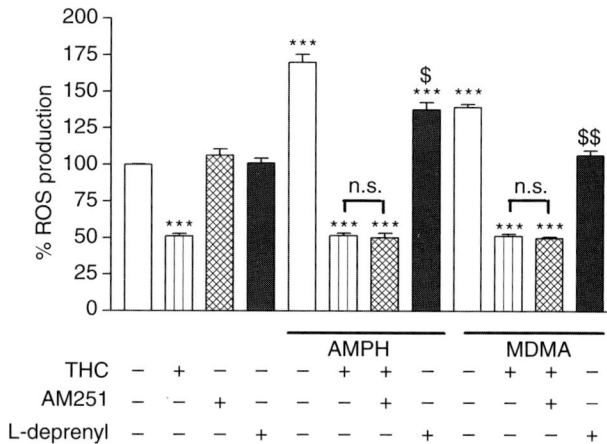

FIG. 5. Effects of pre-treatment with THC (10 μM), AM-251 (0.1 μM) or l-deprenyl (1 μM) on the increase in ROS production in mouse brain synaptosomes induced by AMPH (1 mM) or MDMA (50 μM). Data represent the means $\pm$ S.E.M. of three experiments carried out in triplicate. *** $p < 0.001$ versus control (absence of amphetamine derivative in the medium); $^{\$}p < 0.05$ and $^{\$\$}p < 0.01$ versus AMPH or MDMA alone.

It must be pointed out that AMPH inhibits MAO-B at concentrations higher than 118 μM (Ulus *et al.*, 2000). The same conclusion is valid for MDMA whose K_i for MAO-B is 370 μM (Leonardi and Azmitia, 1994).

Metabolic actions of MAO-B have also been found to participate in the neurotoxic actions of amphetamines. Increased expression of this enzyme results in increased neurotoxicity by METH (Wei *et al.*, 1997). Accordingly, suppression of MAO-B using an antisense oligonucleotide attenuates striatal damage induced by MDMA in rats (Falk *et al.*, 2002).

DA can also form ROS through another pathway besides MAO (Berman and Hastings, 1999) yielding DA quinone, which acts as a dopaminergic neuron-specific neurotoxic factor (Miyazaki *et al.*, 2006). Thus, the incorporation of AMPH into the neuron can stimulate phospholipase A_2 activity via its ability to increase $[pH]_i$ (Giambalvo, 2004). Arachidonic acid and hydrogen peroxide could serve as substrates for the prostaglandin H synthase-catalyzed oxidation of DA to quinones (Hastings, 1995). This can account for its role in the oxidative effect shown at high concentrations of AMPH (which is not prevented by L-deprenyl) and explains the inhibition produced by 3-(4-octadecyl)benzoylacrylic acid (OBAA) (phospholipase A_2 inhibitor) only against the effect of high concentrations of AMPH.

The concomitant abuse of cannabis and amphetamine derivatives is a frequent practice. Tetrahydrocannabinol (THC), the active ingredient in herbal cannabis, acts on specific receptors (especially CB_1). Our results demonstrate that THC fully prevents the oxidative effect of AMPH and MDMA. This preventive effect is not inhibited by the specific CB_1 antagonist, AM251, which points to an unspecific antioxidant action as the factor responsible, and not to a specific interaction with CB_1 receptor. Neuroprotective antioxidant effects of THC have also been reported by other authors (Morley *et al.*, 2004). From our results, it can be deduced that the antioxidant effect of THC prevents the harmful effect of DA oxidation induced by AMPH and MDMA, pointing to a protective effect of THC (Fig. 5). However, this result must be analyzed carefully because chronic cannabis abuse has been reported to cause memory impairment and, therefore, it cannot be proposed as neuroprotective.

The calcium dependence reported with METH and assessed also for AMPH and MDMA, made it advisable to assay the possible implication of NO synthase in the pro-oxidative effect. Surprisingly by preventing the NO effect (by an nNOS inhibitor or an NO trapping agent as carboxy-2-phenyl-4, 4, 5, 5-tetramethyl-imidazoline-1-oxyl-3-oxide(carboxy-PTIO), the effect of MDMA, but not that of AMPH, was inhibited. Consequently, it is likely that, conversely to what happens with AMPH, the initial oxidation induced by MDMA would be potentiated by nNOS, producing $ONOO^-$. Also, another calcium-sensitive enzyme, PKA, showed the same profile. Thus, the implication of nNOS and PKA is null for AMPH and significant for MDMA.

VI. Nicotinic Antagonists Block METH and MDMA-Induced ROS Production

Activation of certain nicotinic receptors enhances DA release induced by amphetamine (Drew and Werling, 2001). Such enhancement is mediated by an increase in Na^+ and Ca^{2+}, as well as by the activation of PKC. Thus, after the initial depolarization mediated by nicotinic receptors and subsequent Ca^{2+} entry, PKC is activated and phosphorylates DAT.

PKC and nNOS require Ca^{2+} to be activated. Nicotine ionotropic homomeric receptors made of $\alpha7$ subunits are permeate to Na^+ and Ca^{2+} and, in a paper of Liu *et al.* (2003), it has been proved that AMPH acts as an agonist on these receptors in chromaffin cells. Although $\alpha7$ neuronal nicotinic receptors are a minority type in the striatum, they are expressed on DA axon terminals (Wonnacott, 1997; Zoli *et al.*, 2002). For this reason, we tested the involvement of nicotinic $\alpha7$ receptors in our preparation, using the specific antagonist methyllycaconitine (MLA). MLA completely inhibited METH-induced ROS production as well as α-bungarotoxin (another prototypic $\alpha7$ antagonist), while dihydro-β-erythroidine (DBE), an antagonist that blocks receptors containing $\beta2$ subunits, was devoid of effect, thus implicating $\alpha7$ receptors (Pubill *et al.*, 2005).

On the basis of these antecedents, we considered the possibility that MDMA may also exert an oxidative effect dependent on nAChR stimulation. Specific $\beta2$-subunit-containing and $\alpha7$ nAChR antagonists fully inhibited the oxidative stress induced by MDMA. Additionally, MLA inhibited the remaining effect of MDMA in the presence of catalase, CAT + SOD (attributed to NO), pointing a role of $\alpha7$ nAChR in the activation of nNOS induced by MDMA.

Moreover, activation of these nAChR could potentiate intracellular calcium increase and activate additional calcium-sensitive transduction processes. As a result, sequestering external calcium with EGTA or pre-incubation of synaptosomes with 2-aminoethoxydiphenyl borate, a cell-permeate IP3 receptor inhibitor, resulted in strong prevention of MDMA-induced oxidative effects. Also nitrendipine, an L-type, voltage-sensitive calcium-channel blocker, and dantrolene, an inhibitor of ryanodine receptor, prevented its oxidative effect.

All these results concluded that METH and MDMA can interact with nAChRs, especially $\alpha7$. The consequent increase in cytosolic calcium could be responsible for PKC activation. Amphetamines act also providing more cytoplasmic DA available to the inward-facing DAT for its further release, and inhibiting DA uptake. Both processes are known to be calcium- and PKC-dependent (Gnegy, *et al.*, 2004; Johnson, *et al.*, 2004; Narita, *et al.*, 2005) (for uptake regulation see Section IX).

When cytoplasmatic DA is higher than that which can be released by DAT reversion, it undergoes intracellular oxidation. The additional calcium that nicotinic interaction produces could be responsible for nNOS activation, leading to ROS production. $ONOO^-$, ROS, and DA quinones could later contribute to long-term inhibition of DAT function (Park *et al.*, 2002), trapping of DA in the terminal, and impairment of initial oxidative stress.

These results allow us to conclude that agents that affect DAT functionality, such as PKC, tyrosine kinase, and intracellular calcium availability, seem also to affect amphetamines-induced ROS production.

For an integrated depiction of the mechanisms leading to this oxidative effect see Fig. 6.

The release of intracellular pools of DA by AMPH has been shown to be a paradigm of amphetamine-derivative effects in experimental models. The calcium dependence of DA release induced by amphetamines is controversial. Thus, a new insight into the mechanism of calcium dependence of amphetamine-induced effects is of interest. The use of low or high concentrations of these drugs and protein substrates that contain or do not contain nAChRs, probably contributes to the contradictory results found in the bibliography.

VII. MLA as a Neuroprotective Substance for METH- and MDMA-Induced Dopaminergic Neurotoxicity

In view of these results, the next goal was to study whether MLA had neuroprotective effects *in vivo*. Assessment of neurotoxicity markers after an *in vivo* treatment with a neurotoxic schedule of METH (7.5 mg/kg s.c., every 2 h,

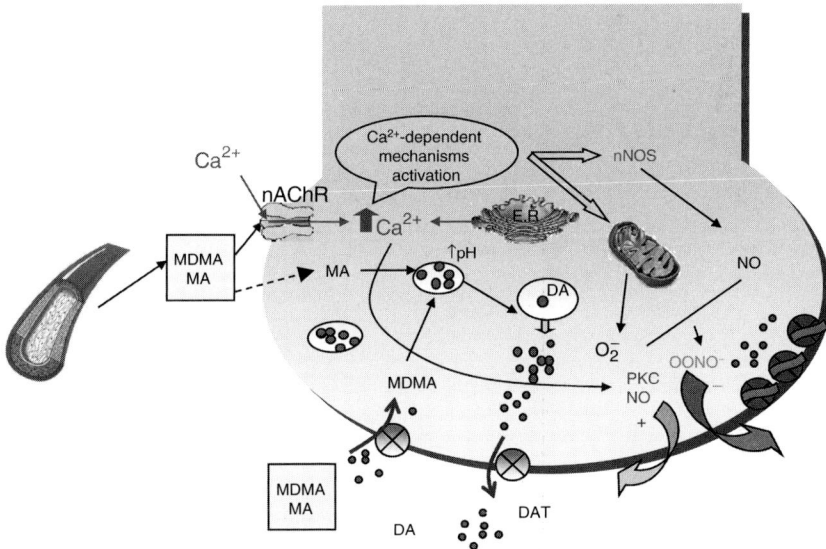

Fig. 6. Integrated picture showing the mechanisms leading to METH/MDMA-induced ROS production in semi-purified striatal synaptosomes. (See Color Insert.)

for a total of four doses) or MDMA (25 mg/kg, s.c., every 3 h, for a total of three doses) was used to determine this neuroprotective effect of MLA (6 mg/kg, i.p., administered 20 min before each dose of METH or MDMA) in mice (Chipana *et al.*, 2006; Escubedo *et al.*, 2005).

METH induced, at 72 h post treatment, a significant loss of striatal DA reuptake sites of about 73%, measured as specific binding of [^{3}H]WIN 35428 in mouse striatum membranes. This dopaminergic injury was attenuated in mice pretreated with MLA (from 73% to 43%) without affecting METH-induced hyperthermia.

Glial activation was assessed by the increase in the peripheral-type benzodiazepine receptor density in striata of the different treatment groups sacrificed at 24 h post treatment. It was measured by [^{3}H]PK 11195 specific binding. In this area, METH-treated animals showed a significant increase of 53% in peripheral-type benzodiazepine receptor density compared with control (CTRL) animals (glial activation). When MLA was administered 30 min before each METH dose, such an increase was fully prevented, pointing to the protective effect of this nicotinic receptor antagonist against METH neurotoxicity in mouse striatum.

The *in vivo* neurotoxic model of MDMA used was characterized by a significant loss of DA terminals (69%) and a decrease of tyrosine hydroxylase levels (65%) in striata from animals sacrificed 7 days post treatment. This terminal loss was prevented by pretreatment with MLA, pointing also to a neuroprotective effect versus this amphetamine derivative (Fig. 7A). MLA reduced the mortality induced by MDMA (14.3% in the MDMA-treated group, but only 7.14% in the

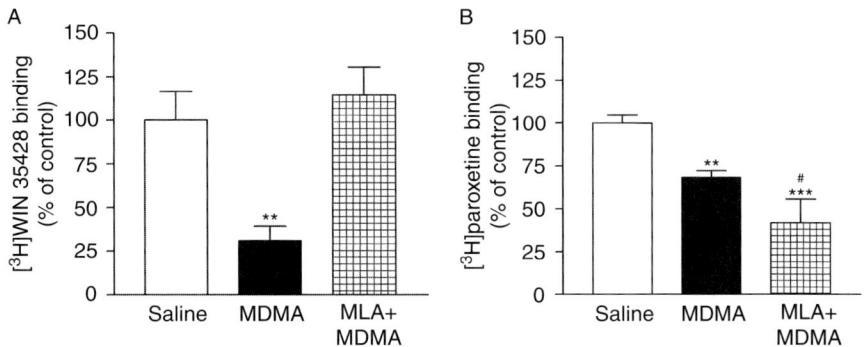

FIG. 7. Panel A: Effect of *in vivo* treatment with MDMA (three injections, 25 mg/kg s.c., 3-h intervals) or in combination with MLA (three injections, 6 mg/kg i.p., 20 min before MDMA) on the density of mouse striatal dopamine reuptake sites at 7 days after treatment, and measured as specific binding of [^{3}H]WIN 35428. **$p < 0.01$ versus saline-treated group. Panel B: Effect of treatment with MDMA alone or in combination with MLA (same dose schedule as above) on the density of mouse striatal serotonin reuptake sites 7 days after treatment, measured as specific binding of [^{3}H]paroxetine. **$p < 0.01$, ***$p < 0.001$ versus saline-treated group; #$p < 0.05$ versus MDMA-treated group. In both panels values are expressed as means ± S.E.M. of those obtained from 5 to 6 animals in each group. One-way ANOVA and Tukey's post hoc test.

MLA + MDMA-treated group). It was not possible to identify a specific mechanism as responsible for this increase in survival, but from results of body temperature, a protective effect based on a hypothermic or antihyperthermic mechanism of MLA can be ruled out.

[^{3}H]Paroxetine binding was measured also in the striatum of these animals, as a marker of degeneration of 5-HT terminals. Conversely, MLA did not prevent the loss in [^{3}H]paroxetine binding sites (Fig. 7B), indicating that its neuroprotective effect is selective for dopaminergic terminals. The more pronounced loss of 5-HT terminals in the MLA + MDMA group may have been due to a protective effect from death of MLA on the mice with more severe losses, rather than to a potentiation of the deleterious effect of MDMA.

Turner (2004) suggested that the α7 nAChR-mediated pathway is tightly and specifically coupled to refilling of the readily releasable pool of vesicles in DA terminals. The nicotine-induced increase in the size of this readily releasable pool was blocked by α-bungarotoxin and by the calmodulin antagonist calmidazolium, suggesting that Ca^{2+} entry through α7 nAChRs specifically enhances synaptic vesicle mobilization at DA terminals. Then, it can be speculated that α7 antagonists can modulate neurotoxicity and some behavioral effects of MDMA or METH.

VIII. Memantine: A New Tool to Prevent Amphetaminic Neurotoxicity

Memantine (MEM), a noncompetitive antagonist of the NMDA receptor, is a drug used to treat moderate to severe Alzheimer's disease; it reduces tonic, but not synaptic, NMDA receptor activity. In 2005, the group of Aracava demonstrated that MEM, at clinically relevant concentrations, can block α7 nAChR in a noncompetitive manner, and more effectively that it does at NMDA receptors (Aracava et al., 2005). Unger et al. (2005) described how treatment with MEM significantly increases the number of α7 nAChR binding sites in the frontal and retrosplenial cortex in mice, suggesting the interaction of MEM with these nicotinic receptors, as upregulation of nAChR is a characteristic effect induced by nicotinic ligands (agonists and antagonists, see Section XI.A for further information).

MEM has shown promising results in the treatment of amphetamine addiction (Levi and Borne, 2002). No drugs are currently approved in the United States or Europe for the treatment of addictions to METH or MDMA. Fluoxetine pretreatment has been recommended as protection from MDMA-induced long-term neurotoxicity, but recently it has been found that fluoxetine decreases the elimination of MDMA and its metabolite, methylenedioxyamphetamine, leading to an increased risk of acute MDMA toxic effects (Upreti and Eddington, 2008).

Our hypothesis derived from previous results with MLA was that MEM could prevent METH- and MDMA-induced neurotoxicity, probably more efficiently than MLA, due to its dual mechanism (Chipana *et al.*, 2008a,c). If it did, MEM could be used not only to treat these addictions or to prevent the effects of these amphetamine derivatives, but also to have a beneficial effect on the memory impairment that abusers of these drugs usually suffer (Simon and Matlick, 2002; Simon *et al.*, 2002).

First, we studied dopaminergic neurotoxicity (characteristic of METH and MDMA in mice) by examining whether MEM prevented ROS production induced by METH or by MDMA *in vitro* in mouse striatal synaptosomes.

In the *in vivo* experiments, we administered MDMA to Dark Agouti rats (18 mg/kg, s.c.) as a model of serotonergic neurotoxicity induced by MDMA. Dark Agouti rats (a strain devoid of some CYP isoforms) suffer a significant serotonergic lesion in response to a single dose of MDMA (Kovacs *et al.*, 2007). In contrast, Sprague-Dawley and Wistar strains do not usually show a similar injury until they have received several doses (Battaglia *et al.*, 1988b).

We also administered METH (the same dose schedule as in the previous section) to mice to study the effect of MEM on dopaminergic METH-induced neurotoxicity. The MEM + MDMA or MEM + METH groups received a dose of MEM (5 mg/kg, i.p.) 30 min before the corresponding dose of MDMA or METH. Hyperthermic response to each amphetamine derivative was not modified by MEM.

A. *In Vitro* Neurotoxicity Studies: ROS Production

MEM had no antioxidant effect versus H_2O_2 (data not shown), but inhibited the ROS production induced by MDMA at all the concentrations tested (Fig. 8A). This inhibition was countered by the presence of PNU 282987, a specific agonist of the $\alpha7$ nAChR, in the incubation medium.

METH induced a concentration-dependent ROS production in mouse striatal synaptosomes yielding an EC_{50} value of about 84 μM, which was significantly increased to 208 μM in the presence of MEM. As can be seen in Fig. 8B, the inhibitory effect of MEM was also countered by the presence in the incubation medium of PNU 282987, and the oxidative effect of METH was reinforced by this $\alpha7$-specific agonist.

It has been suggested that $\alpha7$ nAChR located at striatal glutamatergic terminals and inducing glutamate release could modulate DA release in the striatum (Kaiser and Wonnacott, 2000). Although in synaptosomal preparation the probability of neurochemical cross-talk is very low, we studied the possible involvement of NMDA receptors using glutamic acid (GLU) and phencyclidine (PCP) (an

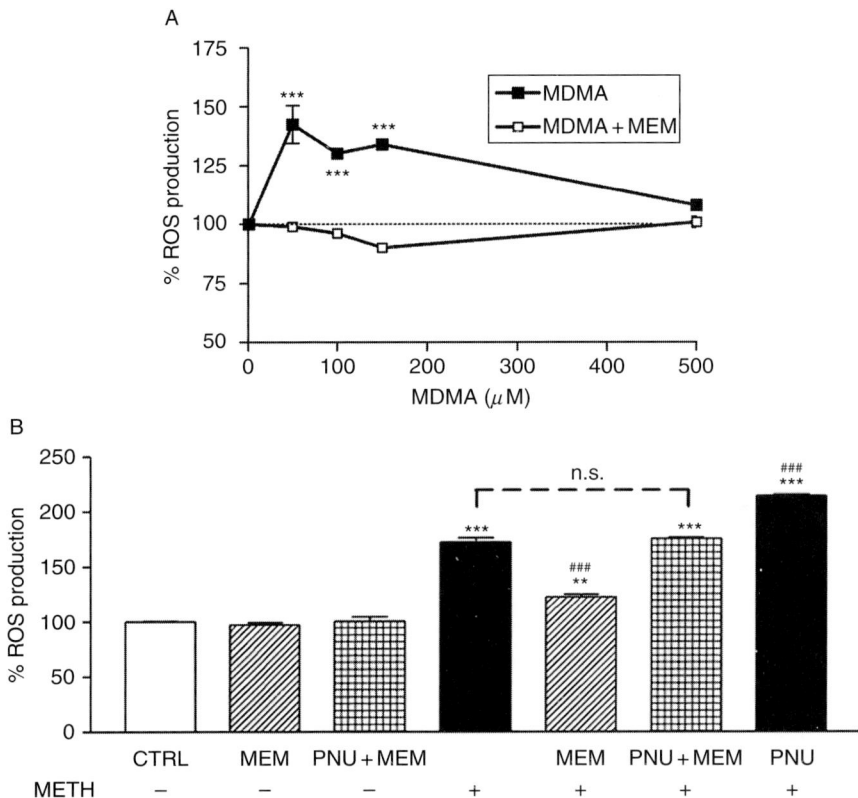

FIG. 8. Panel A. Effect of MEM (0.3 μM) on the ROS production induced by different concentrations of MDMA in mouse striatal synaptosomes, expressed as percentage of control value (dotted line). Panel B. Effect of PNU 282987 (0.5 μM) and MEM (0.3 μM), alone or in combination, on METH (3 mM)-induced ROS production in mouse striatal synaptosomes. Results are expressed as mean $\pm$ S.E.M. from at least three separate experiments run in triplicates. **$p < 0.01$ and ***$p < 0.001$ versus CTRL and ###$p < 0.001$ versus METH n.s.: non significant, one-way ANOVA and Tukey's post hoc test.

NMDA channel blocker). GLU alone induced ROS production that was not prevented by MEM, ruling out a role of NMDA receptors in the oxidative effects of GLU in this preparation, and for extension, NMDA-mediated effect of MEM. Additionally, MEM inhibition of MDMA- or METH-induced oxidative effect was not countered by the presence of GLU in the incubation medium, and PCP did not modify MDMA and METH effects. These results demonstrate a preventive effect of MEM on MDMA- and METH-induced oxidation in mouse striatum as a result of it blocking α7 nAChRs.

B. Memantine Protection Against MDMA-Induced Serotonergic Injury and METH-Induced Dopaminergic Injury

A significant decrease in the density of serotonin transporter (SERT) was observed in both the hippocampus and frontal cortex of MDMA-treated Dark Agouti rats killed 7 days post treatment, but serotonergic injury was already apparent 24 h post treatment (Chipana *et al.*, 2008a). In both cases, MEM significantly prevented the loss of [^{3}H]paroxetine binding sites, suggesting a neuroprotective effect on 5-HT terminals (see Table I).

A characteristic that differentiates this vulnerable rat strain from Sprague-Dawley rats is that we detected not only the rapid appearance of the serotonergic injury (this is the first time that such a rapid MDMA-induced loss in SERT is described in hippocampus, which is probably attributable to the major sensitivity of the rat strain used), but also glial activation 7 days after treatment (characterized by an increase in [^{3}H]PK 11195 binding), which was not found 3 or 7 days after treatment in Sprague-Dawley rats (Pubill *et al.*, 2002). In the present study, a significant increase in [^{3}H]PK 11195 binding was detected in the hippocampus and frontal cortex 7 days post treatment, but not in animals killed 24 h post treatment. Therefore, this effect appears after the neuronal injury and suggests a phagocytic role rather than a pro-inflammatory response for this glial activation. Furthermore, MEM-pretreated animals did not show either serotonergic injury or glial activation.

Increased ROS production can activate different transcription factors such as the nuclear factor kappa B (NF-κB). NF-κB induces the expression of pro-inflammatory and cytotoxic genes and plays a key role in the balance between cell survival and death. To gauge the activation of NF-κB by MDMA, we measured the translocation (from cytosol to the nucleus) of P65, the active subunit of NF-κB, in the hippocampus of rats, detecting a significant p65 nuclear translocation in the hippocampus of MDMA-treated animals (Fig. 9). This translocation was inhibited by MEM pretreatment, suggesting that activation of NF-κB after treatment with this amphetamine derivative participates in the cytotoxic effect, since when this activation is blocked by MEM, neuronal injury is prevented.

In parallel experiments, the extent of METH-induced injury was determined 3 days after ending a dosage regimen that models chronic drug use. Using this schedule, METH induced a dopaminergic injury accompanied by gliosis. Both parameters were prevented in the animals pretreated with MEM (Chipana *et al.*, 2008c), while hyperthermia was not affected. MEM diminished the incidence of death in METH-treated animals, but this difference was not significant.

Although MEM could directly prevent the MDMA- and the METH-induced neurotoxicity through antagonism at NMDA receptors, this is not a feasible hypothesis since antagonists of these receptors fail to prevent oxidative stress and cell death induced by METH and MDMA (Jimenez *et al.*, 2004). Protective effects of

TABLE I

Effect of Memantine on MDMA-Induced Decrease in the Abundance of 5-HT Transporters (Labeled by [³H]Paroxetine) in Rat Hippocampus and Frontal Cortex and Also in the Glial Activation Measured by The [³H]PK11195 Binding

Treatment	Hippocampus				Frontal cortex	
	[³H]Paroxetine binding 1 DPS[a]	[³H]Paroxetine binding 7 DPS[a]	[³H]PK 11195 binding 7 DPS[a]		[³H]Paroxetine binding 7 DPS[a]	[³H]PK 11195 binding 7 DPS[a]
Saline	100.00 ± 2.65	100.00 ± 9.34	100.00 ± 6.35		100.00 ± 5.05	100.00 ± 2.94
MDMA	$62.16 \pm 3.86{***}$	$70.02 \pm 7.85{*}$	$168.22 \pm 20.65{**}$		$56.90 \pm 2.57{***}$	$120.03 \pm 6.70{*}$
Mem	97.31 ± 0.89	116.40 ± 3.95	94.90 ± 7.05		88.93 ± 9.20	90.00 ± 9.12
MDMA + Mem	$89.78 \pm 2.65{\#\#\#}$	94.87 ± 4.96	$104.41 \pm 9.77{\#}$		$74.00 \pm 4.98{**}{\#}$	$97.70 \pm 4.94{*}$

Note: Results (in %) are expressed as mean $\pm$ SEM from six to nine different experiments.

${*} p < 0.05$, ${**} p < 0.01$, ${***} p < 0.001$ versus Saline, ${\#} p < 0.05$, ${\#\#\#} p < 0.001$ versus MDMA group.

[a]DPS: days post treatment.

A

B

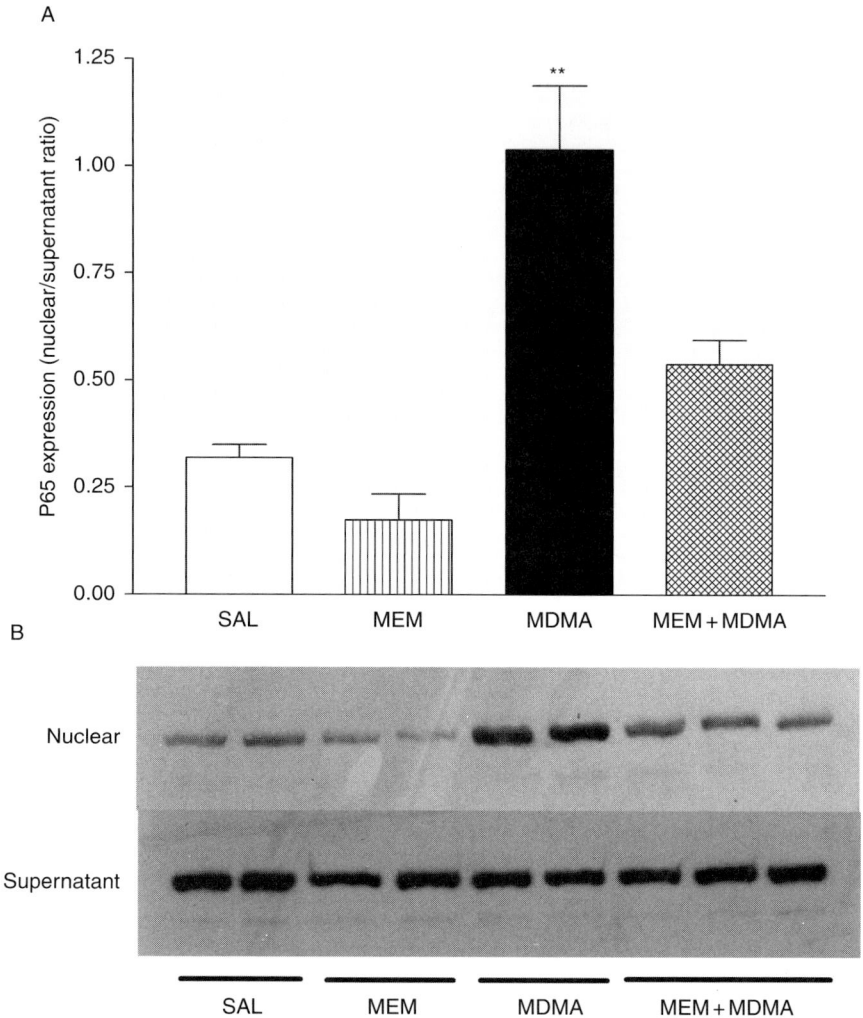

FIG. 9. Activation of NF-κB, measured as the translocation (from cytosol to the nucleus) of P65. P65 expression in rat hippocampus (A) and representative Western blot showing P65 translocation (B). Animals were treated with saline, MDMA, MEM and MEM+MDMA and killed 24 h later. Data are expressed as ratios of nuclear pellet/supernatant. *$p < 0.05$ **$p < 0.01$ versus saline and [#]$p < 0.05$ versus MDMA.

NMDA antagonists are secondary to the blockade of METH- or MDMA-induced hyperthermia, which is not the case for MEM. However, MEM has a better protective effect in comparison to MDMA- and METH-induced neurotoxicity than MLA, a more specific $\alpha7$ nAChR antagonist. The dual antagonism that MEM exerts on NMDA receptor and on $\alpha7$ nAChR probably turns it into a better pharmacological tool to prevent amphetamines-induced damage.

DA neuron firing is modulated by glutamatergic (excitatory) afferents and DA release is evoked by NMDA. If MEM blocks NMDA receptors, a decrease in extracellular DA levels would take place. Following the Sprague *et al.*-integrated hypothesis (Sprague *et al.*, 1998), the probability that released DA might be taken up into the depleted 5-HT terminals will be reduced by MEM. Consequently, the antagonism of NMDA receptors could contribute to the protective effects of MEM. Even so, indirect mechanisms for MEM action based on the interplay between the various neurotransmission systems leading to a modification in basal ACh release should also be taken into account. Nair and Gudelsky (2006) described how MDMA stimulates hippocampal ACh release, involving nondo-paminergic and nonserotonergic mechanisms).

IX. Effects on Serotonin and Dopamine Uptake: Nicotinic Receptor Modulation

It is well-established that the monoamine transporters (i.e., DAT, SERT, and NET) are the main biological targets for stimulants such as AMPH, METH, and MDMA. The functionality of these transporters is affected by these drugs (at low concentrations), promoting the release of neurotransmitters as well as inhibiting their uptake. Consequently, these psychostimulants increase extracellular neuro-transmitter concentrations (especially DA).

In mouse behavioral tests with METH, the increase in extracellular DA concentration is reflected by an increased locomotion observed at low doses, thought to reflect an increased DA transmission in the nucleus accumbent (Ljungberg and Ungerstedt, 1985). If METH dose is increased, it induces stereo-types thought to reflect an increased DA transmission in the neostriatum.

An $\alpha7$ antagonist, MLA, but not a heteromeric nAChR antagonist, DBE, inhibited METH-induced climbing (stereotypic) behavior. However, MLA did not modify either basal locomotor activity or the METH-induced hyperlocomo-tive profile. Thus, it seems that in neostriatum, activation of $\alpha7$ nAChRs is required to permit METH-induced DA release through reverse transport.

At neurochemical level, Khoshbouei *et al.* (2004) proposed a unified model to explain the effect of amphetamines on DA release, suggesting an asymmetrical transporter (DAT) with a conformational property favoring influx over efflux, but with net flux controlled by transmembrane substrate gradients. They also intro-duced a potential second messenger system that may provide the basis for these observations.

Consistent with a regulation of transport direction by second messengers, numerous putative phosphorylation sites for various protein kinases were identified in the intracellular domains of the DAT (Granas *et al.*, 2003). In addition, multiple protein kinases have been shown to regulate DAT function (Carvelli *et al.*, 2002; Granas *et al.*, 2003; Loder and Melikian, 2003; Melikian and Buckley, 1999). Accordingly, PKC activation has been shown to stimulate DAT-mediated release of DA (Giambalvo, 1992a,b), and AMPH-mediated DA release is inhibited by PKC inhibitors and by downregulation of PKC (Kantor and Gnegy, 1998), whereas [^{3}H]DA uptake is unaffected by these manipulations. However, results from our group were contrary to the last statement, as will be explained below.

In contrast to other drugs of abuse, MDMA's action is believed to be more prominent on 5-HT than on DA release/uptake. Additionally, Iravani *et al.* (2000) established that MDMA is a better inhibitor of 5-HT and DA uptake than stimulator of the release of these neurotransmitters.

According to Iravani's results and as a measure of METH and MDMA's acute effects, we studied the mechanisms involved in METH and MDMA inhibition of DA and 5-HT uptake, the role of calcium in this effect, and the modulation of amphetamine derivatives effect by nAChR.

In these studies, we used the semipurified synaptosomal preparation described previously. Results from Sandoval *et al.* (2001) resemble those effects of METH previously reported *in vivo*, which further suggested that the *in vitro* synaptosomal system is an appropriate model for the rapid and reversible changes caused by *in vivo* administration of METH.

In our studies, [^{3}H]5-HT uptake (in hippocampal) and [^{3}H]DA uptake (in striatal) synaptosomes were measured as indicative of the acute serotonergic effect of MDMA and the acute dopaminergic effect of METH, respectively. Pre-incubation of synaptosomes with MDMA (15 μM) induced a significant reduction in [^{3}H]5-HT uptake by 40% (Fig. 10). MDMA (10 μM) and METH (1 μM) also inhibited [^{3}H]DA uptake by 75% and 80%, respectively (Fig. 11). This inhibition remained even after drug washout and therefore cannot be attributed to residual drug presence, but to a persistent alteration of transporters. As incubation of drugs with synaptosomes was carried out in the presence of GSH and MAO-A inhibitor (clorgyline) in the medium, the effect of amphetamines on SERT/DAT cannot be attributed to ROS generation.

The effect of the amphetamine derivatives on these transporters was prevented by EGTA, 7-NI, and PKC inhibitors (GF-109203X or NPC 15437). Some authors had yet described the relationship between PKC and the effect of amphetamines on DAT (Sandoval *et al.*, 2001). Additionally, the physical association of nNOS and SERT has been recently reported, resulting in modulation of SERT activity (Garthwaite, 2007). Also, Cao and Reith (2002) described how NO inhibits DA uptake.

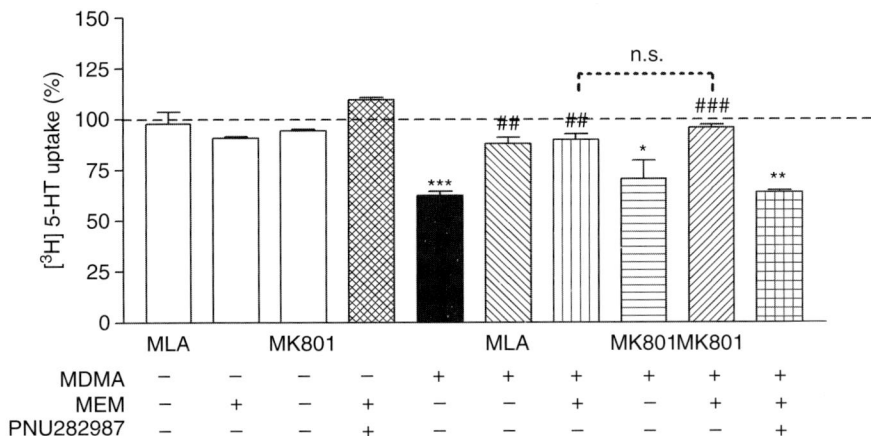

FIG. 10. Effect of MDMA (15 μM), MLA (0.1 μM), MEM (0.15 μM), MK801 (1 μM) and PNU 282987 (0.5 μM), alone or in combination, on [^{3}H]5-HT uptake in synaptosomes from rat hippocampus. Data are presented as mean $\pm$ S.E.M. as percentage of control (dashed line) [^{3}H]5-HT uptake, from at least three separate experiments run in duplicate. 60 Nonspecific [^{3}H]5-HT uptake was determined at 4 °C in parallel samples containing 10 μM fluoxetine and was about 10% of total [^{3}H]5-HT uptake. *$p < 0.05$, **$p < 0.01$ and ***$p < 0.001$ versus control, $^{##}p < 0.01$ and $^{###}p < 0.001$ versus MDMA. n.s.: non significant, one-way ANOVA and Tukey's post hoc test.

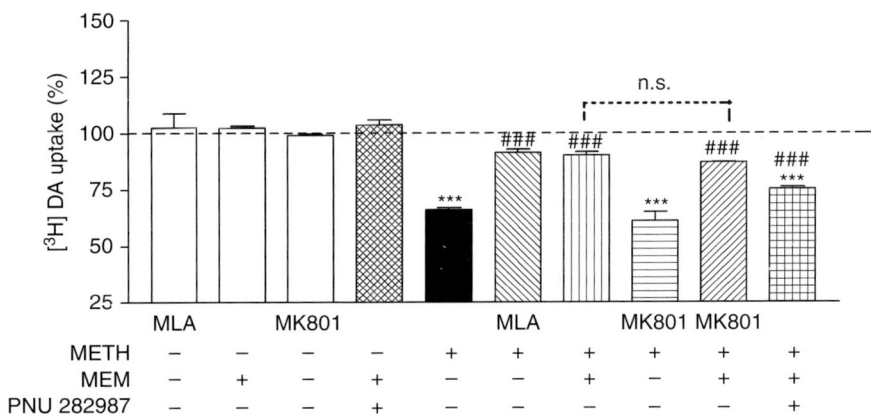

FIG. 11. Effect of METH (1 μM), MLA (0.1 μM), MEM (0.15 μM), MK801 (1 μM) and PNU 282987 (0.5 μM), alone or in combination, on [^{3}H]DA uptake in rat striatal synaptosomes. Data are presented as mean $\pm$ S.E.M. as percentage of control (dashed line) [^{3}H]DA uptake, from at least three separate experiments run in duplicate. Nonspecific [^{3}H]DA uptake was determined at 4 °C in parallel samples containing 100 μM cocaine and was about 7% of total [^{3}H]DA uptake. ***$p < 0.001$ versus control, $^{###}p$.

From our results, we can establish that the inhibition of both 5-HT and DA uptake by MDMA and METH depends on calcium and also on the activity of nNOS and PKC, both of which are calcium-dependent enzymes. If MDMA and METH act as α7 nAChR agonists or positive modulators, calcium entry through this receptor would contribute to their effect on 5-HT or DA uptake. In accordance with this hypothesis, we studied the effects of MLA and MEM in the effects of MDMA and METH on 5-HT and DA uptake. MLA and MEM countered the effect of amphetamines on SERT/DAT. Also, PNU 282987 (α7 nAChR agonist) prevented the effect of MEM, but MK 801 did not modify it, confirming that the effect of MEM on MDMA/METH-induced uptake inhibition is mediated by α7 nAChR and not by NMDA receptor activation (Figs. 10 and 11).

Aznar *et al.* (2005) described the presence of α7 nAChR at 5-HT neurons, in terminals projecting into the hippocampus. Furthermore, in our experiments PNU 282987 alone inhibited 5-HT uptake, which suggests that SERT functionality in the hippocampal serotonergic terminals can be regulated by α7 nAChR. Accordingly, PNU 282987 potentiated the inhibitory effect of MDMA on SERT function.

In order to compare the role of nAChR in the effect of MDMA and AMPH on DAT, [^{3}H]DA uptake inhibition by AMPH and MDMA was measured, in the absence or presence of the specific nAChR antagonists. Nondifferentiated PC 12 cells were chosen for this study because these cells exhibit characteristics of dopaminergic neurons.

Pre-incubation of PC 12 cells with AMPH (10 μM) or MDMA (10 μM) reduced [^{3}H]DA uptake by 50% and 60%, respectively. In our preparation, pre-incubation with DBE or MLA prevented the inhibition of DAT induced by MDMA but not that induced by AMPH. The lack of MLA effect on AMPH-induced DA inhibition correlates with the lack of NOS involvement in AMPH-induced ROS production described by us (see Section V) and with results of AMPH interaction with α7 nAChR (see Section XI.B).

Finally, a functional similarity can be drawn between processes that modulates amphetamines-induced monoamine uptake inhibition and ROS production, meaning that in the MDMA and METH effects, a strong connection must exist between DAT functionality and their oxidative effect leading to neuronal injury.

X. Nicotinic Receptors and the Cognitive Impairment Induced by Amphetamine Derivatives

Although chronic use of METH leads to long-lasting cognitive impairment in humans (Simon *et al.*, 2002, 2004), there are few papers about an animal model that reflects METH-induced alterations of memory processes (Nagai *et al.*, 2007; Schroder *et al.*, 2003).

A deficit in learning and memory is a highly consistent finding of cross-sectional research with experienced adult ecstasy users (Morgan *et al.*, 2002; Rodgers *et al.*, 2006) and forebrain structures that are essential for cognitive function, such as the hippocampus and frontal cortex are highly sensitive to MDMA (Green *et al.*, 2003).

Self-reported ecstasy consumption has been documented as early as at an age of 10 (De Micheli and Formigoni, 2004) and its regular use leads to cognitive and psychiatric disorders, enhanced impulsivity, impaired decision-making capability, anxiety disorder, and depression (Gerra *et al.*, 2005; Lieb *et al.*, 2002). However, results from clinical investigations have been inconsistent and this may be partly due to methodological difficulties that hamper research in this area. The most fundamental of these is the concomitant use of other illicit drugs. In particular, cannabis intake is prevalent among ecstasy users and the combination produces a more deleterious effect on memory than either drug alone (Gouzoulis-Mayfrank and Daumann, 2006).

In addition, McCann *et al.* (1999) found that MDMA users were impaired on measures of sustained or complex attention and incidental learning. Other cognitive functions such as planning and reaction time also appear to be impaired with continued MDMA use (Verkes *et al.*, 2001; Zakzanis and Young, 2001).

MDMA induced cognitive impairment at a dose of 15 mg/kg (Camarasa *et al.*, 2008). Using the interspecies scaling formula, this is equivalent to a dose of 188 mg in a 70 kg human, which may actually be lower than the doses used by chronic abusers (Kouimtsidis *et al.*, 2006; Scholey *et al.*, 2004). Furthermore, it has been shown that administering 10–15 mg/kg to rats creates similar plasma concentrations of MDMA to those of a human who consumes a 150 mg tablet (Kouimtsidis *et al.*, 2006). However, as we wanted to simulate chronic human abuse, maintained MDMA plasma levels had to be reached for a significant period. This can be obtained by using a multiple dose administration schedule (b.i.d. for 4 consecutive days)

Using this schedule of treatment in our laboratory, we have performed some experiments recently to demonstrate a specific effect of MDMA treatment on the object recognition memory test and the Morris water maze in Long Evans rats (Camarasa *et al.*, 2008). The object recognition task is based on the spontaneous tendency of rodents to explore a novel object. The test was performed according to the original description of Ennaceur and Delacour (1988). In the first trial (familiarization), rats were exposed to two identical objects and were allowed to explore for 5 min. In the second trial (testing), one of the objects was replaced by a new, novel object, and animals were also allowed to explore the novel and familiar objects for 5 min. This task is closely analogous to recognition tests that are widely used in humans to test memory and to characterize amnesic syndromes by providing an accurate index of the overall severity of declarative memory impairment (Dix and Aggleton, 1999). Bisagno *et al.* (2002) reported that

METH also induced short-term memory impairments in an object recognition task. Moreover, Morley *et al.* (2001) also evaluated the effects of amphetamines on object recognition, showing that, in rats, a heavy MDMA exposure over 2 days, may lead to memory impairment 3 months later, possibly through a neurotoxic effect.

As can be seen in Fig. 12, the ability to discriminate between the familiar and the novel object was abolished following MDMA treatment. In these experiments, MEM (at a dose of 5 mg/kg) was administered 20 min prior each dose of MDMA. Animals pretreated with MEM recovered the lack of discrimination that appeared in the MDMA-treated animals. This beneficial effect of MEM was illustrated by the partial recovery of the Discrimination Index value to that of saline-treated animals.

MDMA impaired spatial memory when assessed in the water maze (Morris *et al.*, 1982). In the acquisition phase, rats received training sessions, consisting of eight trials per day, and were tested on 4 consecutive days. Rats were allowed to swim to the hidden platform, and the escape latency was determined. The rats were subjected to a single test session (probe trial) after the last training session (free swimming without platform) to assess their memory of the platform location and different parameters of the rat's performance were analyzed.

Animals treated with MDMA (the same dosing schedule as above) showed impaired learning evidenced by a lack of difference between the escape latency times for trials performed on different days. This impairment was abolished by the pretreatment with MEM.

Previous published results with MEM in memory processes were controversial. In a study with normal rats, Mistzal and Danysz (1995) reported that MEM

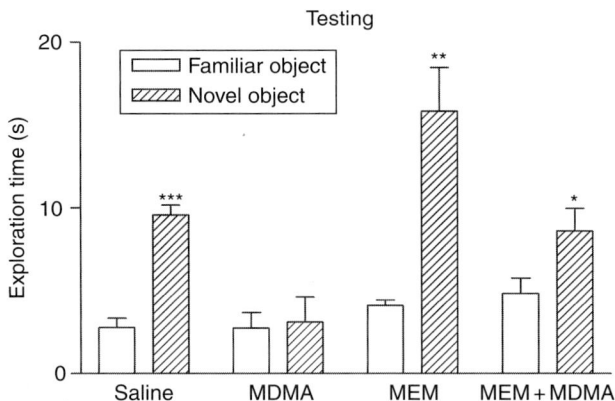

FIG. 12. Comparison of the exploration time obtained in the object recognition task. This figure shows the time spent by rats exploring the old and the new object (Testing Trial).

at high doses (about 20 mg/kg) impaired aversive memory, while at low doses (5 mg/kg), it had no effect on memory performance in a two-choice avoidance task. Conversely, other authors have demonstrated the beneficial effect of MEM in rats displaying memory deficits (Zajaczkowski *et al.*, 1996).

On the other hand, characteristic effects of NMDA receptor antagonists, such as MK-801, in rodents are impairment in water maze performance and increased thigmotaxis (Cain and Saucier, 1996). In our experiments, MEM abolished the memory impairment (see Fig. 13) and also thigmotaxis induced by MDMA, discarding a mechanism of action of MEM based in an antagonism on NMDA receptors.

The neurochemistry involved in solving the Morris water maze can be very complex and is often debated. While acetylcholine is known to be one of the primary neurochemical systems involved in learning and memory, other pathways such as the serotonergic system are also involved (Ricaurte *et al.*, 1993). Moreover, the pathophysiology of memory impairment (mainly Alzheimer's disease) is very complex and involves numerous neurotransmitter systems in the brain. Although the aetiology remains unknown, one of the most significant hallmarks is the early development of cholinergic deficits in the brain. The degree of losses of cholinergic neurons and nAChRs (particularly $\alpha7$ and $\alpha4,\beta2$) correlate well with the magnitude of cognitive decline in Alzheimer's disease patients as the disease progresses from mild to moderate (Nordberg, 2001). Although it has been shown that $\alpha7$ nAChR agonists enhance cognition and are neuroprotective, it is not clear whether these effects are the result of direct receptor activation or activation-induced receptor desensitization. This could explain why $\alpha7$ nAChR agonists and antagonists are neuroprotectives and also why MEM inhibits the MDMA-induced memory impairment while being an $\alpha7$ nAChR antagonist.

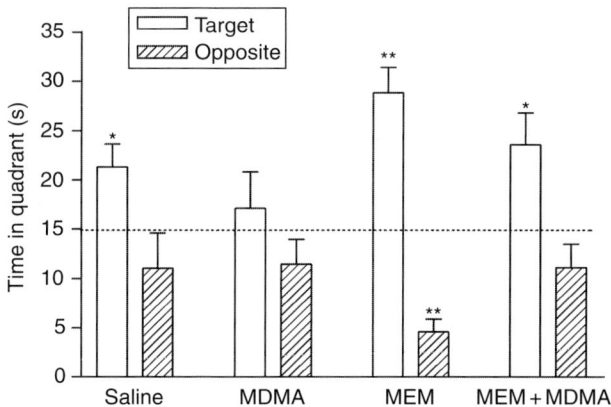

FIG. 13. Comparison of the exploration time obtained in the object recognition task. This figure shows the time spent by rats exploring the old and the new object (Testing Trial).

There have been many studies that demonstrate that activation of $\alpha7$ nAChRs may be involved in processes that contribute to Alzheimer's disease pathophysiology. For example, in rodent hippocampal preparations, β-amyloid peptides impair hippocampal signaling in an $\alpha7$ nAChR-dependent manner and are thought to result from increased Ca^{2+} influx and chronic activation of signal transduction pathways (Dineley *et al.*, 2002). Our group also has demonstrated that, in rats, the serotonergic injury induced by MDMA is $\alpha7$ nicotinic receptor dependent. If memory deficits induced by MDMA are the consequence of its neurotoxicity, it seems that the preventive effect of MEM could be due to its efficacy in preventing MDMA-induced neurotoxicity.

Therefore, MEM by preventing MDMA-induced neuronal injury, contributes to ameliorate cognitive impairment produced by MDMA, and both GLU and $\alpha7$ receptor antagonism are responsible of this beneficial effect. Although the neurochemistry underlying the above-cited effects of MEM must be investigated further, the preventive effect on MDMA-induced impairment suggests a new therapeutic approach to the treatment of long-term adverse effects of amphetamine derivatives also showing promising results in the treatment of amphetamine addiction (Levi and Borne, 2002).

XI. Amphetamine Derivative Interaction with Nicotinic Receptors

A. NICOTINIC ACETYLCHOLINE RECEPTORS

The fact that $\alpha7$ nAChR antagonists protected against METH/MDMA-induced neurotoxicity pointed us to assess the interaction of these drugs with nAChR. Former publications had suggested an interaction between amphetamines and peripheral nAChR: Skau and Gerald (1978) described how AMPH inhibited α-bungarotoxin binding at the neuromuscular junction in mice; Liu *et al.* (2003) reported that AMPH acts as an agonist on nicotinic receptors (probably of $\alpha7$ subtype) in bovine chromaffin cells, inducing catecholamine release; and Klingler *et al.* (2005) more recently identified nAChR as one of the physiological targets of MDMA in the neuromuscular junction. However, there were no reports concerning central nAChRs and amphetamines.

In the nervous system, AChRs participate in several functions including cognition, locomotion, and analgesia (Champtiaux and Changeux, 2002; Drago *et al.*, 2003; Picciotto *et al.*, 2000,2001) and nicotine addiction (Dani and Harris, 2005). nAChRs in the CNS are mainly located presynaptically and modulate the release of almost all neurotransmitters, but they have a postsynaptic localization in some areas, where they mediate fast synaptic transmission (Dajas-Bailador and Wonnacott, 2004; Gotti and Clementi, 2004; Jensen *et al.*, 2005).

Neuronal nAChRs are a heterogeneous class of ionotropic receptor subtypes formed by the association of five subunits encoded by different genes. The genes that have been cloned so far are divided in two subfamilies of nine α ($\alpha2$–$\alpha10$) and three β ($\beta2$–$\beta4$) subunits and are expressed in the nervous system, cochlea, and a number of non-neuronal tissues (Gotti and Clementi, 2004; Hogg et al., 2003). The pentameric assembly of nAChR subunits can be homomeric or heteromeric, forming a central ion pore and have different structural, functional, and pharmacological properties (Le et al., 2002). Two main classes have been identified: the α-bungarotoxin (α-BgTx)-sensitive receptors, which are made up of the $\alpha7$, $\alpha8$, $\alpha9$, and/or $\alpha10$ subunits and can form homomeric or heteromeric receptors, and α-BgTx-insensitive receptors that consist of α 2–6 and (β 2–4) subunits, and bind nicotine and many other nicotinic agonists with high affinity but not α-BgTx (Gotti and Clementi, 2004).

nAChRs are permeable to the monovalent Na^+ and K^+ ions, and also to Ca^{2+}. The cation permeability of the subtypes depends on their subunit composition and has been reviewed (Fucile, 2004). Thus, heteromeric nAChRs comprising α- and β-subunits have in general a low permeability for Ca^{2+} (fractional current of 2–5%). By contrast, homomeric $\alpha7$ subtypes have the highest fractional Ca^{2+} current, which ranges from 6% to 12% depending on the species. Moreover, the fractional Ca^{2+} current through human $\alpha7$ nAChR is the highest reported for homomeric ligand-gated receptors, and matches that of heteromeric NMDA receptors (Jatzke et al., 2002). Also, depolarization induced by entry of Na^+ or Ca^{2+} could open voltage-gated calcium channels and enhance Ca^{2+} influx. These two mechanisms can be physiologically complementary and play important roles in cell signaling by activating different downstream intracellular pathways in neurones (reviewed by Dajas-Bailador and Wonnacott, 2004) such as PKC and nNOS, which have similarly been implicated in the neurotoxicity of amphetamines (Deng and Cadet, 1999; Kramer et al., 1997).

In addition to the ACh binding sites, nAChRs have a number of allosteric binding sites. Various structurally different ligands bind to these sites and modify nAChR function (reviewed by Jensen et al., 2005).

nAChRs exhibit a particular regulation after agonist (i.e., nicotine) treatment, contrarily to what one would expect after continuous stimulation, a downregulation, these receptors develop an increase in ligand binding, that is, upregulation (Flores et al., 1992; Marks et al., 1983). The regulation of nAChRs seems to be complex and a number of work have been focused on the study of the mechanisms involved, which have been recently reviewed (Gaimarri et al., 2007). The explanation for this particular effect of nicotine may be due to the fact that nAChRs, under the constant presence of an agonist, get desensitized (Picciotto, 2003), and it would be a response to restore necessary nicotinic transmission.

Chronic nicotine treatment does not increase mRNA levels of nAChR subunits in different experimental models (Madhok et al., 1995; Marks et al., 1992;

Pauly *et al.*, 1996), which clearly suggests that posttranscriptional mechanisms are involved in upregulation. Most studies have been carried out on the $\alpha4\beta2$ subtype (reviewed by Gaimarri *et al.*, 2007) which, in transfected cells and neurones, is localized on the cell surface and in intracellular pools, where it colocalizes with a number of markers of the endoplasmic reticulum (ER). The synthesized subunits undergo several posttranslational modifications (i.e., glycosilations) and assemble into heteropentameric receptors. Only the correctly assembled receptors exit from the ER to the Golgi (Ren *et al.*, 2005). In the Golgi apparatus, the sugars are trimmed and processed into complex carbohydrates before reaching the cell surface. Finally, the surface receptors are internalized to the endosomal compartment and eventually targeted to the lysosomal compartment to be degraded. The steps previous to surface expression are submitted to a tight quality control that warrants the performance of the new receptors, and many new receptors are degraded/recycled before reaching the cell membrane.

After studying these mechanisms at different levels, a number of models have been proposed to explain the upregulation of $\alpha4\beta2$ receptors, which include increase in receptor transport to the plasma membrane (Harkness and Millar, 2002); a decrease in surface receptor internalization due to nicotine-induced modifications (Peng *et al.*, 1997); isomerization into a more easily activated high-affinity conformation (Buisson and Bertrand, 2001; Vallejo *et al.*, 2005); an increase in subunit assembly and/or decrease in receptor turnover (Darsow *et al.*, 2005; Kuryatov *et al.*, 2005); or a role of nicotine as a molecular chaperone acting as a maturational enhancer toward high-affinity receptors (Sallette *et al.*, 2005). Moreover, not only nicotinic agonists induce nAChR upregulation, but also antagonists have been reported to have such an effect (Gopalakrishnan *et al.*, 1997; Peng *et al.*, 1994). This means that this phenomenon is not related to the efficacy of the ligand to activate the receptor, but that is only binding to either mature or immature forms which stabilizes a high-affinity state, promotes receptor maturation, or prevents degradation.

Then two questions arose about amphetamine derivatives and nAChRs: (i) do amphetamines have affinity for nAChRs? and (ii) if they have, do they induce changes in their density upon prolonged treatment?

B. AFFINITY OF AMPHETAMINE DERIVATIVES FOR nAChRs

When we envisaged the study of the interaction of amphetamine derivatives with nAChRs we chose, in addition to rodent brain membranes, a model that allowed us to study the regulation and test pharmacological modulation as well. For this reason, we chose NGF-differentiated PC 12 cells. This cell line exhibits a phenotype of dopaminergic neurones and has been used by other authors to assess the neurotoxicity of amphetamines (Imam *et al.*, 2002; Wei *et al.*, 1997), and also express nAChRs, including the $\alpha 7$ subtype (Blumenthal *et al.*, 1997; Henderson *et al.*, 1994), and has been used as an *in vitro* model for the upregulation of nAChRs (Jonnala and Buccafusco, 2001; Madhok *et al.*, 1995; Takahashi *et al.*, 1999).

[^{3}H]MLA was used to label $\alpha 7$ receptors, as it has been described to bind α-BgTx sites (Davies *et al.*, 1999) and $\alpha 7$ nAChR are the predominant homomeric type in CNS. On the other hand, the specific radiolabeled agonist [^{3}H] epibatidine ([^{3}H]EB) was used to label the heteromeric type, mainly $\alpha 4 \beta 2$ nAChRs.

In membranes from NGF-differentiated PC 12 cells and in mouse brain, both MDMA and METH displaced competitively the binding of both radioligands, indicating that they can directly interact with nAChRs (Garcia-Rates *et al.*, 2007). MDMA displayed higher affinity than METH for both subtypes of receptors. The resulting K_i values fell in the micromolar range, some being in the low micromolar range (in mouse brain, those of MDMA on both subtypes of nAChR: about 0.7 μM against [^{3}H]EB and 35 μM against [^{3}H]MLA; or that of METH against [^{3}H]EB: 25 μM) and others in the high micromolar range (i.e., METH effects on [^{3}H]MLA binding: about 300 μM in mouse brain). In all the cases, MDMA displayed higher affinity than METH and it was higher for heteromeric than for $\alpha 7$ subtype. As for METH and $\alpha 7$ receptors are concerned, it must be pointed that in mouse brain (unpublished observation) and in rat brain (Chipana *et al.*, 2008c), the displacement curve could be fitted to a two-site model, with the site of higher affinity having a K_i value also in the low micromolar range. In fact, binding to an ion channel can be complex and depending on where the binding site is localized (extracellular domain or within the channel) and the state of activation of the receptor. In addition, the bound radioligand could occlude the access of the competing drug to a site within the ion pore.

Such differences suggest also a different physiological relevance among the effects of METH and MDMA on both subtypes of nAChRs, which would highly depend on the doses administered and resulting brain concentrations. Low micromolar concentrations are achievable in the brain after administration of these drugs (Chu *et al.*, 1996; Johnson *et al.*, 2004; Kitaichi *et al.*, 2003). Special attention must be paid to the affinity of MDMA for heteromeric receptors, which is practically the same that the K_i displayed by MDMA for the 5-HT receptor (0.61 μM, Battaglia *et al.*, 1988a), its main physiological target. Therefore, such an interaction at recreational doses is certainly possible. In addition, it must be taken into account that several factors could increase the amphetamine-derivative concentration in a real noncontrolled ambient (route of administration, association with other drugs, changes in blood–brain barrier permeability, etc.).

By contrast, when the affinity of AMPH was assessed in mouse brain membranes, it was found that its affinity toward $\alpha 7$ nAChRs was out of physiological relevance ($K_i > 870$ μM) (Chipana *et al.*, 2008b), whereas that toward heteromeric receptors was moderate (K_i about 23 μM). This lack of affinity toward α7 nAChRs would explain why AMPH-mediated ROS production is not mediated by nNOS activation and NO, as there would not be an α7-mediated increase in Ca^{2+} necessary to activate nNOS.

The pharmacodynamic consequences of the interaction of METH and MDMA with nAChRs remain to be determined, but according to our previous work, it could be expected that an agonistic or a positive allosteric effect would enhance nicotinic neurotransmission.

C. AMPHETAMINE DERIVATIVES INDUCE nAChR UPREGULATION

Pre-incubation of differentiated PC 12 cells with MDMA or METH induces nAChR upregulation in a concentration- and time-dependent manner (Fig. 14) (Garcia-Rates *et al.*, 2007), that is, they increase binding of [^{3}H]MLA and [^{3}H]EB. This increase was measured after drug-washout, ruling out concomitant allosteric modulations of the binding sites. In general, MDMA induced better increments than METH and these increments were higher in heteromeric than in α7 receptors. Also AMPH induced increases in [^{3}H]EB binding in this cell line of the same magnitude than MDMA (additional unpublished result).

A priori, one would thing that MDMA and METH induce nAChR upregulation by the same mechanism that nicotine. In fact, nicotine has been proved to cross plasma membrane to interact with intracellular forms of nAChR, and METH and MDMA can access inside the cell through the monoamine transporters localized in the plasma membrane.

Saturation binding experiments after MDMA or METH treatment demonstrated that B_{max} for [^{3}H]MLA (an antagonist) binding sites increased without affecting the affinity (K_D) and that in the case of [^{3}H]EB (an agonist), two binding site populations (high and low affinity states) could be detected, increasing the B_{max} of the high-affinity component after MDMA or METH (Garcia-Rates *et al.*, 2007). Agonists bind, at lower concentrations, mainly the receptors in high-affinity state, while antagonists label all the receptor population. These results are in agreement with those of Vallejo *et al.*, (2005), by which nicotine increases/stabilizes the proportion of α4β2 receptors in a high-affinity state and, as for α7 is concerned, it could be deduced that MDMA/METH, by binding to precursor or immature forms of these receptors, would enhance or induce the formation of receptors capable of binding [^{3}H]MLA.

Some additional experiments were performed to clarify the mechanisms involved in MDMA/METH-induced nAChR upregulation (Garcia-Rates *et al.*, 2007). Cycloheximide (an inhibitor of protein synthesis) inhibited such upregulation, indicating that it depends on *de novo* protein synthesis. Similar results were reported for α4β2 (Gopalakrishnan *et al.*, 1997) and α7 nAChR (Kawai and Berg,

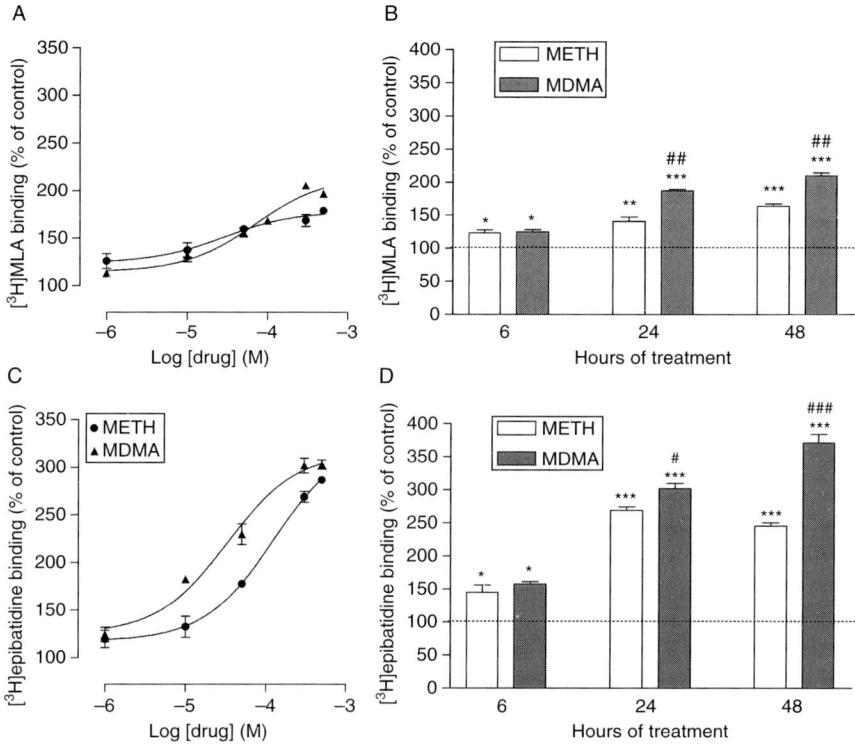

FIG. 14. Increases in nAChR binding densities in NGF-differentiated PC 12 cells as a function of drug concentration (A and C) and incubation time (B and D) (Garcia-Ratés *et al.*, 2007) induced by treatment with MDMA or METH. Homomeric $\alpha 7$ receptors were labeled with [^{3}H]MLA (A and B), while heteromeric receptors were labeled with [^{3}H]epibatidine (C and D). In panels A and B, all the increases are statistically significant versus untreated cells (100%).

2001). These, together by the fact that there is no change in transcriptional activity during nAChR upregulation, point to the involvement of posttranslational mechanisms. Brefeldin A, a drug that inhibits intracellular trafficking of proteins from the Golgi apparatus to the cell surface, although dramatically reduced global surface receptor expression, did no block upregulation induced by MDMA/METH. This indicates that receptor trafficking is not crucial for nAChR upregulation and therefore such regulation starts at early steps in receptor maturation.

Additionally, we demonstrated the participation of cyclophilin A, a chaperone that participates in the maturation of $\alpha 7$ nAChR (Schroeder *et al.*, 2003), and protein tyrosine kinases in MDMA/METH induced upregulation of this subtype of receptor. Another differential factor was the involvement of PKA in the upregulation of heteromeric receptors induced by MDMA/METH and the CTRL by PKC, whose inhibition potentiates such upregulation (Garcia-Rates *et al.*, 2007).

All these results confirm that regulation of nAChRs is a very complex process, with differences between receptor subtypes, host cell type and variations along time that has not been fully elucidated to date.

Further investigation on this interaction is warranted in order to clarify some aspects derived from these findings. Preliminary unpublished results from our group demonstrate that MDMA and METH can induce Ca^{2+} influx in PC 12 cells and that, when taken together, MDMA potentiates the upregulation induced by nicotine.

The direct interaction between MDMA/METH and nAChR and their nicotine-like effects are more likely to occur after high doses of drug, when brain concentrations are close to the K_i, but such an effect is really feasible to occur in the low micromolar range. Such an interaction might account for some of the effects of these drugs, that is, desensitization or stabilization of receptors, or increased neurotoxicity. The direct interaction of these drugs with nAChRs corroborates the hypothesis suggested in previous works of our group (Chipana *et al.*, 2006, 2008c; Escubedo *et al.*, 2005; Pubill *et al.*, 2005) by which binding of amphetamines to nicotinic receptors would enhance their activation and this would in turn activate or potentiate calcium-dependent pathways involved in their neurotoxicity. For this reason, blockade of $\alpha 7$ nAChR was neuroprotective against METH/MDMA. Moreover, the receptor upregulation induced by these drugs could potentiate processes of addiction, dependence and vulnerability to nicotine.

Moreover, taking into account that $\alpha 7$ nAChR are related with cognitive function (Young *et al.*, 2007) and, in addition to $\alpha 4 \beta 2$, are implicated in psychiatric pathologies such as schizophrenia (Ripoll *et al.*, 2004), the interaction of MDMA/METH with these receptors could explain the psychiatric disorders that appear after chronic consumption of these drugs.

References

Ali, S. F., and Itzhak, Y. (1998). Effects of 7-nitroindazole, an NOS inhibitor on methamphetamine-induced dopaminergic and serotonergic neurotoxicity in mice. *Ann. NY Acad. Sci.* **844,** 122–130.

Ali, S. F., Newport, G. D., and Slikker, W. (1996). Methamphetamine-induced dopaminergic toxicity in mice. Role of environmental temperature and pharmacological agents. *Ann. NY Acad. Sci.* **801,** 187–198.

Aracava, Y., Pereira, E. F., Maelicke, A., and Albuquerque, E. X. (2005). Memantine blocks alpha7* nicotinic acetylcholine receptors more potently than *n*-Methyl-d-aspartate receptors in rat hippocampal neurons. *J. Pharmacol. Exp. Ther.* **312,** 1195–1205.

Aznar, S., Kostova, V., Christiansen, S. H., and Knudsen, G. M. (2005). Alpha 7 nicotinic receptor subunit is present on serotonin neurons projecting to hippocampus and septum. *Synapse* **55,** 196–200.

Battaglia, G., Brooks, B. P., Kulsakdinun, C., and De Souza, E. B. (1988a). Pharmacologic profile of MDMA (3,4-methylenedioxymethamphetamine) at various brain recognition sites. *Eur. J. Pharmacol.* **149,** 159–163.

Battaglia, G., Yeh, S. Y., and De Souza, E. B. (1988b). MDMA-induced neurotoxicity: Parameters of degeneration and recovery of brain serotonin neurons. *Pharmacol. Biochem. Behav.* **29,** 269–274.

Bennett, B. A., Hyde, C. E., Pecora, J. R., and Clodfelter, J. E. (1993). Differing neurotoxic potencies of methamphetamine, mazindol, and cocaine in mesencephalic cultures. *J. Neurochem.* **60,** 1444–1452.

Berman, S. B., and Hastings, T. G. (1999). Dopamine oxidation alters mitochondrial respiration and induces permeability transition in brain mitochondria: Implications for Parkinson's disease. *J. Neurochem.* **73,** 1127–1137.

Bisagno, V., Ferguson, D., and Luine, V. N. (2002). Short toxic methamphetamine schedule impairs object recognition task in male rats. *Brain. Res.* **940,** 95–101.

Blumenthal, E. M., Conroy, W. G., Romano, S. J., Kassner, P. D., and Berg, D. K. (1997). Detection of functional nicotinic receptors blocked by alpha-bungarotoxin on pc12 cells and dependence of their expression on post-translational events. *J. Neurosci.* **17,** 6094–6104.

Bowyer, J. F., Davies, D. L., Schmued, L., Broening, H. W., Newport, G. D., Slikker, W., and Holson, R. R. (1994). Further studies of the role of hyperthermia in methamphetamine neurotoxicity. *J. Pharmacol. Exp. Ther.* **268,** 1571–1580.

Buisson, B., and Bertrand, D. (2001). Chronic exposure to nicotine upregulates the human (alpha)4 ((beta)2 nicotinic acetylcholine receptor function. *J. Neurosci.* **21,** 1819–1829.

Cain, D. P., and Saucier, D. (1996). The Neuroscience of spatial navigation: Focus on behavior yields advances. *Rev. Neurosci.* **7,** 215–231.

Camarasa, J., Marimon, J. M., Rodrigo, T., Escubedo, E., and Pubill, D. (2008). Memantine prevents the cognitive impairment induced by 3,4-methylenedioxymethamphetamine in rats.. *Eur. J. Pharmacol.* **589,** 132–139.

Cao, B. J., and Reith, M. E. (2002). Nitric oxide inhibits uptake of dopamine and *N*-Methyl-4-phenylpyridinium (MPP+) but not release of MPP+ in rat C6 glioma cells expressing human dopamine transporter. *Br. J. Pharmacol.* **137,** 1155–1162.

Carvelli, L., Moron, J. A., Kahlig, K. M., Ferrer, J. V., Sen, N., Lechleiter, J. D., Leeb-Lundberg, L. M., Merrill, G., Lafer, E. M., Ballou, L. M., Shippenberg, T. S., Javitch, J. A., *et al.* (2002). PI 3-kinase regulation of dopamine uptake. *J. Neurochem.* **81,** 859–869.

Champtiaux, N., and Changeux, J. P. (2002). Knock-out and knock-in mice to investigate the role of nicotinic receptors in the central nervous system. *Curr. Drug Targets CNS Neurol. Disord.* **1,** 319–330.

Chipana, C., Camarasa, J., Pubill, D., and Escubedo, E. (2006). Protection against MDMA-induced dopaminergic neurotoxicity in mice by methyllycaconitine: Involvement of nicotinic receptors. *Neuropharmacology* **51,** 885–895.

Chipana, C., Camarasa, J., Pubill, D., and Escubedo, E. (2008a). Memantine prevents MDMA-induced neurotoxicity. *Neurotoxicology* **29,** 179–183.

Chipana, C., Garcia-Rates, S., Camarasa, J., Pubill, D., and Escubedo, E. (2008b). Different oxidative profile and nicotinic receptor interaction of amphetamine and 3,4-methylenedioxy-methamphetamine. *Neurochem. Int.* **52,** 401–410.

Chipana, C., Torres, I., Camarasa, J., Pubill, D., and Escubedo, E. (2008c). Memantine protects against amphetamine derivatives-induced neurotoxic damage in rodents. *Neuropharmacology* **54,** 1254–1263.

Chiueh, C. C., Miyake, H., and Peng, M. T. (1993). Role of dopamine autoxidation, hydroxyl radical generation, and calcium overload in underlying mechanisms involved in MPTP-induced Parkinsonism. *Adv. Neurol.* **60,** 251–258.

Chu, T., Kumagai, Y., DiStefano, E. W., and Cho, A. K. (1996). Disposition of methylenedioxymethamphetamine and three metabolites in the brains of different rat strains and their possible roles in acute serotonin depletion. *Biochem. Pharmacol.* **51,** 789–796.

Colado, M. I., Camarero, J., Mechan, A. O., Sanchez, V., Esteban, B., Elliott, J. M., and Green, A. R. (2001). A study of the mechanisms involved in the neurotoxic action of 3,4-methylenedioxymethamphetamine (MDMA, ecstasy') on dopamine neurones in mouse brain. *Br. J. Pharmacol.* **134,** 1711–1723.

Dajas-Bailador, F., and Wonnacott, S. (2004). Nicotinic acetylcholine receptors and the regulation of neuronal signalling. *Trends. Pharmacol. Sci.* **25,** 317–324.

Dani, J. A., and Harris, R. A. (2005). Nicotine addiction and comorbidity with alcohol abuse and mental illness. *Nat. Neurosci.* **8,** 1465–1470.

Darsow, T., Booker, T. K., Pina-Crespo, J. C., and Heinemann, S. F. (2005). Exocytic trafficking is required for nicotine-induced up-regulation of alpha 4 beta 2 nicotinic acetylcholine receptors. *J. Biol. Chem.* **280,** 18311–18320.

Darvesh, A. S., Yamamoto, B. K., and Gudelsky, G. A. (2005). Evidence for the involvement of nitric oxide in 3,4-methylenedioxymethamphetamine-induced serotonin depletion in the rat brain. *J. Pharmacol. Exp. Ther.* **312,** 694–701.

Davidson, C., Gow, A. J., Lee, T. H., and Ellinwood, E. H. (2001). Methamphetamine neurotoxicity: Necrotic and apoptotic mechanisms and relevance to human abuse and treatment. *Brain. Res. Rev.* **36,** 1–22.

Davies, A. R., Hardick, D. J., Blagbrough, I. S., Potter, B. V., Wolstenholme, A. J., and Wonnacott, S. (1999). Characterisation of the binding of [3H]methyllycaconitine: A new radioligand for labelling alpha 7-type neuronal nicotinic acetylcholine receptors. *Neuropharmacology* **38,** 679–690.

De Micheli, D., and Formigoni, M. L. (2004). Drug use by brazilian students: Associations with family, psychosocial, health, demographic and behavioral characteristics. *Addiction* **99,** 570–578.

Demiryurek, A. T., Cakici, I., and Kanzik, I. (1998). Peroxynitrite: A putative cytotoxin. *Pharmacol. Toxicol.* **82,** 113–117.

Deng, X., and Cadet, J. L. (1999). Methamphetamine administration causes overexpression of nNOS in the mouse striatum. *Brain Res.* **851,** 254–257.

Dineley, K. T., Bell, K. A., Bui, D., and Sweatt, J. D. (2002). Beta-amyloid peptide activates alpha 7 nicotinic acetylcholine receptors expressed in xenopus oocytes. *J. Biol. Chem.* **277,** 25056–25061.

Dix, S. L., and Aggleton, J. P. (1999). Extending the spontaneous preference test of recognition: Evidence of object-location and object-context recognition. *Behav. Brain Res.* **99,** 191–200.

Drago, J., McColl, C. D., Horne, M. K., Finkelstein, D. I., and Ross, S. A. (2003). Neuronal nicotinic receptors: Insights gained from gene knockout and knockin mutant mice. *Cell Mol. Life Sci.* **60,** 1267–1280.

Drew, A. E., and Werling, L. L. (2001). Protein kinase C regulation of dopamine transporter initiated by nicotinic receptor activation in slices of rat prefrontal cortex. *J. Neurochem.* **77,** 839–848.

Ellison, G., Eison, M. S., Huberman, H. S., and Daniel, F. (1978). Long-term changes in dopaminergic innervation of caudate nucleus after continuous amphetamine administration. *Science* **201,** 276–278.

Ennaceur, A., and Delacour, J. (1988). A new one-trial test for neurobiological studies of memory in rats. 1: Behavioral data. *Behav. Brain Res.* **31,** 47–59.

Escubedo, E., Guitart, L., Sureda, F. X., Jimenez, A., Pubill, D., Pallas, M., Camins, A., and Camarasa, J. (1998). Microgliosis and down-regulation of adenosine transporter induced by methamphetamine in rats. *Brain. Res.* **814,** 120–126.

Escubedo, E., Chipana, C., Perez-Sanchez, M., Camarasa, J., and Pubill, D. (2005). Methyllycaconitine prevents methamphetamine-induced effects in mouse striatum: Involvement of alpha7 nicotinic receptors. *J. Pharmacol. Exp. Ther.* **315,** 658–667.

Falk, E. M., Cook, V. J., Nichols, D. E., and Sprague, J. E. (2002). An antisense oligonucleotide targeted at MAO-B attenuates rat striatal serotonergic neurotoxicity induced by MDMA. *Pharmacol. Biochem. Behav.* **72,** 617–622.

Flores, C. M., Rogers, S. W., Pabreza, L. A., Wolfe, B. B., and Kellar, K. J. (1992). A subtype of nicotinic cholinergic receptor in rat brain is composed of alpha 4 and beta 2 subunits and is up-regulated by chronic nicotine treatment. *Mol. Pharmacol.* **41,** 31–37.

Foster, J. D., Pananusorn, B., and Vaughan, R. A. (2002). Dopamine transporters are phosphorylated on N-terminal serines in rat striatum. *J. Biol Chem.* **277,** 25178–25186.

Fucile, S. (2004). Ca^{2+} permeability of nicotinic acetylcholine receptors. *Cell Calcium* **35,** 1–8.

Fukami, G., Hashimoto, K., Koike, K., Okamura, N., Shimizu, E., and Iyo, M. (2004). Effect of antioxidant *N*-Acetyl-L-cysteine on behavioral changes and neurotoxicity in rats after administration of methamphetamine. *Brain. Res.* **1016,** 90–95.

Gaimarri, A., Moretti, M., Riganti, L., Zanardi, A., Clementi, F., and Gotti, C. (2007). Regulation of neuronal nicotinic receptor traffic and expression. *Brain Res. Rev.* **55,** 134–143.

Garcia-Rates, S., Camarasa, J., Escubedo, E., and Pubill, D. (2007). Methamphetamine and 3,4-methylenedioxymethamphetamine interact with central nicotinic receptors and induce their up-regulation. *Toxicol. Appl. Pharmacol.* **223,** 195–205.

Garthwaite, J. (2007). Neuronal nitric oxide synthase and the serotonin transporter get harmonious. *Proc. Natl. Acad. Sci. USA* **104,** 7739–7740.

Garthwaite, J., and Boulton, C. L. (1995). Nitric oxide signaling in the central nervous system. *Ann. Rev. Physiol.* **57,** 683–706.

Gerra, G., Garofano, L., Castaldini, L., Rovetto, F., Zaimovic, A., Moi, G., Bussandri, M., Branchi, B., Brambilla, F., Friso, G., and Donnini, C. (2005). Serotonin transporter promoter polymorphism genotype is associated with temperament, personality traits and illegal drugs use among adolescents. *J. Neural. Transm.* **112,** 1397–1410.

Giambalvo, C. T. (1992a). Protein kinase C and dopamine transport-1. Effects of amphetamine *in vivo. Neuropharmacology* **31,** 1201–1210.

Giambalvo, C. T. (1992b). Protein kinase C and dopamine transport-2. Effects of amphetamine *in vitro. Neuropharmacology* **31,** 1211–1222.

Giambalvo, C. T. (2003). Differential effects of amphetamine transport vs. dopamine reverse transport on particulate PKC activity in striatal synaptoneurosomes. *Synapse* **49,** 125–133.

Giambalvo, C. T. (2004). Mechanisms underlying the effects of amphetamine on particulate PKC activity. *Synapse* **51,** 128–139.

Gibb, J. W., Stone, D. M., Johnson, M., and Hanson, G. R. (1989). Role of dopamine in the neurotoxicity induced by amphetamines and related designer drugs. *NIDA Res. Monograph Series* 161–178.

Gnegy, M. E., Khoshbouei, H., Berg, K. A., Javitch, J. A., Clarke, W. P., Zhang, M., and Galli, A. (2004). Intracellular $Ca2^+$ regulates amphetamine-induced dopamine efflux and currents mediated by the human dopamine transporter. *Mol. Pharmacol.* **66,** 137–143.

Gopalakrishnan, M., Molinari, E. J., and Sullivan, J. P. (1997). Regulation of human alpha4beta2 neuronal nicotinic acetylcholine receptors by cholinergic channel ligands and second messenger pathways. *Mol. Pharmacol.* **52,** 524–534.

Gotti, C., and Clementi, F. (2004). Neuronal nicotinic receptors: From structure to pathology. *Prog. Neurobiol.* **74,** 363–396.

Gouzoulis-Mayfrank, E., and Daumann, J. (2006). The confounding problem of polydrug use in recreational ecstasy/MDMA users: A brief overview. *J. Psychopharmacol.* **20,** 188–193.

Granas, C., Ferrer, J., Loland, C. J., Javitch, J. A., and Gether, U. (2003). N-terminal truncation of the dopamine transporter abolishes phorbol ester- and substance P receptor-stimulated phosphorylation without impairing transporter internalization. *J. Biol. Chem.* **278,** 4990–5000.

Green, A. R., Mechan, A. O., Elliott, J. M., O'Shea, E., and Colado, M. I. (2003). The pharmacology and clinical pharmacology of 3,4-methylenedioxymethamphetamine (MDMA, "ecstasy"). *Pharmacol. Rev.* **55,** 463–508.

Guilarte, T. R., Nihei, M. K., McGlothan, J. L., and Howard, A. S. (2003). Methamphetamine-induced deficits of brain monoaminergic neuronal markers: Distal axotomy or neuronal plasticity. *Neuroscience* **122,** 499–513.

Harkness, P. C., and Millar, N. S. (2002). Changes in conformation and subcellular distribution of alpha4beta2 nicotinic acetylcholine receptors revealed by chronic nicotine treatment and expression of subunit chimeras. *J. Neurosci.* **22,** 10172–10181.

Hastings, T. G. (1995). Enzymatic oxidation of dopamine: The role of prostaglandin h synthase. *J. Neurochem.* **64,** 919–924.

Hastings, T. G., Lewis, D. A., and Zigmond, M. J. (1996). Role of oxidation in the neurotoxic effects of intrastriatal dopamine injections. *Proc. Natl. Acad. Sci. USA* **93,** 1956–1961.

Henderson, L. P., Gdovin, M. J., Liu, C., Gardner, P. D., and Maue, R. A. (1994). Nerve growth factor increases nicotinic ACh receptor gene expression and current density in wild-type and protein kinase A-deficient PC12 cells. *J. Neurosci.* **14,** 1153–1163.

Hogg, R. C., Raggenbass, M., and Bertrand, D. (2003). Nicotinic acetylcholine receptors: From structure to brain function. *Rev. Physiol. Biochem. Pharmacol.* **147,** 1–46.

Hom, D. G., Jiang, D., Hong, E. J., Mo, J. Q., and Andersen, J. K. (1997). Elevated expression of glutathione peroxidase in PC12 cells results in protection against methamphetamine but not MPTP toxicity. *Brain Res. Mol. Brain Res.* **46,** 154–160.

Hotchkiss, A. J., and Gibb, J. W. (1980). Long-term effects of multiple doses of methamphetamine on tryptophan hydroxylase and tyrosine hydroxylase activity in rat brain. *J. Pharmacol. Exp. Ther.* **214,** 257–262.

Imam, S. Z., and Ali, S. F. (2000). Selenium, an antioxidant, attenuates methamphetamine-induced dopaminergic toxicity and peroxynitrite generation. *Brain Res.* **855,** 186–191.

Imam, S. Z., Newport, G. D., Islam, F., Slikker, W., and Ali, S. F. (1999). Selenium, an antioxidant, protects against methamphetamine-induced dopaminergic neurotoxicity. *Brain Res.* **818,** 575–578.

Imam, S. Z., el-Yazal, J., Newport, G. D., Itzhak, Y., Cadet, J. L., Slikker, W. Jr., and Ali, S. F. (2001). Methamphetamine-induced dopaminergic neurotoxicity: Role of peroxynitrite and neuroprotective role of antioxidants and peroxynitrite decomposition catalysts. *Ann. NY Acad. Sci.* **939,** 366–380.

Imam, S. Z., Newport, G. D., Duhart, H. M., Islam, F., Slikker, W., Jr., and Ali, S. F. (2002). Methamphetamine-induced dopaminergic neurotoxicity and production of peroxynitrite are potentiated in nerve growth factor differentiated pheochromocytoma 12 cells. *Ann. NY Acad. Sci.* **965,** 204–213.

Iravani, M. M., Asari, D., Patel, J., Wieczorek, W. J., and Kruk, Z. L. (2000). Direct effects of 3,4-methylenedioxymethamphetamine (MDMA) on serotonin or dopamine release and uptake in the caudate putamen, nucleus accumbens, substantia nigra pars reticulata, and the dorsal raphe nucleus slices. *Synapse* **36,** 275–285.

Itzhak, Y., Gandia, C., Huang, P. L., and Ali, S. F. (1998). Resistance of neuronal nitric oxide synthase-deficient mice to methamphetamine-induced dopaminergic neurotoxicity. *J. Pharmacol. Exp. Ther.* **284,** 1040–1047.

Itzhak, Y., Martin, J. L., and Ail, S. F. (2000). NNOS inhibitors attenuate methamphetamine-induced dopaminergic neurotoxicity but not hyperthermia in mice. *Neuroreport* **11,** 2943–2946.

Jatzke, C., Watanabe, J., and Wollmuth, L. P. (2002). Voltage and concentration dependence of Ca(2+) permeability in recombinant glutamate receptor subtypes. *J. Physiol.* **538**, 25–39.

Jayanthi, S., Ladenheim, B., Andrews, A. M., and Cadet, J. L. (1999). Overexpression of human copper/zinc superoxide dismutase in transgenic mice attenuates oxidative stress caused by methylenedioxymethamphetamine (ecstasy). *Neuroscience* **91**, 1379–1387.

Jekabsone, A., Ivanoviene, L., Brown, G. C., and Borutaite, V. (2003). Nitric oxide and calcium together inactivate mitochondrial complex i and induce cytochrome C release. *J. Mol. Cell. Cardiol.* **35**, 803–809.

Jensen, A. A., Frolund, B., Liljefors, T., and Krogsgaard-Larsen, P. (2005). Neuronal nicotinic acetylcholine receptors: Structural revelations, target identifications, and therapeutic inspirations. *J. Med. Chem.* **48**, 4705–4745.

Jimenez, A., Jorda, E. G., Verdaguer, E., Pubill, D., Sureda, F. X., Canudas, A. M., Escubedo, E., Camarasa, J., Camins, A., and Pallas, M. (2004). Neurotoxicity of amphetamine derivatives is mediated by caspase pathway activation in rat cerebellar granule cells. *Toxicol. Appl. Pharmacol.* **196**, 223–234.

Johnson, E. A., O'Callaghan, J. P., and Miller, D. B. (2004). Brain concentrations of d-MDMA are increased after stress. *Psychopharmacology (Berl)* **173**, 278–286.

Jonnala, R. R., and Buccafusco, J. J. (2001). Relationship between the increased cell surface alpha7 nicotinic receptor expression and neuroprotection induced by several nicotinic receptor agonists. *J. Neurosci. Res.* **66**, 565–572.

Kaiser, S., and Wonnacott, S. (2000). Alpha-bungarotoxin-sensitive nicotinic receptors indirectly modulate [(3)H]dopamine release in rat striatal slices via glutamate release. *Mol. Pharmacol.* **58**, 312–318.

Kantor, L., and Gnegy, M. E. (1998). Protein kinase C inhibitors block amphetamine-mediated dopamine release in rat striatal slices. *J. Pharmacol. Exp. Ther.* **284**, 592–598.

Kawai, H., and Berg, D. K. (2001). Nicotinic acetylcholine receptors containing alpha 7 subunits on rat cortical neurons do not undergo long-lasting inactivation even when up-regulated by chronic nicotine exposure. *J. Neurochem.* **78**, 1367–1378.

Khoshbouei, H., Sen, N., Guptaroy, B., Johnson, L., Lund, D., Gnegy, M. E., Galli, A., and Javitch, J. A. (2004). N-terminal phosphorylation of the dopamine transporter is required for amphetamine-induced efflux. *PLoS Biol.* **2**, E78.

Kitaichi, K., Morishita, Y., Doi, Y., Ueyama, J., Matsushima, M., Zhao, Y. L., Takagi, K., and Hasegawa, T. (2003). Increased plasma concentration and brain penetration of methamphetamine in behaviorally sensitized rats. *Eur. J. Pharmacol.* **464**, 39–48.

Klingler, W., Heffron, J. J., Jurkat-Rott, K., O'sullivan, G., Alt, A., Schlesinger, F., Bufler, J., and Lehmann-Horn, F. (2005). 3,4-methylenedioxymethamphetamine (ecstasy) activates skeletal muscle nicotinic acetylcholine receptors. *J. Pharmacol. Exp. Ther.* **314**, 1267–1273.

Kouimtsidis, C., Schifano, F., Sharp, T., Ford, L., Robinson, J., and Magee, C. (2006). Neurological and psychopathological sequelae associated with a lifetime intake of 40,000 ecstasy tablets. *Psychosomatics* **47**, 86–87.

Kovacs, G. G., Ando, R. D., Adori, C., Kirilly, E., Benedek, A., Palkovits, M., and Bagdy, G. (2007). Single dose of MDMA causes extensive decrement of serotoninergic fibre density without blockage of the fast axonal transport in dark agouti rat brain and spinal cord. *Neuropathol. Appl. Neurobiol.* **33**, 193–203.

Kramer, H. K., Poblete, J. C., and Azmitia, E. C. (1997). Activation of protein kinase C (PKC) by 3,4-methylenedioxymethamphetamine (MDMA) occurs through the stimulation of serotonin receptors and transporter. *Neuropsychopharmacology* **17**, 117–129.

Kuryatov, A., Luo, J., Cooper, J., and Lindstrom, J. (2005). Nicotine acts as a pharmacological chaperone to up-regulate human alpha4beta2 acetylcholine receptors. *Mol. Pharmacol.* **68**, 1839–1851.

Le, N. N., Corringer, P. J., and Changeux, J. P. (2002). The Diversity of subunit composition in nAChRs: Evolutionary origins, physiologic and pharmacologic consequences. *J. Neurobiol.* **53,** 447–456.

Leonardi, E. T., and Azmitia, E. C. (1994). MDMA (ecstasy) inhibition of MAO type A and type B: Comparisons with fenfluramine and fluoxetine (prozac). *Neuropsychopharmacology* **10,** 231–238.

Levi, M. S., and Borne, R. F. (2002). A review of chemical agents in the pharmacotherapy of addiction. *Curr. Med. Chem.* **9,** 1807–1818.

Lieb, R., Schuetz, C. G., Pfister, H., von Sydow, K., and Wittchen, H. (2002). Mental disorders in ecstasy users: A prospective-longitudinal investigation. *Drug Alcohol. Depend.* **68,** 195–207.

Liu, P. S., Liaw, C. T., Lin, M. K., Shin, S. H., Kao, L. S., and Lin, L. F. (2003). Amphetamine enhances Ca^{2+} entry and catecholamine release via nicotinic receptor activation in bovine adrenal chromaffin cells. *Eur. J. Pharmacol.* **460,** 9–17.

Ljungberg, T., and Ungerstedt, U. (1985). A rapid and simple behavioural screening method for simultaneous assessment of limbic and striatal blocking effects of neuroleptic drugs. *Pharmacol. Biochem. Behav.* **23,** 479–485.

Loder, M. K., and Melikian, H. E. (2003). The dopamine transporter constitutively internalizes and recycles in a protein kinase C-regulated manner in stably transfected PC12 cell lines. *J. Biol. Chem.* **278,** 22168–22174.

Lyles, J., and Cadet, J. L. (2003). Methylenedioxymethamphetamine (MDMA, ecstasy) neurotoxicity: Cellular and molecular mechanisms. *Brain. Res. Brain. Res. Rev.* **42,** 155–168.

Madhok, T. C., Matta, S. G., and Sharp, B. M. (1995). Nicotine regulates nicotinic cholinergic receptors and subunit mRNAs in PC 12 cells through protein kinase A. *Brain Res. Mol. Brain Res.* **32,** 143–150.

Maragos, W. F., Jakel, R., Chesnut, D., Pocernich, C. B., Butterfield, D. A., St, C. D., and Cass, W. A. (2000). Methamphetamine toxicity is attenuated in mice that overexpress human manganese superoxide dismutase. *Brain Res.* **878,** 218–222.

Marks, M. J., Pauly, J. R., Gross, S. D., Deneris, E. S., Hermans-Borgmeyer, I., Heinemann, S. F., and Collins, A. C. (1992). Nicotine binding and nicotinic receptor subunit RNA after chronic nicotine treatment. *J. Neurosci.* **12,** 2765–2784.

Marks, M. J., Burch, J. B., and Collins, A. C. (1983). Effects of chronic nicotine infusion on tolerance development and nicotinic receptors. *J. Pharmacol. Exp. Ther.* **226,** 817–825.

McCann, U. D., Mertl, M., Eligulashvili, V., and Ricaurte, G. A. (1999). Cognitive performance in $(+/-)$ 3,4-methylenedioxymethamphetamine (MDMA, "ecstasy") users: A controlled study. *Psychopharmacology (Berl)* **143,** 417–425.

Melikian, H. E., and Buckley, K. M. (1999). Membrane trafficking regulates the activity of the human dopamine transporter. *J. Neurosci.* **19,** 7699–7710.

Misztal, M., and Danysz, W. (1995). Comparison of glutamate antagonists in continuous multiple-trial and single-trial dark avoidance. *Behav. Pharmacol.* **6,** 550–561.

Miyazaki, I., Asanuma, M., Diaz-Corrales, F. J., Fukuda, M., Kitaichi, K., Miyoshi, K., and Ogawa, N. (2006). Methamphetamine-induced dopaminergic neurotoxicity is regulated by quinone-formation-related molecules. *FASEB J.* **20,** 571–573.

Monks, T. J., Jones, D. C., Bai, F., and Lau, S. S. (2004). The role of metabolism in 3,4-(+)-methylenedioxyamphetamine and 3,4-(+)-methylenedioxymethamphetamine (Ecstasy) toxicity. *Ther. Drug. Monit.* **26,** 132–136.

Morgan, M. J., McFie, L., Fleetwood, L. H., and Robinson, J. A. (2002). Ecstasy (MDMA): Are the psychological problems associated with its use reversed by prolonged abstinence? *Psychopharmacology* **159,** 294–303.

Morley, K. C., Gallate, J. E., Hunt, G. E., Mallet, P. E., and McGregor, I. S. (2001). Increased anxiety and impaired memory in rats 3 months after administration of 3,4-methylenedioxymethamphetamine (Ecstasy). *Eur. J. Pharmacol.* **433,** 91–99.

Morley, K. C., Li, K. M., Hunt, G. E., Mallet, P. E., and McGregor, I. S. (2004). cannabinoids prevent the acute hyperthermia and partially protect against the 5-HT depleting effects of MDMA ("Ecstasy") in rats. *Neuropharmacology* **46,** 954.

Morris, R. G., Garrud, P., Rawlins, J. N., and O'Keefe, J. (1982). Place navigation impaired in rats with hippocampal lesions. *Nature* **297,** 681–683.

Myhre, O., and Fonnum, F. (2001). The effect of aliphatic, naphthenic, and aromatic hydrocarbons on production of reactive oxygen species and reactive nitrogen species in rat brain synaptosome fraction: The involvement of calcium, nitric oxide synthase, mitochondria, and phospholipase A. *Biochem. Pharmacol.* **62,** 119–128.

Nagai, T., Takuma, K., Dohniwa, M., Ibi, D., Mizoguchi, H., Kamei, H., Nabeshima, T., and Yamada, K. (2007). Repeated methamphetamine treatment impairs spatial working memory in rats: Reversal by clozapine but not haloperidol. *Psychopharmacology (Berl)* **194,** 21–32.

Nair, S. G., and Gudelsky, G. A. (2006). 3,4-methylenedioxymethamphetamine enhances the release of acetylcholine in the prefrontal cortex and dorsal hippocampus of the rat. *Psychopharmacology (Berl)* **184,** 182–189.

Narita, M., Miyatake, M., Shibasaki, M., Tsuda, M., Koizumi, S., Narita, M., Yajima, Y., Inoue, K., and Suzuki, T. (2005). Long-lasting change in brain dynamics induced by methamphetamine: Enhancement of protein kinase C-dependent astrocytic response and behavioral sensitization. *J. Neurochem.* **93,** 1383–1392.

Nash, J. F., and Yamamoto, B. K. (1992). Methamphetamine neurotoxicity and striatal glutamate release: Comparison to 3,4-methylenedioxymethamphetamine. *Brain. Res.* **581,** 237–243.

Nordberg, A. (2001). Nicotinic receptor abnormalities of Alzheimer's disease: Therapeutic implications. *Biol. Psychiatr.* **49,** 200–210.

O'Callaghan, J. P., and Miller, D. B. (1994). Neurotoxicity profiles of substituted amphetamines in the C57BL/6J mouse. *J. Pharmacol. Exp. Ther.* **270,** 741–751.

Park, S. U., Ferrer, J. V., Javitch, J. A., and Kuhn, D. M. (2002). Peroxynitrite inactivates the human dopamine transporter by modification of cysteine 342: Potential mechanism of neurotoxicity in dopamine neurons. *J. Neurosci.* **22,** 4399–4405.

Parrott, A. C., Lees, A., Garnham, N. J., Jones, M., and Wesnes, K. (1998). Cognitive performance in recreational users of MDMA of "Ecstasy": Evidence for memory deficits. *J. Psychopharmacol.* **12,** 79–83.

Pauly, J. R., Marks, M. J., Robinson, S. F., van de Kamp, J. L., and Collins, A. C. (1996). Chronic nicotine and mecamylamine treatment increase brain nicotinic receptor binding without changing alpha4 Or beta2 mRNA levels. *J. Pharmacol. Exp. Ther.* **278,** 361–369.

Peng, X., Gerzanich, V., Anand, R., Whiting, P. J., and Lindstrom, J. (1994). Nicotine-induced increase in neuronal nicotinic receptors results from a decrease in the rate of receptor turnover. *Mol. Pharmacol.* **46,** 523–530.

Peng, X., Gerzanich, V., Anand, R., Wang, F., and Lindstrom, J. (1997). Chronic nicotine treatment up-regulates alpha3 and alpha7 acetylcholine receptor subtypes expressed by the human neuroblastoma cell line SH-SY5Y. *Mol. Pharmacol.* **51,** 776–784.

Picciotto, M. R. (2003). Nicotine as a modulator of behavior: Beyond the inverted U. *Trends. Pharmacol. Sci.* **24,** 493–499.

Picciotto, M. R., Caldarone, B. J., King, S. L., and Zachariou, V. (2000). Nicotinic receptors in the brain. Links between molecular biology and behavior. *Neuropsychopharmacology* **22,** 451–465.

Picciotto, M. R., Caldarone, B. J., Brunzell, D. H., Zachariou, V., Stevens, T. R., and King, S. L. (2001). Neuronal nicotinic acetylcholine receptor subunit knockout mice: Physiological and behavioral phenotypes and possible clinical implications. *Pharmacol. Ther.* **92,** 89–108.

Pubill, D., Verdaguer, E., Sureda, F. X., Camins, A., Pallas, M., Camarasa, J., and Escubedo, E. (2002). Carnosine prevents methamphetamine-induced gliosis but not dopamine terminal loss in rats. *Eur. J. Pharmacol.* **448,** 165–168.

Pubill, D., Canudas, A. M., Pallas, M., Camins, A., Camarasa, J., and Escubedo, E. (2003). Different glial response to methamphetamine- and methylenedioxymethamphetamine-induced neurotoxicity. *Naunyn Schmiedebergs Arch. Pharmacol.* **367,** 490–499.

Pubill, D., Chipana, C., Camins, A., Pallas, M., Camarasa, J., and Escubedo, E. (2005). Free radical production induced by methamphetamine in rat striatal synaptosomes. *Toxicol. Appl. Pharmacol.* **204,** 57–68.

Quinton, M. S., and Yamamoto, B. K. (2006). Causes and consequences of methamphetamine and mdma toxicity. *AAPS J.* **8,** E337–47.

Ren, K., Puig, V., Papke, R. L., Itoh, Y., Hughes, J. A., and Meyer, E. M. (2005). Multiple calcium channels and kinases mediate alpha7 nicotinic receptor neuroprotection in pc12 cells. *J. Neurochem.* **94,** 926–933.

Reneman, L., Endert, E., de Bruin, K., Lavalaye, J., Feenstra, M. G., de Wolff, F. A., and Booij, J. (2002). The acute and chronic effects of MDMA ("Ecstasy") on cortical 5-HT2A receptors in rat and human brain. *Neuropsychopharmacology* **26,** 387–396.

Ricaurte, G. A., Guillery, R. W., Seiden, L. S., Schuster, C. R., and Moore, R. Y. (1982). Dopamine nerve terminal degeneration produced by high doses of methylamphetamine in the rat brain. *Brain Res.* **235,** 93–103.

Ricaurte, G. A., Seiden, L. S., and Schuster, C. R. (1984). Further evidence that amphetamines produce long-lasting dopamine neurochemical deficits by destroying dopamine nerve fibers. *Brain Res.* **303,** 359–364.

Ricaurte, G. A., Markowska, A. L., Wenk, G. L., Hatzidimitriou, G., Wlos, J., and Olton, D. S. (1993). 3,4-methylenedioxymethamphetamine, serotonin and memory. *J. Pharmacol. Exp. Ther.* **266,** 1097–1105.

Riddle, E. L., Fleckenstein, A. E., and Hanson, G. R. (2006). Mechanisms of methamphetamine-induced dopaminergic neurotoxicity. *AAPS J.* **8,** E413–E418.

Ripoll, N., Bronnec, M., and Bourin, M. (2004). Nicotinic receptors and schizophrenia. *Curr. Med. Res. Opin.* **20,** 1057–1074.

Rodgers, J., Buchanan, T., Pearson, C., Parrott, A. C., Ling, J., Hefferman, T. M., and Scholey, A. B. (2006). Differential experiences of the psychobiological sequelae of ecstasy use: Quantitative and qualitative data from an internet study. *J. Psychopharmacol.* **20,** 437–446.

Sallette, J., Pons, S., Devillers-Thiery, A., Soudant, M., Prado, d.C., Changeux, J. P., and Corringer, P. J. (2005). Nicotine upregulates its own receptors through enhanced intracellular maturation. *Neuron* **46,** 595–607.

Sanchez, V., Zeini, M., Camarero, J., O'Shea, E., Bosca, L., Green, A. R., and Colado, M. I. (2003). The nNOS inhibitor, AR-R17477AR, prevents the loss of NF68 immunoreactivity induced by methamphetamine in the mouse striatum. *J. Neurochem.* **85,** 515–524.

Sandoval, V., Riddle, E. L., Ugarte, Y. V., Hanson, G. R., and Fleckenstein, A. E. (2001). Methamphetamine-induced rapid and reversible changes in dopamine transporter function: An *in vitro* model. *J. Neurosci.* **21,** 1413–1419.

Scholey, A. B., Parrott, A. C., Buchanan, T., Heffernan, T. M., Ling, J., and Rodgers, J. (2004). Increased intensity of ecstasy and polydrug usage in the more experienced recreational Ecstasy/MDMA users: A WWW study. *Addict. Behav.* **29,** 743–752.

Schroder, N., O'Dell, S. J., and Marshall, J. F. (2003). Neurotoxic methamphetamine regimen severely impairs recognition memory in rats. *Synapse* **49,** 89–96.

Schroeder, K. M., Wu, J., Zhao, L., and Lukas, R. J. (2003). Regulation by cycloheximide and lowered temperature of cell-surface alpha7-nicotinic acetylcholine receptor expression on transfected SH-EP1 cells. *J. Neurochem.* **85,** 581–591.

Seiden, L. S., Lew, R., and Malberg, J. E. (2001). Neurotoxicity of methamphetamine and methylenedioxymethamphetamine. *Neurotoxicity Res.* **3,** 101–116.

Simon, N. G., and Mattick, R. P. (2002). The impact of regular ecstasy use on memory function. *Addiction* **97,** 1523–1529.

Simon, S. L., Dacey, J., Glynn, S., Rawson, R., and Ling, W. (2004). The effect of relapse on cognition in abstinent methamphetamine abusers. *J. Subst. Abuse. Treat.* **27,** 59–66.

Simon, S. L., Domier, C. P., Sim, T., Richardson, K., Rawson, R. A., and Ling, W. (2002). Cognitive performance of current methamphetamine and cocaine abusers. *J. Addict. Dis.* **21,** 61–74.

Skau, K. A., and Gerald, M. C. (1978). Inhibition of alpha-bungarotoxin binding to rat and mouse diaphragms by amphetamine and related nonquaternary compounds. *J. Pharmacol. Exp. Ther.* **205,** 69–76.

Sprague, J. E., Everman, S. L., and Nichols, D. E. (1998). An integrated hypothesis for the serotonergic axonal loss induced by 3,4-methylenedioxymethamphetamine. *Neurotoxicology* **19,** 427–442.

Stephans, S. E., and Yamamoto, B. K. (1994). Methamphetamine-induced neurotoxicity: Roles for glutamate and dopamine efflux. *Synapse* **17,** 203–209.

Stone, D. M., Hanson, G. R., and Gibb, J. W. (1987). Differences in the central serotonergic effects of methylenedioxymethamphetamine (MDMA) in mice and rats. *Neuropharmacology* **26,** 1657–1661.

Sulzer, D., and Rayport, S. (1990). Amphetamine and other psychostimulants reduce pH gradients in midbrain dopaminergic neurons and chromaffin granules: A mechanism of action. *Neuron* **5,** 797–808.

Takahashi, T., Yamashita, H., Nakamura, S., Ishiguro, H., Nagatsu, T., and Kawakami, H. (1999). Effects of nerve growth factor and nicotine on the expression of nicotinic acetylcholine receptor subunits in PC12 cells. *Neurosci. Res.* **35,** 175–181.

Tata, D. A., Raudensky, J., and Yamamoto, B. K. (2007). Augmentation of methamphetamine-induced toxicity in the rat striatum by unpredictable stress: Contribution of enhanced hyperthermia. *Eur. J. Neurosci.* **26,** 739–748.

Thomas, D. M., Dowgiert, J., Geddes, T. J., Francescutti-Verbeem, D., Liu, X., and Kuhn, D. M. (2004). Microglial activation is a pharmacologically specific marker for the neurotoxic amphetamines. *Neurosci. Lett.* **367,** 349–354.

Turner, T. J. (2004). Nicotine enhancement of dopamine release by a calcium-dependent increase in the size of the readily releasable pool of synaptic vesicles. *J. Neurosci.* **24,** 11328–11336.

Ulus, I. H., Maher, T. J., and Wurtman, R. J. (2000). Characterization of phentermine and related compounds as monoamine oxidase (MAO) inhibitors. *Biochem. Pharmacol.* **59,** 1611–1621.

Unger, C., Svedberg, M. M., Schutte, M., Bednar, I., and Nordberg, A. (2005). Effect of memantine on the alpha 7 neuronal nicotinic receptors, synaptophysin- and low molecular weight MAP-2 levels in the brain of transgenic mice over-expressing human acetylcholinesterase. *J. Neural. Transm.* **112,** 255–268.

Upreti, V. V., and Eddington, N. D. (2008). Fluoxetine pretreatment effects pharmacokinetics of 3,4-methylenedioxymethamphetamine (MDMA, ECSTASY) in rat. *J. Pharm. Sci.* **97,** 1593–1605.

Vallejo, Y. F., Buisson, B., Bertrand, D., and Green, W. N. (2005). Chronic nicotine exposure upregulates nicotinic receptors by a novel mechanism. *J. Neurosci.* **25,** 5563–5572.

Vatassery, G. T. (1996). Oxidation of vitamin E, vitamin C, and thiols in rat brain synaptosomes by peroxynitrite. *Biochem. Pharmacol.* **52,** 579–586.

Vatassery, G. T., Lai, J. C., DeMaster, E. G., Smith, W. E., and Quach, H. T. (2004). Oxidation of vitamin E and vitamin C and inhibition of brain mitochondrial oxidative phosphorylation by peroxynitrite. *J. Neurosci. Res.* **75,** 845–853.

Verkes, R. J., Gijsman, H. J., Pieters, M. S., Schoemaker, R. C., de Visser, S., Kuijpers, M., Pennings, E. J., de Bruin, D., Van de Wijngaart, G., Van Gerven, J. M., and Cohen, A. F. (2001). Cognitive performance and serotonergic function in users of ecstasy. *Psychopharmacology (Berl)* **153,** 196–202.

Virmani, A., Gaetani, F., Imam, S., Binienda, Z., and Ali, S. (2002). The protective role of l-carnitine against neurotoxicity evoked by drug of abuse, methamphetamine, could be related to mitochondrial dysfunction. *Ann. NY Acad. Sci.* **965,** 225–232.

Volkow, N. D., Chang, L., Wang, G. J., Fowler, J. S., Leonido-Yee, M., Franceschi, D., Sedler, M. J., Gatley, S. J., Hitzemann, R., Ding, Y. S., Logan, J., Wong, C., *et al.* (2001). Association of dopamine transporter reduction with psychomotor impairment in methamphetamine abusers. *Am. J. Psychiatr.* **158,** 377–382.

Wang, J. Q., and Lau, Y. S. (2001). Dose-related alteration in nitric oxide synthase mrna expression induced by amphetamine and the full d1 dopamine receptor agonist SKF-82958 in mouse striatum. *Neurosci. Lett.* **311,** 5–8.

Warren, M. W., Zheng, W., Kobeissy, F. H., Cheng Liu, M., Hayes, R. L., Gold, M. S., Larner, S. F., and Wang, K. K. (2007). Calpain- and caspase-mediated alphaii-spectrin and tau proteolysis in rat cerebrocortical neuronal cultures after ecstasy or methamphetamine exposure. *Int. J. Neuropsychopharmacol.* **10,** 479–489.

Wei, Q., Jurma, O. P., and Andersen, J. K. (1997). Increased expression of monoamine oxidase-B results in enhanced neurite degeneration in methamphetamine-treated PC12 cells. *J. Neurosci. Res.* **50,** 618–626.

Wonnacott, S. (1997). Presynaptic nicotinic ACh receptors. *Trends. Neurosci.* **20,** 92–98.

Young, J. W., Crawford, N., Kelly, J. S., Kerr, L. E., Marston, H. M., Spratt, C., Finlayson, K., and Sharkey, J. (2007). Impaired attention is central to the cognitive deficits observed in alpha 7 deficient mice. *Eur. Neuropsychopharmacol.* **17,** 145–155.

Zajaczkowski, W., Quack, G., and Danysz, W. (1996). Infusion of (+)-MK-801 and memantine: Contrasting effects on radial maze learning in rats with entorhinal cortex lesion. *Eur. J. Pharmacol.* **296,** 239–246.

Zakzanis, K. K., and Young, D. A. (2001). Memory impairment in abstinent MDMA ("Ecstasy") users: A longitudinal investigation. *Neurology* **56,** 966–969.

Zhang, L., Coffey, L. L., and Reith, M. E. (1997). Regulation of the functional activity of the human dopamine transporter by protein kinase C. *Biochem. Pharmacol.* **53,** 677–688.

Zoli, M., Moretti, M., Zanardi, A., McIntosh, J. M., Clementi, F., and Gotti, C. (2002). Identification of the nicotinic receptor subtypes expressed on dopaminergic terminals in the rat striatum. *J. Neurosci.* **22,** 8785–8789.

ETHANOL ALTERS THE PHYSIOLOGY OF NEURON–GLIA COMMUNICATION

Antonio González and Ginés M. Salido

Department of Physiology (Cell Physiology Research Group), University of Extremadura, 10071, Cáceres, Spain

In the central nervous system (CNS), both neurones and astrocytes play crucial roles. On a cellular level, brain activity involves continuous interactions within complex cellular circuits established between neural cells and glia. Although it was initially considered that neurones were the major cell type in cerebral function, nowadays astrocytes are considered to contribute to cerebral function too. Astrocytes support normal neuronal activity, including synaptic function, by regulating the extracellular environment with respect to ions and neurotransmitters. There is a plethora of noxious agents which can lead to the development of alterations in organs and functional systems, and that will end in a chronic prognosis. Among the potentially harmful external agents we can find ethanol consumption, whose consequences have been recognized as a major public health concern. Deregulation of cell cycle has devastating effects on the integrity of cells, and has been closely associated with the development of pathologies which can lead to dysfunction and cell death. An alteration of normal neuronal–glial physiology could represent the basis of neurodegenerative processes. In this review we will pay attention on to the recent findings in astrocyte function and their role toward neurons under ethanol consumption.

167

I. Introduction

All living organisms are constantly subjected to a variety of physical, chemical, or biological damaging factors. This situation is commonly referred to as "stress." Since ancient, alcohol has been present in our lives. In the actual society, in which we count with an increased expectance of life, the presence of chronic diseases is increasing, which consequently affects the quality of life. There is a plethora of noxious agents which can lead to the development of alterations in organs and functional systems, and that will end in a chronic prognosis. Among the potentially harmful external agents we can find alcohol consumption, whose consequences have been recognized as a major public health concern.

A large number of studies, from the whole body to both the cellular and molecular levels, and considering acute and chronic administration, have provided important and detailed information about ethanol effects on the physiology of living beings.

Brain activity results in constant accumulation, sorting, analysis, storage, and retrieval of information that determines the most advanced functions of the central nervous system (CNS), represented by cognition and generation of thoughts. These processes involve continuous interactions within complex intercellular circuits mainly established between neurones and glia, but also with microglia and brain vessels. The information is integrated by employing molecules, some of the as simple as ions. Any substance or molecule, either exogenous or endogenously produced, can affect these numerous cell-to-cell interactions with concomitant effects on brain physiology. Following ingestion, the material absorbed from the gut passes into the portal vein and goes to the liver. This applies to ethanol intake too, and therefore the metabolism of ethanol occurs largely in the liver. However, the CNS is able to metabolize ethanol too, and mainly astrocytes.

Ethanol produces a wide array of effects at the neuropsychological, neurophysiological, and morphological levels. The immediate effects of ethanol, which may occur within minutes of consumption, include altered perceptions and emotions, impaired judgment and distorted vision, hearing, and motor skills. On the other side, long-term effects of heavy ethanol consumption may lead to significant brain injury.

Here we will review the recent findings about ethanol effects on the physiology of the CNS. The work starts with a focus onto the types of cells that take part of the cytoarchitecture of the CNS, being actually recognized that some of them are intricately interconnected and thus their correct and equilibrate function is of critical importance for the homeostasis in the CNS. The main focus will be paid to the role of the macroglia, that is, astrocytes, that initially were considered to have passive roles within the CNS function but, conversely, in the recent years have been pointed to as a cellular type with a really active role in the physiology of this tissue.

II. Cytoarchitecture in the CNS

There is a number between 12 and 15 billion neurons in cerebral cortex and about a billion neurons in spinal cord, whereas there is an order of 10–50 times more glial cells than neurons in the CNS (Kuffler, 1984).

Neurons operate in interconnected networks where the information is received, integrated and sent as an output signal. Beside neurons, other cell types in the CNS are oligodendrocytes, which form myelin, astrocytes, with multiple support functions to neurons, ependymal cells, which are epithelial cells that line brain ventricles and central canal of spinal and assist in secretion and circulation of cerebral spinal fluid (CSF) and microglia, which are small cells that proliferate and act as scavengers when tissue is destroyed and play an important role in defense and inflammation (Verkhratsky et al., 1998). Additionally, there are endothelial cells lining the blood–brain barrier, the specialized system of brain microvascular endothelial cells that shields the brain from toxic substances in the blood, supplies the brain tissues with nutrients, and filters harmful compounds from the brain back to the bloodstream, ensuring the proper function of the CNS (Persidsky et al., 2006).

Glial cells act as the "glue" in the CNS. They are located surrounding neurons in order to hold them in place. However, glial cells play a number of other functions which are very important and represent a major support for neuronal functions: supply of nutrients and oxygen to neurons, to insulate one neuron from another, regulate extracellular pH and ion balance (e.g., K^+), regulate glutamate levels, produce myelin, contribute to form the blood-brain barrier, to destroy pathogens, metabolize, and remove dead neurons or, for example, guide migrating neurons during development (radial glia), in addition to a regulation of synaptic transmission.

All these cells, despite having distinguished functions of their own, are remarkably interconnected and demonstrate considerable amount of intercellular signaling between them. Such crosstalk between these different cell types forms the basis of the physiology of the CNS.

Despite sharing several common signaling pathways, sometimes the same ligand induces opposite responses in different CNS cells, suggesting that there are cell-type-specific modulations in cellular signaling pathways. Beside their cytological differences, there is a property that clearly differentiates either type of cell: whereas neurons are electrically excitable cells, that is, they undergo action potentials, the glia is normally considered an electrically nonexcitable cell type, which will be very much related to the array of functions they can develop (Araque et al., 2001; Volterra and Meldolesi, 2005). As a consequence, neurons will employ action potentials to communicate between them, whereas the glia will communicate with each other and neurones by other means. Finally, glia is capable of mitosis, a property absent in neurons.

Attention has been paid to the behavior of the cells that form the CNS subjected to different conditions. Due to the close apposition from the histological point of view, the first problem the researchers find is how to exactly distinguish the cellular type on study, when working with slices or cell cultures.

If cell cultures are to be used, astrocytes and neurones can be distinguished by their morphology. Astrocytes will be observed at the microscope as cells with a wide cell body from which membrane processes irradiate, and normally grow confluently with neighbor astrocytes (Fig. 1). Conversely, neurons will be noticed by their characteristic soma from which dendrites grow (Fig. 2). This property cannot uniquely be used to distinguish both cell types in tissue slices; therefore, the first useful way is to employ immunocytochemical staining techniques for detection of cytoskeleton proteins. Glial fibrillary acidic protein (GFAP) is a specific intermediate filament protein in astrocytes, and is the most common characteristic used to identify this growing cell type in culture (Verkhratsky et al., 1998). S100B, a member of the large S100 Ca^{2+}-binding protein family, is known to be another specific marker for astrocytes (Ogata and Kosaka, 2002). On the other side, microtubule-associated protein-2 (MAP-2) can be used to identify neurons growing in cultures (Liao et al., 2005). A confocal fluorescence image of GFAP- and MAP-2-positive cells can be seen in Fig. 3.

Cell type specific Ca^{2+} responses in hippocampal cells have been shown, which can be employed to distinguish astrocytes and neurons growing in culture (Beck et al., 2004; Dallwig and Deitmer, 2002; González et al., 2006a). Astrocytes have the property to respond to low external KCl concentrations (0.2 mM) by mobilizing Ca^{2+}. This cell type occasionally shows Ca^{2+} mobilization in response to higher KCl concentrations (50 and 100 mM). Conversely, neurons will only show a Ca^{2+} response when challenged with the higher KCl

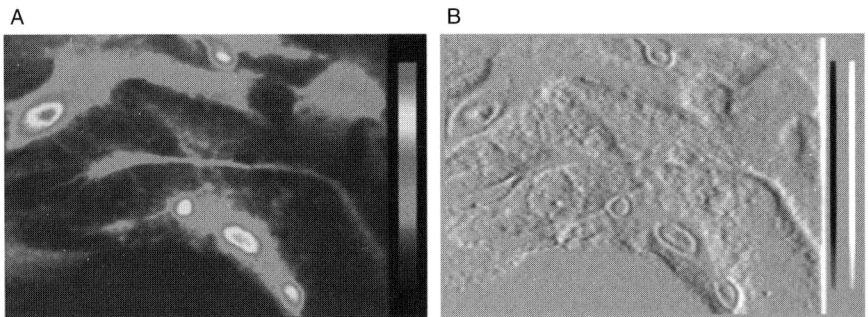

FIG. 1. Fluorescence image of astrocytes in primary culture loaded with the fluorescent probe fura-2. Astrocytes are usually observed as cells with a wide cell body from which membrane processes irradiate, and normally grow confluently with neighbor astrocytes. (A) Fluorescence image of astrocytes in culture. (B) Projection in relief of fura-2-loaded cells. (See Color Insert.)

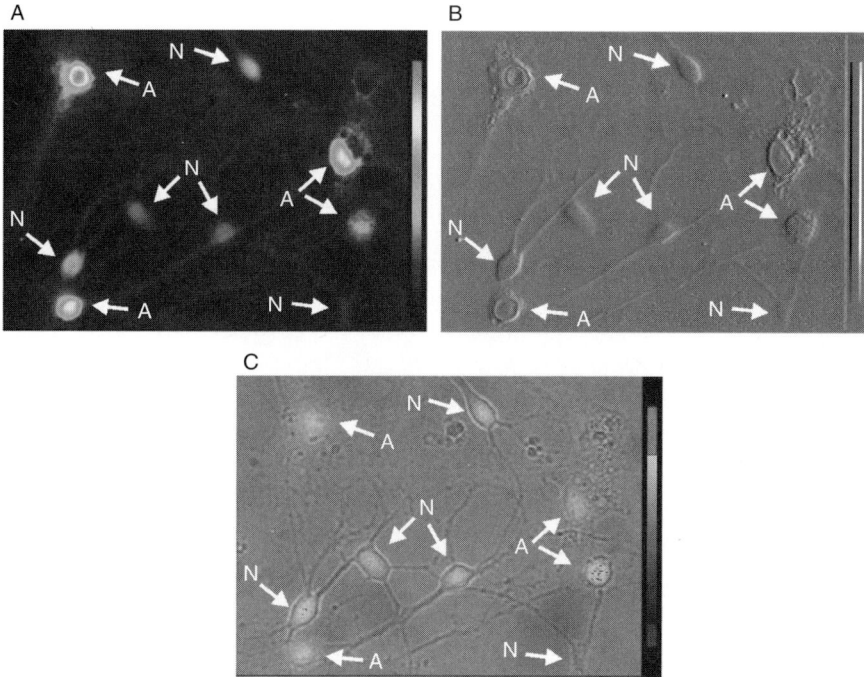

FIG. 2. Fluorescence image of a mixed culture of hippocampal neurons and astrocytes loaded with the fluorescent probe fura-2. (A) Neurons will be noticed by their characteristic soma from which dendrites grow. (B) Projection in relief of fura-2-loaded cells. (C) Transmitted light image of the cell cultures. (A: astrocyte. N: neuron). (See Color Insert.)

concentrations, that is, 50 or 100 mM KCl (Fig. 4). There are other maneuvers to detect the cell type by specifically responding to certain agonists, as for example astrocytes will show responses to 2-Me-SADP, whereas neurons will not (Fam *et al.*, 2000). Conversely, neurons will show responses to apirase, whereas astrocytes will not (Newman, 2006).

III. Focus onto Macroglia: The "Special" Role of Astrocytes in the CNS

Glia was discovered in 1856 by the pathologist Rudolf Virchow in his search for a "connective tissue" in the brain. In 1894, Carl Ludwig Schleich proposed a hypothesis of fully integrated and interconnected neuron–glia circuits as a

A

B

C

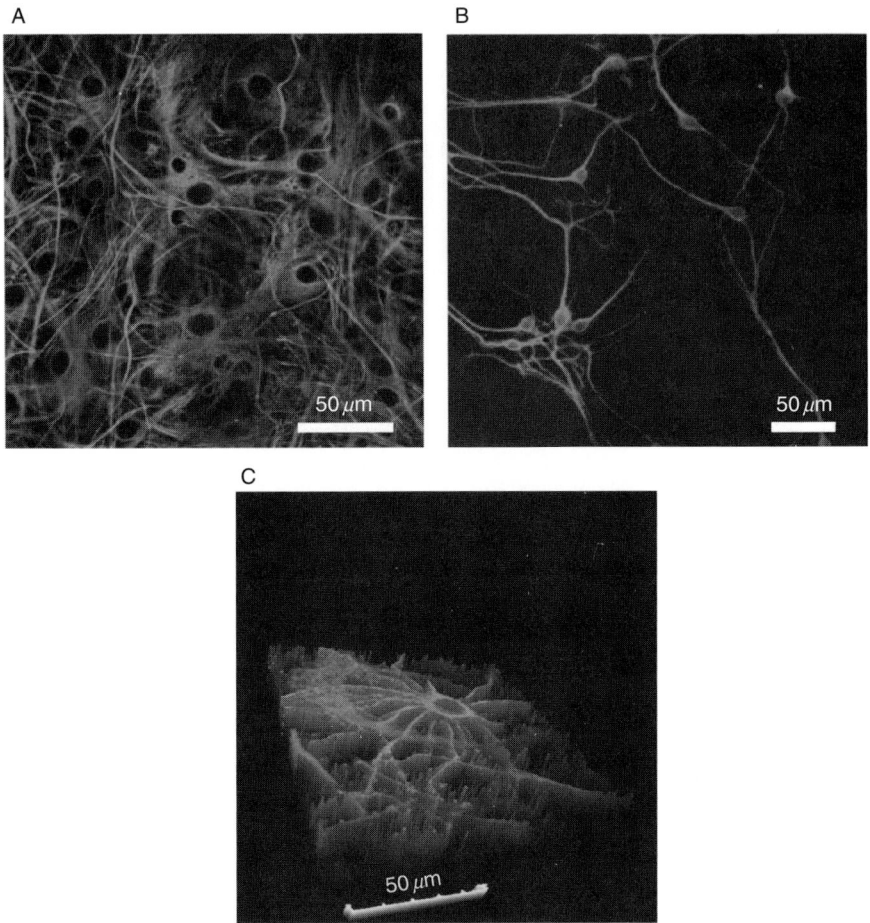

Fig. 3. Immunofluorescence study of hippocampal astrocytes and neurons in culture. (A) Confocal fluorescence image of GFAP-positive cells (astrocytes). (B) Confocal fluorescence image of MAP-2-positive cells (neurons). (C) Mixed cultures of hippocampal neurons and astrocytes. Cultures were incubated with two different fluorescent antibodies directed against GFAP (astrocytes) and MAP-2 (neurons). The figure shows the close apposition of the two cellular types with astrocyte-processes (red) enwrapping one neuron (green). (See Color Insert.)

substrate for brain function. This hypothesis received direct experimental support 100 years later, after several physiological techniques were applied to glial cells (Verkhratsky, 2006).

Astrocytes are the most abundant type of macroglial cell, and present numerous projections that anchor neurons to their blood supply (Braet *et al.*, 2001; Grafstein *et al.*, 2000; Haydon, 2001; Kuffler, 1984). They provide energy supply

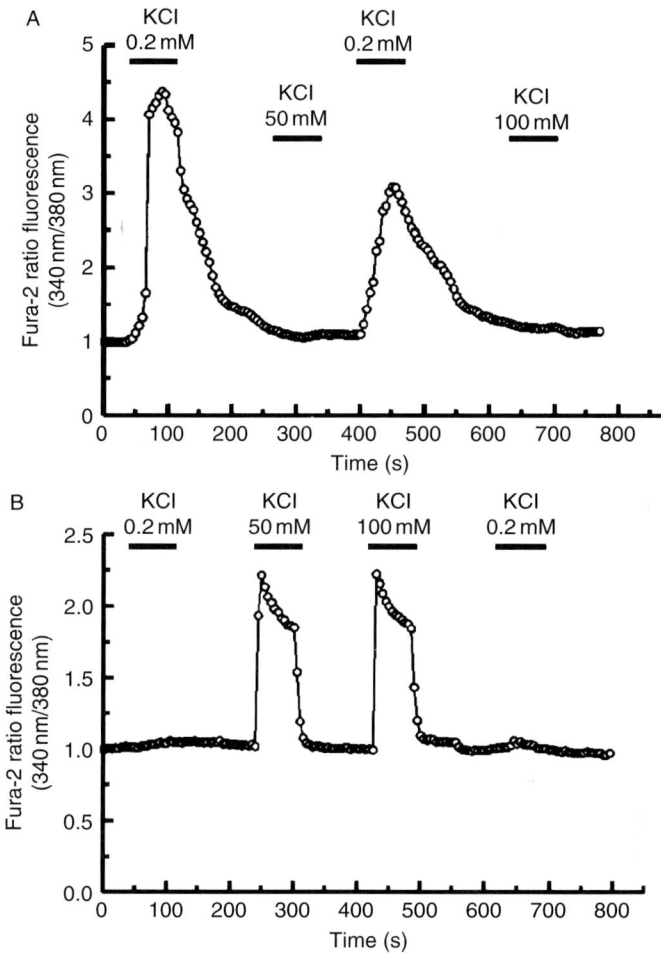

FIG. 4. Cell type specific Ca^{2+} responses in hippocampal cells. (A) Astrocytes have the property to respond to low external KCl concentrations (0.2 mM) by mobilizing Ca^{2+}. (B) Conversely, neurons will only show a Ca^{2+} response when challenged with the higher KCl concentrations, that is, 50 or 100 mM KCl.

to neurons and coordinate metabolic reactions, regulate the external chemical environment by removing excess ions, notably potassium, regulate brain cell volume, participate in neuronal migration during development, recycling neurotransmitters released during synaptic transmission, and in the production of neurotrophic factors (Araque *et al.*, 2001; Eysseric *et al.*, 2000; Haydon, 2000; Ogata and Kosaka, 2002; Phillias and O'Reagan, 2002; Sofroniew *et al*, 1999).

Overall, there is consensus relative to a protective role of astrocytes toward neurons, which will be focused later on in this review.

In addition, a theory suggests that astrocytes may be the predominant "building blocks" of the blood–brain barrier (Bungaard and Abbott, 2008; Choi and Kim, 2008; Paemeleire, 2002; Yang and Aschner, 2003). Astrocytes may regulate endothelial cell metabolism, and vasoconstriction and vasodilation by producing substances with angiogenic properties, such as vascular endothelial growth factor (Proia et al., 2008), ATP (Leybaert et al., 2004), and arachidonic acid (Gabryel et al., 2007; Palomba et al., 2007). Thus, astrocytes play important functions at the level of arterioles where blood flow is controlled, at the level of capillaries where blood-brain barrier is located and at the level of blood immune cells (Leybaert et al., 2004).

Glia retains the ability to undergo cell division in adulthood, while most neurons cannot. The view is based on the general deficiency of the mature nervous system in replacing neurons after an injury, such as a stroke or trauma, while very often there is a profound proliferation of glia, or gliosis near or at the site of damage.

Astrocyte activity is marked by hypertrophy (Da Cunha et al., 1993), resulting in an expression of proteins such as GFAP (Liedtke et al., 1996; McCall et al., 1996), S100B (Marshak et al., 1992; Sheng et al., 1996), adhesion molecules and antigen presenting capabilities, including major histocompatibility antigens (Edwards and Robinson, 2006; Frohman et al., 1989). Reactive astrocytes will represent an obstacle preventing establishment of normal neural contacts and circuitry (Anderson et al., 2003). Related to this, the displacement of presynaptic terminals from the postsynaptic membrane is accompanied by the projection of thin, sheet-like astrocytic processes (Deouiche and Frotsher, 2001). In addition, reactive astrocytes produce a myriad of neurotoxic substances in various brain pathologies (Mori et al., 2006).

Astrocytes signal each other using Ca^{2+} (Braet et al., 2004; Verkhratsky et al., 1998). This quality of cell-to-cell communication has been termed "calcium excitability" (Araque et al., 2001; Cornell-Bell et al., 1990; Verkhratsky et al., 2002). This is manifest as transient or prolonged elevations in intracellular concentration of Ca^{2+} ($[Ca^{2+}]_i$). These can be spontaneous or triggered in response to specific neurotransmitters (Cornell-Bell et al., 1990). In contrast, neurons exhibit electrical excitability, a property absent in astrocytes. The membrane potential of glia is relatively stable, and although they can express voltage-gated channels (Verkhratsky et al., 1998), they exhibit little or no fluctuation in membrane potential.

Astrocytes respond to a variety of extracellular stimuli, including ATP and other nucleotides, by raising $[Ca^{2+}]_i$. Increased $[Ca^{2+}]_i$ modulates various intracellular processes, including differentiation, cytoskeletal reorganization, and the secretion of neuroactive substances (Araque et al., 1998; Neary et al., 1999; Pasti et al., 2001; Verkhratsky and Kettenmann, 1996).

A rise in $[Ca^{2+}]_i$ localized to one part of an astrocyte may propagate throughout the entire cell, and Ca^{2+} responses may be transmitted from one astrocyte to others, leading to regenerative Ca^{2+} signals that spread within astrocyte networks (Cornell-Bell et al., 1990; Dani and Smith, 1995; Fam et al., 2000; Newman and Zahs, 1997). The net effect is a Ca^{2+} wave that propagates from cell to cell, analogous to action potential signal propagation.

This cell-to-cell communication could effectively signal to neurons, endothelial, or other cell type in the CNS. There is emerging evidence that Ca^{2+} signaling in astrocytes is a means for information encoding and transmission, which is complementary to and interacts with signaling in vascular brain cells (Leybaert et al., 2004) and electrical signaling in neurons (Araque et al., 1999; Parpura et al., 1994).

Several pathways have been proposed to mediate Ca^{2+} waves. Ca^{2+} signals could be transmitted from one astrocyte to another via gap junctions. This site of direct cytoplasmic communication between neighboring cells facilitates the transfer of small molecular weight molecules involved in cell signaling and metabolism (Kielian, 2008). Thus, this type of cellular communication would allow messenger molecules to diffuse from one astrocyte to another, for example, Ca^{2+} and/or inositol trisphosphate (IP_3) (Boitano et al., 1992; Paemeleire, 2002; Saez et al., 1989; Scemes et al., 2000).

On the other hand, it has been proposed an alternative mechanism for transmitting signals between astrocytes, such as via a diffusible extracellular messenger, which is released toward the extracellular space. Extracellular release of neurotransmitters like ATP or glutamate, and the consequent activation of specific receptors on neighboring astrocytes, may also mediate Ca^{2+} wave propagation (Bowser and Khakh, 2007; Fellin et al., 2006; Paemeleire, 2002). It is known that ATP evokes Ca^{2+} responses in astrocytes (Salter and Hicks, 1994), and it has been found that ATP is released during Ca^{2+} wave propagation and that propagation of Ca^{2+} waves is blocked by broad spectrum antagonists of receptors activated by ATP (Cotrina et al., 1998; Guthrie et al., 1999). Together, these findings indicate that ATP is a diffusible messenger responsible for transmitting Ca^{2+} waves between astrocytes. The type of response that ATP induces in astrocytes depends on the type of purinoreceptor to which it binds, as well as the cytosolic signaling cascades to which the receptor is coupled (Fam et al., 2003). ADP, formed by enzymatic degradation of ATP, and UTP have been pointed out as putative paracrine messengers within cell-to-cell signalization (Harden and Lazarowski, 1999; Zimmermann et al., 1998).

In addition, it has been proved that astrocytes are able to release glutamate, D-serine and eicosanoids (Araque et al., 2001; Fellin et al., 2004; González et al., 2006a; Montana et al., 2006; Mothet et al., 2005; Palomba et al., 2007; Volterra and Meldolesi, 2005).

More than one of the described mechanisms for neurotransmitter release does operate within astrocytes (Malarkey and Parpura, 2008). Released messengers, in turn, activate Ca^{2+} entry or Ca^{2+} release from intracellular stores by acting on ionotropic and metabotropic receptors, respectively. By this way, ATP and glutamate are the major active neurotransmitters involved in the cell-to-cell communication of Ca^{2+} signals in astrocytes and other cells types in the CNS (Bowser and Khakh, 2007; Perea and Araque, 2007).

Nitric oxide, a signaling molecule synthesized by enzymatic oxidation of L-arginine by nitric oxide synthase, is another putative intercellular signaling molecule for cell-to-cell communication (Willmott et al., 2000). NO activates guanylyl cyclase and increases cytoplasmic cGMP signaling cascades (Galione et al., 1993).

Communication between astrocytes thus seems to rely on many communication systems and signaling molecules, which act in parallel or display regional and cellular specialization. By terms of employing this codex, there is a bidirectional signal communication system within the CNS, which might be mainly carried out by extracellular messengers released from any cell type of the CNS. Because of their close apposition to neurons, signaling molecules released by astrocytes will modulate synaptic transmission and neuronal excitability, as well as neuronal plasticity and survival. Even it could be possible that astrocytes could play roles in higher cognitive functions like learning and memory. It is not therefore estrange that an alteration in Ca^{2+} signaling, and hence in the function of astrocytes, could affect synaptic activity and plasticity and brain homeostasis.

IV. Receptors in Glial Cells Cellular Signalization

Glial cells express a heterogeneous pattern of functional receptors to a variety of chemicals previously known to affect neurons. Astrocytes respond to stimuli, such as neurotransmitters, neuromodulators, and hormones, with an increase in $[Ca^{2+}]_i$ (Braet et al., 2004; Verkhratsky et al., 1998). However, this pattern of communication is not restricted to glial cells, as neurons also exchange Ca^{2+} signals between them as well as with surrounding astrocytes (Charles et al., 1996; Dani et al., 1992; Yuste et al., 1995). These Ca^{2+} signals can be initiated in astrocytes and/or by neuronal stimulation, and can be communicated back to neurons to modulate their synaptic activity (Braet et al., 2004; Verkhratsky, 2006). This could be considered the basis of an active participation of glia in brain function.

$[Ca^{2+}]_i$ responses result from the coordinated activity of a number of molecular cascades responsible for Ca^{2+} movement into or out of the cytoplasm toward the extracellular space or into intracellular stores (Braet et al., 2004; Verkhratsky et al., 1998). The increases in $[Ca^{2+}]_i$ exhibit a variety of temporal and spatial patterns. Such specificity may provide a means for intracellular and intercellular information coding and, in addition, coordinate their responses with those of neurons (Araque *et al.*, 2001), and/or neighboring astrocytes, microvascular cells, microglia, and meningeal cells (Braet *et al.*, 2004).

Glutamate is the major excitatory neurotransmitter in the CNS with important involvement in physiology and pathophysiology within the nervous tissue. Glutamate receptors were probably the first neurotransmitter receptors found in astroglia (Orkand *et al.*, 1981). Nevertheless, it is not the unique molecule that astrocytes employ in cell signaling and in their communication with neurones. Glutamate actions are carried out through the activation of different families of receptors, termed ionotropic, and metabotropic receptors depending on the biochemical and functional consequences derived from their activation.

The ionotropic glutamate receptors are ligand-gated cationic channels for which, according to their pharmacological properties and their response to selective agonists, there have been shown three groups: a-amino-3-hydroxy-5-methylisoxazole-4-propionic acid (AMPA), kainic acid and *N*-methyl-D-aspartate (NMDA) (Buldakova *et al.*, 1999; Coleman *et al.*, 2006; Dingledine *et al.*, 1999; Huettner, 2003; Rodríguez-Moreno and Sihra, 2007; Sommer *et al.*, 1990; Vestergaard *et al.*, 2004).

On the other side, there have been proposed at least eight members of metabotropic glutamate receptors. These are coupled (via G proteins) with PLC, activating therefore the IP_3-mediated intracellular signaling pathway, whereas other metabotropic receptors are connected with adenylate cyclases (Nakanishi, 1994; Nicoletti *et al.*, 2007; Palucha and Pilc, 2007; Pin and Duvoisin, 1995). In astrocytes, Ca^{2+} signaling is mostly involved in activation of either type of membrane receptor.

Adenosine 5'-triphosphate, adenosine and related substances act as neurotransmitters in the CNS to control important physiological functions through the activation of their so named purinoreceptors (Edwards, 1994; Fredholm *et al.*, 2005; Neary *et al.*, 2004; Tran *et al.*, 2008). As for glutamate receptors, purinoreceptors can be subdivided into ionotropic and metabotropic receptors. Numerous experiments on cultured astrocytes demonstrate that they are often associated with PLC and hence IP_3 turnover and, therefore, are involved in the regulation of $[Ca^{2+}]_i$. On the other hand, this kind of receptors can either inhibit or stimulate adenylate cyclase although, in many cases, activation of glial purinoreceptors leads to increases in $[Ca^{2+}]_i$ (Delumeau *et al.*, 1991; Fischer and Krügel, 2007; Lee *et al.*, 2003; Lo *et al.*, 2008; Peakman and Hill, 1995; Pierson *et al.*, 2007). Adenosine may skip Ca^{2+} and cAMP signaling and function as a factor regulating astrocyte differentiation by acting through tyrosine dephosphorylation (Abe and Saito, 1998).

Astrocytes further express adrenoreceptors, on which monoamines epinephrine and norepinephrine exert their physiological action. Via G protein-coupled receptors monoamines produce an increase in $[Ca^{2+}]_i$ through PLC activation, whereas acting on others they can control the activity of adenylate cyclase (Cardinaux and Magistretti, 1996; Chen and Hertz, 1999; Espallergues *et al.*, 2007; Mulligan and MacVicar, 2004; Rosenberg and Li, 1995).

The gamma-aminobutyric acid (GABA) effects in the nervous system are mediated by three distinct subtypes of receptors. In many cases, GABA-induced depolarization of glial cells produces Ca^{2+} influx and measurable $[Ca^{2+}]_i$ transients (Hertz et al., 2006). GABA and glycine coexist in the terminal boutons of spinal interneurons from which they are coreleased to activate glycine and GABA receptors (Örnung et al., 1994; Todd et al., 1996). Some glial cells have been reported to express glycine receptors (Zhang et al., 2008), and activation of these receptors might by linked to $[Ca^{2+}]_i$ (Hou et al., 2008).

Acetylcholine reacts with nicotinic cholinoreceptors that contain an integral cationic channel and metabotropic muscarinic cholinoreceptors that are coupled to PLC by means of G proteins. Therefore, activation of these receptors will induce changes in $[Ca^{2+}]_i$ (Oikawa et al., 2005; Ono et al., 2008; Sharma and Vijayaraghavan, 2001; Shelton and McCarthy, 2000).

Astrocytes also respond to histamine, which acts onto three types of receptors (Li et al., 1999). Measurements of $[Ca^{2+}]_i$ revealed intracellular Ca^{2+} transients mediated via H_1 histamine receptors (Shelton and McCarthy, 2000).

Substance P belongs to a family of neuropeptides, the neurokinins, which acts through stimulation of widely distributed neurokinin receptors in the brain (Boros et al., 2008; Chauhan et al., 2008; Willis et al., 2007). These receptors are G protein-coupled metabotropic receptors linked to phosphoinositide hydrolysis (Marriot et al., 1993) and mediate their actions through an increase in $[Ca^{2+}]_i$ (Bordey et al., 1994; Min et al., 2008).

Glial cells also show bradykinin receptors (Hsieh et al., 2008; Noda et al., 2007; Wang et al., 2007), and endothelin receptors (Gadea et al., 2008; Koyama and Tanaka, 2008; Opitz et al., 2008), which are metabotropic receptors linked to PLC and IP_3 production through G proteins and trigger changes in $[Ca^{2+}]_i$.

Finally, other agonists linked to functional regulation in glial cells are serotonin (Boros et al., 2008; Lee et al., 2006), the neurohormones oxytocin and vasopressin (Parent et al., 2008; Syed et al., 2007), neuropeptide Y (Holmberg and Patterson, 2006), platelet-activating factor (Yoshida et al., 2005) vasoactive intestinal polypeptide (Jozwiak-Bebenista et al., 2007) and angiotensin II (Clark et al., 2008). Furthermore, several groups have reported activation of opioid receptors in astrocytes coupled to intracellular Ca^{2+} signaling (Davis et al., 2007).

These data clearly demonstrate that glial cells can express functional receptors to almost all known neurotransmitters, neuromodulators, and neurohormones. For further information see publications by Verkhratsky et al. (1998) and Bract et al. (2004). In addition, the ability of glial cells to respond to stimulants and/or depressors of the CNS plays a critical role, because of the close apposition of the different cellular types that compose the nervous tissue and, moreover, due to the consequences that a functional impairment would present on either cellular type due to their functional interrelationship.

V. Release of Intercellular Messengers

There are several mechanisms that have been suggested to underline the release of signaling molecules from astrocytes: reverse operation of glutamate transporters, volume-regulated anion channels, gap-junctional hemichannels, diffusional release through purinergic receptors and Ca^{2+}-dependent exocytosis (Araque et al., 2001; Haydon and Carmignoto, 2006; Montana et al., 2006; Parpura et al., 2004; Salazar et al., 2008; Volterra and Meldolesi, 2005). The close physical relationship between astrocytes and neurones provides an opportunity for many functional interactions. Astrocytes have a potential role in integrating synaptic signals and providing feedback responses, in the form of release of neurotransmitters that, as mentioned above, regulate neuronal excitability and synaptic function and brain blood flow. Thus there is a bidirectional signaling pathway between astrocytes and neurons, on one side, and astrocytes and blood vessels on the other, which opens the possibility to an exchange of a huge amount of information in the CNS. Among the different molecules released, the two major signaling messengers released by astrocytes seem to be ATP and glutamate.

The mechanisms by which astrocytes release ATP appear to be diverse, employing vesicular release, connexin hemichannles, cystic fibrosis transmembrane regulator, or the P-glycoprotein (Braet et al., 2004). On the other hand, astrocytic glutamate release can be carried out through connexin hemichannel, excitatory amino acid transporters (EAAT), anion transporter, via $P2X_7$ receptor channels or exocytosis. Depending on the mechanism employed, ATP and/or glutamate release by astrocytes can be Ca^{2+}-dependent or independent (Bowser and Khakh, 2007; Braet et al., 2004; Hazell, 2007).

Among the mechanisms for ATP and/or glutamate release from astrocytes, exocytosis constitutes the mechanism that has recently received special attention, since it was initially considered to occur only in neurons. However, nowadays there exists consistent information favoring its presence in glial cells (Bezzi et al., 2004; Bowser and Khakh, 2007; González et al., 2006a; Jourdain et al., 2007; Ni et al., 2007; Perea and Araque, 2007; Salazar et al., 2008; Volterra and Meldolesi, 2005).

VI. Effects of Ethanol on Developing CNS

Ethanol is a known teratogen and has been implicated in the etiology of human fetal alcohol syndrome, which is characterized by distinct craniofacial abnormalities such as microcephaly, agnathia, and ocular aberrations. Prenatal ethanol exposure induces functional abnormalities during brain development affecting neurogenesis and gliogenesis. Thus, ethanol causes a number of changes in several neurochemical systems. Astrocytes are the predominant source of postnatal retinoic acid synthesis in the cerebellum, and this acid shows teratogenic effects responsible for the fetal alcohol syndrome. McCaffery *et al.* (2004) showed that ethanol could stimulate retinoic acid synthesis leading to abnormal embryonic concentrations of this morphogen and, thus, ethanol could represent a major cause of fetal alcohol syndrome. Additionally, increased sensitivity of glutamate receptors and enhanced transmembrane transport of glutamate has been observed in the presence of ethanol. This was in relationship to the increase in the expression of the excitatory amino acid transporters EAAT1 and EAAT2. Thus, glutamatergic system is affected by ethanol, which can be viewed as a maladaptive process that disposes the developing brain to fetal alcohol syndrome (Zink *et al.*, 2004).

Furthermore, ethanol affects the synthesis, intracellular transport, distribution, and secretion of *N*-glycoproteins in different cell types, including astrocytes and neurons (Braza-Boïls *et al.*, 2006). Glycoproteins, such as adhesion molecules and growth factors, participate in the regulation of nervous system development. Thus, the alterations in the glycosylation process induced by ethanol could be a key mechanism involved in the teratogenic effects of alcohol exposure on brain development. Further studies by Martínez *et al.* (2007) showed that long-term ethanol treatment substantially impairs glycosylation and membrane trafficking in primary cultures of rat astrocytes. Ethanol reduced endogenous levels of active RhoA due to and increase in the activity of small Rho GTPases, reduced phosphoinositides levels and induced changes in the dynamics and organization of the actin cytoskeleton.

Cholesterol is an essential component of cell membranes and plays an important role in signal transduction. There are evidences that cholesterol homeostasis may be affected by ethanol, and this may be involved in neurotoxicity (Guizzetti and Costa, 2007). Indeed, the pathogenesis of Alzheimer's disease has been linked to altered cholesterol homeostasis in the brain. Several functions are carried out by cholesterol and are important for brain development, such as glial cell proliferation, synaptogenesis, neuronal survival and neurite outgrowth. In addition, the brain contains high levels of cholesterol, mostly synthesized *in situ*. Furthermore, astrocytes produce large amounts of cholesterol that can be released by these cells and utilized by neurons to form synapses.

Ethanol presents as well morphological effects on the developing adolescent brain. There were clear effects immediately and long after drinking cessation of a chronic ethanol administration on two neurotransmitter systems (the serotoninergic and nitrergic), which decreased, and the astrocytic cytoskeleton and neuron, which increased and decreased, respectively (Evrard *et al.*, 2006). The authors concluded that drinking cessation can partially ameliorate the ethanol-induced morphological changes on neurons and astrocytes but cannot fully return it to the basal state.

It has been shown that chronic alcohol affects the synaptic structure (Zhou *et al.*, 2007). Following alcohol consumption, the density of dendritic spines was found differentially lower in the nucleus accumbens, and depicted an upregulation of a subunit of the NMDA receptor. The upregulated NMDA receptor subunit is a splice variant isoform which is required for membrane-bound trafficking or anchoring into a spine synaptic site. These changes evoked by ethanol demonstrated an alteration of microcircuitry for glutamate reception.

In a study carried out from a different point of view, ethanol inhibited a Ca^{2+}-insensitive K^+ channel activity, and also affected gap junction coupling, demonstrating that astrocytes play a critical role in brain K^+ homeostasis, and that ethanol effects on astrocytic function could influence neuronal activity (Adermark and Lovinger, 2006).

Finally, despite most of the investigations on the effects of ethanol have been performed following its addition to either tissue or cell cultures, an interesting study has shown excessive activation of glutamatergic neurotransmission in the cerebral cortex following ethanol withdrawal and its contribution to significant behavioral disturbances and to alcohol craving. These effects were related to the activity of the enzyme glutamine synthetase, which converts released glutamate to glutamine (Miguel-Hidalgo, 2006).

VII. Ethanol and Glial Oxidative Stress

The deleterious effects of alcohol in CNS could result either from a direct toxic effect of alcohol or from an indirect effect involving its metabolites and/or reactive oxygen species (ROS) generation. Ethanol can induce several cellular reactions which result in a modification of cellular redox status that can severely affect the cell's capacity to be protected against the endogenous production of ROS (Gonthier *et al.*, 2004). The consequences derived from the effects of ethanol on cellular structures would end in a morphological and functional impairment of cellular physiology.

Among brain cells, astrocytes seem less vulnerable than neurons, but their impairment can dramatically affect neurons because of their protective role toward neurons. Astrocytes have been proposed to represent the major cellular localization of ethanol metabolism within the brain, and have been postulated to protect neurons from ethanol-induced oxidative stress (Watts *et al.*, 2005). The exact enzymatic mechanism responsible for ethanol oxidation in the brain is not clear yet.

Ethanol is normally metabolized in the liver to acetaldehyde by the alcohol dehydrogenase reaction, and acetaldehyde can be further metabolized to acetic acid via aldehyde dehydrogenase reaction. The last step in the pathway is the conversion of acetic acid to acetyl-CoA. Although theoretically the activity of the latter enzyme is high enough to cope with the rate at which alcohol is oxidized by alcohol dehydrogenase, there is a limit to the rate at which the reaction can continue and can therefore lead to accumulation of acetaldehyde, which is toxic for most tissues, including the CNS. Thus, there is always a build up of acetaldehyde which passes out from the liver into the blood, and this acetaldehyde is responsible for some of the unpleasant symptoms of alcohol excess. Once in the bloodstream, the acetaldehyde can also cross the blood–brain barrier and attack the CNS (Haddad, 2004).

In the nervous tissue, ethanol can be metabolized by catalase, cytochrome P450, and alcohol dehydrogenase, with catalase playing a pivotal role among the others (González et al., 2007; Signorini-Allibe et al., 2005; Zimatkin et al., 2006). It also induces upregulation of other antioxidant defenses by increasing, for example, the enzymatic activities of superoxide dismutase, catalase, and glutathione peroxidase (Eysseric et al., 2000; Rathinam et al., 2006). The expression of heat shock proteins like HSP70 (Russo et al., 2001), which have a protective and stabilizing effect on stress-induced injury, is also induced by ethanol. Altogether, this would confer to astrocytes a survival advantage preventing oxidative damage.

A major contribution to the adverse effects of ethanol, as far as interaction with other pathways is concerned, arises from the change in redox state. Alcohol may have several targets in astrocytes and other cell types, impairing for example cellular redox status, cell growth, and differentiation, interfering with the stimulatory effect of trophic factors or altering the expression of cytoskeletal proteins (González et al., 2006b; Guerri et al., 2001). Alterations of glial cells by ethanol would lead to perturbations in neuron–glia interactions generating developmental defects of the brain (Guerri and Renau-Piqueras, 1997; Guerri et al., 2001; Ramachandran et al., 2001).

ROS is a term that describes chemical compounds derived from molecular oxygen (Turrens, 2003). ROS can be produced in the course of different physiological processes and, under normal conditions, are controlled by antioxidant defenses of the cell.

The role of ROS is pivotal for the modulation of critical cellular functions, notably for neurons, astrocytes, and microglia such as apoptosis, ion transport, and Ca^{2+} mobilization (Xu et al., 2003). However, when the capacity of the cell to control the oxidants is overwhelmed, overproduction of ROS results in a potentially toxic oxidative stress state (Taylor and Crack, 2004) that can cause morphological and functional alterations of the cell, including alteration of intracellular Ca^{2+} homeostasis (González et al., 2006a), and has been considered the basis of excitotoxicity injury mechanisms by positive modulation of excitatory amino acid release (Haslew-Layton et al., 2005).

Brain tissue is particularly vulnerable to oxidative damage, possibly due to its high consumption of oxygen and the consequent generation of high quantities of ROS during oxidative phosphorylation. In addition, several regions of the brain are particularly rich in iron, which promotes the production of ROS. Furthermore, the brain counts with relatively poor levels of antioxidant enzymes and antioxidant compounds (Dos Santos *et al.*, 2006; Lamarche *et al.*, 2004a). In a recent paper, González *et al.* (2006a) have shown that ROS lead to an increase in $[Ca^{2+}]_i$, inhibit response of astrocytes to physiological agonists and stimulate glutamate secretion, which in excess can be considered neurotoxic. Although glutamate is the principal excitatory neurotransmitter in the mammalian brain, high levels of this neurotransmitter lead to excitotoxic neuronal death mediated by Ca^{2+} influx, principally through NMDA-gated channels (Bambrick *et al.*, 2004).

As mentioned above, Ca^{2+} signaling is an important medium for neuron–glia interaction, in the sense that neuronal activity can trigger Ca^{2+} signals in glial cells and vice versa. Due to its critical importance for the cellular functions, resting $[Ca^{2+}]_i$ is tightly controlled, and abnormalities in Ca^{2+} regulation will lead to impairment of cellular physiology. Ca^{2+}-ROS interplay can be considered as a push-pull relationship. An elevated level of $[Ca^{2+}]_i$ can lead to excessive ROS production, whereas excessive ROS production can lead to cytosolic Ca^{2+} overload.

It has been shown that an acute exposure of astrocytes to ethanol increased $[Ca^{2+}]_i$ (Allansson *et al.*, 2001; González *et al.*, 2007); inhibition of plasma membrane Ca^{2+}-ATPase activity, the pump that transports Ca^{2+}toward the extracellular space, has been pointed out as one of the molecular mechanisms of action of ethanol (Sepúlveda and Mata, 2004). This would lead to a Ca^{2+} overload of the cytoplasm. Other changes evoked by ethanol are cell swelling and transformation of actin cytoskeleton, consisting of changes in filalamentous actin, appearing as increases in ring formations and a more dispersed appearance of the filalaments (Allansson *et al.*, 2001).

It is well known that mitochondria represent the major source of intracellular ROS, and that Ca^{2+} uptake into the organelle can lead to ROS generation (González *et al.*, 2006b; Granados *et al.*, 2004). Ethanol-evoked ROS production takes place in the mitochondria (González *et al.*, 2006a), and accumulated mitochondrial ROS can be released to the cytoplasm leading to damage of different transport mechanisms, ion channel modification, lipid peroxidation, and DNA damage. Furthermore, damage to mitochondrial metabolism may generate additional damaging radical species, thus activating cellular death pathways (Jacobson and Duchen, 2002).

Despite glutamate functions in normal conditions as a regulator of cell function, it is well known that an excessive and maintained secretion of glutamate can be injurious to brain cells (Emerit *et al.*, 2004; Xu *et al.*, 2006). Furthermore, there are evidences that support the notion that reactive astrocytes produce a myriad of neurotoxic substances in various brain pathologies (Mori *et al.*, 2006).

Salazar *et al.* (2008) have shown that ethanol evokes a dose-dependent increase in glutamate secretion by an exocytosis mechanism, which was dependent on Ca^{2+} mobilization. The secretory effect of ethanol was reduced in the presence of antioxidants, therefore indicating the participation of ROS in ethanol-evoked glutamate secretion by astrocytes.

There are consistent data regarding the role of glutamate and the attendant increase in intracellular Ca^{2+} in triggering excitotoxic cell death in neighboring cells (Lee *et al.*, 2000; Molz *et al.*, 2008; Zipfel *et al.*, 2000). Because astrocytes are the major regulators of glutamate homeostasis, primary and/or secondary glial cell death can cause and/or aggravate human diseases of the CNS.

When the nervous system is subjected to stressful stimuli, reactive gliosis often occurs. Associated with neuronal injuries caused by many CNS insults is an activation of glial cells (particularly astrocytes and microglia), termed gliosis, at the sites of injury. The activation of mechanisms that result in activated glia causes secondary neuronal damage.

De la Monte *et al.* (2008) have recently studied the effects of ethanol on cerebellum. In their study the authors show increased neuronal loss, gliosis, lipid peroxidation, and DNA damage in the presence of ethanol. This drug evoked inhibition of insulin and insulin-like growth factor signaling. Earlier studies showed that ethanol administration to young and aged rats produced an increase in lipid peroxidation and GFAP content, and a decline in glutathione (GSH) levels. In their study, the authors assayed the protective effects of melatonin against ethanol, which exerted its protective effect on injured nervous tissues by scavenging free radicals and stabilizing glial activity (Baydas and Tuzcu, 2005).

In an former study, it was shown that ethanol treatment was able to upregulate both cyclooxygenase 2 (COX-2) and inducible nitric oxide synthase (iNOS) expression, and that these effects were regulated via nuclear factor kappa B (NF-κB), which suggest that ethanol is able to induce inflammatory mediators in astrocytes through the NF-κB activation (Blanco *et al.*, 2004).

Reactive astrocytes demonstrate hypertrophy of cellular processes and then present a stellate morphology, increased GFAP immunoreactivity, and increased number of mitochondria as well as elevated enzymatic and nonenzymatic antioxidant activities.

The specific expression of GFAP suggests that it plays an important role in the function of astrocytes in the CNS. Up regulation of astrocytic intermediate filaments is a crucial step in reactive gliosis (Liedtke *et al.*, 1996; McCall *et al.*, 1996). However, despite its positive and neuroprotective role after CNS injury, abnormal induction of GFAP is considered deleterious for CNS regeneration because these strong GFAP-positive reactive astrocytes will represent an obstacle preventing establishment of normal neural contacts and circuitry (Anderson *et al.*, 2003).

Ethanol exposure significantly and dramatically increased GFAP immunoreactivity and the number of immunoreactive astrocytes (Gonca *et al.*, 2005). More recently, González *et al.* (2007) have shown that ethanol evoked a Ca^{2+}-dependent increase in GFAP expression. Furthermore, the reduction of the effect of ethanol in the presence of the antioxidants catalase and resveratrol indicates that ROS are involved in astrogliosis.

VIII. Ethanol, Inflammation, and Glial Cell Death

Inflammatory processes and cytokine expression have been implicated in the pathogenesis of several neurodegenerative disorders, and could be related to ethanol. Vallés *et al.* (2004) showed that activation of glial cells by ethanol induced stimulation of signaling pathways and inflammatory mediators in brain, and could cause neurotoxicity. Concretely, chronic ethanol treatment upregulates iNOS, COX-2 and IL-1beta in rat cerebral cortex and in cultured astrocytes. The upregulation of these inflammatory mediators concomitantly occurred with the stimulation of MAP kinases, including ERK1/2, p-38, and JNK. These effects were associated with an increased in both caspase-3 and apoptosis.

As mentioned, astroglial cells are involved in the neuropathogenesis of several inflammatory diseases of the brain, where the abnormal activation of inflammatory mediators and cytokines plays an important role. Ethanol upregulates inflammatory mediators in astroglial cells. Astrocytes undergo actin cytoskeleton disorganization, and there is a stimulation of both the IRAK/ERK/NF-κB pathway and the COX-2 expression, which are associated with the inflammatory response (Guasch *et al.*, 2007).

More recently it has been further shown that ethanol-induced inflammatory processes in the brain and glial cells are mediated via the activation of interleukin-1 beta receptor type I (IL-1RI)/toll-like receptor type 4 (TLR4) signaling (Blanco *et al.*, 2008). Ethanol triggers the translocation of IL-1RI and/or TLR4 into lipid rafts of caveolae-enriched fractions, which have been pointed out as platforms for receptor signaling and represent important integrators of signal events and trafficking. This is accompanied by endocytosis of IL-1RI and TLR4 receptors and the stimulation of their downstream signaling pathways, which could be some of the mechanisms for ethanol-induced inflammatory damage in the brain.

Apoptosis or programmed cell death is a form of cell death that occurs in multicellular organisms. Apoptosis is a tightly regulated process which engages multiple cell signaling pathways, and involves the altruistic suicide of individual cells in favor of the organism. This process is desirably during organism development and morphological changes, especially at the embryonic stage, as well as during the activation of the immune system. However, defects in apoptosis can result in cancer, autoimmune diseases and neurodegenerative disorders.

Interesting results have arisen from studies on Ca^{2+} signaling and apoptosis. Ethanol treatment potentiated, in a concentration- and time-dependent manner, apoptotic cell death induction by thapsigargin, caffeine and the protonophore FCCP, which separately caused similar increases in Ca^{2+} levels and also induced similar apoptotic rates (Hirata *et al.*, 2006).

The effect of ethanol on the induction of apoptosis in astrocytes and the formation of ceramide as apoptotic signal was investigated by Schatter *et al.* (2005). Ethanol induced nuclear fragmentation and DNA laddering, and inhibited phospholipase D-mediated formation of phosphatidic acid, which is a mitogenic lipid messenger. The authors concluded that ethanol induced glial apoptosis during brain development via the formation of ceramide.

Further studies have shown that astrocytes exposed to ethanol undergo morphological changes associated with anoikis, a programmed cell death induced by loss of anchorage. Astrocytes depicted peripheral reorganization of both focal adhesions and actin-myosin system, cell contraction, membrane blebbing and chromatin condensation (Miñambres *et al.*, 2006).

Finally, it has been shown that ethanol affects intracellular trafficking. In fact, ethanol could interfere with nucleocytoplasmic transport in astrocytes, in such a way that ethanol induces a delay in both import and export of proteins to the nucleus (Marín *et al.*, 2008).

IX. Astroglia and Neuroprotection

Due to their close apposition to neurons, coupling of astrocytes in glial networks may be functionally important in determining neuronal vulnerability (Benarroch, 2005). Cell survival is a critical issue in the onset and progression of neurodegenerative diseases and following pathological events including ischemia and traumatic brain injury. Oxidative stress is the main cause of cell damage in such pathological conditions.

In the setting of stress, that is, when brain homeostasis is disrupted, astrocytes become activated, a state termed gliosis, in which astrocytes produce a variety of factors such as nitric oxide, neuropeptides and cytokines (Allan and Rothwell, 2001). Nevertheless, astrocytes can release other factors, which show a protective activity onto neurons. The ability of astrocytes to protect neurons depends on their specific energy metabolism, high glutathione level, increased antioxidant enzyme activity (catalase, superoxide dismutase, glutathione peroxidase) and overexpression of antiapoptotic Bcl-2 gene (Gabryel and Trzeciak, 2001).

Growing evidence indicates the prominent role of astrocytes in the defense against oxidative stress-mediated neuronal damage and death (Takuma *et al.*, 2004).

Investigation on the ways the brain reacts to ethanol exposure showed that in astrocytes there was an increase in the activity the antioxidant enzymes catalase, superoxide dismutase (SOD), and glutathione peroxidase (GPX). And this was associated with a reduced amount of radicals derived from ethanol, supporting a role of astrocytes to protect the brain (Eysseric et al., 2000). Li et al. (2008) have recently reported that the absence of astrocyte intermediate filament proteins is linked to changes in glutamate transport, endothelin B receptor-mediated control of gap junctions and plasminogen activator inhibitor-1 expression, which play an especial function in astrocytes-mediated protection in brain ischemia. Furthermore, it has been found an increase in the expression of taurine in reactive astrocytes after ethanol exposure. Since taurine is involved in various neuroprotective functions, this suggests that it might play an important role in neuroprotective processes carried out by astrocytes (Sakurai et al., 2003).

Co-culture studies have shown that astrocytes protect neurons from cell death induced by ROS (Desagher et al., 1996; Skaper et al., 1998). Moreover, astrocytes provide glutathione precursors to neighboring neurons (Dringen et al., 2000; Watts et al., 2005) and are essential in the regulation of released glutamate and its conversion to glutamine through the enzyme glutamine synthetase (Miguel-Hidalgo, 2006). Indeed, the increase in apoptotic neuron death in response to ethanol can be blocked by elevating neuron GSH with N-acetylcysteine or by coculturing neurons with neonatal cortical astrocytes. In this study ethanol increased both gamma-glutamyl transpeptidase expression and activity on astrocytes. The authors concluded that these effects were directed to enhance neuron GSH homeostasis, therefore proposing the participation of ROS in ethanol-induced damage to neurons and the protective role of astrocytes (Rathinam et al., 2006).

Lamarche et al. (2004b) demonstrated that vitamin E, which shows antioxidant properties, and astrocytes can protect neurones from ethanol-induced oxidative stress, notably by contributing to maintaining the intracellular glutathione levels. In addition, selenium also had an interesting neuroprotective effect. In relation to this study, co-culturing neurons with astrocytes prevented GSH depletion observed in the presence of ethanol. The increases in neuron ROS production and apoptosis evoked by ethanol were also prevented by co-culture with astrocytes (Watts et al., 2005).

Additional information regarding the protection of astrocytes toward neurons comes from the studies of Druse et al. (2007), whose investigations demonstrated that treatment with S100B, a protein which is released from activated astrocytes, prevented ethanol-associated apoptosis of fetal rhombencephalic neurons. When S100B is secreted it exerts a regulatory effect on neighboring cells: astrocytes, neurons, and microglia. S100B stimulates glial proliferation, neuronal survival and protects neurons against glutamate excitotoxicity (Nardin et al., 2007). In the work by Druse et al. (2007), the S100B neuroprotective effects were linked to activation of the PI-3K pathways. Moreover, this protein increased the formation of pAkt and the upregulation of two downstream NF-κB-dependent prosurvival genes: XIAP and Bcl-2.

Glial cells are not only essential for maintaining a healthy and well-functioning brain, but they also protect and aid the brain in the functional recovery from injuries. The activation of glial cells in the CNS is the first defense mechanism against pathological abnormalities that occur in neurodegenerative diseases. In fact, ethanol exposure increased astrocytic and oligodendrocytic differentiation of neural stem cells as a compensatory mechanism to repair the impaired neural network (Tateno *et al.*, 2005).

X. Conclusion

Astrocytes face the synapses, send end-foot processes that enwrap the brain capillaries, and form an extensive network interconnected by gap junctions. Astrocytes express several membrane proteins and enzymes that are critical for uptake of glutamate at the synapses, ammonia detoxification, buffering of extracellular K^+, and volume regulation. They also participate in detection, propagation, and modulation of excitatory synaptic signals, provide metabolic support to the active neurons, and contribute to functional hyperemia in the active brain tissue.

Although until recently no specific role has been identified for these cells in the pathogenesis of brain diseases, alterations in the neuron–astrocyte relationship have begun to emerge. A dysfunction of glial cell receptors, which alters the glial cells' sense of their environment, can lead to the development of neurological diseases as varied as brain tumors, AIDS-related neuropathology, Alzheimer's disease, and amyotrophic lateral sclerosis.

Ethanol has an extensive array of actions in the body. Due to its effects on astroglial physiology, impairment in the function of this "specialized" cell type could alter their protective role toward neurons, transforming these "pacemaker" cells into an activated potentially injurious cell with negative consequences to neuronal function and survival, and to brain function.

Acknowledgment

This work was supported by Junta de Extremadura-FEDER (PRI08A018).

References

Abe, K., and Saito, H. (1998). Adenosine stimulates stellation of cultured rat cortical astrocytes. *Brain Res.* **804,** 63–71.

Adermark, L., and Lovinger, D. M. (2006). Ethanol effects on electrophysiological properties of astrocytes in striatal brain slices. *Neuropharmacology* **51,** 1099–1108.

Allan, S. M., and Rothwell, N. J. (2001). Cytokines and acute neurodegeneration. *Nat. Rev. Neurosci.* **2,** 734–744.

Allansson, L., Khatibi, S., Olsson, T., and Hansson, E. (2001). Acute ethanol exposure induces $[Ca^{2+}]_i$ transients, cell swelling and transformation of actin cytoskeleton in astroglial primary cultures. *J. Neurochem.* **76,** 472–479.

Anderson, M. F., Blomstrand, F., Blomstrand, C., Eriksson, P. S., and Nilsson, M. (2003). Astrocytes and stroke: Networking for survival? *Neurochem. Res.* **28,** 293–305.

Araque, A., Parpura, V., Sanzgiri, R. P., and Haydon, P. G. (1998). Glutamate dependent astrocyte modulation of synaptic transmission between cultured hippocampal neurons. *Eur. J. Neurosci.* **10,** 2129–2142.

Araque, A., Parpura, V., Sanzgiri, R. P., and Haydon, P. G. (1999). Tripartite synapses: Glia, the unacknowledged partner. *Trends Neurosci.* **22,** 208–215.

Araque, A., Carmignoto, G., and Haydon, P. G. (2001). Dinamic signaling between astrocytes and neurons. *Ann. Rev. Physiol.* **63,** 795–813.

Bambrick, L., Kristian, T., and Fiskum, G. (2004). Astrocyte mitochondrial mechanisms of ischemic brain injury and neuroprotection. *Neurochem. Res.* **29,** 601–608.

Baydas, G., and Tuzcu, M. (2005). Protective effects of melatonin against ethanol-induced reactive gliosis in hippocampus and cortex of young and aged rats. *Exp. Neurol.* **194,** 175–181.

Beck, A., Nieden, R. Z., Schneider, H. P., and Deitmer, J. W. (2004). Calcium release from intracellular stores in rodent astrocytes and neurons *in situ*. *Cell Calcium* **35,** 47–58.

Benarroch, E. E. (2005). Neuron-astrocyte interactions: Partnership for normal function and disease in the central nervous system. *Mayo Clin. Proc.* **80,** 1326–1338.

Bezzi, P., Gundersen, V., Galbete, J. L., Seifert, G., Steinhäuser, C., Pilati, E., and Volterra, A. (2004). Astrocytes contain a vesicular compartment that is competent for regulated exocytosis of glutamate. *Nat. Neurosci.* **7,** 613–620.

Blanco, A. M., Pascual, M., Valles, S. L., and Guerri, C. (2004). Ethanol-induced iNOS and COX-2 expression in cultured astrocytes via NF-kappa B. *Neuroreport* **15,** 681–685.

Blanco, A. M., Perez-Arago, A., Fernandez-Lizarbe, S., and Guerri, C. (2008). Ethanol mimics ligand-mediated activation and endocytosis of IL-1RI/TLR4 receptors via lipid rafts caveolae in astroglial cells. *J. Neurochem.* **106,** 625–639.

Boitano, S., Dirksen, E. R., and Sanderson, M. J. (1992). Intercellular propagation of calcium waves mediated by inositol trisphosphate. *Science* **258,** 292–295.

Bordey, A., Feltz, P., and Trousland, J. (1994). Mobilization of intracellular calcium by substance P in a human astrocytoma cell line (U-373 MG). *Glia* **11,** 277–283.

Boros, C., Lukácsi, E., Horváth-Oszwald, E., and Réthelyi, M. (2008). Neurochemical architecture of the filum terminale in the rat. *Brain Res.* **1209,** 105–114.

Bowser, D. N., and Khakh, B. S. (2007). Vesicular ATP is the predominant cause of intercellular calcium waves in astrocytes. *J. General Physiol.* **129,** 485–491.

Braet, K., Paemeleire, K., D'Herde, K., Sanderson, M. J., and Leybaert, L. (2001). Astrocyte-endothelial cell calcium signals conveyed by two signalling pathways. *Eur. J. Neurosci.* **13,** 79–91.

Braet, K., Cabooter, L., Paemeleire, K., and Leybaert, L. (2004). Calcium signal communication in the central nervous system. *Biol. Cell* **96,** 79–91.

Braza-Boïls, A., Tomás, M., Marín, M. P., Megías, L., Sancho-Tello, M., Fornas, E., and Renau-Piqueras, J. (2006). Glycosylation is altered by ethanol in rat hippocampal cultured neurons. *Alcohol Alcoholism* **41,** 494–504.

Buldakova, S. L., Vorobjev, V. S., Sharonova, I. N., Samoilova, M. V., and Magazanik, L. G. (1999). Characterization of AMPA receptor populations in rat brain cells by the use of subunit-specific open channel blocking drug, IEM-1460. *Brain Res.* **94,** 261–268.

Bungaard, M., and Abbott, N. J. (2008). All vertebrates started out with a glial blood-brain barrier 4–500 million years ago. *Glia* **56,** 699–708.

Cardinaux, J. R., and Magistretti, P. J. (1996). Vasoactive intestinal peptide, pituitary adenylate cyclase-activating peptide, and noradrenaline induce the transcription factors CCAAT/enhancer binding protein (C/EBP)-beta and C/EBP delta in mouse cortical astrocytes: Involvement in cAMP-regulated glycogen metabolism. *J. Neurosci.* **16,** 919–929.

Charles, A. C., Kodali, S. K., and Tyndale, R. F. (1996). Intercellular calcium waves in neurons. *Mol. Cell. Neurosci.* **7,** 337–353.

Chauhan, V. S., Sterka, Jr., D. G., Gray, D. L., Bost, K. L., and Marriott, I. (2008). Neurogenic exacerbation of microglial and astrocyte responses to *Neisseria meningitidis* and *Borrelia burgdorferi*. *J. Immunol.* **180,** 8241–8249.

Chen, Y., and Hertz, L. (1999). Noradrenaline effects on pyruvate decarboxylation: Correlation with calcium signaling. *J. Neurosci. Res.* **58,** 599–606.

Choi, Y. K., and Kim, K. W. (2008). AKAP12 in astrocytes induces barrier functions in human endothelial cells through protein kinase Czeta. *FEBS J.* **275,** 2338–2353.

Clark, M. A., Guillaume, G., and Pierre-Louis, H. C. (2008). Angiotensin II induces proliferation of cultured rat astrocytes through c-Jun N-terminal kinase. *Brain Res. Bull.* **75,** 101–106.

Coleman, S. K., Möykkynen, T., Cai, C., von Ossowski, L., Kuismanen, E., Korpi, E. R., and Keinänen, K. (2006). Isoform-specific early trafficking of AMPA receptor flip and flop variants. *J. Neurosci.* **26,** 11220–11229.

Cornell-Bell, A. H., Finkbeiner, S. M., Cooper, M. S., and Smith, S. J. (1990). Glutamate induces calcium waves in cultured astrocytes: Long-range glial signalling. *Science* **247,** 470–473.

Cotrina, M. L., Lin, J. H., and Nedergaard, M. (1998). Cytoskeletal assembly and ATP release regulate astrocytic calcium signalling. *J. Neurosci.* **18,** 8794–8804.

Da Cunha, A., Jefferson, J. J., Tyor, W. R., Glass, J. D., Jannotta, F. S., and Vitkovic, L. (1993). Gliosis in human brain: Relationship to size but not other properties of astrocytes. *Brain Res.* **600,** 161–165.

Dallwig, R., and Deitmer, J. W. (2002). Cell-type specific calcium responses in acute rat hippocampal slices. *J. Neurosci. Method* **116,** 77–87.

Dani, J. W., and Smith, S. J. (1995). The triggering of astrocytic calcium waves by NMDA-induced neuronal activation. *Ciba Foundation Symp.* **188,** 195–205.

Dani, J. W., Chernjavski, A., and Smith, S. J. (1992). Neuronal activity triggers calcium waves in hippocampal astrocyte netrworks. *Neuron* **8,** 429–440.

Davis, R. L., Buck, D. J., Saffarian, N., and Stevens, C. W. (2007). The opioid antagonist, beta-funaltrexamine, inhibits chemokine expression in human astroglial cells. *J. Neuroimmunol.* **186,** 141–149.

De la Monte, S. M., Tong, M., Cohen, A. C., Sheedy, D., Harper, C., and Wands, J. R. (2008). Insulin and insulin-like growth factor resistance in alcoholic neurodegeneration. *Alcoholism Clin. Exp. Res.* **32,** 1630–1644.

Delumeau, J. C., Petitet, F., Cordier, J., Glowinski, J., and Premont, J. (1991). Synergistic regulation of cytosolic Ca^{2+} concentration in mouse astrocytes by NK1 tachykinin and adenosine agonists. *J. Neurochem.* **57,** 2026–2035.

Deouiche, A., and Frotsher, M. (2001). Peripheral astrocyte processes: Monitoring by selective immunostaining for the actin-binding ERM proteins. *Glia* **36,** 330–341.

Desagher, S., Glowinski, J., and Premont, J. (1996). Astrocytes protect neurons from hydrogen peroxide toxicity. *J. Neurosci.* **16,** 2553–2562.

Dingledine, R., Borges, K., Bowie, D., and Traynelis, S. F. (1999). The glutamate receptor ion channels. *Pharmacol. Rev.* **51,** 7–61.

Dos Santos, A. Q., Nardin, P., Funchal, C., de Almeida, L. M., Jacques-Silva, M. C., Wofchuk, S. T., Gonçalves, C. A., and Gottfried, C. (2006). Resveratrol increases glutamate uptake and glutamine synthetase activity in C6 glioma cells. *Arch. Biochem. Biophys.* **453,** 161–167.

Dringen, R., Gutterer, J. M., and Hirrlinger, J. (2000). Glutathione metabolism in brain metabolic interaction between astrocytes and neurons in the defense against reactive oxygen species. *Eur. J. Biochem.* **267,** 4912–4916.

Druse, M. J., Gillespie, R. A., Tajuddin, N. F., and Rich, M. (2007). S100B-mediated protection against the pro-apoptotic effects of ethanol on fetal rhombencephalic neurons. *Brain Res.* **1150,** 46–54.

Edwards, F. A. (1994). ATP receptors. *Curr. Opin. Neurobiol.* **4,** 347–352.

Edwards, M. M., and Robinson, S. R. (2006). TNF alpha affects the expression of GFAP and S100B: Implications for Alzheimer's disease. *J. Neural Trans.* **113,** 1709–1715.

Emerit, J., Edeas, M., and Bricaire, F. (2004). Neurodegenerative diseases and oxidative stress. *Biomed. Pharmacother.* **58,** 39–46.

Espallergues, J., Solovieva, O., Técher, V., Bauer, K., Alonso, G., Vincent, A., and Hussy, N. (2007). Synergistic activation of astrocytes by ATP and norepinephrine in the rat supraoptic nucleus. *Neuroscience* **148,** 712–723.

Evrard, S. G., Duhalde-Vega, M., Tagliaferro, P., Mirochnic, S., Caltana, L. R., and Brusco, A. (2006). A low chronic ethanol exposure induces morphological changes in the adolescent rat brain that are not fully recovered even after a long abstinence: An immunohistochemical study. *Exp. Neurol.* **200,** 438–459.

Eysseric, H., Gonthier, B., Soubeyran, A., Richard, M. J., Daveloose, D., and Barret, L. (2000). Effects of chronic ethanol exposure on acetaldehyde and free radical production by astrocytes in culture. *Alcohol* **21,** 117–125.

Fam, S. R., Gallagher, C. J., and Salter, M. W. (2000). P2Y1 Purinoceptor-mediated Ca^{2+} signaling and Ca^{2+} wave propagation in dorsal spinal cord astrocytes. *J. Neurosci.* **20,** 2800–2808.

Fam, S. R., Gallagher, C. G., Kalia, L. V., and Salter, M. W. (2003). Differential frequency dependence of P2Y1- and P2Y2-ediated Ca^{2+} signaling in astrocytes. *J. Neurosci.* **23,** 4437–4444.

Fellin, T., Pascual, O., Gobbo, S., Pozzan, T., Haydon, P. G., and Carmignoto, G. (2004). Neuronal synchrony mediated by astrocytic glutamate through activation of extrasynaptic NMDA receptors. *Neuron* **43,** 729–743.

Fellin, T., Sul, J. Y., D'Ascenzo, M., Takano, H., Pascual, O., and Haydon, P. G. (2006). Bidirectional astrocyte-neuron communication: The many roles of glutamate and ATP. *Novartis Foundation Symp.* **276,** 208–217; discussion 217–221, 233–207, 275–281.

Fischer, W., and Krügel, U. (2007). P2Y receptors: Focus on structural, pharmacological and functional aspects in the brain. *Curr. Med. Chem.* **14,** 2429–2455.

Fredholm, B. B., Chen, J. F., Cunha, R. A., Svenningsson, P., and Vaugeois, J. M. (2005). Adenosine and brain function. *Int. Rev. Neurobiol.* **63,** 191–270.

Frohman, E. M., van den Noort, S., and Gupta, S. (1989). Astrocytes and intracerebral immune responses. *J. Clin. Immunol.* **9,** 1–9.

Gabryel, B., and Trzeciak, H. I. (2001). Role of astrocytes in pathogenesis of ischemic brain injury. *Neurotox. Res.* **3,** 205–221.

Gabryel, B., Chalimoniuk, M., Stolecka, A., and Langfort, J. (2007). Activation of cPLA2 and sPLA2 in astrocytes exposed to simulated ischemia *in vitro*. *Cell Biol. Int.* **31,** 958–965.

Gadea, A., Schinelli, S., and Gallo, V. (2008). Endothelin-1 regulates astrocyte proliferation and reactive gliosis via a JNK/c-Jun signaling pathway. *J. Neurosci.* **28,** 2394–2408.

Galione, A., White, A., Willmott, N., Turner, M., Potter, B. V., and Watson, S. P. (1993). cGMP mobilizes intracellular Ca^{2+} in sea urchin eggs by stimulating cyclic ADP-ribose synthesis. *Nature* **365,** 456–459.

Gonca, S., Filiz, S., Dalçik, C., Yardimoğlu, M., Dalçik, H., Yazir, Y., and Erden, B. F. (2005). Effects of chronic ethanol treatment on glial fibrillary acidic protein expression in adult rat optic nerve: An immunocytochemical study. *Cell Biol. Int.* **29,** 169–172.

Gonthier, B., Signorini-Allibe, N., Soubeyran, A., Eysseric, H., Lamarche, F., and Barret, L. (2004). Ethanol can modify the effects of certains free radical-generating systems on astrocytes. *Alcoholism* **28,** 526–533.

González, A., Granados, M. P., Pariente, J. A., and Salido, G. M. (2006a). H2O2 mobilizes Ca^{2+} from agonist- and thapsigargin-sensitive and insensitive intracellular stores and stimulates glutamate secretion in rat hippocampal astrocytes. *Neurochem. Res.* **31,** 741–750.

González, A., Núñez, A. M., Granados, M. P., Pariente, J. A., and Salido, G. M. (2006b). Ethanol impairs CCK-8-evoked amylase secretion through Ca^{2+}-mediated ROS generation in mouse pancreatic acinar cells. *Alcohol* **38,** 51–57.

González, A., Pariente, J. A., and Salido, G. M. (2007). Ethanol stimulates ROS generation by mitochondria through Ca^{2+} mobilization and increases GFAP content in rat hippocampal astrocytes. *Brain Res.* **1178,** 28–37.

Grafstein, B., Liu, S., Cotrina, M. L., Goldman, S. A., and Nedergaard, M. (2000). Meningeal cells can communicate with astrocytes by calcium signalling. *Ann. Neurol.* **47,** 18–25.

Granados, M. P., Salido, G. M., Pariente, J. A., and González, A. (2004). Generation of ROS in response to CCK-8 stimulation in mouse pancreatic acinar cells. *Mitochondrion* **3,** 285–296.

Guasch, R. M., Blanco, A. M., Pérez-Aragó, A., Miñambres, R., Talens-Visconti, R., Peris, B., and Guerri, C. (2007). RhoE participates in the stimulation of the inflammatory response induced by ethanol in astrocytes. *Exp. Cell Res.* **313,** 3779–3788.

Guerri, C., and Renau-Piqueras, J. (1997). Alcohol, astroglia and brain development. *Mol. Neurobiol.* **15,** 65–81.

Guerri, C., Pascual, M., and Reanau-Piqueras, J. (2001). Glial and fetal alcohol syndrome. *Neurotoxicology* **22,** 593–599.

Guizzetti, M., and Costa, L. G. (2007). Cholesterol homeostasis in the developing brain: A possible new target for ethanol. *Human Exp. Toxicol.* **26,** 355–360.

Guthrie, P. B., Knappenberger, J., Segal, M., Bennett, M. V. L., Charles, A. C., and Kater, S. B. (1999). ATP released from astrocytes mediates glial calcium waves. *J. Neurosci.* **19,** 520–528.

Haddad, J. J. (2004). Alcoholism and neuro-immune-endocrine interactions: Physiochemical aspects. *Biochem. Biophys. Res. Commun.* **323,** 361–371.

Harden, T. K., and Lazarowski, E. R. (1999). Release of ATP and UTP from astrocytoma cells. *Prog. Brain Res.* **120,** 135–143.

Haslew-Layton, R. E., Mongin, A., and Kimelberg, H. K. (2005). Hydrogen peroxide potentiates volume-sensitive excitatory amino acid release via a mechanism involving Ca^{2+}/calmodulin-dependent protein kinase II. *J. Biol. Chem.* **280,** 3548–3554.

Haydon, P. G. (2000). Neuroglial networks: Neurons and glia talk to each other. *Curr. Biol.* **10,** R712–714.

Haydon, P. G. (2001). GLIA: Listening and talking to the synapse. *Nat. Rev. Neurosci.* **2,** 185–193.

Haydon, P. G., and Carmignoto, G. (2006). Astrocyte control of synaptic transmission and neurovascular coupling. *Physiol. Rev.* **86,** 1009–1031.

Hazell, A. S. (2007). Excitotoxic mechanisms in stroke: An update of concepts and treatment strategies. *Neurochem. Int.* **50,** 941–953.

Hertz, L., Zhao, Z., and Chen, Y. (2006). The astrocytic GABA(A)/benzodiazepine-like receptor: The Joker receptor for benzodiazepine-mimetic drugs? *Recent Patents CNS Drug Discovery* **1,** 93–103.

Hirata, H., Machado, L. S., Okuno, C. S., Brasolin, A., Lopes, G. S., and Smaili, S. S. (2006). Apoptotic effect of ethanol is potentiated by caffeine-induced calcium release in rat astrocytes. *Neurosci. Lett.* **393,** 136–140.

Holmberg, K. H., and Patterson, P. H. (2006). Leukemia inhibitory factor is a key regulator of astrocytic, microglial and neuronal responses in a low-dose pilocarpine injury model. *Brain Res.* **1075,** 26–35.

Hou, M., Duan, L., and Slaughter, M. M. (2008). Synaptic inhibition by glycine acting at a metabotropic receptor in tiger salamander retina. *J. Physiol.* **586,** 2913–2926.

Huettner, J. E. (2003). Kainate receptors and synaptic transmission. *Prog. Neurobiol.* **70,** 387–407.

Hsieh, H. L., Wu, C. Y., and Yang, C. M. (2008). Bradykinin induces matrix metalloproteinase-9 expression and cell migration through a PKC-delta-dependent ERK/Elk-1 pathway in astrocytes. *Glia* **56,** 619–632.

Jacobson, J., and Duchen, M. R. (2002). Mitochondrial oxidative stress and cell death in astrocytes-requirement for stored Ca^{2+} and sustained opening of the permeability transition pore. *J Cell Sci.* **115,** 1175–1188.

Jourdain, P., Bergersen, L. H., Bhaukaurally, K., Bezzi, P., Santello, M., Domercq, M., Matute, C., Tonello, F., Gundersen, V., and Volterra, A. (2007). Glutamate exocytosis from astrocytes controls synaptic strength. *Nat. Neurosci.* **10,** 331–339.

Jozwiak-Bebenista, M., Dejda, A., and Nowak, J. Z. (2007). Effects of PACAP, VIP and related peptides on cyclic AMP formation in rat neuronal and astrocyte cultures and cerebral cortical slices. *Pharmacol. Rep.* **59,** 414–420.

Kielian, T. (2008). Glial connexins and gap junctions in CNS inflammation and disease. *J. Neurochem.* **106,** 1000–1016.

Koyama, Y., and Tanaka, K. (2008). Endothelins stimulate the production of stromelysin-1 in cultured rat astrocytes. *Biochem. Biophys. Res. Commun.* **371,** 659–663.

Kuffler, S. W. (1984). Physiology of neuroglia cells. *In* "From neuron to brain," (S. W. Kuffler, J. G. Nicholls, and A. R. Martin, Eds.), 2nd ed. p. 324. Sinauer, Sunderland, MA.

Lamarche, F., Gonthier, B., Signorini, N., Eysseric, H., and Barret, L. (2004a). Impact of ethanol and acetaldehyde on DNA and cell viability of cultured neurones. *Cell Biol. Toxicol.* **20,** 361–374.

Lamarche, F., Signorini-Allibe, N., Gonthier, B., and Barret, L. (2004b). Influence of vitamin E, sodium selenite, and astrocyte-conditioned medium on neuronal survival after chronic exposure to ethanol. *Alcohol* **33,** 127–138.

Lee, J. M., Grabb, M. C., Zipfel, G. J., and Choi, D. W. (2000). Brain tissue responses to ischemia. *J. Clin. Invest.* **106,** 723–7731.

Lee, Y. C., Chien, C. L., Sun, C. N., Huang, C. L., Huang, N. K., Chiang, M. C., Lai, H. L., Lin, Y. S., Chou, S. Y., Wang, C. K., Tai, M. H., Liao, W. L., *et al.* (2003). Characterization of the rat A2A adenosine receptor gene: A 4.8-kb promoter-proximal DNA fragment confers selective expression in the central nervous system. *Eur. J. Neurosci.* **18,** 1786–1796.

Lee, M. Y., Song, H., Nakai, J., Ohkura, M., Kotlikoff, M. I., Kinsey, S. P., Golovina, V. A., and Blaustein, M. P. (2006). Local subplasma membrane Ca^{2+} signals detected by a tethered Ca^{2+} sensor. *Proc. Natl Acad. Sci. USA* **103,** 13232–13237.

Leybaert, L., Cabooter, L., and Braet, K. (2004). Calcium signal communication between glial and vascular brain cells. *Acta Neurologica Belgica* **104,** 51–56.

Li, Z., Miyata, S., and Hatton, G. I. (1999). Inositol 1,4,5-trisphosphate-sensitive Ca^{2+} stores in rat supraoptic neurons: Involvement in histamine-induced enhancement of depolarizing afterpotentials. *Neuroscience* **93,** 667–674.

Li, L., Lundkvist, A., Andersson, D., Wilhelmsson, U., Nagai, N., Pardo, A. C., Nodin, C., Ståhlberg, A., Aprico, K., Larsson, K., Yabe, T., Moons, L., *et al.* (2008). Protective role of reactive astrocytes in brain ischemia. *J. Cereb. Blood Flow Metabol.* **28,** 468–481.

Liao, C. H., Ho, C. T., and Lin, J. K. (2005). Effects of garcinol on free radical generation and NO production in embryonic rat cortical neurons and astrocytes. *Biochem. Biophys. Res. Commun.* **329,** 1306–1314.

Liedtke, W., Edelmann, W., Bieri, P. L., Chiu, F. C., Cowan, N. J., Kucherlapati, R., and Raine, C. S. (1996). GFAP is necessary for the integrity of CNS white matter architecture and long-term maintenance of myelination. *Neuron* **17,** 607–615.

Lo, J. C., Huang, W. C., Chou, Y. C., Tseng, C. H., Lee, W. L., and Sun, S. H. (2008). Activation of P2X(7) receptors decreases glutamate uptake and glutamine synthetase activity in RBA-2 astrocytes via distinct mechanisms. *J. Neurochem.* **105,** 151–164.

Malarkey, E. B., and Parpura, V. (2008). Mechanisms of glutamate release from astrocytes. *Neurochem. Int.* **52,** 142–154.

Marín, M. P., Tomas, M., Esteban-Pretel, G., Megías, L., López-Iglesias, C., Egea, G., and Renau-Piqueras, J. (2008). Chronic ethanol exposure induces alterations in the nucleocytoplasmic transport in growing astrocytes. *J. Neurochem.* **106,** 1914–1928.

Marriot, D. R., Wilkin, G. P., and Wood, J. N. (1993). Substance P-induced release of prostaglandins from astrocytes: Regional sperecialisation and correlation with phosphoinositol metabolism. *J. Neurochem.* **56,** 259–265.

Marshak, D. R., Pesce, S. A., Stanley, L. C., and Griffin, W. S. (1992). Increased S100 beta neurotrophic activity in Alzheimer's disease temporal lobe. *Neurobiol. Aging* **13,** 1–7.

Martínez, S. E., Lázaro-Diéguez, F., Selva, J., Calvo, F., Piqueras, J. R., Crespo, P., Claro, E., and Egea, G. (2007). Lysophosphatidic acid rescues RhoA activation and phosphoinositides levels in astrocytes exposed to ethanol. *J. Neurochem.* **102,** 1044–1052.

McCaffery, P., Koul, O., Smith, D., Napoli, J. L., Chen, N., and Ullman, M. D. (2004). Ethanol increases retinoic acid production in cerebellar astrocytes and in cerebellum. *Brain Res. Develop. Brain Res.* **153,** 233–241.

McCall, M. A., Gregg, R. G., Behringer, R. R., Brenner, M., Delaney, C. L., Galbreath, E. J., Zhang, C. L., Pearce, R. A., Chiu, S. Y., and Messing, A. (1996). Targeted deletion in astrocyte intermediate filament (Gfap) alters neuronal physiology. *Proc. Natl Acad. Sci. USA* **93,** 6361–6366.

Miguel-Hidalgo, J. J. (2006). Withdrawal from free-choice ethanol consumption results in increased packing density of glutamine synthetase-immunoreactive astrocytes in the prelimbic cortex of alcohol-preferring rats. *Alcohol Alcoholism* **41,** 379–385.

Min, M. Y., Wu, Y. W., Shih, P. Y., Lu, H. W., Lin, C. C., Wu, Y., Li, M. J., and Yang, H. W. (2008). Physiological and morphological properties of, and effect of substance P on, neurons in the A7 catecholamine cell group in rats. *Neuroscience* **153,** 1020–1033.

Miñambres, R., Guasch, R. M., Perez-Aragó, A., and Guerri, C. (2006). The RhoA/ROCK-I/MLC pathway is involved in the ethanol-induced apoptosis by anoikis in astrocytes. *J. Cell Sci.* **119,** 271–282.

Molz, S., Decker, H., Dal-Cim, T., Cremonez, C., Cordova, F. M., Leal, R. B., and Tasca, C. I. (2008). Glutamate-induced toxicity in hippocampal slices involves apoptotic features and p38(MAPK) signalling. *Neurochem. Res.* **33,** 27–36.

Montana, V., Malarkey, E. B., Verderio, C., Matteoli, M., and Parpura, V. (2006). Vesicular transmitter release from astrocytes. *Glia* **54,** 700–715.

Mori, T., Town, T., Tan, J., Yada, N., Horikoshi, Y., Yamamoto, J., Shimoda, T., Kamanaka, Y., Tateishi, N., and Asano, T. (2006). Arundic acid ameliorates cerebral amyloidosis and gliosis in Alzheimer transgenic mice. *J. Pharmacol. Exp. Therap.* **318,** 571–578.

Mothet, J. P., Pollegioni, L., Ouanounou, G., Martineau, M., Fossier, P., and Baux, G. (2005). Glutamate receptor activation triggers a calcium-dependent and SNARE protein-dependent release of the gliotransmitter D-serine. *Proc. Natl Acad. Sci. USA* **102,** 5606–5611.

Mulligan, S. J., and MacVicar, B. A. (2004). Calcium transients in astrocyte endfeet cause cerebrovascular constrictions. *Nature* **431,** 195–199.

Nakanishi, S. (1994). Metabotropic glutamate receptors: Synaptic transmission, modulation, and plasticity. *Neuron* **13,** 1031–1037.

Nardin, P., Tramontina, F., Leite, M. C., Tramontina, A. C., Quincozes-Santos, A., Vieira de Almeida, L. M., Battastini, A. B., Gottfried, C., and Gonçalves, C. A. (2007). S100B content and secretion decrease in astrocytes cultured in high-glucose medium. *Neurochem. Int.* **50,** 774–782.

Neary, J. T., Kang, Y., Bu, Y., Yu, E., Akong, K., and Peters, C. M. (1999). Mitogenic signaling by ATP/P2Y purinergic receptors in astrocytes: Involvement of a calcium-independent protein kinase C, extracellular signal-regulated protein kinase pathway distinct from the phosphatidylinositol-specific phospholipase C/calcium pathway. *J. Neurosci.* **19,** 4211–4220.

Neary, J. T., Kang, Y., and Shi, Y. F. (2004). Signaling from nucleotide receptors to protein kinase cascades in astrocytes. *Neurochem. Res.* **29,** 2037–2042.

Newman, E. A. (2006). A purinergic dialogue between glia and neurons in the retina. *Novartis Found. Symp.* **276,** 193–202.

Newman, E. A., and Zahs, K. R. (1997). Calcium waves in retinal glial cells. *Science* **275,** 844–847.

Ni, Y., Malarkey, E. B., and Púrpura, V. (2007). Vesicular release of glutamate mediates bidirectional signaling between astrocytes and neurons. *J. Neurochem.* **103,** 1273–1284.

Nicoletti, F., Arcella, A., Iacovelli, L., Battaglia, G., Giangaspero, F., and Melchiorri, D. (2007). Metabotropic glutamate receptors: New targets for the control of tumor growth? *Trends Pharmacol. Sci.* **28,** 206–213.

Noda, M., Sasaki, K., Ifuku, M., and Wada, K. (2007). Multifunctional effects of bradykinin on glial cells in relation to potential anti-inflammatory effects. *Neurochem. Int.* **51,** 185–191.

Ogata, K., and Kosaka, T. (2002). Structural and quantitative analysis of astrocytes in the mouse hippocampus. *Neuroscience* **113,** 221–233.

Ono, K., Toyono, T., and Inenaga, K. (2008). Nicotinic receptor subtypes in rat subfornical organ neurons and glial cells. *Neuroscience* **154,** 994–1001.

Opitz, C. F., Ewert, R., Kirch, W., and Pittrow, D. (2008). Inhibition of endothelin receptors in the treatment of pulmonary arterial hypertension: Does selectivity matter? *Eur. Heart J.* **29,** 1936–1948.

Orkand, R. K., Orkand, P. M., and Tang, C. M. (1981). Membrane properties of neuroglia in the optic nerve of Necturus. *J. Exp. Biol.* **95,** 49–59.

Örnung, G., Shupliakov, O., Ottersen, O. P., Storm-Mathisen, J., and Cullheim, S. (1994). Immuno-histochemical evidence for coexistence of glycine and GABA in nerve terminals on cat spinal motoneurones: An ultrastructural study. *Neuroreport* **5,** 889–892.

Oikawa, H., Nakamichi, N., Kambe, Y., Ogura, M., and Yoneda, Y. (2005). An increase in intracellular free calcium ions by nicotinic acetylcholine receptors in a single cultured rat cortical astrocytes. *J. Neurosci. Res.* **79,** 535–544.

Paemeleire, K. (2002). Calcium signaling in and between brain astrocytes and endothelial cells. *Acta Neurol. Belg.* **102,** 137–140.

Palomba, L., Amadori, A., and Cantoni, O. (2007). Early release of arachidonic acid prevents an otherwise immediate formation of toxic levels of peroxynitrite in astrocytes stimulated with lipopolysaccharide/interferon-gamma. *J. Neurochem.* **103,** 904–913.

Palucha, A., and Pilc, A. (2007). Metabotropic glutamate receptor ligands as possible anxiolytic and antidepressant drugs. *Pharmacol. Therap.* **115,** 116–147.

Parent, A. S., Rasier, G., Matagne, V., Lomniczi, A., Lebrethon, M. C., Gérard, A., Ojeda, S. R., and Bourguignon, J. P. (2008). Oxytocin facilitates female sexual maturation through a glia-to-neuron signaling pathway. *Endocrinology* **149,** 1358–1365.

Parpura, V., Basarsky, T. A., Liu, F., Jeftinija, K., Jeftinija, S., and Haydon, P. G. (1994). Glutamate-mediated astrocyte-neuron signalling. *Nature* **369,** 744–747.

Parpura, V., Scemes, E., and Spary, D. C. (2004). Mechanisms of glutamate release from astrocytes: Gap junction "hemichanneles," purinergic receptors and exocytotic release. *Neurochem. Int.* **45**, 259–264.

Pasti, L., Zonta, M., Pozzan, T., Vicini, S., and Carmignoto, G. (2001). Cytosolic calcium oscillations in astrocytes may regulate exocytotic release of glutamate. *J. Neurosci.* **21**, 477–484.

Peakman, M. C., and Hill, D. S. J. (1995). Adenosine A1 receptor-mediated changes in basal and histamine-stimulated levels of intracellular calcium in primary rat astrocytes. *Brit. J. Pharmacol.* **115**, 801–810.

Perea, G., and Araque, A. (2007). Astrocytes potentiate transmitter release at single hippocampal synapses. *Science* **317**, 1083–1086.

Persidsky, Y., Ramirez, S. H., Haorah, J., and Kanmogne, G. D. (2006). Blood-brain barrier: Structural components and function under physiologic and pathologic conditions. *J. Neuroimmun. Pharmacol.* **1**, 233–236.

Phillias, J. W., and O'Reagan, M. H. (2002). Evidence for swelling-induced adenosine and adenine nucleotide release in rat cerebral cortex exposed to monocarboxylate-containing of hypotonic artificial cerebrospinal fkluids. *Neurochem. Int.* **40**, 629–635.

Pierson, P. M., Peteri-Brunbäck, B., Pisani, D. F., Abbracchio, M. P., Mienville, J. M., and Rosso, L. (2007). A(2b) receptor mediates adenosine inhibition of taurine efflux from pituicytes. *Biol. Cell* **99**, 445–554.

Pin, J. P., and Duvoisin, R. (1995). The metabotropic glutamate receptors: Structure and functions. *Neuropharmacology* **34**, 1–26.

Proia, P., Schiera, G., Mineo, M., Ingrassia, A. M., Santoro, G., Savettieri, G., and Di Liegro, I. (2008). Astrocytes shed extracellular vesicles that contain fibroblast growth factor-2 and vascular endothelial growth factor. *Int. J. Mol. Med.* **21**, 63–67.

Ramachandran, V., Perez, A., Chen, J., Senthil, D., Schenker, S., and Henderson, G. I. (2001). In utero ethanol exposure causes mitochondrial dysfunction, which can result in apoptotic cell death in fetal brain: A potential role for 4-hydroxynonenal. *Alcoholism Clin. Exp. Res.* **25**, 862–871.

Rathinam, M. L., Watts, L. T., Stark, A. A., Mahimainathan, L., Stewart, J., Shenker, S., and Henderson, G. I. (2006). Astrocyte control of fetal cortical neuron glutathione homeostasis: Up-regulation by ethanol. *J. Neurochem.* **96**, 1289–1300.

Rodríguez-Moreno, A., and Sihra, T. S. (2007). Metabotropic actions of kainate receptors in the CNS. *J. Neurochem.* **103**, 2121–2135.

Rosenberg, P. A., and Li, Y. (1995). Adenylyl cyclase activation underlies intracellular cyclic AMP accumulation, cyclic AMP transport, and extracellular adenosine accumulation evoked by beta-adrenergic receptor stimulation in mixed cultures of neurons and astrocytes derived from rat cerebral cortex. *Brain Res.* **692**, 227–232.

Russo, A., Palumbo, M., Scifo, C., Cardile, V., Barcellona, M. L., and Renis, M. (2001). Ethanol-induced oxidative stress in rat astrocytes: Role of HSP70. *Cell Biol. Toxicol.* **17**, 153–168.

Saez, J. C., Connor, J. A., Spray, D. C., and Bennett, M. V. (1989). Hepatocyte gap juntions are permeable to the second messenger, inositol 1,4,5-trsiphosphate, and to calcium ions. *Proc. Natl Acad. Sci. USA* **86**, 2708–2712.

Sakurai, T., Miki, T., Li, H. P., Miyatake, A., Satriotomo, I., and Takeuchi, Y. (2003). Colocalization of taurine and glial fibrillary acidic protein immunoreactivity in mouse hippocampus induced by short-term ethanol exposure. *Brain Res.* **959**, 160–164.

Salazar, M., Pariente, J. A., Salido, G. M., and González, A. (2008). Ethanol induces glutamate secretion by Ca^{2+} mobilization and ROS generation in rat hippocampal astrocytes. *Neurochem. Int.* **52**, 1061–1067.

Salter, M. W., and Hicks, J. L. (1994). ATP-evoked increases intracellular calcium in cultured neurons and glia from the dorsal spinal cord. *J. Neurosci.* **14**, 1563–1575.

Scemes, E., Suadicani, S. O., and Spray, D. C. (2000). Intercellular communication in spinal cord astrocytes: Fine tuning between gap junctions and P2 nucleotide receptors in calcium wave propagation. *J. Neurosci.* **20,** 1435–1445.

Schatter, B., Jin, S., Löffelholz, K., and Klein, J. (2005). Cross-talk between phosphatidic acid and ceramide during ethanol-induced apoptosis in astrocytes. *BMC Pharmacol.* **5,** 3.

Sepúlveda, M. R., and Mata, A. M. (2004). The interaction of ethanol with reconstituted synaptosomal plasma membrane Ca^{2+}-ATPase. *Biochim. Biophys. Acta* **1665,** 75–80.

Sharma, G., and Vijayaraghavan, S. (2001). Nicotinic cholinergic signaling in hippocampal astrocytes involves calcium-induced calcium release from intracellular stores. *Proc. Natl Acad. Sci. USA* **98,** 4148–4153.

Shelton, M. K., and McCarthy, K. D. (2000). Hippocampal astrocytes exhibit Ca^{2+}-elevating muscarinic cholinergic and histaminergic receptors *in situ. J. Neurochem.* **74,** 555–563.

Sheng, J. G., Ito, K., Skinner, R. D., Mrak, R. E., Rovnaghi, C. R., Van Eldik, L. J., and Griffin, W. S. (1996). *In vivo* and *in vitro* evidence supporting a role for the inflammatory cytokine interleukin-1 as a driving force in Alzheimer pathogenesis. *Neurobiol. Aging* **17,** 761–766.

Signorini-Allibe, N., Gonthier, B., Lamarche, F., Eysseric, H., and Barret, L. (2005). Chronic consumption of ethanol leads to substantial cell damage in cultured rat astrocytes in conditions promoting acetaldehyde accumulation. *Alcohol Alcoholism* **40,** 163–171.

Skaper, S. D., Ancona, B., Facci, L., Franceschini, D., and Giusti, P. (1998). Melatonin prevents the delayed death of hippocampal neurons induced by enhanced excitatory neurotransmission and the nitridergic pathway. *FASEB J* **12,** 725–731.

Sofroniew, M. V., Bush, T. G., Blumauer, N., Lawrence, K., Mucke, L., and Johnson, M. H. (1999). Genetically-targeted and conditionally-regulated ablation of astroglial cells in the central, enteric and peripheral nervous systems in adult transgenic mice. *Brain Res.* **835,** 91–95.

Sommer, B., Keinanen, K., Verdoorn, T. A., Wisden, W., Burnashev, N., Herb, A., Kohler, M., Takagi, T., Sakmann, B., and Seeburg, P. H. (1990). Flip and flop: A cell-specific functional switch in glutamate-operated channels of the CNS. *Science* **249,** 1580–1585.

Syed, N., Martens, C. A., and Hsu, W. H. (2007). Arginine vasopressin increases glutamate release and intracellular Ca^{2+} concentration in hippocampal and cortical astrocytes through two distinct receptors. *J. Neurochem.* **103,** 229–237.

Takuma, K., Baba, A., and Matsuda, T. (2004). Astrocyte apoptosis: Implications for neuroprotection. *Prog. Neurobiol.* **72,** 111–127.

Tateno, M., Ukai, W., Yamamoto, M., Hashimoto, E., Ikeda, H., and Saito, T. (2005). The effect of ethanol on cell fate determination of neural stem cells. *Alcoholism Clin. Exp. Res.* **29,** 225S–229S.

Taylor, J. M., and Crack, P. J. (2004). Impact of oxidative stress on neuronal survival. *Clin. Exp. Pharmacol. Physiol.* **31,** 397–406.

Tran, M. D., Wanner, I. B., and Neary, J. T. (2008). Purinergic receptor signaling regulates N-cadherin expression in primary astrocyte cultures. *J. Neurochem.* **105,** 272–286.

Todd, A. J., Watt, C., Spike, R. C., and Sieghart, W. (1996). Colocalization of GABA, glycine, and their receptors at synapses in the rat spinal cord. *J. Neurosci.* **16,** 974–982.

Turrens, J. F. (2003). Mitochondrial formation of reactive oxygen species. *J. Physiology* **522,** 335–344.

Vallés, S. L., Blanco, A. M., Pascual, M., and Guerri, C. (2004). Chronic ethanol treatment enhances inflammatory mediators and cell death in the brain and in astrocytes. *Brain Pathol.* **14,** 365–371.

Verkhratsky, A. (2006). Patching the glia reveals the functional organisation of the brain. *Pflügers Archives: Eur. J. Physiol.* **453,** 411–420.

Verkhratsky, A., and Kettenmann, H. (1996). Calcium signalling in glial cells. *Trends Neurosci.* **19,** 346–352.

Verkhratsky, A., Orkand, R. K., and Kettenmann, H. (1998). Glial calcium: Homeostasis and signalling function. *Physiological Rev.* **78,** 99–141.

Verkhratsky, A., Solovyeva, N., and Toescu, E. C. (2002). Calcium excitability in glial cells. *In* "The Tripartite Synapse: Glia in Synaptic Transmission," (A. Volterra, P. Magistreti, and P. Haydon, Eds.), pp. 99–109. Oxford University Press, Oxford.

Vestergaard, H. T., Vogensen, S. B., Madsen, U., and Ebert, B. (2004). Analogues of homoibotenic acid show potent and selective activity following sensitization by quisqualic acid. *Eur. J. Pharmacol.* **488,** 101–109.

Volterra, A., and Meldolesi, J. (2005). Astrocytes, from brain glue to communication elements: The revolution continues. *Nat. Rev. Neurosci.* **6,** 626–640.

Wang, Y. B., Peng, C., and Liu, Y. H. (2007). Low dose of bradykinin selectively increases intracellular calcium in glioma cells. *J. Neurol. Sci.* **258,** 44–51.

Watts, L. T., Rathinam, M. L., Schenker, S., and Henderson, G. I. (2005). Astrocytes protect neurons from ethanol-induced oxidative stress and apoptotic death. *J. Neurosci. Res.* **80,** 655–666.

Willis, M., Hutter-Paier, B., Wietzorrek, G., Windisch, M., Humpel, C., Knaus, H. G., and Marksteiner, J. (2007). Localization and expression of substance P in transgenic mice overexpressing human APP751 with the London (V717I) and Swedish (K670M/N671L) mutations. *Brain Res.* **1143,** 199–207.

Willmott, N. J., Wong, K., and Strong, A. J. (2000). A fundmental role for the nitric oxide-G-kinase signaling pathway in mediating intercellular Ca(2+) waves in glia. *J. Neurosci.* **20,** 1767–1779.

Xu, J., Yu, S., Sun, A. Y., and Sujn, G. Y. (2003). Oxidant-mediated AA release from astrocytes involves cPLA(2) and iPLA(2). *Free Radical Biol. Med.* **34,** 1531–1543.

Xu, J. H., Hu, H. T., Liu, Y., Qian, Y. H., Liu, Z. H., Tan, Q. R., and Zhang, Z. J. (2006). Neuroprotective effects of ebselen are associated with the regulation of Bcl-2 and Bax proteins in cultured mouse cortical neurons. *Neurosci. Lett.* **399,** 210–214.

Yang, J., and Aschner, M. (2003). Developmental aspects of blood-brain barrier (BBB) and rat brain endothelial (RBE4) cells as in vitro model for studies on chlorpyrifos transport. *Neurotoxicology* **24,** 741–745.

Yoshida, H., Imaizumi, T., Tanji, K., Sakaki, H., Metoki, N., Hatakeyama, M., Yamashita, K., Ishikawa, A., Taima, K., Sato, Y., Kimura, H., and Satoh, K. (2005). Platelet-activating factor enhances the expression of nerve growth factor in normal human astrocytes under hypoxia. *Brain Res. Mol. Brain Res.* **133,** 95–101.

Yuste, R., Nelson, D. A., Rubin, W. W., and Katz, L. C. (1995). Neuronal domains in development neocortex: Mechanisms of coactivation. *Neuron* **14,** 7–17.

Zhang, L. H., Gong, N., Fei, D., Xu, L., and Xu, T. L. (2008). Glycine uptake regulates hippocampal network activity via glycine receptor-mediated tonic inhibition. *Neuropsychopharmacology* **33,** 701–711.

Zimatkin, S. M., Pronko, S. P., Vasiliou, V., Gonzalez, F. J., and Deitrich, R. A. (2006). Enzymatic mechanisms of ethanol oxidation in the brain. *Alcoholism* **30,** 1500–1505.

Zimmermann, H., Braun, N., Kegel, B., and Heine, P. (1998). New insights into molecular structure and function of ectonucleotidases in the nervous system. *Neurochem. Int.* **32,** 421–425.

Zink, M., Schmitt, A., Vengeliene, V., Henn, F. A., and Spanagel, R. (2004). Ethanol induces expression of the glutamate transporters EAAT1 and EAAT2 in organotypic cortical slice cultures. *Alcoholism Clin. Exp. Res.* **28,** 1752–1757.

Zipfel, G. J., Babcock, D. J., Lee, J. M., and Choi, D. W. (2000). Neuronal apoptosis after CNS injury: The roles of glutamate and calcium. *J. Neurotrauma* **17,** 857–869.

Zhou, F. C., Anthony, B., Dunn, K. W., Lindquist, W. B., Xu, Z. C., and Deng, P. (2007). Chronic alcohol drinking alters neuronal dendritic spines in the brain reward center nucleus accumbens. *Brain Res.* **1134,** 148–161.

THERAPEUTIC TARGETING OF "DARPP-32": A KEY SIGNALING MOLECULE IN THE DOPIMINERGIC PATHWAY FOR THE TREATMENT OF OPIATE ADDICTION

Supriya D. Mahajan,* Ravikumar Aalinkeel,* Jessica L. Reynolds,*
Bindukumar B. Nair,* Donald E. Sykes,* Zihua Hu,[†] Adela Bonoiu,[‡] Hong Ding,[‡]
Paras N. Prasad,[‡] and Stanley A. Schwartz*

*Department of Medicine, Division of Allergy, Immunology, and Rheumatology, Buffalo
General Hospital, State University of New York, Buffalo, New York 14203, USA
[†]Department of Biostatistics, Center for Computational Research, State University of New York
(SUNY), Buffalo, New York 14260, USA
[‡]Institute for Lasers Photonics and Biophotonics, State University of New York (SUNY), Buffalo,
New York 14260, USA

INTERNATIONAL REVIEW OF
NEUROBIOLOGY, VOL. 88
DOI: 10.1016/S0074-7742(09)88008-2

199

The 32-kDa dopamine- and adenosine 3′,5′-monophosphate-regulated phosphoprotein (DARPP-32) is recognized to be critical to the pathogenesis of drug addiction. Opiates via the μ-receptor act on the dopaminergic system in the brain and modulates the expression of DARPP-32 phosphoprotein which is an important mediator of the activity of the extracellular signal-regulated kinase (ERK) signaling cascades, the activation of which represents an exciting nexus for drug-induced changes in neural long-term synaptic plasticity. Silencing of DARPP-32 using an siRNA against DARPP-32 may provide a novel gene therapy strategy to overcome drug addiction. In this study, we investigated the effect of the opiate (heroin) on D1 receptor (D1R) and DARPP-32 expression and additionally, evaluated the effects of DARPP-32-siRNA gene silencing on protein phosphatase-1 (PP-1), ERK, and cAMP response element-binding (CREB) gene expression in primary normal human astrocytes (NHA) cells *in vitro*. Our results indicate that heroin significantly upregulated both D1R and DARPP-32 gene expression, and that DARPP-32 silencing in the NHA cells resulted in the significant modulation of the activity of downstream effector molecules such as PP-1, ERK, and CREB which are known to play an important role in opiate abuse-induced changes in long-term neural plasticity. These findings have the potential to facilitate the development of DARPP32 siRNA-based therapeutics against drug addiction.

I. Introduction

Drug addiction is a chronic, relapsing disease, characterized by compulsive drug-seeking behavior that is caused by neurochemical and molecular changes in the brain. Substance abuse is a major public health concern in the United State and several factors contribute to the development and persistence of drug addiction. The key among them is alteration in the individual's neurophysiological functioning induced by the addictive drugs. It is believed that the neurochemical/neurophysiological alterations in the brain caused by addictive drugs have a cellular and molecular basis and may be persistent and contribute to changes in behavior leading to drug addiction. Drugs like opiates induce dopamine-receptor stimulation leading to induction of transcription factors and phosphorylation of many substrate proteins involved in neuronal excitability (Colvis *et al.*, 2005; Fienberg *et al.*, 1998, 2000; Greengard *et al.*, 1998; Guitart-Masip *et al.*, 2006; Nestler, 2005; Nestler *et al.*, 1994; Self *et al.*, 1998). Disruptions in the dynamic balance of dopamine-receptor-mediated phosphorylation and dephosphorylation cascades leads to impaired integration of synaptic inputs that can cause altered neuronal communication and disruption of this balance. This, in turn, may result in the induction of transcription factors and their downstream targets causing long-lasting plastic changes in the brain (Nairn *et al.*, 2004; Takahashi *et al.*, 2005). This process may be the principal *molecular mechanism* underlying drug addiction. Mechanisms by which neuroimmune signaling affects the brain under normal conditions or during pathologic processes remain largely unexplored.

A. MOLECULAR MECHANISMS UNDERLYING OPIATE ADDICTION

Long-term use of opiates leads to tolerance, sensitization, and physical dependence. High rates of relapse to drug use following prolonged withdrawal periods characterize the behavior of experienced heroin users (O'Brien, 1997). In heroin-free individuals, drug craving and relapse to drug use can be triggered by stress (Kreek and Koob, 1998) and by stimuli previously associated with drug use (Carter and Tiffany, 1999; Childress *et al.*, 1992). Many drugs of abuse, administered repeatedly over time, cause tolerance and physical dependence. Tolerance and physical dependence have been correlated with changes in the intracellular cAMP signal transduction cascade. Elements of the cascade found to be altered include G-proteins, adenylate cyclase, protein kinase A (PKA), and its target CREB (cAMP response element-binding) (Nestler *et al.*, 1994; Wang and Gintzler, 1997; Wang *et al.*, 1996). The activity of CREB is regulated by phosphorylation and opiates have been shown to alter phosphorylation of CREB protein. CREB may serve as a final common mediator in the ultimate expression of both positive and negative reinforcing properties of drugs of abuse like opiates (Widnell *et al.*, 1996). Elegant studies by Walters *et al.* (2001) demonstrated that activation of CREB plays a complex role in the neuroadaptive processes associated with opiate addiction. Repeated opiate exposure induces biochemical changes in specific brain regions, predominantly the mesolimbic dopamine system which mediates the reinforcing actions of the opiates.

B. OPIATE-INDUCED DOPAMINE-RECEPTOR STIMULATION ACTIVATES THE cAMP-DEPENDENT PKA–DARPP-32 SIGNALING PATHWAY

Opiate dependence is characterized by enhanced neuronal excitability associated with upregulation of the cAMP second messenger system. Opiate-induced dopamine-receptor stimulation activates the cAMP-dependent PKA pathway, leading to induction of transcription factors and phosphorylation of many substrate proteins involved in neuronal excitability. Opiate administration can also alter gene expression in different regions of the brain, which may contribute to the plastic changes associated with addictive behavior (Calabresi *et al.*, 2000). An extensive bidirectional communication takes place between the nervous and the immune systems. Both immune cells and neuroreactive molecules may modulate brain function through multiple signaling pathways. DARPP-32 is recognized as critical to the pathogenesis of drug addiction (Greengard *et al.*, 1998, 1999). Opiates act on the dopaminergic system in the brain and perturb DARPP-32 function. DARPP-32 is involved in mediating the actions of virtually all neurotransmitters in all parts of the brain (Greengard *et al.*, 1998, 1999; Nairn *et al.*, 2004; Svenningsson *et al.*, 2004, 2005). Neurotransmitters such as dopamine, through direct or indirect pathways, regulate the phosphorylation of DARPP-32. DARPP-32, when phosphorylated at Thr^{34}, acts as an amplifier of PKA-mediated signaling through its ability to potently inhibit protein phosphatase-1 (PP-1). This amplifying property of DARPP-32 is critical for dopaminergic signaling, but it is also utilized by multiple other neurotransmitters, including glutamate in various regions of the brain. In addition to its role as a PP-1 inhibitor, DARPP-32 when phosphorylated at Thr^{75} inhibits PKA. Upon dopaminergic neurotransmission, the phosphorylation state at Thr^{75} is decreased allowing disinhibition of PKA and further increasing phosphorylation at Thr^{34}. This complex positive feedback loop potentiates dopaminergic signaling (Greengard *et al.*, 1998, 1999; Nairn *et al.*, 2004; Nishi *et al.*, 1997; Svenningsson *et al.*, 2004, 2005).

Thus, modulation of the expression of DARPP-32 can cause an inhibition of PP-1 and a concomitant dysregulation of its downstream effector proteins, glycogen synthesis kinase-3 (GSK-3), CREB, and c-Fos. CREB mediates morphine-induced upregulation of specific components of the cAMP pathway that contributes to opiate dependence (Chao and Nestler, 2004; Mahajan *et al.*, 2005b). Thus, DARPP-32, a major regulatory hub of neurosignaling networks, is a unique potential target for new therapeutic approaches to modulate neural plasticity within the brain reward circuitry and break the vicious cycle of addiction.

Studies of the biochemical and molecular basis of drug addiction have several important clinical implications. A better understanding of the neurobiological mechanisms underlying the addictive action of drugs of abuse will help in the development of pharmacological agents that prevent or reverse drug abuse. The availability of such therapeutic agents would represent a significant advancement in our fight against drug addiction. Additionally, these studies will pave the way for further investigations that may ultimately yield designer drugs that may be taken orally that target DARPP-32 for the prevention or therapy of drug addiction.

II. Study Design

In the current study, we used primary normal human astrocytic cell cultures. Interactions between neurons and astrocytes are critical for signaling, energy metabolism, and neuroprotection in the central nervous system (CNS). Astrocytes face the synapses, send end-foot processes that enwrap the brain capillaries, and form an extensive network interconnected by gap junctions. Astrocytes express several membrane proteins and enzymes that are critical for uptake of neuro-transmittors such as dopamine, glutamate, etc. at the synapses and participate in detection, propagation, and modulation of excitatory synaptic signals, provide metabolic support to the active neurons. Normal human astrocytes (NHA), 1–3 million cells/ml, were treated with heroin (10^{-7}–10^{-11}) molar concentrations for 48 h, and the effect of heroin on D1 receptor (D1R) and DARPP-32 gene expression was evaluated. An initial gene microarray analysis is done using NHA cells treated with heroin (10^{-7} M) to identify genes that might play a key role in opiate addiction. Data from our gene microarray analysis of NHA treated with heroin (10^{-7} M) showed a significant modulation of the DARPP-32 gene and other members of this signaling cascade indicating the involvement and the significance of this pathway in the opiate addiction process. Since, NHA are among these several neuronal cells that expresses DARPP-32, we believe that silencing the expression of DARPP-32 in these cells, may result in the modulation of the expression activation of the extracellular signal-regulated kinase (ERK) mitogen-activated protein (MAP) kinase cascade that causes drug-induced changes in neural plasticity and thereby neurological dysfunction in opiate addicts. Therefore, using commercially available transfection reagent such as siPROT (Ambion Inc., Austin, TX), we did a transient transfection that silenced DARPP-32 gene expression in NHA cells by about 80% for a time period of up to 1 week. Since, optimal transfection efficiency was observed at 48 h, further experiments that evaluated the expression of secondary messengers PP-1, ERK, and CREB in DARPP32-siRNA transfected, scrambled siRNA transfected, and untransfected controls are done at 48 h posttransfection. These transfected cells were treated with/without heroin (10^{-7} M) for a period of 24 h posttransfection.

III. Material and Methods

A. CELL CULTURE: IN VITRO TREATMENT OF CELLS

We will use primary cultures of NHA obtained from Applied Cell Biology Research Institute (ACBRI), Kirkland, WA. Astrocytes were >95% GFAP posi-tive, as characterized by immunocytochemistry, and were >98% viable by trypan blue exclusion criteria. NHA are cultured in Astrocyte Growth Medium kit that includes serum-free basal medium with attachment factors (ABCRI, Cat # 4Z0-210) and Passage Reagent Group™ (ABCRI, Cat # 4Z0-800) and were passaged 1:4 at 80–90% confluence. NHA are obtained at passage 2 for each experiment and are used for all experimental paradigms between within the 8–20 cumulative population doublings. Previous kinetics (12–96 h) and dose (heroin [10^{-7}, 10^{-9}, and 10^{-11}]) response studies indicate that a 48-h time period and a dose of 10^{-7} M concentration heroin were optimal. The opiate (heroin), 10^{-7}–10^{-11} M, concentrations that we use in our study are well within the physiological range and have been used by us and other investigators in prior studies (Baritaki *et al.*, 2005; Mahajan *et al.*, 2005a,b; Peterson *et al.*, 2004; Singh *et al.*, 2004). Heroin hydrochloride (Cat # H5144; Sigma, St Loius, MO) is obtained as a 100 μg/ml stock solution.

B. PRODUCTION OF cDNA MICROARRAYS

The microarrays used in this experiment were produced at the Roswell Park Cancer Institute (RPCI) Microarray and Genomics Core Facility and contain approximately 5043 cDNA clones (Research Genetics).

C. PREPARATION AND HYBRIDIZATION OF FLUORESCENT LABELED cDNA

A total of six RNA samples (i.e., six slides or six cDNA arrays) that included the three separate NHA samples treated with heroin (10^{-7} M) for 48 h and the corresponding untreated samples were screened for gene expression. cDNA was synthesized from each of those RNA samples and the heroin (10^{-7} M) treated samples were labeled with Cy5 dye, while the corresponding untreated controls were labeled with the Cy3 dye using the Atlas Powerscript Fluorescent Labeling Kit (BD BioSciences). To minimize intra-array variations, two hybridizations for each sample were performed, details of hybridization procedures have been described earlier (Mahajan *et al.*, 2006). Additionally, the labeling with the Cy dye was interchanged, so as to have technical replicates of the matched pairs of heroin (10^{-7} M) treated and untreated sample. These standard normalization approaches correct the labeling bias between the Cy3 and Cy5 dyes (Mahajan *et al.*, 2006).

D. IMAGE ANALYSIS

The hybridized slides were scanned using a GenePix 4200A scanner to generate high-resolution (10 μm) images for both Cy3 and Cy5 channels. Image analysis was performed on the raw image files using ImaGene (current version 6.0.1) from BioDiscovery Inc. Each cDNA spot was defined by a circular region. The size of the region was programmatically adjusted to match the size of the spot. Local background for a spot was determined by ignoring a 2–3 pixel buffer region around the spot and then measuring signal intensity in a 2–3 pixel wide area outside the buffer region. Raw signal intensity values for each spot and its background region were segmented using a proprietary optimized segmentation algorithm, which excludes pixels that were not representative of the majority pixels in that region. The background corrected signal for each cDNA spot was the mean signal (of all the pixels in the region)—mean local background. The output of the image analysis was two tab delimited files, one for each channel, containing all of the raw fluorescence data.

E. Microarray Data Processing and Analysis

Expression data extracted from image files were first checked by a M $(\log_2(\text{Cy3}/\text{Cy5}))$ versus A $[(\log_2(\text{Cy3}) + \log_2(\text{Cy5}))/2)]$ plot to see whether intensity-dependent expression bias existed between spots (genes) labeled with Cy3 and Cy5 on each individual slide. After finding that intensity-dependent expression bias existed for all slides, we first performed a Lowess data normalization to correct the observed intensity-dependent expression bias. We then performed a global normalization to bring the median expression values of Cy3 and Cy5 on all three slides to the same scale. This was done by selecting a baseline array (e.g., Cy3) from one of the three slides, followed by scaling expression values of the remaining six arrays to the median value of the baseline array $(\tilde{m}_{\text{base}})$:

$$x'_i = \frac{\tilde{m}_{\text{base}}}{\tilde{m}_i} x_i.$$

After data normalization, the average intensity of individual gene from multiple spots on each slide was computed using an in-house developed PERL script. A total of 5043 average expression values were obtained, including empty, dry, null, and DMSO control spots. Paired t-test on normalized intensity with p-values <0.05 was used to generate a list of genes with significant change in expression between normal and heroin (10^{-7} M)-treated samples. The false positive rate (FDR) of the significant genes was estimated by using the SAM algorithm (Tusher et al., 2001). Quality control measures including ratios of housekeeping genes G3PDH and β-actin, scaling factors, background, and Q-values were within acceptable limits.

F. siRNA Transfection

NHA are transiently transfected with DARPP-32 siRNA for 24–96 h. siRNA (stealth[TH] siRNA) is obtained from Invitrogen (Carlsbad, CA). The siRNA sequences for DARPP-32 (Accession no. AF464196) are sense—ACA CAC

CAC CUU CGC UGA AAG CUG U and antisense—ACA GCU UUC AGC GAA GGU GGU GUG U. The appropriate scrambled control siRNA sequences are sense—ACA CCC AUC CUC GGU AAG ACA CUG U and antisense— ACA GUG UCU UAC CGA GGA UGG GUG U. Twenty-four hours before siRNA transfection, 1×10^4 NHA are seeded onto six-well plates in OPTI minimal essential medium containing 4% FBS with no antibiotics to give 30–50% confluence at the time of transfections. The siRNAs are transfected to a final concentration of 200 pM using siPROT (Ambion Inc.) according to the manufacturer's recommendations. The final concentration of siRNA used for transfections was optimized (data not shown) and are consistent with concentrations used by other investigators (Hassani *et al.*, 2005). We obtained a significant knockdown (~80–90%) of DARPP-32 gene expression using stealthTH siRNA against DARPP-32. Optimal gene knockdown effect is observed up to 1 week posttransfection. The efficiency of gene silencing was determined by measuring the percentage inhibition of the expression of DARPP-32 using quantitative real time Q-PCR. Since these are transient transfections, all experiments using transfected cells are carried out at 48 h posttransfection. Transfected NHA are treated with heroin (10^{-7} M) for 48 h. Appropriate scrambled siRNA control and a nontransfected control are used. The gene expression levels of DARPP-32, PP-1, ERK, and CREB are determined by real time, quantitative PCR.

G. Cell Viability Assay

Cell viability is measured using the MTT assay, that is, based on the reduction of a tetrazolium component 3-(4,5-dimethylthiazol-2-yl)-2,5-diphenyltetrazolium bromide (MTT) into an insoluble formazan product by the mitochondria of viable cells (Promega, Madison, WI). The MTT assay is a quantitative, sensitive detection of cell proliferation since it measures the growth rate of cells by virtue of a linear relationship between cell activity and absorbance.

H. RNA Extraction

Cytoplasmic RNA is extracted using the Trizol reagent (Invitrogen) (Chomczynski and Sacchi, 1987). RNA concentrations are determined using a Nanodrop spectrophotometer. Isolated RNA is stored at −80 °C.

I. Real Time, Quantitative RT-PCR

The relative abundance of each mRNA species is determined by real time, quantitative PCR. To provide precise quantification of initial target in each PCR reaction, the amplification plot is examined at a point during the early log phase of product accumulation. This is accomplished by assigning a fluorescence threshold above background and determining the time point at which each sample's amplification plot reaches the threshold (defined as the threshold cycle number or C_T). Differences in threshold cycle number are used to quantify the relative amount of PCR target contained within each tube. Relative expression of mRNA species is calculated using the comparative C_T method (Higuchi et al., 1993). All data are controlled for quantity of RNA input by performing measurements on an endogenous reference gene, β-actin. Results on RNA from treated samples are normalized to results obtained on RNA from the control, untreated sample. Briefly, the analysis is performed as follows: for each sample, a difference in C_T values (ΔC_T) is calculated for each mRNA by taking the mean C_T of duplicate tubes and subtracting the mean C_T of the duplicate tubes for the reference RNA (β-actin) measured on an aliquot from the same RT reaction. The ΔC_T for the treated sample is then subtracted from the ΔC_T for the untreated control sample to generate a $\Delta\Delta C_T$. The mean of these $\Delta\Delta C_T$ measurements is then used to calculate expression of the test gene relative to the reference gene and normalized to the untreated control as follows: relative expression / transcript accumulation index = $2^{-\Delta\Delta C_T}$. This calculation assumes that all PCR reactions are working with 100% efficiency. All PCR efficiencies were found to be >95%; therefore, this assumption introduces minimal error into the calculations (Bustin, 2002; Radonic et al., 2004).

J. Immunofluorescent Staining

DARPP-32 siRNA transfected and untranfected (control) NHA were grown to 70% confluence in a Petri dish with a glass bottom. Both transfected and untransfected cells were treated with heroin (10^{-7} M) for a period of 48 h. Cells were fixed 10 min at 37 °C in 2% formaldehyde followed by permabilization with ice-cold 90% methanol. Cells were then washed in 1× phosphate buffer saline (PBS) and blocking done with 10% normal goat serum and 1% BSA in PBS was followed by incubation with of anti-ERK rabbit polyclonal antibody (Cat # sc-94; Santa Cruz Biotechnologies, Santa Cruz, CA) overnight at 4 °C. Samples were then incubated with a mixture of antirabbit secondary antibodies conjugated to Alexa Fluor® 647 (Cat # A21244; Molecular Probes, Eugene, OR), washed with PBS counter, stained with DAPI, and mounted on glass slides. Standard immunofluoresecent staining procedures were followed. Imaging was done using a Leica Confocal Laser Scanning Microscope (TCS SP2 AOBS, Leica Microsystems Heidelberg GmbH) with an Oil Immersion objective lens 63×. HeNe 633 nm laser was applied to excite Alexa Fluor® 647, argon ion laser was applied to excite DAPI.

IV. Results

A. Microarray Data Preprocessing

To obtain meaningful data from these microarray analyses, preprocessing of the microarray data is very important. This involves data normalization procedures, which corrects systematic differences such as intensity-dependent expression bias and different dye efficiency between and across datasets. Normalization procedures used have been described in details in our previous studies (Mahajan *et al.*, 2006). In microarray studies, a large proportion of genes are usually not expressed across all samples to be compared and, as a common practice, are filtered out before performing statistical analysis. In this study, genes whose expression values were less than $2\times$ the DMSO control expression values across the six array samples were filtered out before performing statistical analysis. These led to an $\sim13.3\%$ reduction (671 genes) in gene sample size.

To account for variation between replicates, we performed data filtering by measuring the repeatability of gene expression using coefficient of variation (CV). This was done by computing the CV of Cy3/Cy5 ratios for all individual genes from six slides, followed by constructing a 99% confidence interval for CV values of all genes. Genes with CVs outside the upper 99% confidence interval bound were regarded as unreliable measurements and were removed from further analysis. This process eliminated another $\sim3.0\%$ of genes. The remaining 4372 genes were used for the detection of the differentially expressed genes.

B. mRNA Gene Expression Patterns in Heroin-Treated NHA

To compare gene expression levels between heroin (10^{-7} M)-treated NHA and their respective untreated controls, we used a two-step statistical data analyses. First, we used the regularized paired *t*-test to detect differentially expressed genes, which resulted in 1403 ($p < 0.05$) significantly modulated genes. For multiple test correction to control the FDR, these genes were further subjected to SAM analysis, which resulted in a shorter list of 214 genes with an estimated FDR of 4.9%. Of the 214 differentially expressed genes, 173 (81.6%) were upregulated and 41 (18.3%) were downregulated on treatment with heroin (10^{-7} M). The differentially expressed genes, showing their fold changes with respect to the housekeeping controls were further subdivided into different categories based on their biological function. An interrogation of our microarray datasets of the significantly and reproducibly changed genes showed that many of the genes could be classified into distinct functional groups. Our data (Table I) suggest that heroin (10^{-7}M) treatment significantly modulated genes involved in

TABLE I

LIST OF GENES THAT ARE SIGNIFICANTLY MODULATED BY HEROIN TREATMENT IN NHA

Gene accession no.	Genes grouped into different functional categories		Fold change	p-Value
		24 h		
	Cytokines and chemokines			
AA969475 | AI733305	Interferon gamma receptor 2	▲	1.27	3.43E−02
AA683550	Interleukin-1 receptor-associated kinase 1	▲	1.64	3.53E−02
AA463497	MCP	▲	1.4	1.36E−02
AA931884	Small inducible cytokine A1	▼	0.3	1.27E−02
AA487034	Transforming growth factor, beta	▲	1.25	3.47E−02
AA504211	Tumor necrosis factor (ligand) Superfamily	▲	1.91	9.86E−02
AA630120	Vascular endothelial growth factor B	▲	1.43	3.39E−02
	Signaling molecules			
W68281	Mitogen-activated protein kinase-activated protein kinase 3	▲	1.37	1.09E−01
R50953	Mitogen-activating protein kinase kinase kinase kinase 2	▲	1.59	7.77E−03
AA018980	**Protein kinase, cAMP dependent**	▲	**1.37**	**9.92E−03**
R26186 |	**Protein phosphatase 1, catalytic subunit, beta isoform**	▲	**1.37**	**1.69E−03**
AA460827	**Protein phosphatase 1, regulatory (inhibitor) subunit 1A (DARPP-32)**	▲	**1.55**	**5.52E−03**
R51209	**Protein phosphatase 2A**	▲	**1.33**	**4.67E−02**
	Cell cycle protein			
AA401479	**Cyclin-dependent kinase 5**	▲	**1.45**	**4.41E−02**
	Apoptosis regulation			
H74208	BCL2	▼	0.63	2.30E−02
AA459263	BCL2-related protein A1	▼	0.24	2.54E−02
R42530	Caspase 3, apoptosis-related cysteine protease	▲	1.46	9.45E−03

The genes highlighted in boldface are those that are involved in the cAMP-dependent PKA–DARPP-32 pathway.

cytokine and chemokine regulation (INF-γ, MCP-1, TGF-β, v-EGF, IL-1β, TNF-α); signal transduction (IP3, JNK, PI3 kinase, MAPK, DARRP-32, PP-1); cell cycle regulation (Cdk5); and apoptosis regulation (Caspases; Bcl-2). Since several of the molecules implicated in the DARPP-32 signaling pathway were significantly modulated by heroin in NHA as indicated gene microarray data analysis, the current study will focus on evaluating the role of DARPP-32 in the regulation of secondary messengers downstream of DARPP-32 that play a role in transcriptional regulation and consequently bring about changes in neuronal plasticity observed in drug addiction (Fig. 1).

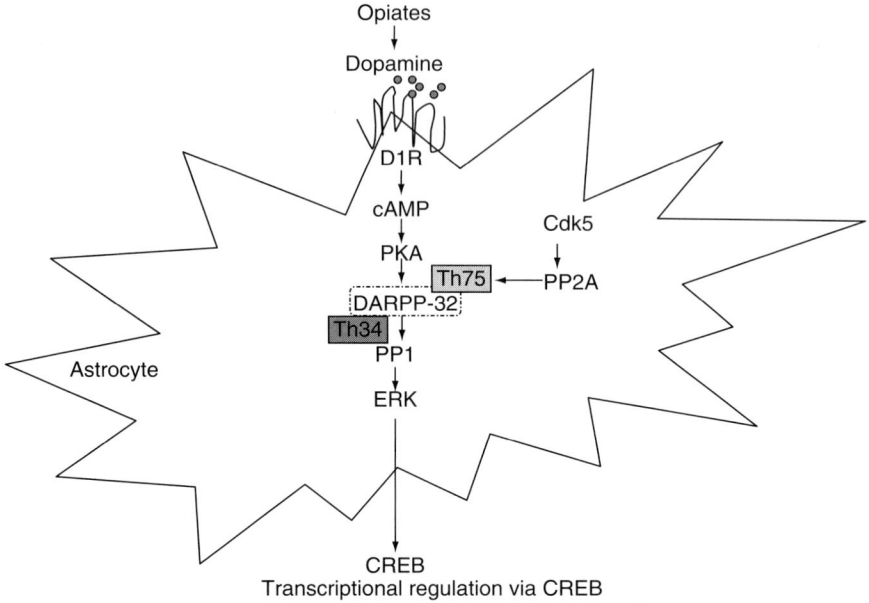

Fig. 1. Schematic of a signaling cascade activated by opiates. Opiates via the μ-opioid receptor activates the dopiminergic pathway leading to the increased DARPP-32 phosphorylation at Thr34 and increased PP-1 inhibition which modulates ERK activation and consequently the expression of the downstream transcriptional molecule CREB.

C. Opiates Increase D1R Gene Expression by NHA

Opiates activate the dopaminergic system in the brain via dopamine D1Rs. Our results show that heroin significantly increase D1R gene expression by NHA (Fig. 2). These results indicate that the dopaminergic pathway is triggered by opiates. NHA were treated with heroin (10^{-7} and 10^{-9} M) for 48 h, RNA was extracted, reverse transcribed, followed by quantitation of gene expression using quantitative, real time PCR. Heroin treatment significantly upregulated D1R gene expression at 10^{-7} M (TAI = 1.45, 45% increase, $p < 0.01$) and 10^{-9} M (TAI = 1.37, 37% increase, $p < 0.01$). The results are representative of three independent experiments and statistical significance was determined using ANOVA where the comparisons were performed between morphine treated and untreated controls.

D. Opiates Modulate the DARPP-32 Gene Expression in NHA Cells

As reported previously by Greengard *et al.*, DARPP-32 occupies a unique position whereby it modulates both dopaminergic and glutamatergic signaling depending upon which Thr residue within the protein is phosphorylated

Effect of heroin on D1 receptor gene expression in BMVEC

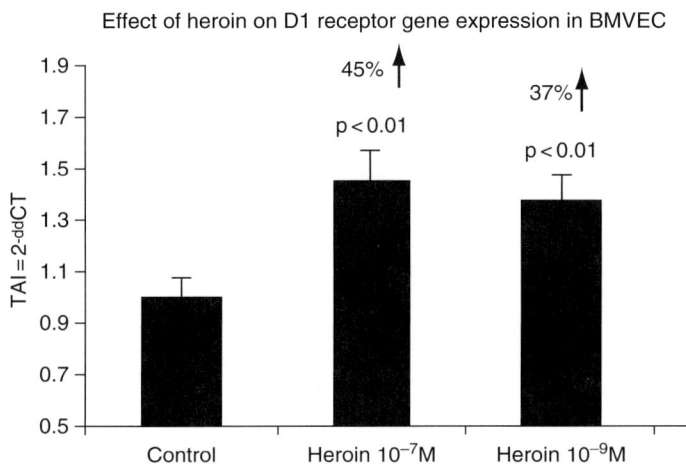

FIG. 2. Effect of heroin on dopamine D1 receptor gene expression by NHA. NHA were treated with 10^{-7}–10^{-9} M heroin for 48 h. RNA was extracted, reverse transcribed, and the D1 receptor gene expression was quantitated using real time PCR.

(Greengard et al., 1998, 1999; Nairn et al., 2004; Svenningsson et al., 2004, 2005). Phospho-Thr^{34}-DARPP-32 amplifies the D1/PKA pathway, whereas phospho-Thr^{75} DARPP-32 inhibits it, thereby shifting the balance toward dephosphorylation of target substrates via PP-1. DARPP-32 thus represents a "molecular switch" that regulates and fine-tunes the phosphorylation state of PP-1 and downstream target proteins. In an in vivo mouse model, acute administration of morphine has been shown to increase the phosphorylation of DARPP-32 at Thr^{34}, without affecting phosphorylation at Thr^{75} in striatal medium spiny neurons (Borgkvist et al., 2007). NHA were treated with 10^{-11}–10^{-7} M heroin for 48 h, RNA was extracted followed by quantitation of DARPP-32 gene expression using quantitative, real time PCR. As shown in Fig. 3, heroin at 10^{-11}, 10^{-9}, and 10^{-7} M concentrations significantly upregulated DARPP-32 gene expression by 43% (TAI $= 1.75 \pm 0.17$, $p < 0.0001$), 58% (TAI $= 1.58 \pm 0.08$, $p < 0.001$), and 43% (TAI $= 1.43 \pm 0.04$, $p < 0.001$), respectively as compared to control cultures (TAI $= 1.00 \pm 0.03$). These results are representative of three separate experiments and statistical significance was determined using Students' t-test, where the sample comparisons were performed between heroin treated and untreated controls.

We have previously shown that NHA treated with opiate (morphine) significantly downregulated CREB gene expression and also decreased endogenous CREB protein expression (Mahajan et al., 2005). It is believed that morphine activation of the μ-opioid receptor increases intracellular cAMP levels in NHA that triggers the D1/DARPP-32/PP-1 signaling cascade. This, in turn, modulates the activity of CREB, which initiates the transcription of immediate-early genes.

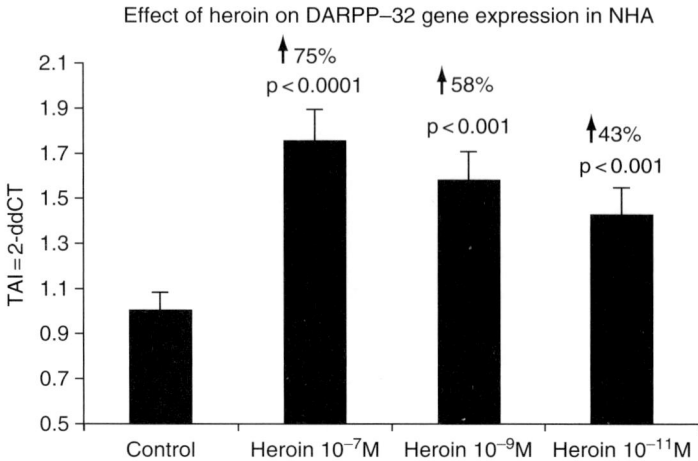

FIG. 3. Effect of heroin on DARPP-32 gene expression by NHA. NHA were treated with 10^{-7}–10^{-11} heroin for 48 h, RNA was extracted, reverse transcribed, and DARPP-32 gene expression was quantitated using real time PCR. Morphine significantly increased DARPP-32 gene expression in a dose-dependent manner.

E. EFFECT OF DARPP-32 siRNA TRANSFECTION ON CELL VIABILITY

Approximately 10,000 cells/ml transfected NHA cells were incubated with the MTT reagent for approximately 3 h, followed by addition of a detergent solution to lyse the cells and solubilize the colored crystals. The samples are read using an ELISA plate reader at a wavelength of 570 nm. Figure 4 shows >90% viability in both the transfected and untransfected NHA cells up to 96 h posttransfection.

F. TRANSFECTION EFFICIENCY OF DARPP-32 siRNA

The efficiency of gene silencing was determined by measuring the percentage inhibition of the DARPP-32 gene expression using quantitative real time Q-PCR. The commercially available transfection reagent siPROT was used to transfect the cells using a final concentration of 200 pmol siRNA-DARPP-32 in culture. Our results (Fig. 5) show that the percentage decrease over a time period of 24 h up to 1 week in DARPP-32 gene expression in transfected NHA as compared to the scrambled siRNA used as a control. The percentage decrease in DARPP-32 gene expression was 40% ($p < 0.01$), 85% ($p < 0.0001$), and 82% ($p < 0.0001$) at 24, 48, and 1 week posttransfection, respectively. Significant knockdown of DARPP-32 gene expression was observed even 1 week posttransfection.

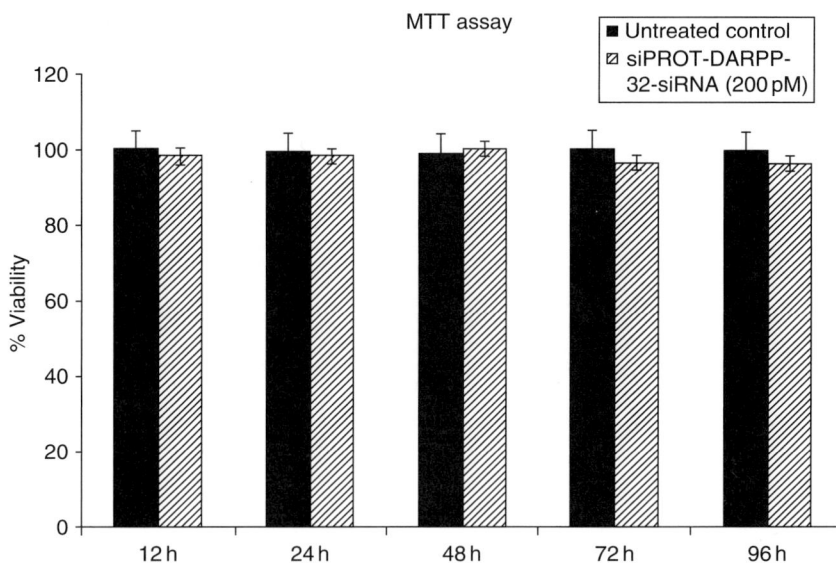

FIG. 4. Effect of the DARPP-32-siRNA transfection on cell viability in NHA. Cell viability was measured using the MTT assay. NHA were transfected with DARPP-32-siRNA (200 pM) using siPROT transfection reagent from Ambion Inc. (Austin, TX), for a time period of up to 96 h. Data were expressed as a percentage of viable cells calculated with respect to the untreated control and represented as the mean ± S.D. of three separate experiments done in duplicate. Our results showed a >90% viability over the range of time periods tested.

G. EFFECT OF DARPP-32 SILENCING ON PP-1, ERK, AND CREB GENE EXPRESSION

Both acute and chronic opiate exposure induce biochemical changes in specific regions in the brain and specifically the dopaminergic pathway which mediates the reinforcing actions of the opiates. We evaluated the effect of DARPP-32 gene silencing on PP1, ERK, and CREB gene expression in both heroin treated and control cultures. Our data (Fig. 6A–D) suggest that DARPP-32 gene silencing resulted in a significant decrease in PP-1, ERK, and CREB gene expression, when comparison were made between the DARPP-32 siRNA transfected cells and the untransfected controls or the scrambled siRNA transfected cells. Heroin treatment of the DARPP-32 siRNA transfected cells resulted in a significant decrease ($p = 0.02$) in ERK gene expression. Using immunofluorescence techniques, we further examined the effect of heroin on ERK protein expression in DARPP-32 siRNA transfected and untransfected cells. Our results (Fig. 7) show a significant decrease in ERK protein expression in DARPP-32 siRNA transfected cells.

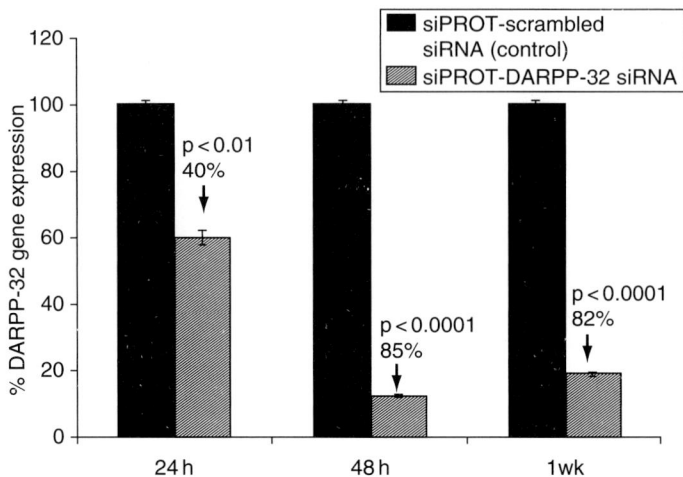

FIG. 5. Effect of the DARPP-32-siRNA transfection on DARPP-32 gene silencing in NHA. NHA were transfected with DARPP-32-siRNA (200 pM) using siPROT transfection reagent from Ambion Inc. (Austin, TX), for a time period of up to 1 week. NHA were treated *in vitro* with the DARPP-32-siRNA (200 pM) and scrambled-siRNA (transfection control), RNA was extracted, reverse transcribed, cDNA amplified, and the DARPP-32 gene expression was determined by real time quantitative PCR. Relative expression of mRNA species was calculated using the comparative C_T method. Data are the mean ± S.D. of three separate experiments done in duplicate. Statistical significance was determined using ANOVA based on comparison between, the positive control, DARPP-32 siRNA and the scrambled-siRNA samples. Our results show a >80% suppression in DARPP-32 gene expression in NHA that were transfected with DARPP-32 siRNA.

V. Discussion

The addictive process occurs as a result of impairments in the three functional systems, namely the motivation-reward, affect regulation, and behavioral inhibition. Nobel laureate Dr. Paul Greengard and his team have extensively studied the phosphoprotein-DARPP-32. DARPP-32 is regulated by many different neurotransmitters and may either inhibit cellular phosphatases or inhibit cAMP-dependent kinases. These opposing activities of DARPP-32 affect the overall sensitivity of the neurons to neurotransmitter signals. The neurotransmitter, dopamine, plays a key role in the development of drug abuse and dependence. Dopamine has been associated with the reinforcing effects of various types of drugs of abuse including opiates.

DARPP-32 is localized in many regions of the brain, and its high enrichment is observed in the dopaminoceptive neurons in the striatum, particularly caudate-putamen and nucleus accumbens. DARPP-32 is also localized in the olfactory

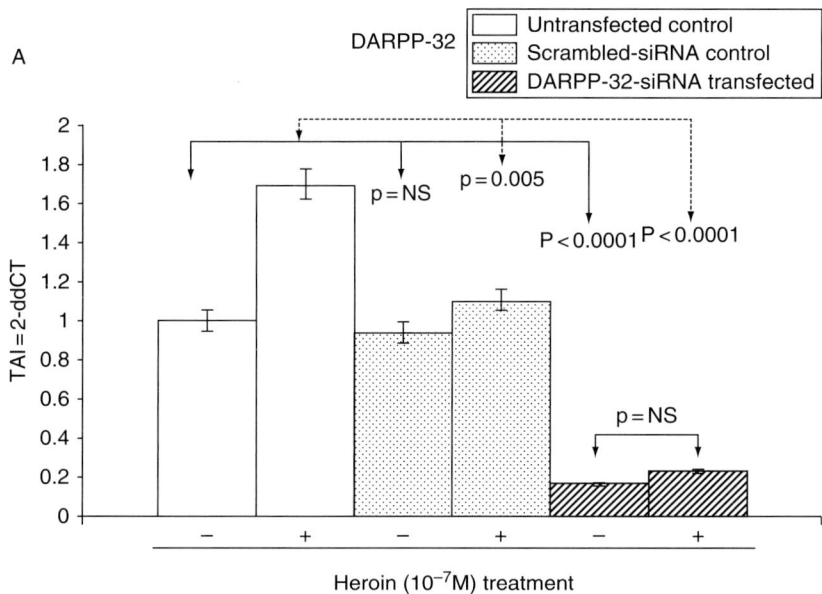

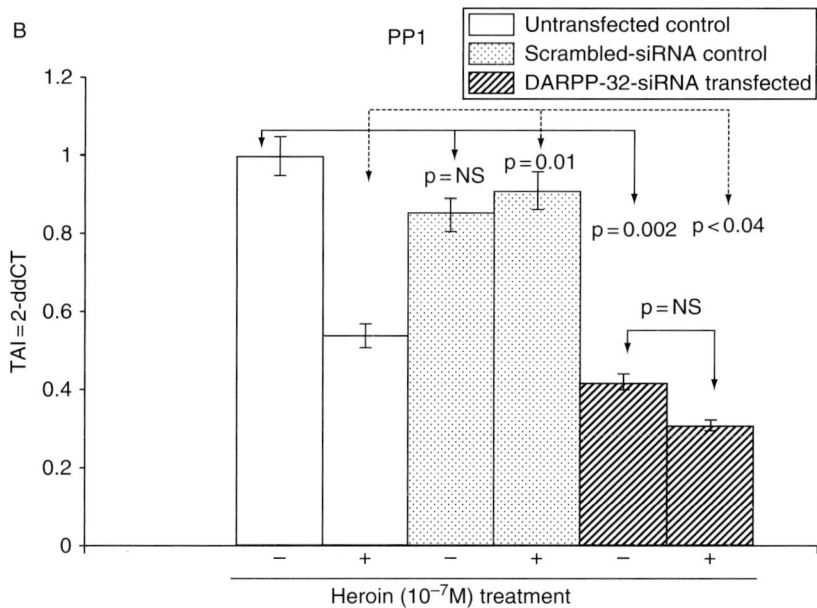

FIG. 6. (*Cont.*)

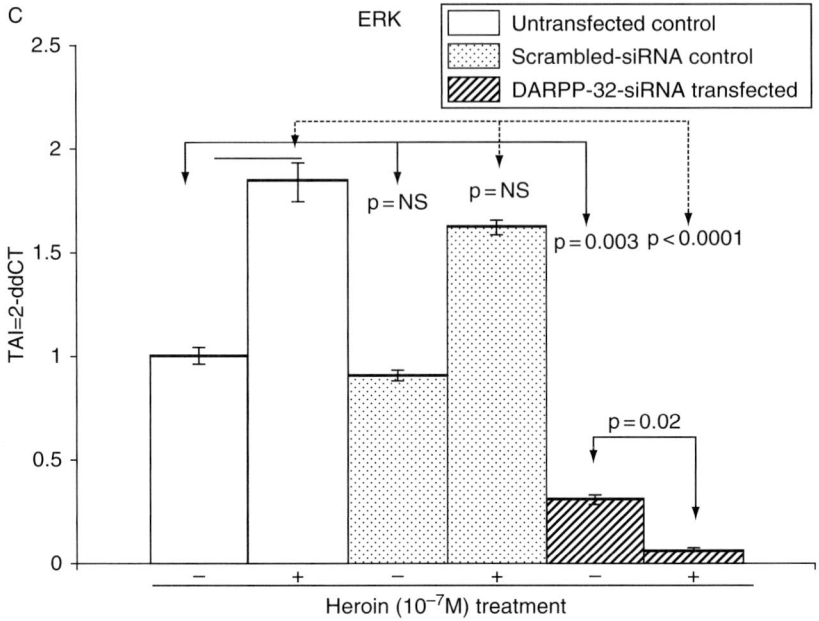

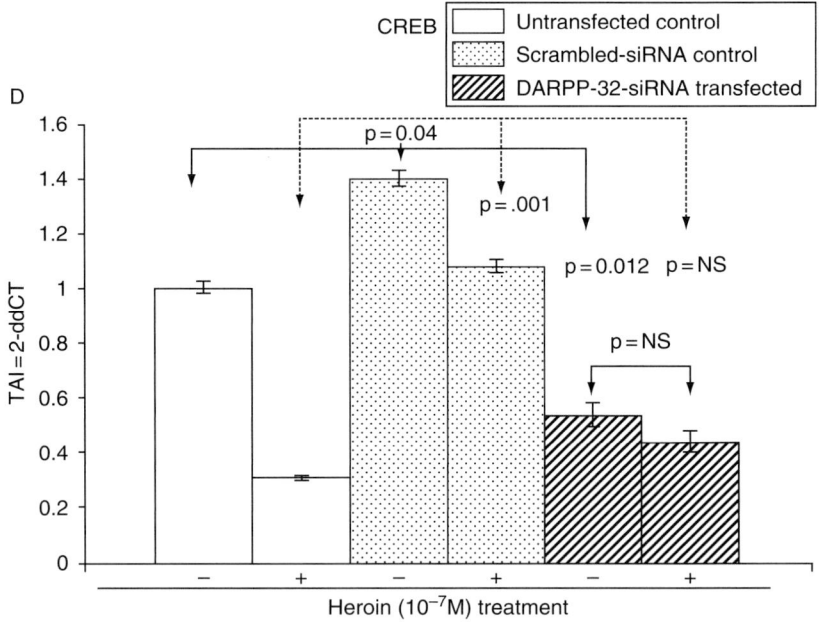

tubercle, bed nucleus of stria terminalis, and portions of the amygdaloid complex and throughout the neocortex, with particular enrichment in layers II, III, and VI, in the dentate gyrus of the hippocampus, choroid plexus, hypothalamus, and cerebellum. Within the striatum, DARPP-32 is found in medium-sized spiny neurons (Ouimet et al., 1984, 1992; Yoshida et al. 1971). Thus, DARPP32 is widely expressed in both neurons and glial cells. Drug abuse results in damage to brain cells, which is evident long after drug abuse has ceased. The damage to the brain cells, neurodegeneration due to drugs of abuse may be irreversible. In response to neuroinflammation and apoptotic stress as a consequence of chronic opiate abuse, neuropathological abnormalities such as increase in the number of astrocytes or astrogliosis, astrocyte hypertrophy, that involves extension of their processes, and increased production of glial fibrillary acidic protein (GFAP), an intermediate filament protein located within their cytoplasm have been observed. GFAP expression is believed to alter the vulnerability of neurons to excitotoxic and metabolic insults and confer some degree of neuroprotection after an excito-toxic or metabolic insult (Hanbury et al., 2003). Additionally, an association between DARRP-32 expression and astrogliosis after an ischemic insult has also been observed (Han and Holtzman, 2000). These studies lead us to speculate that morphine via the DARPP-32 signaling pathway may alter neuronal plasticity via modulation of signaling intermediates such as ERK and CREB and thereby transcriptional regulation in astrocytes. Astrocytes are important cells in the CNS as they face the synapses, send end-foot processes that enwrap the brain capillaries, and form an extensive network interconnected by gap junctions. Astrocytes express several membrane proteins and enzymes that are critical for uptake of neurotransmittors at the synapses and participate in detection, propagation, and modulation of excitatory synaptic signals and provide metabolic support to the active neurons.

The current treatment regimens in opiate addiction therapy involve the use of drugs such as methadone, naltrexone, and levo-alpha-acetylmethadol (LAAM) and no single treatment is appropriate for all individuals. Relapses to drug use can

---

FIG. 6. Effect of DARPP-32 gene silencing on PP-1, ERK, and CREB gene expression in transfected NHA cells treated with/without heroin. NHA were transfected with DARPP-32-siRNA (200 pM) using siPROT transfection reagent from Ambion Inc. (Austin, TX), for 48 h followed by a 24-h treatment with/without heroin (10^{-7} M). Appropriate scrambled-siRNA (transfection control) and untransfected cells were used as controls. RNA was extracted, reverse transcribed, cDNA amplified, and the (A) DARPP-32, (B) PP-1, (C) ERK, (D) CREB gene expression was determined by real time quantitative PCR. Relative expression of mRNA species was calculated using the comparative C_T method. Data are the mean ± S.D. of three separate experiments done in duplicate. Statistical significance was determined using ANOVA based on comparison between, the DARPP-32 siRNA transfected, scrambled-siRNA transfected, and the untransfected samples in the heroin treated and untreated groups. Our results show a significant decrease in PP-1 and ERK gene expression in heroin treated NHA that were transfected with DARPP-32 siRNA.

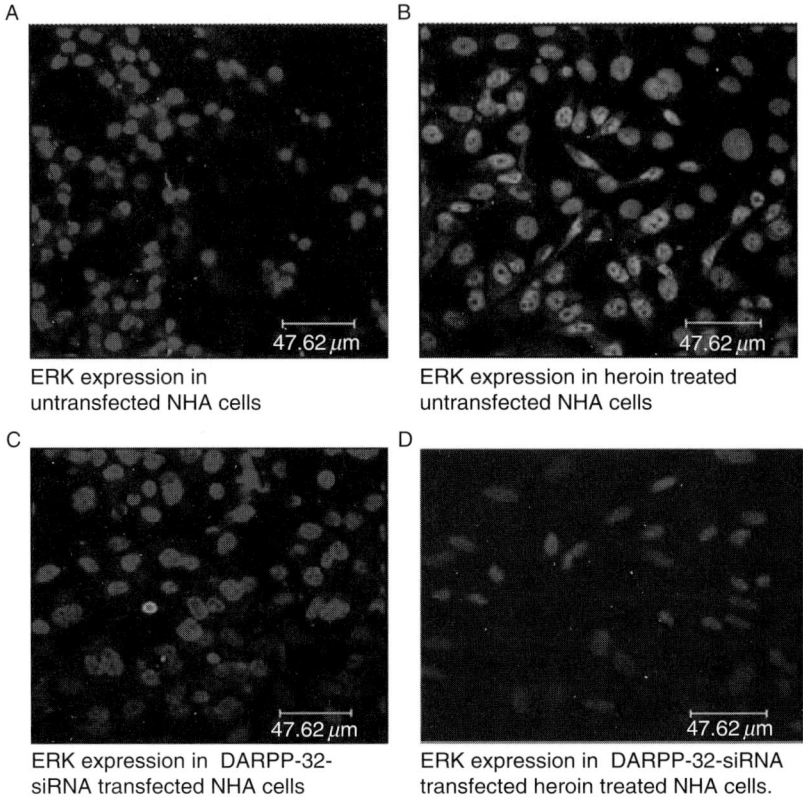

FIG. 7. Immunofluorescent staining for ERK in DARPP-32-siRNA transfected and untransfected NHA cells treated with/without heroin. (A) ERK expression in untransfected NHA cells; (B) ERK expression in heroin (10^{-7} M) treated untransfected NHA cells; (C) ERK expression in DARPP-32-siRNA transfected NHA cells; (D) ERK expression in heroin (10^{-7} M)-treated DARPP-32-siRNA transfected NHA cells. Our results show a significant decrease in ERK expression in heroin-treated DARPP-32-siRNA transfected NHA cells. Data shown are representative images from three separate experiments. (See Color Insert.)

occur during or after successful treatment episodes. Addicted individuals may require prolonged treatment and multiple episodes of treatment to achieve long-term abstinence and complete cure from drug addiction. Silencing the gene expression of DARPP-32 by RNA interference (RNAi) mechanism may thus be an effective therapeutic strategy toward the treatment of drug addiction.

The results of our study suggest that DARPP-32 gene silencing resulted in a significant decrease in PP-1, ERK, and CREB expression in NHA cells. Treatment of NHA with heroin must result in the release of dopamine, via the activation of the D1R, which activates adenylate cyclase resulting in

PKA-mediated phosphorylation of DARPP-32 and increased expression of DARPP-32 gene expression levels that we observed in our study (Fig. 6A). DARPP-32 is a potent inhibitor of PP-1, and therefore it is not surprising that we observed a significant decrease of PP-1 gene expression (Fig. 6B). PP-1 is known to play a central role in both dopaminergic and glutamatergic signaling (Greengard *et al.*, 1998, 1999). Our results showed that DARPP-32 gene silencing also resulted in the significant downregulation of ERK and CREB gene expression. ERK activity has been known to be important in neuronal plasticity and its pharmacologic blockade prevents the transcriptional and behavioral effects of various drugs of abuse. The control of the ERK pathway by DARPP-32 is common to most drugs of abuse (Valjent *et al.*, 2001, 2005). ERKs have emerged as an important target in neuronal signal transduction and are believed to participate in diverse processes, including neuronal maturation and survival, synaptic function (Kuroki *et al.*, 2001; Runden, *et al.*, 1998; Xia, *et al.*, 1995). Chronic activation and nuclear retention of ERK may be a critical factor in triggering proapoptotic signals and neuronal cell death. Runden *et al.* (1998) have shown sustained activation of ERK brought about by PP-1 inhibition induces neuronal cell death via the activation of the caspase-dependent pathway in hippocampal slices (Stanciu, 2000; Stanciu and DeFranco, 2002). Our gene microarray data have shown a significant increase in caspase activity and a significant decrease in the expression of the antiapoptotic protein bcl-2 indicating heroin-induced proapoptotic stimulus in the NHA. The opiate induced increase in ERK expression in NHA and suggests that ERK may be required for activation of proapoptotic pathways.

In the current study, heroin induced D1R-mediated regulation of ERK activity via the DARPP-32 signaling pathway and suggests that the DARPP-32 phosphoprotein may be important in regulating the duration of ERK activation and the subsequent downstream signaling cascades that can modulate the induction of transient immediate early genes and transcription factors such as CREB that modulate transcription and neuronal function and are implicated in the effects of opiates on motor sensitization and reward. The DARPP-32 signaling pathway is thus the central molecular mechanism that underlies the neurobiological alterations due to abuse of opiate drugs. The ERK pathway, as well as DARPP-32 mediated inhibition of PP-1, is critical to the modulation of behavioral responses which are a consequence of changes in neuronal plasticity with structural modification of neural networks in the CNS during drug abuse.

A gene therapy approach such as the silencing of DARPP-32 may provide an effective strategy to overcome drug addiction, particularly because it may also help modify cognitive behavior, thereby preventing a relapse. While we certainly do not anticipate the clinical use of intracerebral DARPP-32 siRNA for the treatment of drug abuse, we expect that results from these studies will support the premise that blocking this pathway can eliminate addictive, drug-seeking behavior, and prevent neurologic dysfunction in drug abusing subjects.

Although chemically synthesized, siRNA can be delivered *in vitro* and *in vivo* to achieve therapeutic gene silencing, without interfering with the cell's endogenous microRNA. A major challenge associated with siRNA therapy and its delivery into the cell is the issue of increased cellular toxicity. Innovative methodologies using nanotechnology, that can provide efficient gene delivery that not only provides sustained release of siRNA molecules which can result in effective DARPP-32 gene silencing but also show no significant toxicity over a wide range of dose and time conditions are currently being explored.

Acknowledgments

The authors acknowledge the Cameron Troup Foundation, Kaleida Health, Buffalo General Hospital, Buffalo, NY, for the grant support provided for this research work.

References

Baritaki, S., Dittmar, M. T., Spandidos, D. A., and Krambovitis, E. (2005). *In vitro* inhibition of R5 HIV-1 infectivity by X4 V3-derived synthetic peptides. *Int. J. Mol. Med.* **16**(2), 333–336.

Borgkvist, A., and Fisone, G. (2007). Psychoactive drugs and regulation of the cAMP/PKA/DARPP-32 cascade in striatal medium spiny neurons. *Neurosci. Biobehav. Rev.* **31**(1), 79–88.

Bustin, S. A. (2002). Quantification of mRNA using real-time reverse transcription PCR (RT-PCR): Trends and problems. *J. Mol. Endocrinol.* **1**, 23–39.

Calabresi, P., Gubellini, P., Centonze, D., Picconi, B., Bernardi, G., Chergui, K., Svenningsson, P., Fienberg, A. A., and Greengard, P. (2000). Dopamine and cAMP-regulated phosphoprotein 32 kDa controls both striatal long-term depression and long-term potentiation, opposing forms of synaptic plasticity. *J. Neurosci.* **20**(22), 8443–8451.

Carter, B. L., and Tiffany, S. T. (1999). Meta-analysis of cue-reactivity in addiction research. *Addiction* **94**, 327–340.

Chao, J., and Nestler, E. J. (2004). Molecular neurobiology of drug addiction. *Annu. Rev. Med.* **55**, 113–132.

Childress, A. R., Ehrman, R., Rohsenow, D. J., Robbins, S. J., and O'Brien, C. P. (1992). Classically conditioned factors in drug dependence. *In* "Substance Abuse: A Comprehensive Textbook" (J. W. Lowinson, P. Luiz, R. B. Millman, and G. Langard, Eds.), pp. 56–69. Williams and Wilkins, Baltimore, MD.

Chomczynski, P., and Sacchi, N. (1987). Single-step method of RNA isolation by acid guanidinium thiocyanate-phenol-chloroform extraction. *Anal. Biochem.* **162**, 156–159.

Colvis, C. M., Pollock, J. D., Goodman, R. H., Impey, S., Dunn, J., Mandel, G., Champagne, F. A., Mayford, M., Korzus, E., Kumar, A., Renthal, W., Theobald, D. E., *et al.* (2005). Epigenetic mechanisms and gene networks in the nervous system. *J. Neurosci.* **25**(45), 10379–10389.

Fienberg, A. A., and Greengard, P. (2000). The DARPP-32 knockout mouse. *Brain Res. Rev.* **31**(2–3), 313–319.

Fienberg, A. A., Hiroi, N., Mermelstein, P. G., Song, W., Snyder, G. L., Nishi, A., Cheramy, A., O'Callaghan, J. P., Miller, D. B., Cole, D. G., Corbett, R., Haile, C. N., *et al.* (1998). DARPP-32: Regulator of the efficacy of dopaminergic neurotransmission. *Science* **281**(5378), 838–842.

Greengard, P., Nairn, A. C., Girault, J. A., Ouimet, C. C., Snyder, G. L., Fisone, G., Allen, P. B., Fienberg, A., and Nishi, A. (1998). The DARPP-32/protein phosphatase-1 cascade: A model for signal integration. *Brain Res. Brain Res. Rev.* **26**(2–3), 274–284.

Greengard, P., Allen, P. B., and Nairn, A. C. (1999). Beyond the dopamine receptor: The DARPP-32/ protein phosphatase-1 cascade. *Neuron* **23**(3), 435–447.

Guitart-Masip, M., Johansson, B., Fernandez-Teruel, A., Canete, T., Tobena, A., Terenius, L., and Gimenez-Llort, L. (2006). Divergent anatomical pattern of D1 and D3 binding and dopamine- and cyclic AMP-regulated phosphoprotein of 32 kDa mRNA expression in the Roman rat strains: Implications for drug addiction. *Neuroscience* **142**(4), 1231–1243.

Han, B. H., and Holtzman, D. M. (2000). BDNF protects the neonatal brain from hypoxic-ischemic injury *in vivo* via the ERK pathway. *J. Neurosci.* **20**, 5775–5781.

Hanbury, R., Ling, Z. D., Wuu, J., and Kordower, J. H. (2003). GFAP knockout mice have increased levels of GDNF that protect striatal neurons from metabolic and excitotoxic insults. *J. Comp. Neurol.* **461**(3), 307–316.

Hassani, Z., Lemkine, G. F., Erbacher, P., Palmier, K., Alfama, G., Giovannangeli, C., Behr, J. P., and Demeneix, B. A. (2005). Lipid-mediated siRNA delivery down-regulates exogenous gene expression in the mouse brain at picomolar levels. *J. Gene Med.* **7**(2), 198–207.

Higuchi, R., Fockler, C., Dollinger, G., and Watson, R. (1993). Kinetic PCR analysis: Real-time monitoring of DNA amplification reactions. *Biotechnology (NY)* **11**, 1026.

Kreek, M. J., and Koob, G. F. (1998). Drug dependence: Stress and dysregulation of brain reward systems. *Drug Alcohol Depend.* **51**, 23–47.

Kuroki, Y., Fukushima, K., Kanda, Y., Mizuno, K., and Watanabe, Y. (2001). Neuroprotection by estrogen via extracellular signal-regulated kinase against quinolinic acid-induced cell death in the rat hippocampus. *Eur. J. Neurosci.* **13**, 472–476.

Mahajan, S. D., Aalinkeel, R., Reynolds, J. L., Nair, B. B., Fernandez, S. F., Schwartz, S. A., and Nair, M. P. (2005a). Morphine exacerbates HIV-1 viral protein gp120 induced modulation of chemokine gene expression in U373 astrocytoma cells. *Curr. HIV Res.* **3**(3), 277–288.

Mahajan, S. D., Schwartz, S. A., Aalinkeel, R., Chawda, R. P., Sykes, D. E., and Nair, M. P. (2005b). Morphine modulates chemokine gene regulation in normal human astrocytes. *Clin. Immunol.* **115**(3), 323–332.

Mahajan, S. D., Hu, Z., Reynolds, J. L., Aalinkeel, R., Schwartz, S. A., and Nair, M. P. N. (2006). Methamphetamine modulates gene expression patterns in Monocyte derived mature Dendritic cells: Implications for HIV-1 pathogenesis. *Mol. Diagn. Ther.* **10**(4), 257–269.

Nairn, A. C., Svenningsson, P., Nishi, A., Fisone, G., Girault, J. A., and Greengard, P. (2004). The role of DARPP-32 in the actions of drugs of abuse. *Neuropharmacology* **47**(Suppl. 1), 14–23.

Nestler, E. J. (2005). Is there a common molecular pathway for addiction? *Nat. Neurosci.* **8**(11), 1445–1449.

Nestler, E. J., Alreja, M., and Aghajanian, G. K. (1994). Molecular and cellular mechanisms of opiate action: Studies in the rat locus coeruleus. *Brain Res. Bull.* **35**, 521–528.

Nishi, A., Snyder, G. L., and Greengard, P. (1997). Bidirectional regulation of DARPP-32 phosphorylation by dopamine. *J. Neurosci.* **17**(21), 8147–8155.

O'Brien, C. P. (1997). A range of research-based pharmacotherapies for addiction. *Science* **278**, 66–70.

Ouimet, C. C., Miller, P. E., Hemmings, H. C. Jr., Walaas, S. I., and Greengard, P. (1984). DARPP-32, a dopamine- and adenosine 3′:5′-monophosphate-regulated phosphoprotein enriched in dopamine-innervated brain regions. III. Immunocytochemical localization. *J. Neurosci.* **4**, 111–124.

Ouimet, C. C., LaMantia, A. S., Goldman-Rakic, P., Rakic, P., and Greengard, P. (1992). Immunocytochemical localization of DARPP-32, a dopamine and cyclic-AMP-regulated phosphoprotein, in the primate brain. *J. Comp. Neurol.* **323**(2), 209–218.

Peterson, P. K., Gekker, G., Hu, S., Cabral, G., and Lokensgard, J. R. (2004). Cannabinoids and morphine differentially affect HIV-1 expression in CD4(+) lymphocyte and microglial cell cultures. *J. Neuroimmunol.* **147**(1–2), 123–126.

Radonić, A., Thulke, S., Mackay, I.M., Landt, O., Siegert, W., Nitsche, A. (2004). Guideline to reference gene selection for quantitative real-time PCR. *Biochem Biophys Res Commun.* **313**(4), 856–862.

Runden, E., Seglen, P. O., Haug, F. M., Ottersen, O. P., Wieloch, T., Shamloo, M., and Laake, J. H. (1998). Regional selective neuronal degeneration after protein phosphatase inhibition in hippocampal slice cultures: Evidence for a MAP kinase-dependent mechanism. *J. Neurosci.* **18**, 7296–7305.

Self, D. W., and Nestler, E. J. (1998). Relapse to drug-seeking: Neural and molecular mechanisms. *Drug Alcohol Depend.* **51**(1–2), 49–60.

Singh, I. N., Goody, R. J., Dean, C., Ahmad, N. M., Lutz, S. E., Knapp, P. E., Nath, A., and Hauser, K. F. (2004). Apoptotic death of striatal neurons induced by human immunodeficiency virus-1 Tat and gp120: Differential involvement of caspase-3 and endonuclease G. *J. Neurovirol.* **10**(3), 141–151.

Stanciu, M. (2000). Persistent activation of ERK contributes to glutamate-induced oxidative toxicity in a neuronal cell line and primary cortical neuron cultures. *J. Biol. Chem.* **275**, 12200–12206.

Stanciu, M., and DeFranco, D. B. (2002). Prolonged nuclear retention of activated extracellular signal-regulated protein kinase promotes cell death generated by oxidative toxicity or proteasome inhibition in a neuronal cell line. *J. Biol. Chem.* **277**, 4010–4017.

Svenningsson, P., Nishi, A., Fisone, G., Girault, J. A., Nairn, A. C., and Greengard, P. (2004). DARPP-32: An integrator of neurotransmission. *Annu. Rev. Pharmacol. Toxicol.* **44**, 269–296.

Svenningsson, P., Nairn, A. C., and Greengard, P. (2005). DARPP-32 mediates the actions of multiple drugs of abuse. *AAPS J.* **7**(2), E353–E360.

Takahashi, S., Ohshima, T., Cho, A., Sreenath, T., Iadarola, M. J., Pant, H. C., Kim, Y., Nairn, A. C., Brady, R. O., Greengard, P., and Kulkarni, A. B. (2005). Increased activity of cyclin-dependent kinase 5 leads to attenuation of cocaine-mediated dopamine signaling. *Proc. Natl. Acad. Sci. USA* **102**(5), 1737–1742.

Tusher, V. G., Tibshirani, R., and Chu, G. (2001). Significance analysis of microarrays applied to the ionizing radiation response. *Proc. Natl. Acad. Sci. USA* **98**(9), 5116–5121.

Valjent, E., Caboche, J., and Vanhoutte, J. (Eds.) (2001). *In* "Mitogen-Extracellular Signal-R/Extracellular Signal-Regulated Kinase Induced Gene Regulation in Brain: A Molecular Substrate for Learning and Memory?" Vol. 23. Humana Press, Totowa, NJ, USA.

Valjent, E., Herve, D., and Girault, J. A. (2005). Drugs of abuse, protein phosphatases, and ERK pathway. *Med. Sci. (Paris)* **21**, 453–454.

Walters, C. L., and Blendy, J. A. (2001). Different requirements for cAMP response element binding protein in positive and negative reinforcing properties of drugs of abuse. *J. Neurosci.* **21**(23), 9438–9444.

Wang, L., and Gintzler, A. R. (1997). Altered mu-opiate receptor-G protein signal transduction following chronic morphine exposure. *J. Neurochem.* **68**, 248–254.

Wang, L., Medina, V. M., Rivera, M., and Gintzler, A. R. (1996). Relevance of phosphorylation state to opioid responsiveness in opiate naive and tolerant/dependent tissue. *Brain Res.* **723**, 61–69.

Widnell, K. L., Self, D. W., Lane, S. B., Russell, D. S., Vaidya, V. A., Miserendino, M. J., Rubin, C. S., Duman, R. S., and Nestler, E. J. (1996). Regulation of CREB expression: *In vivo* evidence for a functional role in morphine action in the nucleus accumbens. *J. Pharmacol. Exp. Ther.* **276**(1), 306–315.

Xia, Z., Dickens, M., Raingeaud, J., Davis, R. J., and Greenberg, M. E. (1995). Opposing effects of ERK and JNK-p38 MAP kinases on apoptosis. *Science* **270**, 1326–1331.

Yoshida, M., and Precht, W. (1971). Monosynaptic inhibition of neurons of the substantia nigra by caudato-nigral fibers. *Brain Res.* **32**, 225–228.

PHARMACOLOGICAL AND NEUROTOXICOLOGICAL ACTIONS MEDIATED BY BUPROPION AND DIETHYLPROPION

Hugo R. Arias,* Abel Santamaría,[†] and Syed F. Ali[‡]

*Department of Pharmaceutical Sciences, College of Pharmacy, Midwestern University, Glendale, Arizona 85308, USA
[†]Laboratorio de Aminoácidos Excitadores, Instituto Nacional de Neurología y Neurocirugía, México City, Mexico
[‡]Neurochemistry Laboratory, Division of Neurotoxicology, National Center of Toxicological Research, Food and Drug Administration, Jefferson, Arkansas 72079, USA

The antiappetite agent diethylpropion (DEP), and the antidepressant and antismoking aid compound bupropion (BP), not only share the same structural motif but also present similar mechanisms of action in the CNS. For example, both drugs induce the release as well as inhibit the reuptake of neurotransmitters such as a dopamine (DA) and norepinephrine (NE). In general, they produce mild side effects, including reversible psychomotor alterations mostly in geriatric patients (by BP), or moderate changes in neurotransmitter contents linked to oxidative damage (by DEP). Therefore, attention must be paid during any therapeutic use of these agents. Regarding the interaction of BP with the DA transporter, residues S359, located in the middle of TM7, and A279, located close to the extracellular end of TM5, contribute to the binding and blockade of translocation mediated by BP, respectively. Additional mechanisms of action have also been determined for each compound. For example, BP is a

INTERNATIONAL REVIEW OF
NEUROBIOLOGY, VOL. 88
DOI: 10.1016/S0074-7742(09)88009-4

noncompetitive antagonist (NCA) of several nicotinic acetylcholine receptors (AChRs). Based on this evidence, the dual antidepressant and antinicotinic activity of BP is currently considered to be mediated by its stimulatory action on DA and NE systems as well as its inhibitory action on AChRs. Considering the results obtained in the archetypical mouse muscle AChR, a sequential mechanism can be hypothesized to explain the inhibitory action of BP on neuronal AChRs: (1) BP first binds to AChRs in the resting state, decreasing the probability of ion channel opening, (2) the remnant fraction of open ion channels is subsequently decreased by accelerating the desensitization process, and finally (3) BP interacts with a binding domain located between the serine (position 9′) and valine (position 13′) rings that is shared with the NCA phencyclidine and other tricyclic antidepressants. The homologous location in the $\alpha3\beta4$ AChR is between the serine and valine/phenylalanine rings. This new evidence opens a window for further investigation using AChRs as targets for the action of safer antidepressants and novel antiaddictive compounds.

I. Introduction

The communication between individual neurons in the central nervous system (CNS) is performed at a morphologically and functionally highly specialized region named the synapse (reviewed by Ziv and Garner, 2004). A proper neuronal communication is essential for the correct functioning of the CNS. The synapse is structurally shaped by the pre- and postsynaptic membranes, both separated by the synaptic cleft. Figure 1 shows a simple scheme of the synapse. An important concept on the synaptic mechanism is that the transference of information between cells is achieved neither by a direct contact nor electrically but chemically.

Hundreds of molecules bearing chemical information, so-called neurotransmitters, have been identified so far as responsible for chemical signaling. In addition, it has been found that a single neuron can synthesize and liberate several of these substances at the same time. Different neurotransmitters operate at different parts of the nervous system, having excitatory and inhibitory actions. Neuroscientists have set up a few guidelines or criteria to prove that an endogenous ligand is really a neurotransmitter (reviewed by Arias, 2006a,b): (1) the endogenous ligand must be produced and found within a neuron; (2) the endogenous chemical must be released by the neuron after the neuron is stimulated; (3) when the ligand is released, it must act on a postsynaptic receptor and cause a biological effect; (4) if the endogenous ligand is applied on the postsynaptic

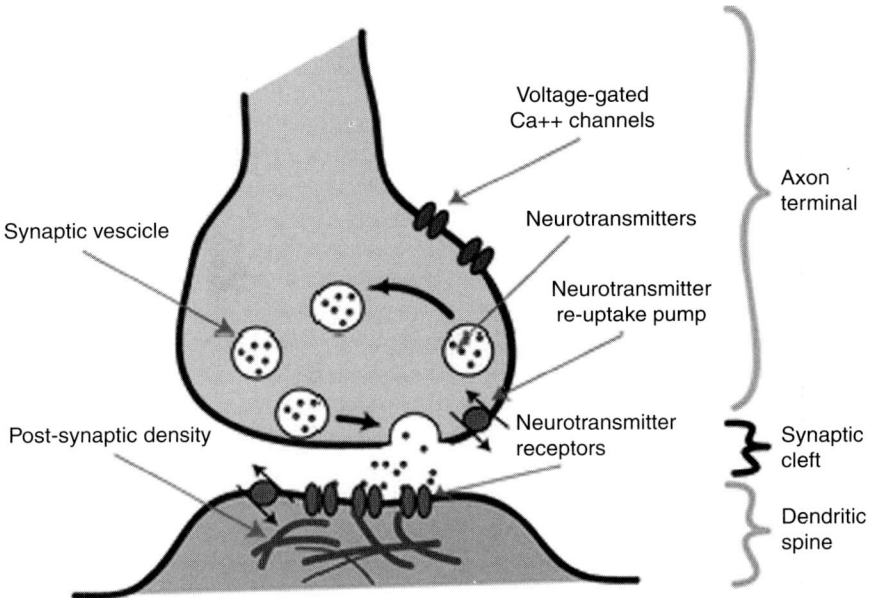

FIG. 1. Diagram showing the most important components of the synapse, including the pre- (axon terminal) and postsynaptic cells (dendritic spine) and the synaptic cleft (modified from Arias, 2006b). Voltage-gated Ca^{2+} channels involved in NR, vesicles for neurotransmitter storage, and neurotransmitter reuptake transporters are depicted in the presynaptic cell. Neurotransmitters are released from the presynaptic membrane to the synaptic cleft, reaching the postsynaptic membrane by diffusion, and finally activating neurotransmitter receptors.

membrane, it should have the same biological effect as when it is released by a neuron; and finally (5) after the chemical activates a postsynaptic receptor, the action of the ligand must be stopped or slowed to prevent clogging of cell communication. At least three mechanisms modulating the neurotransmitter concentration in the synaptic cleft are well known: (a) enzymatic inactivation. For instance, the neurotransmitter acetylcholine (ACh) is degraded to choline and acetate by acetylcholinesterases; (b) presynaptic membrane reuptake by specific neurotransmitter transporters (see Fig. 1). This mechanism is used for the reuptake of monoamine neurotransmitters such as norepinephrine (NE), epinephrine, dopamine (DA), and serotonin (5-hydroxytryptamine; 5-HT), as well as for other neurotransmitters; and (c) soluble binding proteins acting as carriers to buffer the neurotransmitter concentration in the synapse and/or to transport neurotransmitters to extrasynaptic sites, allowing the interaction with, for instance, glial cells or other adjacent neurons (reviewed by Sixma and Smit, 2003).

The chemoelectrical communication mechanism starts with the elicited action potential on the nerve ending that after several steps activates voltage-gated

Ca^{2+} channels enhancing Ca^{2+} permeation (reviewed by Reid *et al.*, 2003). Subsequently, the raised intracellular Ca^{2+} concentration triggers a series of lipid membrane-vesicle fusion processes (reviewed by Stevens, 2004) (see Fig. 1). A specialized region of the presynaptic membrane called "active zone" fuses with the neurotransmitter-containing vesicles by means of different fusion proteins (reviewed by Stevens, 2004). Most steps after vesicle-membrane fusion are actually rationalized (reviewed by Farrant and Nusser, 2005). For example, it is known that after the vesicle content is released into the synaptic cleft the time course of neurotransmitter clearance is in the submillisecond-to-millisecond time regime. The neurotransmitter molecules diffuse through the synaptic space in less than 0.2 ms reaching concentrations of 1–5 mM. Nevertheless, these parameters vary between synapses on a single neuron as well as among synapses from different neurons. Neurotransmitter release (NR) can be modulated by presynaptic receptors (reviewed by Hogg *et al.*, 2003; Kullmann *et al.*, 2005). The location of presynaptic receptors is shown in Fig. 1. The modulation of NR may lead to short- or long-term adaptations of the synapse to altered environmental signals (reviewed by Kullmann *et al.*, 2005).

The chemical information is finally converted into electrical currents on the postsynaptic neuron, where the recognition and binding of neurotransmitters take place by means of highly specific lipid-embedded protein receptors. Among postsynaptic receptors, ionotropic receptors can be stimulated by neurotransmitters inducing a fast opening of the ion channel intrinsically coupled to its receptor. Figure 1 shows the location of postsynaptic ionotropic receptors.

II. Neurotransmitter Release Modulation by Bupropion and Diethylpropion

Neurotransmitter release and changes in neurotransmission are major actions by drugs exerting neuroactive effects in the brain. For instance, NR creates a scenario of enhanced neurotransmission where several brain functions may be transiently exacerbated or inhibited, and these events precede a series of toxic episodes where major neurotransmitter systems are irreversibly affected in a time-dependent manner. Indeed, whether a neurotransmitter system is enhanced or depleted in time by the action of some drugs, the initial step of induced release will determine the course of toxic events in a near future. Thus, NR and changes in neurotransmission are of major relevance for the study of how addictive drugs may modify the primary functions of the CNS to exert pharmacological and/or neurotoxic actions.

In this minireview we will focus on those molecular and cellular mechanisms exerted by the psychostimulant bupropion (BP) to induce specific patterns of neurotransmission, ultimately leading to the therapeutic properties described for this molecule. In order to establish theoretical basis for the characterization of the pharmacological properties of drugs with CNS activity, the many features evoked by BP are contrasted with that for diethylpropion (DEP), another related psychostimulant of moderate activity. The chemical structure of BP [(±)-2-(*tert*-butylamino)-1-(3-chlorophenyl) propan-1-one; see Fig. 2] differs from all other antidepressants on the market. It belongs to the aminoketone class of compounds, sharing the category with the anorectic drug DEP [2-(diethylamino)-1-phenyl-propan-1-one; see Fig. 2]. The main difference between the two aminoketones is that DEP has two ethyl groups connected to its nitrogen, whereas BP has a single butyl group. Because both drugs have a phenyl group attached to the aminoketone part, they look like phenylethylamines including amphetamine, methamphetamine, and 3,4-methylenedioxymethamphetamine (i.e., MDMA or ecstacy).

A. DIETHYLPROPION ACTION ON NEUROTRANSMISSION

Diethylpropion (Tenuate®), also known as amphepramone or diethylcathinone, is an amphetamine-like psychostimulant and anorectic drug (Galván-Arzate and Santamaría, 2002). This drug exerts its effects in the brain in a similar manner to amphetamine, stimulating the release as well as inhibiting the reuptake of neurotransmitters such as DA, NE, and 5-HT (Ollo *et al.*, 1996; Garcia-Mijares *et al.*, 2008). In addition, DEP has an active role on opioid receptor stimulation when combined with ethanol, potentiating motor disturbances in mice (Gevaerd *et al.*, 1999). Given the centrally acting appetite-suppressive action of DEP, it has been employed for the treatment of obesity (Poston *et al.*, 1998). Despite its moderate actions as a psychostimulant and its many potential clinical applications, studies using DEP are very limited, especially in regard to the implications of neurotransmitter modulation in the brain producing side effects, and because DEP is considered a Schedule IV drug by the DEA.

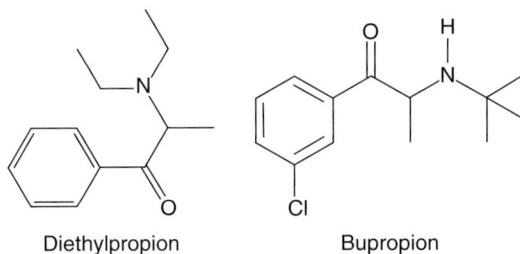

Diethylpropion   Bupropion

FIG. 2. Chemical structures of diethylpropion [2-(diethylamino)-1-phenylpropan-1-one] and bupropion [(±)-2-(*tert*-butylamino)-1-(3-chlorophenyl)propan-1-one]. The main difference between these two aminoketones is that diethylpropion has two ethyl groups connected to its nitrogen, whereas bupropion has a single butyl group, which gives its name.

Surprisingly, there is limited information regarding the molecular effects of DEP on NR from different neurotransmitter systems. An early study from the 1970s (Safta *et al.*, 1976) is probably one of the first suggestions that DEP exerts its effects at a central level through modulation/modification of neurotransmitter systems in the brain. The authors described locomotor and electroencephalographic alterations that can be antagonized by neuroleptics. The authors concluded that these changes can be correlated with the amphetamine-like psychosis observed in humans. Obviously, the observed changes imply the concept of major alterations in neurotransmitter systems that, in turn, might be accounting for the observed changes.

Shortly thereafter, Friström *et al.* (1977) described changes in 5-HT release from rabbit platelets to plasma by different known sympathomimetic or anorectic phenethylamines and derivatives. The aim of the study was to find and describe possible structure–activity relationships of anorectic agents producing pulmonary hypertension. Drugs like amphetamine and ephedrine produced high 5-HT release, followed in magnitude by other related drugs. Indeed, DEP was among these molecules, producing moderate 5-HT release. These findings served to suggest that pulmonary hypertension could be associated to the high levels of free 5-HT produced by these drugs, although the moderate effect exhibited by DEP is more suggestive of a lack of toxicity in this paradigm. In this regard, Samanian and Garattini (1993) stated that the anorectic effects produced by (+)-amphetamine, phentermine, DEP, and phenylpropanolamine are likely to be the result of their ability to enhance the release of NE and/or DA from nerve terminals as well as to inhibit their reuptake. This double pharmacological action will produce an augment in the concentration of these neurotransmitters in the synaptic cleft and subsequently, a persistent action on postsynaptic receptors. Previously, Borsini *et al.* (1979) explored the roles of NE and DA in the anorectic activity of DEP in rats. In this study, the role of NE in DEP's anorectic actions was evidenced by preventing the effects of both DEP and D-amphetamine by means of a lesion of the ventral noradrenergic bundle which selectively decreases brain NE. Their findings showed that the integrity of central noradrenergic neurons is essential for DEP and D-amphetamine to exert their anorectic actions.

Yu *et al.* (2000) addressed a very interesting issue by evaluating the effects of DEP and three synthesized metabolites on the uptake of biogenic amines. The findings of this study provided relevant mechanistic information on the pharmacological action of DEP as it was revealed that the *in vivo* activity of DEP on biogenic amine transporters is mostly due to its endogenous metabolite (±)-2-ethylamino-1-phenyl-propan-1-one, while the other metabolites $(1R,2S)$- and $(1S,2R)$-$(-)$-N,N-diethylnorephedrine had moderate or no effect on the transporters. In addition, (±)-2-ethylamino-1-phenyl-propan-1-one was shown to act as a substrate at both NE ($IC_{50} = 99$ nM) and 5-HT ($IC_{50} = 2118$ nM) transporters (NET and SERT, respectively), as well as an uptake inhibitor ($IC_{50} = 1014$ nM) at the DA transporter (DAT). These findings allowed the authors to suggest not only that (±)-2-ethylamino-1-phenyl-propan-1-one is the most potent neuroactive metabolite of DEP but also that amphetamine-type subjective effects may be partially mediated by brain levels of NE.

The effects of DEP treatment and withdrawal on adaptive responses in aorta reactivity, endothelial factors, and rat behavior were described by Bispo-Da Silva and Cordellini (2003). DEP treatment produced a decreased noradrenergic reactivity of aorta, but withdrawal showed reversibility to this process. Interestingly, locomotor activity was clearly enhanced by DEP treatment and decreased by withdrawal. Results of this investigation evidenced chronological differences in the vascular and neuroadaptative responses, where the neuroadaptative responses were evident even after DEP treatment. On the basis of these and other evidences, DEP and other sympathomimetic drugs mimicking central NE effects remain under current investigation in regard to their mechanistic effects on neurochemical transmission (Bray, 2005).

Although serotonergic, noradrenergic, and dopaminergic systems are likely to be actively participating in DEP's central effects, some other neurotransmitter systems might be involved as well. Taking into consideration that many of the neurotransmitter systems involved in DEP's actions could be modulated either by excitatory or inhibitory signals, our group was interested in characterize a possible participation of glutamatergic and GABAergic systems in its actions. As a first approach to this issue, we tested the effects of DEP on the brain regional levels of excitatory (Asp and Glu) and inhibitory [γ-aminobutyric acid (GABA) and glycine (Gly)] amino acids in rats (Galván-Arzate and Santamaría, 2002). Our results indicate an active role of excitatory and inhibitory transmissions in DEP's activity. Whether these changes were due to primary alterations in the dopaminergic system remain to be tested in further studies. These findings are relevant because they suggest that DEP recruits different neurotransmission systems to exert its effects.

B. Bupropion Action on Neurotransmission

Bupropion has been used for long time as an antidepressant, as well as in the pharmacotherapy for smoking cessation (Wilkes, 2006). It has also recently been used "off-label" for treatment of deficit hyperactivity disorder (Covey et al., 2008). These therapeutic properties have been discovered and adopted on the basis of the many pharmacological properties that this molecule exhibits. For instance, BP has been pressumed to be a dual NE and DA reuptake inhibitor (NDRI) (see Table I), a vesicular monoamine transport enhancer (Rau et al., 2005), an anti-inflammatory agent against the actions of cytokines such as tumor necrosis factor-α, a cytochrome P450 CYT2D6 inhibitor (Dhillon et al., 2008;

TABLE I

Interaction of Bupropion and Its Endogenous Metabolites With Different Neurotransmitter Transporters

Used technique	Tissue or cell	IC$_{50}$ or K$_i$ (μM)		IC$_{50}$ (μM)		References
		DAT	NET	SERT		
[^{3}H]DA uptake	DAT or NET expressed in COS-7 cells	0.95 ± 0.25	4.63 ± 0.58			Mortensen and Amara (2006)
[^{3}H]Nisoxetine competition binding	DAT or NET expressed in COS-7 cells	1.51 ± 0.32	3.42 ± 0.48			Mortensen and Amara (2006)
[^{3}H]NE uptake	Human NET expressed in C6 glial cells		1.37 ± 0.14			Foley and Cozzi (2002)
[^{3}H]WIN 35,428 competition binding	Rat caudate putamen membranes	0.373				Katz et al. (2000)
In vivo [^{3}H]WIN 35,428 competition binding	Mouse striatum	0.199				Stathis et al. (1995)
[^{3}H]WIN 35,428 competition binding	Rabbit caudate membranes	0.055 ± 0.004				Aloyo et al. (1995)
[^{3}H]Mazindol competition binding	Rat striatal synaptosomes	0.178 ± 0.038				Dersch et al. (1994)
[^{3}H]DA uptake	Rat DAT expressed in Ltk$^-$ cells	0.52				Giros et al. (1992)
[^{3}H]DA uptake	Human DAT expressed in Ltk$^-$ cells	0.33				Giros et al. (1992)
[^{3}H]DA uptake	Rat striatal synaptosomes	2.0 ± 0.6 23 ± 8[a] 47 ± 2[b]				Ascher et al. (1995)
[^{3}H]DA uptake	Mouse striatal synaptosomes	2.0 ± 0.8 17 ± 9[a] 23 ± 13[b]				Ascher et al. (1995)
[^{3}H]NE uptake	Rat hypothalamic synaptosomes		5 ± 1 7 ± 3[a] 16 ± 7[b]			Ascher et al. (1995)

[³H]NE uptake	Mouse hypothalamic synaptosomes	4 ± 1 4 ± 1[a] 10 ± 13[b]	Ascher et al. (1995)
[³H]5-HT uptake	Rat hypothalamic synaptosomes	58 ± 15 105 ± 11[a] 67 ± 2[b]	Ascher et al. (1995)
[³H]5-HT uptake	Mouse hypothalamic synaptosomes	36 ± 21 100 ± 11[a] 92 ± 15[b]	Ascher et al. (1995)

DAT, dopamine; NET, norepinephrine; SERT, serotonin transporters.
IC_{50}, concentration of bupropion to produce 50% inhibition of binding or uptake at the neurotransmitter transporter; K_i, inhibition constant.
[³H]WIN 35,428, (−)-2-β-carbomethoxy-3-β-(4-fluorophenyl)tropane.
[a]Hydroxybupropion; [b]Threohydrobupropion.

231

Wilkes, 2006), and as a noncompetitive antagonist (NCA) of several nicotinic receptors (for more details see Section V). Currently, BP constitutes an important pharmacological tool for biomedical research given its proved capacity to inhibit DA and NE reuptake. Despite that the main actions of BP are mostly occurring at the dopaminergic and noradrenergic systems, additional levels of actions of this agent will be considered in this section.

It was in the early 1980s when neurochemical studies on the central mechanisms of BP started to appear. Ferris *et al.* (1982) described experiments suggesting that the antidepressant activity of BP is not merely due to monoamine oxidase (MAO) inhibition or biogenic amines release from nerve endings. In this study, authors showed that BP possesses weak properties as catecholaminergic pumps inhibitor, but its selective blockade on dopaminergic pumps *in vivo* could be correlated with mild stimulation of the CNS in rodents. These and other properties, such as the lack of desensitization of β-adrenergic receptors in rat cerebral cortex, served to postulate BP as an "atypical" antidepressant with different modes of action of those for MAO inhibitors (MAOIs) and tricyclic antidepressants (TCAs) (e.g., imipramine, amitriptyline, and doxepin). Shortly thereafter, the same group (Ferris *et al.*, 1983) reinforced these observations by discriminating the antidepressant activity of BP from MAO inhibition. Although in this study it was clear that DA neurons have to be present for BP to exert its effects on the CNS, they found that at antidepressant doses of BP, the DA turnover is decreased. Moreover, the antidepressant properties of BP were also dissociated from downregulation of postsynaptic β-adrenoceptors. Based on these findings, the authors concluded that the antidepressive properties of BP cannot be merely explained on the basis of alterations in either presynaptic or receptor-mediated postsynaptic events in ACh or 5-HT pathways.

A relevant issue addressed by Gandolfini *et al.* (1983) revealed important features of BP's modulatory actions at a central level in rats chronically administered with this drug: its ability to downregulate β-adrenergic receptors (a trademark of most antidepressants with NE activity) in the frontal cortex, as well as to decrease NE-induced stimulation of adenylate cyclase. Other effects, such as MAO inhibition, release of biogenic amines, or monoamine reuptake inhibition, were discarded in this study. Furthermore, in consideration to these observations, an even when a role of DA as a central mediator of BP actions was not discarded at all, a question raised on how this antidepressant exerts its effects.

Nomikos *et al.* (1989) studied the acute effects of BP on the extracellular concentration of DA and its metabolites by means of microdialysis in freely moving rats. The dose- and time-dependent increase in DA release (peaking at 20 min) and the changes in metabolites produced by BP in the striatum and in the nucleus accumbens lasted for 2 h. Interestingly, the increase in extracellular DA levels resulted action potential-dependent since it was blocked by tetrodotoxin, and was correlated with stereotyped behavior during the first hour. The authors concluded that BP-induced DA release contributed to the observed behavioral effects through complex mechanisms still unknown, and is likely involved in the antidepressant actions of this molecule. These findings were soon corroborated in another report of the same group (Brown *et al.*, 1991) in which BP, apomorphine, D-amphetamine and other drugs were shown to induce DA and 3-methoxytyramine (a DA metabolite and biomarker of NR) in the rat striatum. In this study, BP was assumed to act as a DA reuptake inhibitor. Indeed, further studies simply accepted the DA reuptake inhibition effect of BP as its main feature (Beauregard *et al.*, 1991). Furthermore, when BP was given to rats before and after the iontophoretical DA application, it resulted in a reduced duration of the inhibitory responses evoked by DA on spontaneous neuronal firing in anterior cingulate, while lengthening them in prefrontal cortex, neostriatum, and nucleus accumbens, thus supporting, together with other evidences from the same study, the existence of a presynaptic, positive-feedback mechanism in the anterior cingulate triggered by DA which in turn favors DA release upon its reuptake in DA nerve terminals.

The dopaminergic component of BP actions was reinforced in a model of canine cataflexy (Mignot *et al.*, 1993). This event is typically controlled by adrenergic mechanisms, and therefore the BP-induced DA reuptake inhibition was nonsensitive. In contrast, the cytoprotective effect of BP, in its character of DA reuptake blocker, was evidenced when tested as a tool against two neurotoxic molecules, 6-hydroxydopamine and hydrogen peroxide (H_2O_2), typically affecting the dopaminergic system (Abad *et al.*, 1995). BP significantly increased cell viability, suggesting that the induction of high extracellular levels of DA counteracts the deleterious selective effects of a disrupted DA transmission. In addition, it has been shown that DA-induced NE is not affected by BP when tested in cardiac β-adrenoceptors, suggesting that the cardiac effects of DA result from the indirect action of β-adrenoceptor stimulation and the subsequent NE release, with no further involvement of pharmacologically enhanced extracellular levels of DA (Habuchi *et al.*, 1997).

BP has also been employed as a tool to characterize the kinetics of DA elimination in rats. BP and other DA reuptake blockers significantly enhanced the decay time of DA oxidation in the striatum (Suaud-Chagny *et al.*, 1995). The authors concluded that BP and other reuptake inhibitors are able to potentiate dopaminergic transmission enhanced by action potential-mediated DA release. Moreover, Holson *et al.* (1998) employed BP in an experimental design in rats using microdialysis in order to evaluate the mechanisms by which striatal DA release declines in time. These authors found that the use of BP is also effective to induce DA release, and this event is significantly decreased in time, when for example, DA release is compared at 2, 6, or 26 h postinsertion of the probe. This is probably due to a limited capacity of the striatum to release and synthesize DA over time in a system that has been implanted with a probe and challenged with this and other drugs. These findings are particularly relevant when considering the use of microdialysis designs to evaluate the response of BP.

On the other hand, Heal *et al.* (1992) demonstrated that some effects of BP can be different to those of other drugs affecting DA function, such as sibutramine and metamphetamine. For instance, BP, in contrast to the other drugs, was unable to prevent reserpine ptosis in rats. In addition, BP was able to induce 6-hydroxydopamine-mediated circling behavior only at high doses. Although BP shared some effects with these molecules, those contrasting effects emphasize the involvement of differential mechanistic actions for this drug.

Another relevant issue in the mechanisms of action of BP is its proved ability to increase the firing rate of serotonergic neurons, possibly by a NE-facilitated mechanism. In this regard, mechanisms such as the spontaneous firing of NE, 5-HT, and DA neurons induced by BP have been investigated in rats (Dong and Blier, 2001). In this study, BP attenuated the firing activity of NE in a mechanism clearly dependent on α-adrenoceptors stimulation. In contrast, the mean firing of 5-HT neurons was drastically increased by BP's treatment, whereas firing of DA neurons was unaffected. It was concluded that the decreased firing rate of NE neurons induced by BP was due to an increased activation of their inhibitory somatodendritic α_2-adrenoceptors, which in turn was attributable to increased NE release, an issue supported by the observation that NE reuptake blockers did not change the firing rate of 5-HT neurons. No DA activity mediated by BP was evidenced on mesolimbic/cortical DA neurons.

Along this line of studies dealing with the involvement of other neurotransmitter systems in BP's actions, this agent has also been employed to enhance the potentiated Glu receptor-mediated excitatory postsynaptic potentials induced by clozapine in the rat prefrontal cortex (Chen and Yang, 2002). Although BP was used once again as a tool on the basis of its capacity to induce DA reuptake inhibition, its effect on a typical Glu-mediated event might serve to suggest a possible involvement of this neurotransmission system in its actions through an interaction with the DA system. Of further relevance in this study was the demonstration that BP readily acts as a neuroactive molecule. These and other findings have served to reconsider the term by which BP is recognized in regards to its activity. Thus, in clinical trials, BP is more accepted now as a dual NDRI, and that it is more effective using a sustained-release preparation (Paul *et al.*, 2002).

Since this drug may be acting in different manners between animal models and humans, special attention was paid to the neurochemical effects of BP in humans (Gobbi *et al.*, 2003). Recognized as a weak NDRI and lacking direct action on 5-HT transporters in the rat brain (see Table I), BP is also known to suppress NE neuronal firing activity. In healthy male subjects, these authors found no evidence for BP to act as a NE or 5-HT reuptake inhibitor when assessing the effect of these markers on blood pressure response to tyramine, although it may increase NE availability by increasing its release. It was concluded that the neurochemical properties of BP in humans are complex and still remain undetermined.

Another relevant contribution to the understanding of BP's mechanisms of action in humans was made by Learned-Coughlin's group (Learned-Coughlin *et al.*, 2003). These authors evaluated the *in vivo* activity of BP in the human brain measuring, by positron emission tomography (PET), the extent of duration of DAT occupancy by BP and its metabolites in human volunteers under sustained-release oral treatment. As expected, BP and its metabolites inhibited the striatal reuptake of the specific DAT radioligand [^{11}C]-βCIT-FE {[^{11}C]-N-ω-fluoroalkyl-2-β-carboxy-3-β-(4-iodophenyl) nortropane ester}, thus inducing a low occupancy of striatal DAT. This evidence supports the hypothesis that DA reuptake inhibition by BP is partially responsible for its therapeutic effects on human subjects. However, considering that BP potently inhibits CYP26D activity, special care should be taken when patients are taking drugs metabolized by CYP26D during clinical trials (Kotlyar *et al.*, 2005).

When used again as a tool to evaluate the amphetamine-induced increase in protein kinase C (PKC) activity in rat striatal synaptoneurosomes, BP and other DA reuptake blockers attenuated this effect, but when tested alone these drugs increased PKC activity by themselves (Giambalvo, 2003). This is relevant since several metabolic pathways might be eventually activated by BP through PKC for its central actions. These initial findings were soon confirmed by the same group (Giambalvo and Price, 2003).

Recently, BP has also been shown to increase brain and core temperatures in freely moving rats in a microdialysis trial (Hasegawa *et al.*, 2005). The study demonstrated that the acute inhibition of DA and NE reuptake, but not of 5-HT reuptake, is responsible for the elevation of brain and core temperatures, an issue needing further consideration in clinical trials. In addition, Sidhpura's group compared the effects of BP on AChR-mediated DA release in rat striatal synaptosomes and slices (Sidhpura *et al.*, 2007a,b). Although addressing classical aspects of BP, the contribution of this study is based on the proposal that a modest blockade of DAT by low concentrations of BP generates a feedback inhibition *via* D2 DA presynaptic autoreceptors, and that this is overcome at higher concentrations of BP. However, at this higher concentration, BP does not directly inhibit AChRs. Therefore, the BP-induced inhibition of DAT and AChRs are two events clearly dependent on BP's concentration. In this regard, studies on BP's pharmacokinetic represent an issue deserving further attention, specially in the brain, since they establish the basis for the understanding of the many interactions of this drug with receptors and transporters in the brain. Indeed, several reports addressed this aspect by describing these interactions, mostly supporting the concepts that we have described in this review. As examples, we will mention a couple of reports evaluating the concentrations of BP in the brain, as this issue is crucial for the effects evoked by this agent in the CNS. One of the first reports describing BP kinetics was performed by Schroeder (1983). The author, using rats, dogs, and normal volunteers, demonstrated that BP was rapidly and completely absorbed, widely distributed in tissues, and metabolized extensively prior to its excretion. BP metabolism in rats and dogs was produced predominantly by side chain oxidative cleavage, while reduction of the intact parent aminoketone to an aminoalcohol was an additional major pathway in men. The author also showed that BP, but not its metabolites, was concentrated in many tissues, with a brain to plasma ratio of about 25:1. This investigation was followed soon by a study performed in three different species including, rat, mouse, and guinea pig (Suckow *et al.*, 1986). The pharmacokinetic profiles of BP and its major endogenous metabolites, hydroxybupropion and threohydrobupropion, were determined following the intraperitoneal administration of 40 mg/kg BP to these animals. The authors also carried out further investigation on the reduced BP metabolite threohydrobupropion. The pharmacokinetic analysis showed that rat quickly metabolized BP, but not its accumulated metabolites, while mouse metabolized BP predominantly to hydroxybupropion, and guinea pig converted BP to both threohydrobupropion and hydroxybupropion. Although brain/plasma ratios of BP among these species did not vary significantly, both metabolites exhibited differences in their brain/plasma ratios. The authors demonstrated important differences in BP's metabolism among animal species, where guinea pig, when compared to rat or mouse, constitutes a model that most closely resembles that of human BP metabolism. Altogether, these studies demonstrated two major issues: (a) BP is rapidly absorbed and distributed in organs from the studied species, including man. Once into the body, BP reaches the brain in considerable amounts, attaining those concentrations needed to exert its therapeutic effects at the central level; and (b) all tested species exhibit differences in BP metabolism. Particularly, men show alternative routes to metabolize BP. Interestingly, hydroxybupropion has almost the same affinity as BP for NET (Ascher *et al.*, 1995). However, during clinical treatment, hydroxybupropion reaches concentrations ∼10-fold higher than that of the parent drug (Ascher *et al.*, 1995; Hsyu *et al.*, 1997). This increase in concentration would counteract for the observed low affinity for NET compared to that for DAT, causing meaningful inhibition of NE reuptake, even at therapeutic dosages. Finally, both (*S*)- and (*R*)-BP isomers have the same affinity for monoamine transporters, indicating lack of enantioselectivity (Musso *et al.*, 1993). These aspects should be considered for future design of pharmacological therapeutic strategies.

More recent evidence describes a role of neuromodulatory receptors in BP-induced behavioral changes. In this study, Dhir and Kulkarni (2008) present pharmacological evidence that sigma-1 opioid receptors are involved in the anti-immobility effects elicited by BP. These receptors are known to modulate catecholamine release in the brain, and therefore, they can be involved in the modes of action of several antidepressants. Using the forced swim test, the authors found that BP increased locomotor activity *per se*, and that this effect was reversed by the use of different specific sigma-1 receptor antagonists. Thus, this is another modulatory pathway by which BP might be exerting its central actions, and provides an additional explanation to the complex pharmacological and behavioral patterns evoked by this molecule.

C. ANTINICOTINIC ACTION OF BUPROPION

Bupropion (Wellbutrin®, Zyban®) is recommended as a first aid for smoking cessation, and its efficacy can be attributed to two reasons: BP acts as a nicotine substitute (Young and Glennon, 2002; Wiley *et al.*, 2002) and as a nicotinic antagonist (for more details see Section V). In this regard, BP shows a positive response in different animal models (reviewed by Dwoskin *et al.*, 2006). For example, BP attenuates nicotine-induced unconditioned behaviors, sharing or enhancing discriminative stimulus properties of nicotine, and affects nicotine self-administration in a complex manner, that is, low doses augments whereas high doses attenuates self-administration. Current studies show that BP facilitates the acquisition of nicotine conditioned place preference in rats, further suggesting that BP enhances the rewarding properties of nicotine. BP has been shown to attenuate the expression of nicotine withdrawal symptoms in both animal models and human subjects. With respect to relapse, current studies show that BP attenuates nicotine-induced reinstatement in rats. Recent studies by Paterson *et al.* (2007) explored the hypothesis that BP ameliorates, at least partially, nicotine withdrawal by a DA-dependent mechanism. For this purpose, the authors investigated the effects of a chronic infusion of BP on behavioral aspects of nicotine withdrawal including brain reward thresholds and somatic signs of withdrawal. The authors concluded—based on the findings that BP lowers reward thresholds, increases K^+-evoked DA release, and blocks withdrawal-associated somatic signs—that the BP-induced increase in DA extracellular levels in the nucleus accumbens shell decreases the anhedonic component of nicotine withdrawal, an issue clearly in favor of smoking cessation facilitation.

Other interesting findings of BP for antinicotinic therapy have been recently described. For instance, it has been shown that BP inhibits the cellular effects of nicotine in the rat ventral tegmental area (an area involved in brain reward) and reduces GABAergic transmission to DA neurons, thereby increasing DA neuron excitability, which in turn contributes to its antidepressant effects (Mansvelder *et al.*, 2007). Moreover, BP is responsible for increased locomotor activity in freely moving rats in mechanisms typically involving DA reuptake inhibition and blockade of AChRs, potentiating the behavioral effects of nicotine (Sidhpura *et al.*, 2007a,b).

In summary, whether the actions of BP are mediated by DA and NE reuptake inhibition, by direct actions on dopaminergic, adrenergic, serotonergic, and/or cholinergic receptors, or all of them, the central actions of this drug remain to be elucidated with priority as it represents a suitable strategy for different depression-related disorders.

III. Neurotoxicity of DEP and BP

Several mechanisms have been linked to the neurotoxic events evoked by different psychostimulants, including amphetamine analogs, drugs of abuse, and mild stimulants. Among the most prominent of these potential mechanisms are major changes in DA levels in corticolimbic and striatal pathways (Itzhak and Ali, 2006). Further studies have also given particular emphasis to major alterations in dopaminergic and serotonergic systems, and the potential role that the nitrergic system (NO and nitrocompounds) may exert on these neurochemical pathways (Itzhak and Ali, 2006). In contrast to well-described amphetamine analogs and other potent psychostimulants, agents like BP and DEP seem to be less potent in regard to their toxic effects, an issue pointing to a relatively safe use of these agents for clinical purposes. Nonetheless, some reports in literature describe toxic events associated with these two drugs.

A. NEUROTOXICITY ELICITED BY DEP

To our knowledge, the only report describing the neurotoxic effects of DEP so far appeared 7 years ago (Galván-Arzate and Santamaría, 2002). Our group described changes in brain regional neurochemical alterations and behavioral markers of toxicity in rats assessed after 15 days of daily DEP administration (5 mg/kg), a scheme looking to simulate those conditions at which the drug is administered to humans for clinical purposes. These amphetamine derivatives produced significant changes in total levels of Asp in the hypothalamus and cortex, Glu in hypothalamus, cortex, midbrain, and striatum, and Gln in cortex, whereas the regional levels of GABA remained unchanged. In addition, lipid peroxidation, a current marker of oxidative stress, was increased in the hippo-campus, midbrain, and striatum. In regard to the behavioral markers explored, the actions of DEP on both mercaptopropionic acid-induced seizures (mercapto-propionic acid is a well-known inhibitor of glutamate decarboxylase, a key enzyme for GABA synthesis) and kainic acid-induced wet-dog body shakes (kainic acid is a specific Glu receptor agonist with excitatory activity) were explored to investigate whether DEP induces behavioral sensitization to the effects of agents affecting the central activity of neuroactive amino acids. The latency to the first seizure induced by mercaptopropionic acid (1.2 mmol/kg; i.p.) and the total number of wet-dog body shakes induced by kainic acid (10 mg/kg; i.p.) were significantly affected by DEP treatment. Based on our findings, we concluded that low doses of DEP might be selectively affecting neurochemical substrates, therefore inducing changes in neuroactive amino acids along the brain regions, in a mechanism probably involving DA release. Moreover, the observed behavioral alterations could be the result of excitotoxic events related to Glu changes or the lack of inhibitory processes, and these sensitizations might, in turn, be related to free radical formation and oxidative stress. After all, it seems that DEP is not as innocuous as it has been considered in the literature, an issue to take into consideration in future clinical trials.

B. Neurotoxicity Elicited by BP

The same as DEP, BP has also been mostly considered as a safe agent for clinical purposes. As an antidepressant, this agent has been reported to produce fewer adverse effects than those for TCAs (Van Wyck-Fleet et al., 1983). Although it possesses minimal effects on the reuptake of 5-HT (see Table I) and no proved effects on MAO-A and MAO-B activity, some clinical trials have revealed its toxic features thus, revealing an evident lack of experimental assessments, turning this section more into a description of BP's side effects.

As an example of those studies validating the safeness of the use of this drug in humans, Cato et al. (1983) described a strategy for identifying and classifying adverse events of BP treatment, in direct relation to therapy. BP was studied in four double-blind, placebo-controlled trials. The authors found that the incidence and frequency of adverse events associated with BP were minimal, and correlated well with the known pharmacologic and clinical properties of this antidepressant. By the same year, Van Wyck-Fleet et al. (1983) supported these observations in a study addressing the clinical development of Wellbutrin® in 1153 depressed patients and 157 normal volunteers. Safety measures during the clinical trial program included adverse event symptoms, vital signs, clinical laboratory examinations, and EEGs. The authors found that there were no severe BP-related changes in vital signs, clinical laboratory, or EEG results supporting treatment discontinuation. The only adverse experience considered of medical significance in BP-treated patients was major motor seizure. However, this event occurred with a very limited incidence, comparable with that for TCAs.

Szuba and Leuchter (1992) reported the case histories of two geriatric patients treated with BP for major depression. They found that both patients experienced a previously unreported side effect associated with BP, the so-called "falling backward." Interestingly, orthostatic hypotension and vertigo were not consistent with patients' histories. In addition, both patients manifested other symptoms consistent with parkinsonian syndrome. Their conclusions pointed out to consider that this side effect, apparently unique to BP among antidepressants, is likely to be mediated by dopaminergic effects in the basal ganglia, apparently affecting the geriatric population preferentially. Shortly thereafter, Stoll *et al.* (1994) established further support for the clinical use of BP since they contrasted its effects with those of TCAs, fluoxetine, and MAOIs in patients with antidepressant-associated mania. Maniac criteria included delusions, hallucinations, psychomotor agitation, and bizarre behavior. In general terms, those patients with MAOI- and BP-associated mania exhibited a slightly lower overall rating of psychopathological severity at admission than those subgroups with fluoxetine- and TCAs-associated mania. The authors concluded that MAOI and BP may be associated with milder manic states than either TCAs or fluoxetine.

Together with reversible dyskinesia (Gardos, 1997), BP has also shown to produce seizures, mostly when overdosed (Pesola and Avasarala, 2002). An investigation was performed to determine the importance of BP medication as a possible etiologic factor of new-onset seizures relative to other drugs and new-onset seizures in general (Pesola and Avasarala, 2002). By means of a retrospective study, all new onset generalized seizures were evaluated over a 4-year period in subjects 16 years of age and older. Neurologic consultation, chemistry screening, and neuroimaging served to establish the etiologic diagnosis. The incidence of drug related new-onset seizures was 6% of cases, and all of them presented additional linked components, such as sleep deprivation, attention deficit disorder, bulimia, and alcohol abuse. Moreover, all BP-related seizures occurred in patients taking what was considered to be a therapeutic dose (i.e., 450 mg/day or less). The authors concluded that BP might be a more common cause of drug-related new-onset generalized seizures than previously assumed, thus suggesting an exacerbated excitatory component. Exclusion criteria should be considered for therapeutic dosages of BP in light of this evidence. de Graaf *et al.* (2003) addressed the issue of an extrapyramidal disorder possibly related to the use of BP. As described by the authors, the patient received a dose of 150 mg a day, increasing to 150 mg twice a day on the 4th day. At day 8 of treatment, unvoluntary movements with no cognitive disturbances started to appear all over the body in short but frequent attacks and were diagnosed as ballism. Once therapy was discontinued, treatment with haloperidol diminished the extrapyramidal symptoms. A role of BP-induced ballism by interaction with 5-HT receptors was discarded. These and other extrapyramidal signs of BP such as dyskinesia, might be, according to the authors, attributable to the dopaminergic activity of this drug. Other cases of extrapyramidal disorders associated with BP support these observations. In those cases, characterized symptoms have been orofacial dyskinesia, tremor, and retropulsion, as well as rigidity of the trunk and extremities, and roving eye movements, all in geriatric patients using BP as antidepressant.

The ability of other dopaminergic drugs, such as levodopa and bromocriptine, to induce acute dyskinesia is also substantiated in the literature (de Graaf *et al.*, 2003). Although extrapyramidal disorders have been described in relation to amphetamines and other CNS stimulants, whether the structural relationship between BP and amphetamines contributes to the occurrence of dyskinesia still remains unclear.

In summary, although limited, the adverse and toxic effects of BP are present and deserve intense experimental and clinical characterization in order to support the use of this drug under specific schedule of dosage conditions, or to discard it under other clinical conditions.

IV. Molecular Interaction of BP with DAT and NET

Both DAT and NET belong to the family of Na^+/Cl^--dependent transporters found in the nervous system (reviewed by Iversen, 2006; Mortensen and Amara, 2003). These proteins are expressed in synaptic regions close to the active zone as well as in dendrites and soma of presynaptic neurons (see Fig. 1). These transporters have 12 transmembrane segments (TM1–TM12), where regions comprising TM5–TM8 contain structural elements important for several classes of inhibitors. More specifically, S359, located in the middle of TM7, contributes to the initial binding of BP, whereas A279, located close to the extracellular end of TM5, is responsible for the subsequent steps in the blockade of translocation in DAT (Mortensen and Amara, 2006). These two amino acids, and probably F367, forming a cluster of residues that contribute to BP sensitivity, are located far from the substrate binding pocket located between TM1 and TM6. In addition, the homologous residues V276 and I356 in NET interact specifically with BP.

V. Role of AChRs in Depression and in the Pharmacological Action of Antidepressants

Nicotinic acetylcholine receptors (AChRs) are the paradigm of the Cys-loop ligand-gated ion channel superfamily. This genetically linked superfamily includes types A and C GABA, type 3 5-HT, and Gly receptors (reviewed by Arias, 2006a,b). The malfunctioning of these receptors has been considered as the origin of several neurological disorders (reviewed by Hogg *et al.*, 2003; Lloyd and Williams, 2000). For example, the evidence showing a higher rate of smokers in depressed patients than in the general population supports a possible role of AChRs in depression mechanisms (reviewed by Picciotto *et al.*, 2002). In this regard, it has been reported that hypercholinergic neurotransmission, which is associated with depressed mood states, may be mediated through excessive neuronal AChR activation and that the therapeutic actions of many antidepressants may be mediated in part through inhibition of one or more AChRs (reviewed by Shytle *et al.*, 2002).

Structurally different antidepressansts behave pharmacologically as NCAs of several AChRs (Gumilar *et al.*, 2003; Sanghvi *et al.*, 2008; reviewed by Arias *et al.*, 2006a). *In vivo* and *in vitro* results indicate that BP also inhibits several AChRs in a noncompetitive manner (Arias *et al.*, 2009; Bondarev *et al.*, 2003; Damaj *et al.*, 2004; Fryer and Lukas, 1999; Slemmer *et al.*, 2000; reviewed by Arias *et al.*, 2006a). Table II shows the inhibitory potency and affinity of BP and its endogenous metabolites on several AChR types in different conformational states. Of particular importance is the fact that BP inhibits the $\alpha 3 \beta 4$ AChR, since this receptor type has been considered as the main target for the antiaddictive actions mediated by BP and other NCAs (Glick *et al.*, 2002; Maissonueve and Glick, 2003; Taraschenko *et al.*, 2005). An interesting study performed by Miller *et al.* (2002) showed the capability of BP to inhibits nicotine-evoked [^{3}H]DA overflow in rat striatal slices preloaded with [^{3}H]DA and in rat hippocampal slices preloaded with [^{3}H]NE. It was concluded that, besides of its known ability to inhibit DA and NE transporters, BP also acts as an NCA of $\alpha 3 \beta 2$ and $\alpha 3 \beta 4$ AChRs in both brain regions.

The most accepted mechanism of action for BP is that this antidepressant inhibits the catecholamine reuptake in presynaptic neurons, modulating the concentrations of DA and NE in the synaptic cleft (see Section II.B). However, the affinity of BP for these neurotransmitter transporters is only moderate (see Table I), and there is not clear-cut evidence explaining the dual antidepressant and antinicotinic modes of action elicited by BP. In this regard, the combined inhibition of AChRs and neurotransmitter transporters produced by BP might account for its clinical efficacy in smoking cessation therapy. Thus, the contribution of the BP-induced AChR inhibition to its clinical action could be twofold important: as part of the side effects elicited by BP, and/or as part of its clinical outcome. Thus, a better understanding of the interaction of BP with the AChR in different conformational states to determine its noncompetitive inhibitory mechanism is crucial to develop safer antidepressants and/or antinicotinic drugs. In this regard, the interaction of BP with different conformational states of muscle-type AChRs, as a model of neuronal AChRs, was determined by functional and structural approaches (Arias *et al.*, 2009).

A. MOLECULAR MECHANISMS OF AChR INHIBITION MEDIATED BY BP

In a first attempt to determine the effect of BP on epibatidine-activated Ca^{2+} influx in TE671 cells expressing the $\alpha 1 \beta 1 \varepsilon \delta$ AChR, preincubation and coinjection protocols were used (Arias *et al.*, 2009). The inhibitory potency of BP was in the 10–20 μM concentration range as was previously determined by $^{86}Rb^+$ efflux experiments using the same cell type (Fryer and Lukas, 1999) (see Table II).

TABLE II

INTERACTION OF BUPROPION AND ITS ENDOGENOUS METABOLITES WITH SEVERAL AChRs IN DIFFERENT CONFORMATIONAL STATES

AChR type	Used technique	Resting $IC_{50}{}^{a}$ or $K_i{}^{b}$ (μM)	Activated $IC_{50}{}^{a}$ (μM)	Desensitized $K_i{}^{b}$ (μM)	References
Human $\alpha 1\beta 1\gamma\delta$	Ca^{2+} influx fluorimetry		20.5 ± 3.9^{c} 10.5 ± 2.1^{c}		Arias et al. (2009)
Human $\alpha 1\beta 1\gamma\delta$	$^{86}Rb^{+}$ efflux		10.5 ± 1.0		Fryer and Lukas (1999)
Torpedo	[^{3}H]TCP competition binding[b]	5.1 ± 0.3^{b}		2.0 ± 0.1^{b}	Arias et al. (2009)
Torpedo	[^{3}H]Imipramine competition binding[b]	21.4 ± 3.4^{b}		11.6 ± 1.3^{b}	Arias et al. (2009)
Mouse $\alpha 1\beta 1\varepsilon\delta$	Patch-clamp	0.40 ± 0.04^{d}	40.1 ± 5.1		Arias et al. (2009)
Human $\alpha 3\beta 4^{e}$	Ca^{2+} influx fluorimetry		0.82 ± 0.10		Sidhpura et al. (2007)
Human $\alpha 3\beta 4^{e}$	$^{86}Rb^{+}$ efflux		1.51 ± 0.32		Fryer and Lukas (1999)
Rat $\alpha 4\beta 2$	Voltage-clamp	8.0 ± 2.6^{d}	55		Slemmer et al. (2000)
Rat $\alpha 3\beta 2$	Voltage-clamp	1.3 ± 0.6^{d}			Slemmer et al. (2000)
Rat $\alpha 7$	Voltage-clamp	$\sim 60^{d}$			Slemmer et al. (2000)
Human $\alpha 3\beta 4^{*}$	$^{86}Rb^{+}$ efflux		1.8 ± 1.1		Damaj et al. (2004)
Human $\alpha 4\beta 2$	$^{86}Rb^{+}$ efflux		12.0 ± 1.1		Damaj et al. (2004)
Human $\alpha 4\beta 4$	$^{86}Rb^{+}$ efflux		14.0 ± 1.1		Damaj et al. (2004)
Human $\alpha 1\beta 1\gamma\delta$	$^{86}Rb^{+}$ efflux		7.9 ± 1.1		Damaj et al. (2004)
Human $\alpha 3\beta 4^{*}$	Ca^{2+} influx fluorimetry		0.82 ± 0.10		Sidhpura et al. (2007)
Rat $\alpha 3\beta 4^{*}$	Nicotine-evoked [^{3}H]NE overflow in hippocampal slices		0.32		Miller et al. (2002)
Rat $\alpha 3\beta 2^{*}$	Nicotine-evoked [^{3}H]DA overflow in striatal slices		1.27		Miller et al. (2002)
Rat $\alpha 3\beta 4$	$^{86}Rb^{+}$ efflux		7		Bondarev et al. (2003)
Rat $\alpha 3\beta 4$	$^{86}Rb^{+}$ efflux		14^{f}		Bondarev et al. (2003)

(continued)

TABLE II *(continued)*

AChR type	Used technique	Resting IC$_{50}$^{a} or K$_i$^{b} (μM)	Activated IC$_{50}$^{a} (μM)	Desensitized K$_i$^{b} (μM)	References
Rat α3β4	^{86}Rb$^+$ efflux		20^g 18^h		Bondarev *et al.* (2003)
Rat α3β4	HPLC-immobilized α3β4 AChRs		0.18		Jozwiak *et al.* (2004)
Human α4β2	^{86}Rb$^+$ efflux		31.0 ± 1.1^g 3.3 ± 1.1^h		Damaj *et al.* (2004)
Human α3β4^e	^{86}Rb$^+$ efflux		6.5 ± 1.2^g 10.0 ± 1.5^h		Damaj *et al.* (2004)
Human α4β4	^{86}Rb$^+$ efflux		41.0 ± 1.1^g 30.0 ± 1.1^h		Damaj *et al.* (2004)
Human α1β1γδ	^{86}Rb$^+$ efflux		7.6 ± 1.1^g 28.0 ± 1.4^h		Damaj *et al.* (2004)

^{a}IC$_{50}$ is the required drug concentration to produce 50% inhibition of agonist-activated AChRs; ^{b}K$_i$ values were obtained in the presence of α-bungarotoxin (resting state) or carbamylcholine (desensitized state), respectively; ^{c}These values were obtained by preincubating the cells with BP for 5 min or by conjecting BP and epibatidine for several seconds, respectively; ^{d}These values were obtained by a brief preincubation with BP before agonist activity was measured; ^{e}The native AChR can have additional subunits; ^{f}These values correspond to either (−)- or (+)-threohydrobupropion; ^{g}These values correspond to (2R,3R)-hydroxybupropion; ^{h}These values correspond to (2S,3S)-hydroxybupropion.

Comparing these values with that found in other AChR types (see Table II), we can conclude that the BP specificity follows the sequence: $\alpha3$- > $\alpha4$- ~ $\alpha1$- > $\alpha7$-containing AChRs. Considering the serum levels of BP attained after its oral administration (~0.5–1 μM; Hsyu *et al.*, 1997), the blockade of non-$\alpha3$-containing AChRs is unlikely. However, hydroxybupropion, which has practically the same activity as BP for the $\alpha4\beta2$ AChR (Damaj *et al.*, 2004; see Table II), can attain ~10-fold the plasma concentration of the parent compound (Ascher *et al.*, 1995; Hsyu *et al.*, 1997), increasing the possibility that $\alpha4\beta2$ AChRs contribute to the clinical efficacy of BP.

In order to distinguish the effect of BP on a particular conformational state, macroscopic currents of agonist-activated receptors rapidly pre- (mainly in the resting state) or coincubated (mainly in the open state) with BP were recorded (Fig. 3). Macroscopic currents show that the interaction of BP with the open state decreases the decay time constant of currents activated by 300 μM ACh. At this ACh concentration, the decay rate equals the rate of desensitization from the double liganded open state (Spitzmaul *et al.*, 2001). Thus, our results reveal that BP increases the desensitization rate from the open state. The current decay was adequately fitted using a single exponential function, which discards the possibility of a fast open-channel blockade mechanism. However, the occurrence of slow channel blockade, which has been described for other NCAs (Papke *et al.*, 1989), may result in similar changes in macroscopic currents. If the antidepressant blocked the AChR and its unblocking process were quite slow, distinguishing between this process and the increase in the desensitization rate would be difficult (Gumilar *et al.*, 2003).

Macroscopic current recordings also show a decrease in the peak current when BP interacts with the resting channel. Thus, together with the acceleration of desensitization from the open state, BP inhibits the opening of channels by binding to the resting state. The latter effect might be due to either an increase in desensitization of the resting state or a direct blockade of unliganded channels. A similar effect is observed with TCAs (Gumilar *et al.*, 2003). The potency of inhibition is greater when BP acts on the resting state than on the open state. A similar finding was reported for the inhibition of different Cys-loop receptors by TCAs (Choi *et al.*, 2003; Gumilar and Bouzat, 2008).

The two different effects detected for each different conformational state, that is, reduced peak current and increased decay rate, are additive when the receptor is exposed continuously to the drug. Thus, the mechanism by which BP inhibits AChR function is selective for each conformational state.

B. LOCALIZATION OF THE BP BINDING SITE IN THE AChR ION CHANNEL

To characterize the BP binding sites in different receptor conformational states, a combination of several methods were used on the *Torpedo* AChR (Arias *et al.*, 2009). The results from the radioligand competition binding and

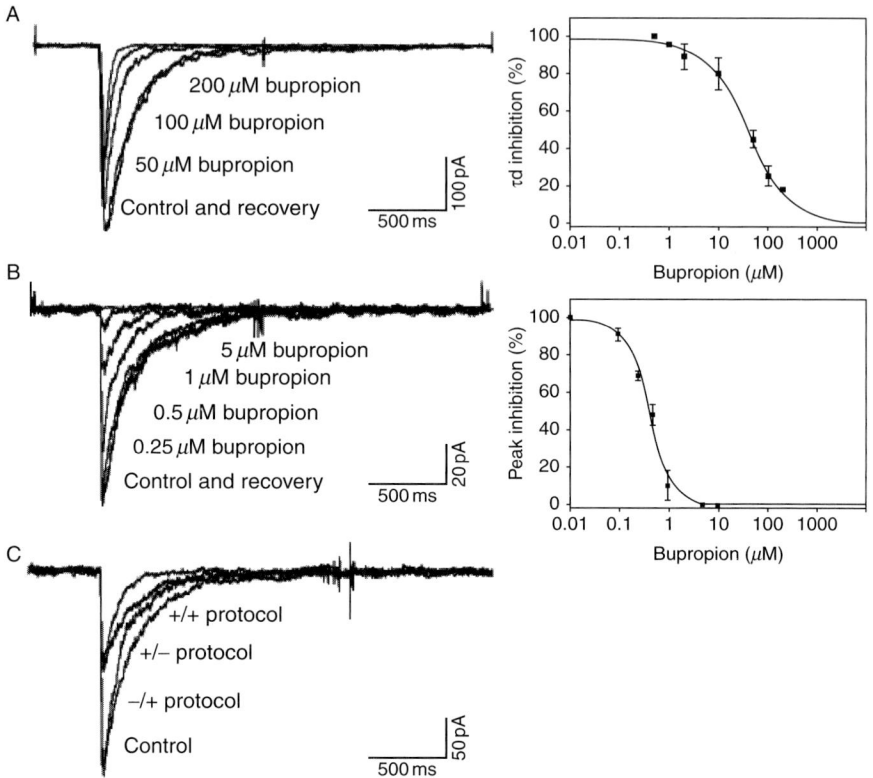

FIG. 3. Effects of bupropion on agonist-induced macroscopic currents in HEK293 cells expressing mouse $\alpha 1\beta 1\varepsilon\delta$ (adult) AchRs (modified from Arias *et al.*, 2009). (A) Bupropion effects from the open state. *Left*: Ensembled mean currents obtained from outside-out patches activated in absence (control) or simultaneous application of ACh and BP, without preincubation with BP (protocol $-/+$; open state). Each trace represents the average of 4–8 applications of agonist. Curves from right to left correspond to: control and recovery, 50, 100, and 200 μM bupropion. The calculated decay time constants (τ_d) are 25 ms for control and recovery curves, and 11.3, 6.3, and 4.5 ms, for 50, 100, and 200 μM BP, respectively. Membrane potential: -50 mV. *Right*: Concentration–response curve for the decrease in the decay time constant ($n = 5$). (B) Effects of bupropion from the resting/activatable state. *Left*: Effect of BP application protocol $+/-$ (resting/activatable state): 2 min preincubation of BP following ACh application. Each trace represents the average of 4–8 applications of agonist. Curves from right to left correspond to: control and recovery, 0.25, 0.5, 1, and 5 μM BP, respectively. The peak current decreases with increased BP concentrations. Membrane potential: -50 mV. The calculated IC_{50} values are summarized in Table II. *Right*: Concentration–response curve for the decrease in the peak current on the resting/activatable state. (C) Effects of bupropion application protocol on macroscopic currents. Superimposed currents responses to 300 μM ACh and BP concentrations correspond to IC_{50} for the resting/activatable state ($\sim$0.4 μM) and for the open state ($\sim$40 μM) using different protocols. From right to left curves correspond to control condition ($-/-$ protocol; $\tau_d = 13.4$ ms), simultaneous 300 μM ACh/40 μM BP application without preincubation with bupropion ($-/+$ protocol; $\tau_d = 6.5$ ms; peak current 99% of the control), ACh application following

column-immobilized AChR experiments indicate that BP binds with higher affinity to the desensitized AChR compared to the resting AChR (Arias et al., 2009). The observed higher affinity of BP for the desensitized state is due to a slower rate of dissociation from the Torpedo AChR ion channel.

The radioligand competition experiments also indicate that BP inhibits the binding of both [^{3}H]imipramine and [piperidyl-3,4-^{3}H(N)]-N-(1-(2-thienyl)cyclo-hexyl)-3,4-piperidine ([^{3}H]TCP) (the structural and functional analog of the NCA phencyclidine (PCP)) to both desensitized and resting AChRs with Hill coefficients (n_H) close to unity (Arias et al., 2009). Hill coefficients close to unity indicate a noncooperative interaction between BP and PCP or imipramine, respectively, suggesting that BP interacts with only one binding site on each AChR conformational state. This hypothesis is supported by the Schild-type analysis of imipramine-induced inhibition of [^{3}H]TCP binding to Torpedo AChRs at different initial concentrations of PCP (Sullivan et al., 2008). Although this evidence suggests a steric mode of competition between BP and PCP or imipramine, we cannot predict whether the location of the BP binding site is the same or distinct in each conformational state.

The location of the PCP binding site depends on the conformational state of the AChR ion channel (reviewed by Arias et al., 2006a). For instance, photoaffinity labeling studies using [^{3}H]ethidium diazide, which binds with high affinity to the PCP locus, helped to determine the structural components of this site in the desensitized state (Pratt et al., 2000). The results indicated that [^{3}H]ethidium diazide mainly labeled residues at and close to the leucine ring (position 9′) from the α1-M2 transmembrane segment. In addition, new photoaffinity labeling results using the hydrophobic probe 3-trifluoromethyl-3-(m-[^{125}I]iodophenyl)diazirine suggest that the PCP binding site is located between the threonine (position 2′) and serine (position 6′) rings, closer to the cytoplasmic end of the desensitized ion channel (Hamouda et al., 2008). Our photoaffinity labeling studies using [^{3}H]2-azidoimipramine supports the idea that TCAs bind to the PCP locus in the desensitized ion channel (Sanghvi et al., 2008). On the other hand, site-directed mutagenesis studies determined that the PCP binding site in the open ion channel includes residues located between the serine (position 6′) and leucine (position 9′) rings (Eaton et al., 2000). Finally, we suggested that the PCP binding site in the resting state is located closer to the external mouth than that in the desensitized and open states (Arias et al., 2002, 2003, 2006b). We also speculated that the aromatic tertiary amino group from the PCP (or TCP) molecule might interact with acidic residues (i.e., α1-Glu^{262}) located at position 20′ (e.g., the outer or

---

2 min preincubation with 0.4 μM BP (+/− protocol; τ_d = 13.9 ms; peak current = 50.4% of the control), and simultaneous 300 μM ACh/40μM BP application after preincubation with 0.4 μM BP (+/+ protocol; τ_d = 6.3 ms; peak current = 47% of the control). The calculated IC$_{50}$ values are summarized in Table II.

extracellular ring) (Arias *et al.*, 2003, 2006b; reviewed by Arias *et al.*, 2006a). Considering our previous findings, we suggest that the secondary amino group from the BP molecule can also be responsible for the binding to one of the two Glu^{262} residues by charge interactions, when the receptor is in the resting state.

Proposed mechanisms of BP binding were additionally studied by molecular modeling. Docking simulations followed by molecular dynamics of BP were performed on $\alpha 3\beta 4$ (Jozwiak *et al.*, 2004) and on *Torpedo* and mouse muscle AChRs (Arias *et al.*, 2009). Figure 4 shows the interaction of BP with the *Torpedo* AChR ion channel. The results from the docking simulations using muscle-type AChRs indicate no enantioselectivity for BP. Although there is no experimental evidence of bupropion enantioselectivity on AChRs, the $(2R,3R)$-hydroxybupropion isomer inhibits the muscle-type AChR with 3.7-fold higher potency than the $(2S,3S)$-isomer (Damaj *et al.*, 2004) (see Table II), indicating the possibility of bupropion enantioselectivity. In this regard, further experiments need to be performed to address this question.

The analysis of the obtained molecular complexes clearly indicates that BP in either the neutral or protonated form binds to the middle portion of muscle-type AChR ion channels between the serine (position 6') and valine rings (position 13') (see Fig. 4). Exactly the same locus was observed for PCP and imipramine in the *Torpedo* AChR ion channel (Sanghvi *et al.*, 2008). However, a distinction was observed in the $\alpha 3\beta 4$ ion channel: BP in the protonated state interacted with the polar region of the intermediate ring (position 1'), whereas the neutral form was positioned between the valine/phenylalanine (position 15') and serine (position 8') rings (Jozwiak *et al.*, 2004). The pocket formed by the cleft between the phenyl ring provided by $\beta 4$-Phe and the isopropyl moiety from $\alpha 3$-Val interacts with the hydrophobic portion of BP, whereas hydrogen bonds are formed between polar residues at the serine ring and the polar region of BP. Considering this difference, distinct BP binding site locations may exist on each AChR ion channel. Nevertheless, we have to take into consideration that the $\alpha 3\beta 4$ AChR ion channel was constructed by homology with the model of the 23 mer peptide imitating the M2 sequence of the *Torpedo* AChR δ subunit (Jozwiak *et al.*, 2004), whereas the *Torpedo* AChR ion channel model (Arias *et al.*, 2009) was based on the cryo-electron microscopy structure determined at ~ 4 Å resolution (PDB ID 2BG9; Unwin, 2005).

The evidence obtained using the *Torpedo* AChR model also supports the [³H] TCP competition experiments, indicating that there is a binding site for antidepressants that overlaps the PCP binding sites in the resting and desensitized *Torpedo* AChR ion channels (Arias *et al.*, 2009; Sanghvi *et al.*, 2008). The thermodynamic parameters also indicate that BP interacts with the AChR ion channels mainly by an entropy-driven process (Arias *et al.*, 2009). In this regard, BP may induce local or global conformational changes or solvent reorganization in the binding pocket. Negative enthalpic values also suggest the existence of attractive forces (e.g., van der Waals, hydrogen bond, and polar interactions) forming the stable complex.

A

B

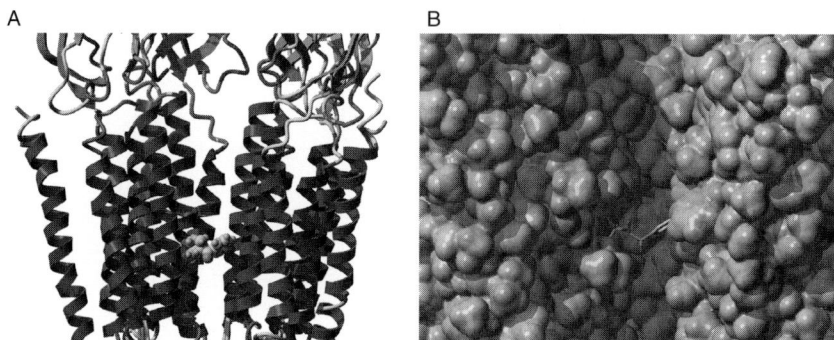

FIG. 4. Complex formed between (*S*)-BP and the *Torpedo* AChR ion channel obtained by molecular docking. The same results were obtained using the (*R*)-isomer and the mouse muscle AChR. Computational simulations were performed using the same protocol as recently reported (Arias *et al.*, 2009; Sanghvi *et al.*, 2008). Briefly, BP molecules were docked into the *Torpedo* ion channel model (PDB ID 2BG9; Unwin, 2005) using the following settings (Molegro Virtual Docker): numbers of runs = 100; maximal number of iterations = 10,000; maximal number of poses = 10. Molecular dynamics (1.0 ns) were subsequently performed (Yasara 6.10.18) using the following parameters: temperature = 298 K; multiple timesteps = 1 fs for intramolecular and 2 fs for intermolecular forces; pressure control: water probe (0.99 g/ml) ensemble. Snapshots of the simulations were saved every 5 ps. (A) Side view of the lowest energy complex showing four subunits rendered in secondary structure mode, whereas the ligand in the neutral form is rendered in element color coded ball mode. (B) Side view of the target protein rendered in semitransparent surface with visible secondary structure and explicit CPK atoms of residues forming the valine (position 13′ in green) and serine (position 6′ in red) rings. The ligand in the neutral form is rendered in stick mode with hydrogen atoms not shown explicitly. On both pictures, the δ subunit was removed for clarity, and the order of remaining subunits is (from left to right) $\alpha 1$, γ, $\alpha 1$, and $\beta 1$. (See Color Insert.)

Considering the above results, we envision a dynamic process where BP binds first to the resting AChR, probably close to the mouth of the ion channel, decreasing the probability of ion channel opening. The remnant fraction of open ion channels is subsequently decreased by accelerating the desensitization process, but not by an open-channel block mechanism. Finally, the dissociation rate of BP from the desensitized ion channel is decreased by stronger interactions with residues from a binding domain located between the serine and valine rings that is shared by antidepressants and PCP.

VI. Conclusions

Different components involved in the process of neurotransmission are important targets for the action of drugs with CNS activity. The aminoketones BP and DEP are in this category of drugs. Both drugs present structural and functional similitude. However, whereas DEP is clinically used as an anorectic agent, BP is used in depression and nicotine addiction therapies. The current evidence indicates that both DEP and BP stimulate the release as well as inhibit the reuptake of NE and DA, whereas BP also blocks different AChRs. This triple action might be the basis of the particular mechanisms by which BP mediates its clinical actions. In this regard, AChRs can be envisioned as targets for the pharmacological action of new and safer antidepressants as well as for novel antiaddictive compounds. Finally, although mostly considered as "safe" drugs for clinical purposes, BP and DEP produce reversible noxious effects at the central level. Thus, these side effects should be taken into consideration in further clinical uses.

Acknowledgments

This research was supported by grants from the Science Foundation Arizona and Stardust Foundation and the Office of Research and Sponsored Programs, Midwestern University (to H. R.), and from the Consejo Nacional de Ciencia y Tecnología (CONACYT 48370-Q), Mexico (to A. S.).

References

Abad, F., Maroto, R., López, M. G., Sánchez-García, P., and García, A. G. (1995). Pharmacological protection against the cytotoxicity induced by 6-hydroxydopamine and H_2O_2 in chromaffin cells. *Eur. J. Pharmacol.* **293,** 55–64.

Aloyo, V. J., Ruffin, J. S., Pazdalski, P. S., Kirifides, A. L., and Harvey, J. A. (1995). [^{3}H]WIN 35,428 binding in the caudate nucleus of the rabbit: Evidence for a single site on the dopamine transporter. *J. Pharmacol. Exp. Ther.* **273,** 435–444.

Arias, H. R. (2006a). Ligand-gated ion channel receptor superfamilies. *In* "Biological and Biophysical Aspects of Ligand-Gated Ion Channel Receptor Superfamilies" (H. R. Arias, Ed.), pp. 1–25. Research Signpost, India.

Arias, H. R. (2006b). Marine toxins targeting ion channels. *Mar. Drugs (Special Issue: Marine Toxins and Ion Channels)* **4,** 37–69.

Arias, H. R., Bhumireddy, P., and Bouzat, C. (2006a). Molecular mechanisms and binding site locations for noncompetitive antagonists of nicotinic acetylcholine receptors. *Int. J. Biochem. Cell Biol.* **38,** 1254–1276.

Arias, H. R., Bhumireddy, P., Spitzmaul, G., Trudell, J. R., and Bouzat, C. (2006b). Molecular mechanisms and binding site location for the noncompetitive antagonist crystal violet on nicotinic acetylcholine receptors. *Biochemistry* **45,** 2014–2026.

Arias, H. R., Gumilar, F., Rosenberg, A., Targowska-Duda, K. M., Feuerbach, D., Jozwiak, K., Moaddel, R., Weiner, I. W., and Bouzat, C. (2009). Interaction of bupropion with muscle-type nicotinic acetylcholine receptors in different conformational states. *Biochemistry* **48,** 4506–4518.

Arias, H. R., McCardy, E. A., Bayer, E. Z., Gallagher, M. J., and Blanton, M. P. (2002). Allosterically linked noncompetitive antagonist binding sites in the resting nicotinic acetylcholine receptor ion channel. *Arch. Biochem. Biophys.* **403,** 121–131.

Arias, H. R., Trudell, J. R., Bayer, E. Z., Hester, B., McCardy, E. A., and Blanton, M. P. (2003). Noncompetitive antagonist binding sites in the *Torpedo* nicotinic acetylcholine receptor ion channel. Structure-activity relationship studies using adamantane derivatives. *Biochemistry* **42,** 7358–7370.

Ascher, J. A., Cole, J. O., Colin, J. N., Feighner, J. P., Ferris, R. M., Fibiger, H. C., Golden, R. N., Martin, P., Potter, W. Z., Richelson, E., and Sulser, F. (1995). Bupropion: A review of its mechanism of antidepressant activity. *J. Clin. Psychiatry* **56,** 395–401.

Beauregard, M., Ferron, A., and Descarries, L. (1991). Comparative analysis of the effects of iontophoretically applied dopamine in different regions of the rat brain, with special reference to the cingulate cortex. *Synapse* **9,** 27–34.

Bispo-Da Silva, L. B., and Cordellini, S. (2003). Effects of diethylpropion treatment and withdrawal on aorta reactivity, endothelial factors and rat behavior. *Toxicol. Appl. Pharmacol.* **190,** 170–176.

Bondarev, M. L., Bondareva, T. S., Young, R., and Glennon, R. A. (2003). Behavioral and biochemical investigations of bupropion metabolites. *Eur. J. Pharmacol.* **474,** 85–93.

Borsini, F., Bendotti, C., Carli, M., Pogessi, E., and Samanian, R. (1979). The roles of brain noradrenaline and dopamine in the anorectic activity of diethylpropion in rats: A comparison with D-amphetamine. *Res. Commun. Chem. Pathol. Pharmacol.* **26,** 3–11.

Bray, G. A. (2005). Drug insight: Appetite suppressants. *Nat. Clin. Pract. Gastroenterol. Hepatol.* **2,** 89–95.

Brown, E. E., Damsma, G., Cumming, P., and Fibiger, H. C. (1991). Interstitial 3-methoxytyramine reflects striatal dopamine release: An *in vivo* microdialysis study. *J. Neurochem.* **57,** 701–707.

Cato, A. E., Cook, L., Starbuck, R., and Heatherington, D. (1983). Methodologic approach to adverse events applied to bupropion clinical trials. *J. Clin. Psychiatry* **44,** 187–190.

Chen, L., and Yang, C. R. (2002). Interaction of dopamine D1 and NMDA receptors mediates acute clozapine potentiation of glutamate EPSPs in rat prefrontal cortex. *J. Neurophysiol.* **87,** 2324–2336.

Choi, J. S., Choi, B. H., Ahn, H. S., Kim, M. J., Rhie, D. J., Yoon, S. H., Min, D. S., Jo, Y. H., Kim, M. S., Sung, K. W., and Hahn, S. J. (2003). Mechanism of block by fluoxetine of 5-hydroxytryptamine3 (5-HT$_3$)-mediated currents in NCB-20 neuroblastoma cells. *Biochem. Pharmacol.* **66,** 2125–2132.

Covey, L. S., Manubay, J., Jiang, H., Nortick, M., and Palumbo, D. (2008). Smoking cessation and inattention or hyperactivity/impulsivity: A post hoc analysis. *Nicotine Tob. Res.* **10,** 1717–1725.

Damaj, M. I., Carroll, F. I., Eaton, J. B., Navarro, H. A., Blough, B. E., Mirza, S., Lukas, R. J., and Martin, B. R. (2004). Enantioselective effects of hydroxy metabolites of bupropion on behavior and on function of monoamine transporters and nicotinic receptors. *Mol. Pharmacol.* **66,** 675–682.

de Graaf, L., Admiraal, P., and van Puijenbroek, E. P. (2003). Ballism associated with bupropion use. *Ann. Pharmacotherapy* **37,** 302–303.

Dersch, C. M., Akunne, H. C., Partilla, J. S., Char, G. U., de Costa, B. R., Rice, K. C., Carroll, F. I., and Rothman, R. B. (1994). Studies of the biogenic amine transporters. 1. Dopamine reuptake blockers inhibit [^{3}H]mazindol binding to the dopamine transporter by a competitive mechanism: Preliminary evidence for different binding domains. *Neurochem. Res.* **19,** 201–208.

Dhillon, S., Yang, L. P., and Curran, M. P. (2008). Bupropion: A review of its use in the management of major depressive disorder. *Drugs* **68,** 653–689.

Dhir, A., and Kulkarni, S. K. (2008). Possible involvement of sigma-1 receptors in the anti-immobility action of bupropion, a dopamine reuptake inhibitor. *Fundam. Clin. Pharmacol.* **22,** 387–394.

Dong, J., and Blier, P. (2001). Modification of norepinephrine and serotonin, but not dopamine, neuron firing by sustained bupropion treatment. *Psychopharmacology* **155,** 52–57.

Dwoskin, L. P., Rauhut, A. S., King-Pospisil, K. A., and Bardo, M. T. (2006). Review of the pharmacology and clinical profile of bupropion, an antidepressant and tobacco use cessation agent. *CNS Drug Rev* **12,** 178–207.

Eaton, M. J., Labarca, C., and Eterović, V. A. (2000). M2 Mutations of the nicotinic acetylcholine receptor increase the potency of the non-competitive inhibitor phencyclidine. *J. Neurosci. Res.* **61,** 44–51.

Farrant, M., and Nusser, Z. (2005). Variations on an inhibitory theme: Phasic and tonic activation of GABA$_A$ receptors. *Nat. Rev. Neurosci.* **6,** 215–229.

Ferris, R. M., Maxwell, R. A., Cooper, B. R., and Soroko, F. E. (1982). Neurochemical and neuropharmacological investigations into the mechanisms of action of bupropion-HCl—A new atypical antidepressant agent. *Adv. Biochem. Psychopharmacol.* **31,** 277–286.

Ferris, R. M., Cooper, B. R., and Maxwell, R. A. (1983). Studies of bupropion's mechanism of antidepressant activity. *J. Clin. Psychiatry* **44,** 74–78.

Foley, K. F., and Cozzi, N. V. (2002). Inhibition of transport function and desipramine binding at the human noradrenaline transporter by N-ethylmaleimide and protection by substrate analogs. *Naunyn Schmiedebergs Arch. Pharmacol.* **365,** 457–461.

Friström, S., Airaksinen, M. M., and Halmekoski, J. (1977). Release of platelet 5-hydroxytryptamine by some anorexic and other sympathomimetics and their acetyl derivatives. *Acta Pharmacol. Toxicol.* **41,** 218–224.

Fryer, J. D., and Lukas, R. J. (1999). Noncompetitive functional inhibition at diverse, human nicotinic acetylcholine receptor subtypes by bupropion, phencyclidine, and ibogaine. *J. Pharmacol. Exp. Ther.* **288,** 88–92.

Galván-Arzate, S., and Santamaría, A. (2002). Neurotoxicity of diethylpropion: Neurochemical and behavioral findings in rats. *Ann. NY Acad. Sci.* **965,** 214–224.

Gandolfini, O., Barbaccia, M. L., Chuang, D. M., and Costa, E. (1983). Daily bupropion injections for 3 weeks attenuate the NE stimulation of adenylate cyclase and the number of β-adrenergic recognition sites in rat frontal cortex. *Neuropharmacology* **22,** 927–929.

Garcia-Mijares, M., Bernardes, A. M., and Silva, M. T. (2008). Diethylpropion produces psychostimulant and reward effects. *Pharmacol. Biochem. Behav.* (Epub ahead of print).

Gardos, G. (1997). Reversible dyskinesia during bupropion therapy. *J. Clin. Psychiatry* **58,** 218.

Gevaerd, M. S., Sultowski, E. T., and Takahashi, R. N. (1999). Combined effects of diethylpropion and alcohol on locomotor activity of mice: Participation of the dopaminergic and opioid systems. *Braz. J. Med. Biol. Res.* **32,** 1545–1550.

Giambalvo, C. T. (2003). Differential effects of amphetamine vs. dopamine reverse transport on particulate PKC activity in striatal synaptoneurosomes. *Synapse* **49,** 125–133.

Giambalvo, C. T., and Price, L. H. (2003). Effects of fenfluramine and antidepressants on protein kinase C activity in rat cortical synaptoneurosomes. *Synapse* **50,** 212–222.

Giros, B., El Mestikawy, S., Godinot, N., Zheng, K., Han, H., Yang-Feng, T., and Caron, M. G. (1992). Cloning, pharmacological characterization, and chromosome assignment of the human dopamine transporter. *Mol. Pharmacol.* **42,** 383–390.

Glick, S. D., Maisonneuve, I. M., and Kitchen, B. A. (2002). Modulation of nicotine self-administration on rats by combination therapy with agents blocking $\alpha 3\beta 4$ nicotinic receptors. *Eur. J. Pharmacol.* **448,** 185–191.

Gobbi, G., Slater, S., Boucher, N., Debonnel, G., and Blier, P. (2003). Neurochemical and psychotropic effects of bupropion in healthy male subjects. *J. Clin. Psychopharmacol.* **23,** 233–239.

Gumilar, F., Arias, H. R., Spitzmaul, G., and Bouzat, C. (2003). Molecular mechanisms of inhibition of nicotinic acetylcholine receptors by tricyclic antidepressants. *Neuropharmacology* **45,** 964–976.

Gumilar, F., and Bouzat, C. (2008). Tricyclic antidepressant inhibit homomeric Cys-Loop receptors by acting at different conformational states. *Eur. J. Pharmacol.* **584,** 30–39.

Habuchi, Y., Tanaka, H., Nishio, M., Yamamoto, T., Komori, T., Morikawa, J., and Yoshimura, M. (1997). Dopamine stimulation of cardiac β-adrenoceptors: The involvement of sympathetic amine transporters and the effect of SKF38393. *Br. J. Pharmacol.* **122,** 1669–1678.

Hamouda, A. K., Chiara, D. C., Blanton, M. P., and Cohen, J. B. (2008). Probing the structure of the affinity-purified and lipid-reconstituted *Torpedo* nicotinic acetylcholine receptor. *Biochemistry* **47,** 12787–12794.

Hasegawa, H., Meeusen, R., Sarre, S., Diltoer, M., Piacentini, M. F., and Michotte, Y. (2005). Acute dopamine/norepinephrine reuptake inhibition increases brain and core temperature in rats. *J. Appl. Physiol.* **99,** 1397–1401.

Heal, D. J., Frankland, A. T., Gosden, J., Hutchins, L. J., Prow, M. R., Luscombe, G. P., and Buckett, W. R. (1992). A comparison of the effects of sibutramine hydrochloride, bupropion and metamphetamine on dopaminergic function: Evidence that dopamine is not a pharmacological target form sibutramine. *Psychopharmacology* **107,** 303–309.

Hogg, R. C., Raggenbass, M., and Bertrand, D. (2003). Nicotinic acetylcholine receptors: From structure to brain function. *Physiol. Biochem. Pharmacol.* **147,** 1–46.

Holson, R. R., Gazzara, R. A., and Gough, B. (1998). Declines in stimulated striatal dopamine release over the first 32 h following microdialysis probe insertion: Generalization across releasing mechanisms. *Brain Res* **808,** 182–189.

Hsyu, P. H., Singh, A., Giargiari, T. D., Dunn, J. A., Ascher, J. A., and Johnston, J. A. (1997). Pharmacokinetics of bupropion and its metabolites in cigarette smokers versus nonsmokers. *J. Clin. Pharmacol.* **37,** 737–743.

Itzhak, Y., and Ali, S. F. (2006). Role of nitrergic system in behavioral and neurotoxic effects of amphetamine analogs. *Pharmacol. Ther.* **109,** 246–262.

Iversen, L. (2006). Neurotransmitter transporters and their impact on the development of psychopharmacology. *Br. J. Pharmacol.* **147,** S82–S88.

Jozwiak, K., Ravichandran, S., Collins, J. R., and Wainer, I. W. (2004). Interaction of noncompetitive inhibitors with an immobilized $\alpha 3\beta 4$ nicotinic acetylcholine receptor investigated by affinity chromatography, quantitative-structure activity relationship analysis, and molecular docking. *J. Med. Chem.* **47,** 4008–4021.

Katz, J. L., Izenwasser, S., and Terry, P. (2000). Relationships among dopamine transporter affinities and cocaine-like discriminative-stimulus effects. *Psychopharmacology* **148,** 90–98.

Kotlyar, M., Brauer, L. H., Tracy, T. S., Hatsukami, D. K., Harris, J., Bronars, C. A., and Adson, D. E. (2005). Inhibition of CYP2D6 activity by bupropion. *J. Clin. Psychopharmacol.* **25,** 226–229.

Kullmann, D. M., Ruiz, A., Rusakov, D. M., Scott, R., Scmyanov, A., and Walker, M. C. (2005). Presynaptic, extrasynaptic and axonal GABAA receptors in the CNS: Where and why? *Prog. Biophys. Mol. Biol.* **87,** 33–46.

Learned-Coughlin, S. M., Bergström, M., Sativcheva, I., Ascher, J., Schmith, V. D., and Langstrom, B. (2003). *In vivo* activity of bupropion at the human dopamine transporter as measured by positron emission tomography. *Biol. Psychiatry* **54,** 800–805.

Lloyd, G. K., and Williams, M. (2000). Neuronal nicotinic acetylcholine receptors as novel drug targets. *J. Pharmacol. Exp. Ther.* **292,** 461–467.

Maissonueve, I. M., and Glick, S. D. (2003). Anti-addictive actions of an *iboga* alkaloid congener: A novel mechanism for a novel treatment. *Pharmacol. Biochem. Behav.* **75,** 607–618.

Mansvelder, H. D., Fagen, Z. M., Chang, B., Mitchum, R., and McGehee, D. S. (2007). Bupropion inhibits the cellular effects of nicotine in the ventral tegmental area. *Biochem. Pharmacol.* **74,** 1283–1291.

Mignot, E., Renaud, A., Nishino, S., Arrigoni, J., Guilleminault, C., and Dement, W. C. (1993). Canine cataplexy is preferentially controlled by adrenergic mechanisms: Evidence using monoamine selective uptake inhibitors and release enhancers. *Psychopharmacology* **113,** 76–82.

Miller, D. K., Sumithran, S. P., and Dwoskin, L. P. (2002). Bupropion inhibits nicotine-evoked [^{3}H]dopamine overflow from rat striatal slices preloaded with [^{3}H]dopamine and from rat hippocampal slices preloaded with [^{3}H]norepinephrine. *J. Pharmacol. Exp. Ther.* **302,** 1113–1122.

Mortensen, O. V., and Amara, S. G. (2003). Dynamic regulation of the dopamine transporter. *Eur. J. Pharmacol.* **479,** 159–170.

Mortensen, O. V., and Amara, S. G. (2006). Gain of function mutants reveal sites important for the interaction of the atypical inhibitors benztropine and bupropion with monoamine transporters. *J. Neurochem.* **98,** 1531–1540.

Musso, D. L., Mehta, N. B., Soroko, F. E., Ferris, R. M., Hollingsworth, E. B., and Kenney, B. T. (1993). Synthesis and evaluation of the antidepressant activity of the enantiomers of bupropion. *Chirality* **5,** 495–500.

Nomikos, G. G., Damsma, G., Wenkstern, D., and Fibiger, H. C. (1989). Acute effects of bupropion on extracellular dopamine concentrations in rat striatum and nucleus accumbens studied by *in vivo* microdyalisis. *Neuropsychopharmacology* **2,** 273–279.

Ollo, C., Alim, T. N., Rosse, R. B., Lindquist, T., Green, T., Gillis, T., Ricci, J., Khan, M., and Deutsch, S. I. (1996). Lack of neurotoxic effect of diethylpropion in crack-cocaine abusers. *Clin. Neuropharmacol.* **19,** 52–58.

Papke, R. L., and Oswald, R. E. (1989). Mechanisms of noncompetitive inhibition of acetylcholine-induced single-channel currents. *J. Gen. Physiol.* **93,** 785–811.

Paterson, N. E., Balfour, D. J., and Markou, A. (2007). Chronic bupropion attenuated the anhecdonic component of nicotine withdrawal in rats via inhibition of dopamine reuptake in the nucleus accumbens shell. *Eur. J. Neurosci.* **25,** 3099–3108.

Paul, M. A., Gray, G., Kenny, G., and Lange, M. (2002). The impact of bupropion on psychomotor performance. *Aviat. Space Environ. Med.* **73,** 1094–1099.

Pesola, G. R., and Avasarala, J. (2002). Bupropion seizure proportion among new-onset generalized seizures and drug related seizures presenting to an emergency department. *J. Emerg. Med.* **22,** 235–239.

Picciotto, M. R., Brunzell, D. H., and Caldarone, B. J. (2002). Effect of nicotine and nicotinic receptors on anxiety and depression. *Neuroreport* **13,** 1097–1106.

Poston, W. S. 2nd, Foreyt, J. P., Borrell, L., and Haddock, C. K. (1998). Challenges in obesity management. *South Med. J.* **91,** 710–720.

Pratt, M. B., Pedersen, S. E., and Cohen, J. B. (2000). Identification of the sites of incorporation of [^{3}H] ethidium diazide within the *Torpedo* nicotinic acetylcholine receptor ion channel. *Biochemistry* **39,** 11452–11462.

Rau, K. S., Birdsall, E., Hanson, J. E., Johnson-Davis, K. L., Carroll, F. I., Wilkins, D. G., Gibb, J. W., Hanson, G. R., and Fleckenstein, A. E. (2005). Bupropion increases striatal vesicular monoamine transport. *Neuropharmacology* **49,** 820–830.

Reid, C. A., Bekkers, J. M., and Clements, J. D. (2003). Presynaptic Ca^{2+} channels: A functional patchwork. *Trends Neurosci* **26,** 683–687.

Safta, L., Cuparencu, B., Sirbu, A., and Secareanu, A. (1976). Experimental observations on the effect of amphepramone on the behavior, locomotion, pentretrazol seizures and electroencephalogram. *Psycopharmacology* **50,** 165–169.

Samanian, R., and Garattini, S. (1993). Neurochemical mechanisms of action of anorectic drugs. *Pharmacol. Toxicol.* **73,** 63–68.

Sanghvi, M., Hamouda, A. K., Jozwiak, K., Blanton, M. P., Trudell, J. R., and Arias, H. R. (2008). Identifying the binding site(s) for antidepressants on the *Torpedo* nicotinic acetylcholine receptor: [^{3}H]2-Azidoimipramine photolabeling and molecular dynamics studies. *Biochem. Biophys. Acta* **1778,** 2690–2699.

Schroeder, D. H. (1983). Metabolism and kinetics of bupropion. *J. Clin. Psychiatry* **44,** 79–81.

Shytle, R. D., Silver, A. A., Lukas, R. J., Newman, M. B., Sheehan, D. V., and Sanberg, P. R. (2002). Nicotinic receptors as targets for antidepressants. *Mol. Psychiatry* **7,** 525–535.

Sidhpura, N., Redfern, P., and Wonnacott, S. (2007a). Comparison of the effects of bupropion on nicotinic receptor-evoked [³H]dopamine release from rat striatal synaptosomes and slices. *Eur. J. Pharmacol.* **567,** 102–109.

Sidhpura, N., Redfern, P., Rowley, H., Heal, D., and Wonnacott, S. (2007b). Comparison of the effects of bupropion and nicotine on locomotor activation and dopamine release *in vivo*. *Biochem. Pharmacol.* **74,** 1292–1298.

Sixma, T. K., and Smit, A. B. (2003). Acetylcholine binding protein (AChBP): A secreted glial protein that provides a high-resolution model for the extracellular domain of pentameric ligand-gated ion channels. *Annu. Rev. Biophys. Biomol. Struct.* **32,** 311–334.

Slemmer, J. E., Martin, B. R., and Damaj, M. I. (2000). Bupropion is a nicotinic antagonist. *J. Pharmacol. Exp. Ther.* **295,** 321–327.

Spitzmaul, G., Dilger, J. P., and Bouzat, C. (2001). The noncompetitive inhibitor quinacrine modifies the desensitization kinetics of muscle acetylcholine receptors. *Mol. Pharmacol.* **60,** 235–243.

Stathis, M., Scheffel, U., Lever, S. Z., Boja, J. W., Carroll, F. I., and Kuhar, M. J. (1995). Rate of binding of various inhibitors at the dopamine transporter *in vivo*. *Psychopharmacology* **119,** 376–384.

Stevens, C. F. (2004). Presynaptic function. *Curr. Opin. Neurobiol.* **14,** 341–345.

Stoll, A. L., Mayer, P. V., Kolbrener, M., Goldstein, E., Suplit, B., Lucier, J., Cohen, B. M., and Tohen, M. (1994). Antidepressant-associated mania: A controlled comparison with spontaneous mania. *Am. J. Psychiatry* **151,** 1642–1645.

Sullivan, C. L., Crundsen, M., and Arias, H. R. (2008). Characterization of Antidepressant Binding Sites on the Nicotinic Acetylcholine Receptor. 2008 ACCP Spring Practice and Research Forum Phoenix, AZ, USA, April 5–9.

Suaud-Chagny, M. F., Dugast, C., Chergui, K., Msghina, M., and Gonon, F. (1995). Uptake of dopamine released by impulse flow in the rat mesolimbic and striatal systems *in vivo*. *J. Neurochem.* **65,** 2603–2611.

Suckow, R. F., Smith, T. M., Perumal, A. S., and Cooper, T. B. (1986). Pharmacokinetics of bupropion and metabolites in plasma and brain of rats, mice, and guinea pigs. *Drug Metab. Dispos.* **14,** 692–697.

Szuba, M. P., and Leuchter, A. F. (1992). Falling backward in two elderly patients taking bupropion. *J. Clin. Psychiatry* **53,** 157–159.

Taraschenko, O. D., Panchal, V., Maisonneuve, I. M., and Glick, S. D. (2005). Is antagonism of α3β4 nicotinic receptors a strategy to reduce morphine dependence? *Eur. J. Pharmacol.* **513,** 207–218.

Unwin, N. (2005). Refined structure of the nicotinic acetylcholine receptor at 4 Å resolution. *J. Mol. Biol.* **346,** 967–989.

Van-Wyck-Fleet, J., Manberg, P. J., Miller, L. L., Harto-Truax, N., Sato, T., Fleck, R. J., Stern, W. C., and Cato, A. E. (1983). Overview of clinically significant adverse reactions to bupropion. *J. Clin. Psychiatry* **44,** 191–196.

Wiley, J. L., Lavecchia, K. L., Martin, B. R., and Damaj, M. I. (2002). Nicotine-like discriminative stimulus effects of bupropion in rats. *Exp. Clin. Psychopharmacol.* **10,** 129–135.

Wilkes, S. (2006). Bupropion. *Drugs Today* **42,** 671–681.

Young, R., and Glennon, R. A. (2002). Nicotine and bupropion share a similar discriminative stimulus effect. *Eur. J. Pharmacol.* **443,** 113–118.

Yu, H., Rothman, R. B., Dersch, C. M., Partilla, J. S., and Rice, K. C. (2000). Uptake and release of diethylpropion and its metabolites with biogenic amine transporters. *Bioorg Med. Chem.* **8,** 2689–2692.

Ziv, N. E., and Garner, C. C. (2004). Cellular and molecular mechanisms of presynaptic assembly. *Nature Rev. Neurosci.* **5,** 385–399.

NEURAL AND CARDIAC TOXICITIES ASSOCIATED WITH 3,4-METHYLENEDIOXYMETHAMPHETAMINE (MDMA)

Michael H. Baumann and Richard B. Rothman

Clinical Psychopharmacology Section, Intramural Research Program (IRP), National Institute on Drug Abuse (NIDA), National Institutes of Health (NIH), Baltimore, Maryland 21224, USA

($\pm$)-3,4-Methylenedioxymethamphetamine (MDMA) is a commonly abused illicit drug which affects multiple organ systems. In animals, high-dose administration of MDMA produces deficits in serotonin (5-HT) neurons (e.g., depletion of forebrain 5-HT) that have been viewed as neurotoxicity. Recent data implicate MDMA in the development of valvular heart disease (VHD). The present paper reviews several issues related to MDMA-associated neural and cardiac toxicities. The hypothesis of MDMA neurotoxicity in rats is evaluated in terms of the effects of MDMA on monoamine neurons, the use of scaling methods to extrapolate MDMA doses across species, and functional consequences of MDMA exposure. A potential treatment regimen (L-5-hydroxytryptophan plus carbidopa) for MDMA-associated neural deficits is discussed. The pathogenesis of MDMA-associated VHD is reviewed with specific reference to the role of valvular 5-HT$_{2B}$ receptors. We conclude that pharmacological effects of MDMA occur at the same doses in rats and humans. High doses of MDMA that produce 5-HT depletions in rats are associated with tolerance and impaired 5-HT release. Doses of MDMA that fail to deplete 5-HT in rats can cause persistent behavioral dysfunction, suggesting even moderate doses may pose risks. Finally, the MDMA metabolite, 3,4-methylenedioxyamphetamine (MDA), is a potent 5-HT$_{2B}$ agonist which could contribute to the increased risk of VHD observed in heavy MDMA users.

INTERNATIONAL REVIEW OF
NEUROBIOLOGY, VOL. 88
DOI: 10.1016/S0074-7742(09)88010-0

0074-7742/09 $35.00

I. Introduction

3,4-Methylenedioxymethamphetamine (MDMA or *Ecstasy*) is a popular illicit drug in the United States, Europe, and elsewhere. The allure of MDMA is likely related to its unique profile of psychotropic effects which includes amphetamine-like mood elevation, coupled with feelings of increased emotional sensitivity and closeness to others (Liechti and Vollenweider, 2001; Vollenweider *et al.*, 1998). MDMA misuse among adolescents is widespread in the United States (Landry, 2002; Yacoubian, 2003), and a recent sampling of high school seniors found 10% reported using MDMA at least once (Banken, 2004). The incidence of MDMA-related medical complications has risen in parallel with the increasing popularity of the drug. Serious adverse effects of MDMA intoxication include hyperthermia, serotonin (5-HT) syndrome, cardiac arrhythmias, hypertension, hyponatremia, liver problems, seizures, coma, and death (Schifano, 2004). Accumulating evidence suggests that heavy MDMA use is associated with cognitive impairments, mood disturbances, and cardiac valve dysfunction, in some cases lasting for months after cessation of drug intake (Morgan, 2000; Parrott, 2002). Despite the risks of illicit MDMA use, a number of clinicians believe the drug has therapeutic value in the treatment of psychiatric disorders, such as posttraumatic stress disorder (Doblin, 2002), and clinical studies with MDMA are underway (see http://clinicaltrials.gov). In controlled research settings, MDMA has been administered safely to humans, but adverse effects have been documented in certain individuals (Harris *et al.*, 2002; Mas *et al.*, 1999).

These considerations provide compelling reasons to evaluate the pharmacology and toxicology of MDMA. The present paper several issues related to MDMA-associated toxicities. The hypothesis of MDMA neurotoxicity in rats is evaluated in terms of MDMA effects on monoamine neurons, the use of scaling methods to adjust MDMA doses across species, and functional consequences of MDMA-induced 5-HT depletions. A potential treatment regimen (L-5-hydroxytryptophan plus carbidopa) for MDMA-associated neural deficits is discussed. The pathogenesis of MDMA-associated valvular heart disease (VHD) is reviewed with particular reference to the role of $5-HT_{2B}$ receptors. Previously published data from our laboratory at the National Institute on Drug Abuse (NIDA) will be included to supplement literature reports. Clinical findings will be mentioned in specific instances to note comparisons between rats and humans.

The neurotoxicity aspect of this review will focus on data obtained from rats since most preclinical MDMA research has been carried out in this animal model. Data from mice will not be considered because this species displays the unusual characteristic of long-term dopamine (DA) depletions (i.e., DA neurotoxicity) in response to MDMA, rather than long-term 5-HT depletions observed in rats, nonhuman primates, and most other animals (reviewed by Colado *et al.*, 2004).

We will not address possible molecular mechanisms underlying MDMA-induced 5-HT deficits, as several excellent reviews have covered this subject (Lyles and Cadet, 2003; Monks *et al.*, 2004; Sprague *et al.*, 1998). All experiments in our laboratory utilized male Sprague–Dawley rats (Wilmington, MA) weighing 300–350 g. Rats were maintained in facilities accredited by the Association for the Assessment and Accreditation of Laboratory Animal Care, and procedures were carried out in accordance with the Animal Care and Use Committee of the NIDA Intramural Research Program (IRP).

II. Effects of MDMA on Monoamine Neurons

In order to critically evaluate the hypothesis of MDMA-induced 5-HT neurotoxicity, a brief review of MDMA pharmacology is necessary. Figure 1 shows that MDMA is a ring-substituted analog of methamphetamine. *Ecstasy* tablets ingested by humans contain a racemic mixture of (+) and (−) isomers of MDMA, and both isomers are bioactive (Johnson *et al.*, 1986; Schmidt, 1987). *Ecstasy* tablets often contain other psychoactive substances such as substituted amphetamines, caffeine, or ketamine, which can contribute to the overall effects of the ingested preparation (Parrott, 2004). Upon systemic administration, *N*-demethylation of MDMA occurs via first-pass metabolism to yield the ring-substituted amphetamine analog 3,4-methylenedioxyamphetamine (MDA) (de la Torre *et al.*, 2004). Initial studies carried out in the 1980s showed that MDMA and MDA stimulate efflux of preloaded [^{3}H]5-HT, and to a lesser extent [^{3}H]DA, in nervous tissue

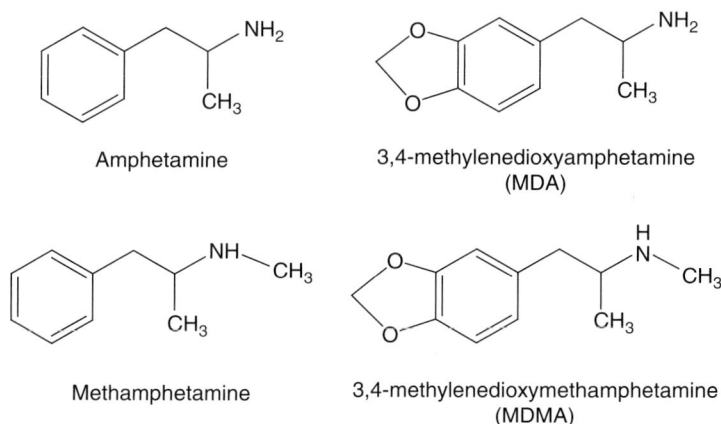

FIG. 1. Chemical structures of MDMA and related amphetamines.

(Johnson *et al.*, 1986; Nichols *et al.*, 1982; Schmidt *et al.*, 1987). Subsequent findings revealed that MDMA interacts with monoamine transporter proteins to stimulate nonexocytotic release of 5-HT, DA, and norepinephrine (NE) in rat brain (Berger *et al.*, 1992; Crespi, 1997; Fitzgerald and Reid, 1993).

Like other substrate-type releasers, MDMA and MDA bind to transporter proteins and are subsequently transported into the cytoplasm of nerve terminals, releasing neurotransmitters via a process originally described as carrier-mediated exchange (for review see Rothman and Baumann, 2006; Rudnick and Clark, 1993). However, the mechanism by which substrates induce the release of neurotransmitters is more complicated than a simple exchange process. Recent studies of the DA transporter (DAT) have shown that inward transport of a substrate like amphetamine is accompanied by an inward sodium current, which increases the concentration of intracellular sodium at the transporter, thereby facilitating reverse transport of DA (Goodwin *et al.*, 2008; Pifl *et al.*, 2009). Determining the precise molecular mechanism of transporter-mediated release of 5-HT is a topic of ongoing research.

Table I summarizes published data from our laboratory showing structure–activity relationships for stereoisomers of MDMA, MDA, and related drugs as monoamine releasers in rat brain synaptosomes (Partilla *et al.*, 2000; Rothman *et al.*, 2001; Setola *et al.*, 2003). Stereoisomers of MDMA and MDA are substrates for 5-HT transporters (SERT), NE transporters (NET), and DAT, with (+) isomers exhibiting greater potency as releasers. In particular, (+) isomers of MDMA and MDA are much more effective DA releasers than their corresponding (−) isomers. It is noteworthy that (+) isomers of MDMA and

TABLE I

Profile of MDMA and Related Compounds as Monoamine Transporter Substrates in Rat Brain Synaptosomes

Drug	5-HT release EC_{50} (nM ± S.D.)	NE release EC_{50} (nM ± S.D.)	DA release EC_{50} (nM ± S.D.)
(+)-Methamphetamine	736 ± 45	12 ± 0.7	24 ± 2
(−)-Methamphetamine	4640 ± 240	29 ± 3	416 ± 20
(±)-MDMA	74.3 ± 5.6	136 ± 17	278 ± 12
(+)-MDMA	70.8 ± 5.2	110 ± 16	142 ± 6
(−)-MDMA	337 ± 34	564 ± 60	3682 ± 178
(+)-Amphetamine	1765 ± 94	7.1 ± 1.0	25 ± 4
(±)-MDA	159 ± 12	108 ± 12	290 ± 10
(+)-MDA	99.6 ± 7.4	98.5 ± 6.1	50.0 ± 8.0
(−)-MDA	313 ± 21	287 ± 23	900 ± 49

Data are taken from Partilla *et al.* (2000), Rothman *et al.* (2001), and Setola *et al.* (2003). Details concerning *in vitro* methods can be found in these papers. Substrate activity at SERT, NET, and DAT is reflected as release efficacy for the corresponding transmitter.

MDA are rather nonselective in their ability to stimulate monoamine release *in vitro*. When compared to amphetamine and methamphetamine, the major effect of methylenedioxy ring-substitution is enhanced potency for 5-HT release and reduced potency for DA release. For example, (+)-MDMA releases 5-HT (EC_{50} = 70.8 nM) about 10 times more potently than (+)-methamphetamine (EC_{50} = 736 nM), whereas (+)-MDMA releases DA (EC_{50} = 142 nM) about six times less potently than (+)-methamphetamine (EC_{50} = 24 nM).

Consistent with *in vitro* results, *in vivo* microdialysis experiments demonstrate that MDMA increases extracellular 5-HT and DA in rat brain, with effects on 5-HT being greater in magnitude (Baumann *et al.*, 2005; Gudelsky and Nash, 1996; Kankaanpaa *et al.*, 1998; Yamamoto *et al.*, 1995). Figure 2 depicts data from

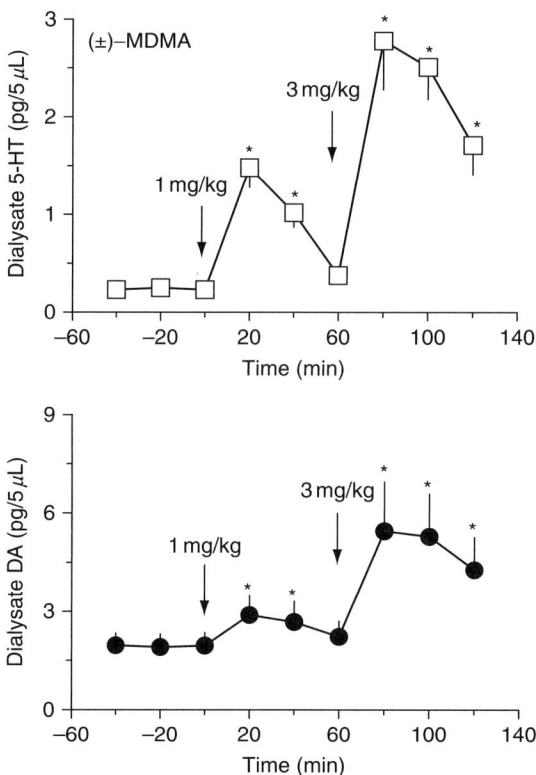

FIG. 2. Effects of i.v. MDMA on concentrations of 5-HT (top panel) and DA (bottom panel) in dialysate samples obtained from rat n. accumbens. Conscious rats undergoing *in vivo* microdialysis received i.v. injections of 1 mg/kg MDMA at time zero followed by 3 mg/kg 60 min later. Dialysate samples obtained at 20 min intervals were assayed for 5-HT and DA by HPLC-ECD. Data are mean ± S.E.M., expressed as pg/5 μl sample for N = 7 rats/group. *P < 0.05 versus preinjection baseline. (Modified from Baumann *et al.*, 2005).

our laboratory showing the stimulatory effects of MDMA on extracellular 5-HT and DA in rat n. accumbens (Baumann et al., 2005). In these experiments, rats undergoing microdialysis received i.v. injections of 1 mg/kg MDMA at time zero, followed by 3 mg/kg MDMA 60 min later. Importantly, these doses of MDMA are comparable to those used in rat self-administration paradigms (Ratzenboeck et al., 2001; Schenk et al., 2003). In all of our microdialysis studies, samples are collected at 20 min intervals beginning 1 h before injections, until 1 h after the second dose of drug; samples are assayed for 5-HT and DA by high-performance liquid chromatography coupled to electrochemical detection (HPLC-ECD) (Baumann and Rutter, 2003). Neurochemical data are expressed as pg/5 μl sample. MDMA causes significant dose-related increases in dialysate 5-HT [$F_{8,45} = 33.04, P < 0.0001$] and DA [$F_{8,45} = 5.56, P < 0.001$] in n. accumbens. At both doses, elevations in extracellular 5-HT are much greater than corresponding elevations in DA. For example, the 1 mg/kg dose of MDMA produces a 7.5-fold rise in 5-HT but only a 1.5-fold rise in DA.

Acute central nervous system (CNS) effects of MDMA are presumably mediated by release of monoamine transmitters, with the subsequent activation of pre- and postsynaptic receptors (reviewed by Cole and Sumnall, 2003; Green et al., 2003). As a specific example in rats, MDMA elicits a unique spectrum of motor actions characterized by forward locomotion (i.e., ambulation) and elements of the 5-HT behavioral syndrome such as forepaw treading and head weaving (i.e., stereotypy) (Gold et al., 1988; Slikker et al., 1989; Spanos and Yamamoto, 1989). It is well established that hyperactivity caused by MDMA is dependent upon activation of multiple 5-HT and DA receptor subtypes in the brain (reviewed by Bankson and Cunningham, 2001; Geyer, 1996). We examined the role of brain monoamines in mediating locomotor effects of MDMA by combining in vivo microdialysis with automated analysis of motor behaviors (Baumann et al., 2008b). Microdialysis guide cannulae were aimed at various brain regions implicated in motor stimulation, including n. accumbens, striatum, and prefrontal cortex. Conscious rats undergoing microdialysis were housed in chambers equipped with photobeams to assess ambulation and stereotypy. Neurochemical and locomotor effects of MDMA administration were determined simultaneously in the same individual rats, thereby allowing the use of Pearson correlation analysis to examine relationships between pg amounts of transmitter, distance traveled and stereotypic events. Figure 3 depicts representative data from the n. accumbens where MDMA-induced elevations in dialysate 5-HT are positively correlated with stereotypy [$r = 0.5276, P < 0.0001$] but not ambulation [$r = 0.3241$, NS]. On the other hand, Fig. 4 shows that increases in accumbens DA are highly correlated with ambulation [$r = 0.6409, P < 0.0001$] and somewhat less so with stereotypy [$r = 0.4490, P < 0.01$]. Our correlation data from various brain regions indicate that motor effects produced by MDMA involve transporter-mediated release of 5-HT and DA in a region- and modality-specific manner.

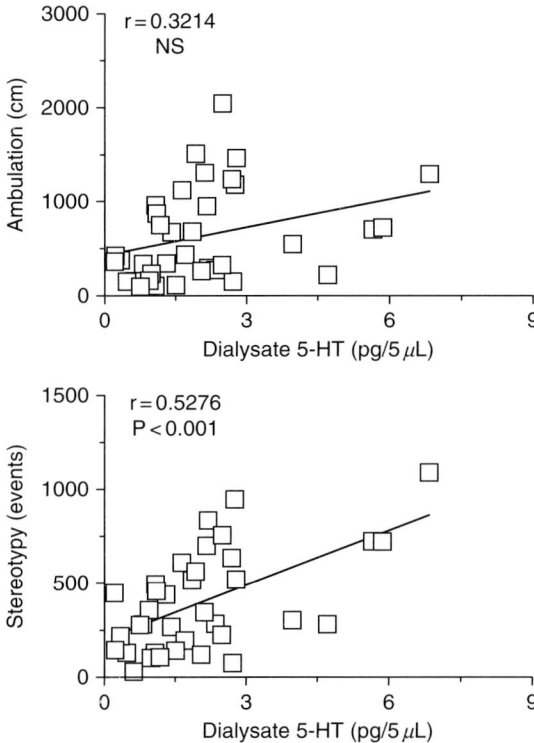

FIG. 3. Correlations between dialysate 5-HT in the n. accumbens versus ambulation (top panel) and stereotypy (bottom panel) produced by MDMA. Conscious rats undergoing *in vivo* microdialysis received i.v. injections of 1 mg/kg MDMA at time zero followed by 3 mg/kg 60 min later. Data points represent raw values obtained at 20 min intervals, from +20 min (first sample after 1 mg/kg) through +120 min (last sample after 3 mg/kg) for $N = 6$ rats. Pearson correlation coefficients (r) are shown with corresponding P values. (Modified from Baumann *et al.*, 2008b).

Additionally, the role of NE in modulating behavioral effects of MDMA is largely unexplored, and this issue warrants further study (Selken and Nichols, 2007; Starr *et al.*, 2008).

Adverse effects of acute **MDMA** administration, including cardiovascular stimulation and elevated body temperature, are thought to involve monoamine release from sympathetic nerves in the periphery or nerve terminals in the CNS. MDMA increases heart rate and mean arterial pressure in conscious rats (O'Cain *et al.*, 2000); this cardiovascular stimulation is likely mediated by MDMA-induced release of peripheral NE stores, similar to the effects of amphetamine (Fitzgerald and Reid, 1994). MDMA has weak agonist actions at α2-adrenoreceptors and

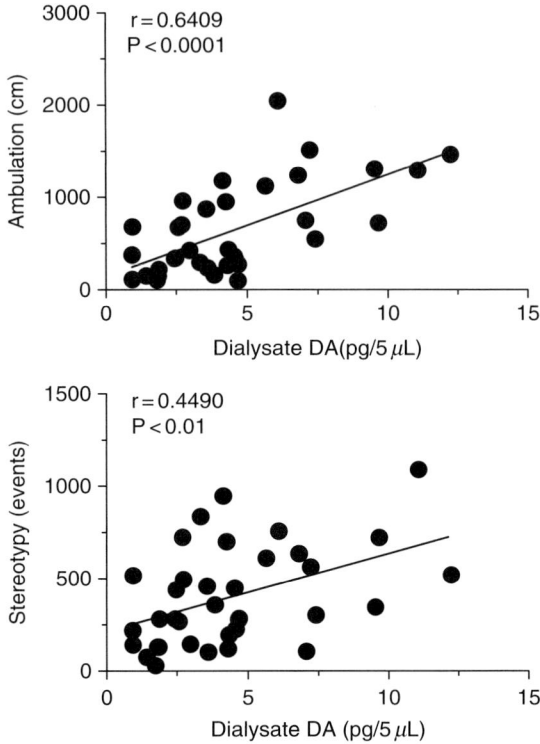

FIG. 4. Correlations between dialysate DA in the n. accumbens versus ambulation (top panel) and stereotypy (bottom panel) produced by MDMA. Conscious rats undergoing *in vivo* microdialysis received i.v. injections of 1 mg/kg MDMA at time zero followed by 3 mg/kg 60 min later. Data points represent raw values obtained at 20 min intervals, from +20 min (first sample after 1 mg/kg) through +120 min (last sample after 3 mg/kg) for $N = 6$ rats. Pearson correlation coefficients (r) are shown with corresponding P values. (Modified from Baumann *et al.*, 2008b).

5-HT$_2$ receptors that might influence its cardiac and pressor effects (Battaglia and De Souza, 1989; Lavelle *et al.*, 1999; Lyon *et al.*, 1986). Moreover, as will be described in more detail in a subsequent part of this review, MDA is a potent 5-HT$_{2B}$ agonist and this property could contribute to adverse cardiovascular effects (Setola *et al.*, 2003). The ability of MDMA to elevate body temperature is well characterized in rats (Dafters, 1995; Dafters and Lynch, 1998; Nash *et al.*, 1988), and this response has been considered a 5-HT-mediated process. However, the data of Mechan *et al.* (2002) provide convincing evidence that MDMA-induced hyperthermia involves the activation of postsynaptic D1 receptors by released DA.

III. Long-Term Effects of MDMA on 5-HT Systems

The long-term adverse effects of MDMA on 5-HT systems have attracted substantial interest, since studies in rats and nonhuman primates show that high-dose MDMA administration produces persistent reductions in markers of 5-HT nerve terminal integrity (reviewed by Lyles and Cadet, 2003; Sprague *et al.*, 1998). Table II summarizes the findings of investigators who first demonstrated that MDMA causes long-term (>1 week) inactivation of tryptophan hydroxylase activity, depletions of brain tissue 5-HT, and reductions in SERT binding and function (Battaglia *et al.*, 1987; Commins *et al.*, 1987; Schmidt, 1987; Stone *et al.*, 1987). These serotonergic deficits are observed in various regions of rat forebrain, including frontal cortex, striatum, hippocampus, and hypothalamus. Immunohistochemical analysis of 5-HT in cortical and subcortical areas reveals an apparent loss of 5-HT axons and terminals in MDMA-treated rats, especially the fine-diameter projections arising from the dorsal raphe nucleus (O'Hearn *et al.*, 1988). Moreover, 5-HT axons and terminals remaining after MDMA treatment appear swollen and fragmented, suggesting structural damage.

Time-course studies indicate that MDMA-induced 5-HT depletions occur in a biphasic manner, with a rapid acute phase followed by a delayed long-term phase (Schmidt, 1987; Stone *et al.*, 1987). In the acute phase, which lasts for the first few h after drug administration, massive depletion of brain tissue 5-HT is accompanied by inactivation of tryptophan hydroxylase. By 24 h later, tissue

TABLE II

LONG-TERM EFFECTS OF (±)-MDMA ON 5-HT NEURONAL MARKERS IN RATS

5-HT deficit	Dose	Survival interval	References
Depletions of 5-HT in cortex, as measured by HPLC-ECD	10 mg/kg, s.c., single dose	1 week	Schmidt (1987)
Depletions of 5-HT in forebrain regions as measured by HPLC-ECD	10–40 mg/kg, s.c., twice daily, 4 days	2 weeks	Commins *et al.* (1987)
Reductions in tryptophan hydroxylase activity in forebrain regions	10 mg/kg, s.c., single dose	2 weeks	Stone *et al.* (1987)
Loss of [^{3}H]-paroxetine-labeled SERT binding sites in forebrain regions	20 mg/kg, s.c., twice daily, 4 days	2 weeks	Battaglia *et al.* (1987)
Deceased immunoreactive 5-HT in fine axons and terminals in forebrain regions	20 mg/kg, s.c., twice daily, 4 days	2 weeks	O'Hearn *et al.* (1988)

5-HT recovers to normal levels but hydroxylase activity remains diminished. In the long-term phase, which begins within 1 week and lasts for months, marked depletion of 5-HT is accompanied by sustained inactivation of tryptophan hydroxylase and loss of SERT binding and function (Battaglia et al., 1988; Scanzello et al., 1993). The findings summarized in Table II have been replicated by many investigators, and the spectrum of decrements is typically described as 5-HT neurotoxicity (Baumann et al., 2007; Ricaurte et al., 2000). Most of the studies designed to examine MDMA neurotoxicity in rats have employed i.p. or s.c. injections of 10 mg/kg or higher, either as single or repeated treatments. These MDMA dosing regimens are known to produce significant hyperthermia, which can exacerbate 5-HT depletions caused by the drug (Green et al., 2004; Malberg and Seiden, 1998). Most investigations have involved non-contingent administration of MDMA rather than self-administration, and this factor could significantly influence effects of the drug.

There are caveats to the hypothesis that MDMA produces 5-HT neurotoxicity. O'Hearn et al. (1988) showed that high-dose MDMA administration has no effect on 5-HT cell bodies in the dorsal raphe, despite profound loss of 5-HT in forebrain projection areas. Thus, the effects of MDMA on 5-HT neurons are sometimes referred to as "axotomy," to account for the fact that perikarya are not damaged (Molliver et al., 1990; O'Hearn et al., 1988). MDMA-induced reductions in 5-HT levels and SERT binding eventually recover (Battaglia et al., 1988; Scanzello et al., 1993), suggesting that 5-HT terminals are not destroyed. Many drugs used clinically produce effects that are similar to those produced by MDMA. For instance, reserpine causes sustained depletions of brain tissue 5-HT, yet reserpine is not considered a neurotoxin (Carlsson, 1976). Chronic administration of 5-HT selective reuptake inhibitors (SSRIs), like paroxetine and sertraline, leads to a marked loss of SERT binding and function comparable to MDMA, but these agents are therapeutic drugs rather than neurotoxins (Benmansour et al., 1999; Frazer and Benmansour, 2002). Finally, high-dose administration of SSRIs produces swollen, fragmented, and abnormal 5-HT terminals which are indistinguishable from the effects of MDMA and other substituted amphetamines (Kalia et al., 2000).

The caveats mentioned above raise a number of questions with respect to MDMA neurotoxicity. Of course, the most important question is whether MDMA abuse causes neurotoxic damage in humans. This complex issue is a matter of ongoing debate which has been addressed by a number of recent papers (Gouzoulis-Mayfrank et al., 2002; Kish, 2002; Reneman, 2003). Clinical studies designed to critically evaluate the long-term effects of MDMA are hampered by a number of factors including comorbid psychopathology and poly-drug abuse among MDMA users. Animal models afford the unique opportunity to evaluate the effects of MDMA without many of these complicating factors, and the main focus here will be to review the evidence pertaining to MDMA-induced 5-HT neurotoxicity in rats.

IV. Scaling Methods and MDMA Dosing

A major point of controversy relates to the relevance of MDMA doses administered to rats when compared to doses taken by humans (see Cole and Sumnall, 2003). As noted earlier, MDMA regimens that produce 5-HT depletions in rats involve administration of single or multiple injections of 10–20 mg/kg, whereas the typical dose of *Ecstasy* abused by humans is one or two tablets of 80–100 mg MDMA, or 1–3 mg/kg administered orally (Green *et al.*, 2003; Schifano, 2004). Based on principles of allometric scaling (i.e., interspecies scaling), some investigators have proposed that neurotoxic doses of MDMA in rats correspond to recreational doses in humans (Ricaurte *et al.*, 2000). In order to critically evaluate this claim, a brief discussion of interspecies scaling is warranted. The concept of interspecies scaling is based upon shared biochemical mechanisms among eukaryotic cells (e.g., aerobic respiration), and was initially developed to describe variations in basal metabolic rate (BMR) between animal species of different sizes (reviewed by White and Seymour, 2005). In the 1930s, Kleiber (1932) derived what is now called the "allometric equation" to describe the relationship between body mass and BMR. The generic form of the allometric equation is: $Y = aW^b$, where "Y" is the variable of interest, "W" is body weight, "a" is the allometric coefficient, and "b" is the allometric exponent. In the case where "Y" is BMR, b is accepted to be 0.75. West *et al.* (2002) have shown that most biological phenomena scale according to a universal quarter-power law, as illustrated by the space-filling fractal networks of branching tubes used by the circulatory system.

Given that the allometric equation is grounded in fundamental commonalities across organisms, it is not surprising this equation can describe the relationship between body mass and physiological variables, such as BMR, heart rate, and circulation time (e.g., Noujaim *et al.*, 2004). Because circulation time and organ blood flow strongly influence drug pharmacokinetics, the allometric equation has been used in the medication development process to "scale-up" dosages from animal models to man (reviewed by Mahmood, 1999). In general, smaller animals have faster heart rates and circulation times, leading to faster clearance of exogenous drugs. This relationship does not hold true for all classes of drugs, however, especially those that are subject to complex metabolism (Lin, 1998).

MDMA is extensively metabolized in humans, as depicted in Fig. 5, and the major pathway of biotransformation involves: (1) *O*-demethylenation catalyzed by cytochrome P450 2D6 (CYP2D6) and (2) *O*-methylation catalyzed by catechol-*O*-methyltransferase (COMT) (reviewed by de la Torre *et al.*, 2004). CYP2D6 and COMT are both polymorphic in humans, and differential expression of CYP2D6 isoforms leads to interindividual variations in the metabolism of 5-HT medications (e.g., SSRIs) (Charlier *et al.*, 2003). Interestingly, CYP2D6 is not present in rats,

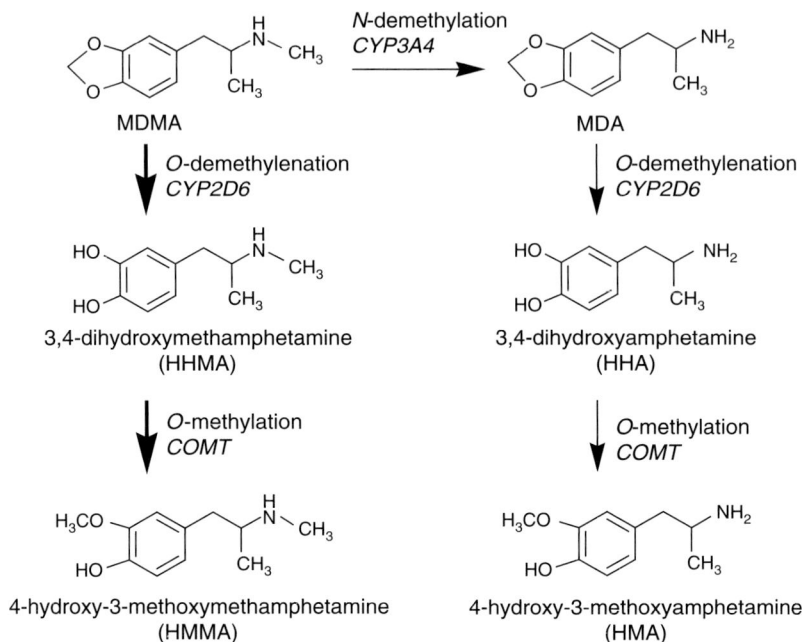

Fig. 5. Metabolism of MDMA in humans. The main cytochrome P450 (CYP) isoforms responsible for particular biotransformation reactions are indicated in italics. Thick arrows highlight major pathways, whereas thin arrows represent minor pathways.

which express a homologous but functionally distinct cytochrome P450 2D1 (Malpass *et al.*, 1999; Maurer *et al.*, 2000). A minor pathway of MDMA biotransformation in humans involves *N*-demethylation of MDMA to form MDA, which is subsequently *O*-demethylenated and *O*-methylated. *N*-Demethylation of MDMA represents a more important pathway for rats when compared to humans (de la Torre *et al.*, 2004). The metabolism of MDMA and MDA generates a number of metabolites, some of which may be bioactive (e.g., Escobedo *et al.*, 2005; Forsling *et al.*, 2002). Determining the potential neurotoxic properties of the various metabolites of MDMA is an important area of research (reviewed by Baumgarten and Lachenmayer, 2004; Monks *et al.*, 2004).

To complicate matters further, de la Torre *et al.* (2000) have shown that MDMA displays nonlinear kinetics in humans such that administration of increasing doses, or multiple doses, leads to unexpectedly high plasma levels of the drug. Enhanced plasma and tissue levels of MDMA are most likely related to autoinhibition of MDMA metabolism, mediated via formation of a metabolite–enzyme complex that irreversibly inactivates CYP2D6 (Wu *et al.*, 1997). Because MDMA displays nonlinear kinetics, repeated drug dosing could produce serious adverse consequences due to unusually high blood and tissue levels of the drug (Parrott, 2002; Schifano, 2004). The existing database of MDMA pharmacokinetic studies represents a curious situation where clinical findings are well documented, while preclinical data are lacking. Specifically, only a handful of animal studies have assessed the relationship between pharmacodynamic effects of MDMA and pharmacokinetics of the drug after administration of single or repeated doses (Chu *et al.*, 1996; Mechan *et al.*, 2006). Few studies have systemically characterized the nonlinear kinetics of MDMA in animal models (Mueller *et al.*, 2008). Collectively, the available data demonstrate that potential species differences in tissue drug uptake, variations in metabolic enzymes and their activities, and the phenomenon of nonlinear kinetics, preclude the use of allometric scaling to extrapolate MDMA doses across different species.

The uncertainties and limitations of allometric scaling led us to investigate the method of "effect scaling" as an alternative strategy for matching equivalent doses of MDMA in rats and humans (see Winneke and Lilienthal, 1992). In this approach, the lowest dose of drug that produces a specific pharmacological response is determined for rats and humans, and subsequent dosing regimens in rats are calculated with reference to the predetermined threshold dose. In the case of MDMA, this strategy is simplified because CNS drug effects, such as neuroendocrine and behavioral changes, have already been investigated in different species. Theoretically, equivalent drug effects *in vivo* should reflect similar drug concentrations reaching active sites in tissue, suggesting the method of effect scaling can account for differences in drug absorption, distribution and metabolism across species (at least for low drug doses). Table III shows the doses of

TABLE III

COMPARATIVE NEUROBIOLOGICAL EFFECTS OF (±)-MDMA ADMINISTRATION IN RATS AND HUMANS

CNS effect	Dose in rats	Dose in humans
In vivo release of 5-HT and DA	2.5 mg/kg, i.p. (Gudelsky and Nash, 1996); 1 mg/kg, s.c. (Kankaanpaa *et al.*, 1998)	1.5 mg/kg p.o. (Liechti *et al.*, 2000; Liechti and Vollenweider, 2001)[a]
Secretion of prolactin and glucocorticoids	1–3 mg/kg, i.p. (Nash *et al.*, 1988)	1.67 mg/kg, p.o. (Mas *et al.*, 1999); 1.5 mg/kg, p.o. (Harris *et al.*, 2002)
Drug discrimination	1.5 mg/kg, i.p. (Oberlender and Nichols, 1988; Schechter, 1988)	1.5 mg/kg, p.o. (Johanson *et al.*, 2005)
Drug reinforcement	1 mg/kg, i.v. (Wakonigg *et al.*, 2003)	1–2 mg/kg, p.o. (Tancer and Johanson, 2003)[b]

[a]Subjective effects were attenuated by 5-HT uptake blockers, suggesting the involvement of transporter-mediated 5-HT release.

[b]Reinforcing effects were determined based on a multiple choice procedure.

MDMA that produce comparable CNS effects in rats and humans. Remarkably, the findings reveal that MDMA doses in the range of 1–2 mg/kg produce pharmacological effects that are equivalent in both species. It is noteworthy that MDMA is typically administered to rats via the i.p. or s.c. route whereas humans take the drug orally. Given the similar effects of MDMA in rats and humans at the same doses, it appears that drug bioavailability is comparable after i.p., s.c., or oral administration (e.g., Finnegan *et al.*, 1988), but verification of this hypothesis awaits further investigation.

Administration of MDMA at i.p. doses of 1–3 mg/kg causes marked elevations in extracellular 5-HT and DA in rat brain, as determined by *in vivo* microdialysis (Baumann *et al.*, 2005; Gudelsky and Nash, 1996; Kankaanpaa *et al.*, 1998). Recall the data from Fig. 2 which demonstrate that 1 mg/kg i.v. MDMA increases extracellular 5-HT and DA rat n. accumbens. Although it is impossible to directly measure 5-HT and DA release in living human brain, clinical studies indicate that subjective effects of recreational doses of MDMA (1.5 mg/kg, p.o.) are mediated via transporter-mediated release of 5-HT (Liechti and Vollenweider, 2001; Liechti *et al.*, 2000). Nash *et al.* (1988) showed that i.p. injections of 1–3 mg/kg of MDMA stimulate prolactin and corticosterone secretion in rats, and similar oral doses increase plasma prolactin and cortisol in human drug users (Harris *et al.*, 2002; Mas *et al.*, 1999). The dose of MDMA discriminated by rats and humans is identical: 1.5 mg/kg, i.p., for rats (Glennon and Higgs, 1992; Oberlender and Nichols, 1988; Schechter, 1988) and 1.5 mg/kg, p.o., for humans (Johanson *et al.*, 2005). A few studies have shown that rats will self-administer MDMA at doses ranging from 0.25 to 1.0 mg/kg i.v., indicating these doses possess reinforcing efficacy (Ratzenboeck *et al.*, 2001; Schenk *et al.*, 2003). Wakonigg *et al.* (2003) demonstrated that a single i.v. injection of 1 mg/kg MDMA serves as a powerful reinforcer in an operant runway procedure, and MDMA displays similar reinforcing potency in Sprague–Dawley and Long–Evans rat strains. Tancer and Johanson (2003) reported that 1 and 2 mg/kg of MDMA have reinforcing properties in humans which resemble those of (+)-amphetamine. The findings summarized in Table III suggest there is no scientific justification for using allometric scaling to "adjust" MDMA doses between rats and humans.

Based on this analysis, we devised an MDMA dosing regimen in rats which attempts to mimic binge use of MDMA in humans. Male Sprague–Dawley rats were double-housed in plastic cages, under conditions of constant ambient temperature (22 °C) and humidity (70%) in a vivarium. In our most recent studies, three i.p. injections of 1.5 or 7.5 mg/kg MDMA were administered, one dose every 2 h, to yield cumulative doses of 4.5 or 22.5 mg/kg (Baumann *et al.*, 2008a). Control rats received saline vehicle according to the same schedule. Rats were removed from their cages to receive i.p. injections, but were otherwise confined to their home cages. The 1.5 mg/kg dose was used as a low "behavioral" dose whereas the 7.5 mg/kg dose was used as a high "toxic" dose (i.e., a dose fivefold greater than threshold). Our repeated dosing regimen was designed to account for the common practice of sequential dosing (i.e., "bumping") used by humans during rave parties (Parrott, 2002). During the binge dosing procedure, body temperatures were measured by a rectal thermometer probe, and 5-HT-mediated behaviors were scored for 6 h after the first dose. Rats were decapitated 2 weeks after dosing, brain regions were dissected, and tissue levels of 5-HT and DA were determined by HPLC-ECD as described previously (Baumann *et al.*, 2001).

Figure 6 demonstrates that our binge MDMA dosing regimen increases core body temperature in rats [$F_{2,105} = 218.11, P < 0.0001$]. Specifically, repeated i.p. doses of 7.5 mg/kg MDMA elicit persistent hyperthermia on the day of treatment, whereas doses of 1.5 mg/kg do not. The 7.5 mg/kg dose causes temperature increases which are about 2 °C above normal, and such elevations are present for at least 6 h. The data in Fig. 7 demonstrate that binge MDMA treatment significantly decreases tissue 5-HT levels in the n. accumbens [$F_{2,17} = 50.52, P < 0.0001$], striatum [$F_{2,17} = 25.49, P < 0.001$], and mediobasal hypothalamus [$F_{2,17} = 14.31, P < 0.0001$] when assessed 2 weeks later. *Post hoc* tests reveal that high-dose MDMA produces long-term depletions of tissue 5-HT (~50% reductions) in all three regions examined, but the low-dose group has 5-HT concentrations similar to saline controls. Transmitter depletion is selective

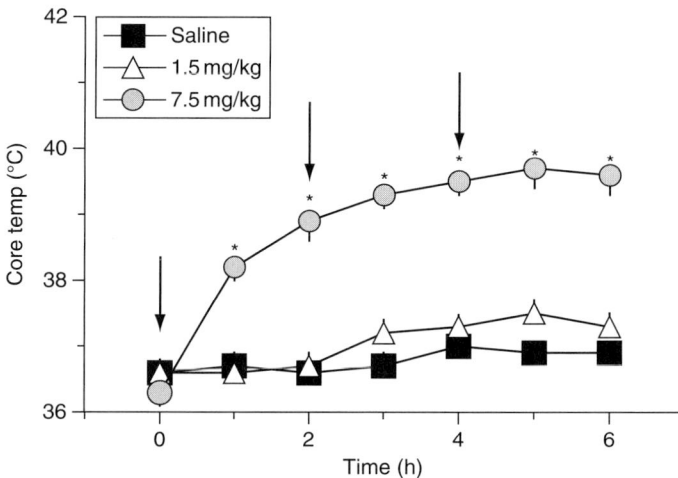

FIG. 6. Effects of MDMA binge administration on body temperature in rats. Male rats received three i.p. injections of saline, 1.5 mg/kg (low dose) or 7.5 mg/kg MDMA (high dose); injections were given at time 0, 2, and 4 h. Core body temperature was monitored hourly via insertion of a rectal temperature probe. Data are degrees Celsius expressed as mean ± S.E.M. for $N = 6$ rats/group. *$P < 0.05$ compared to saline control at a given time point.

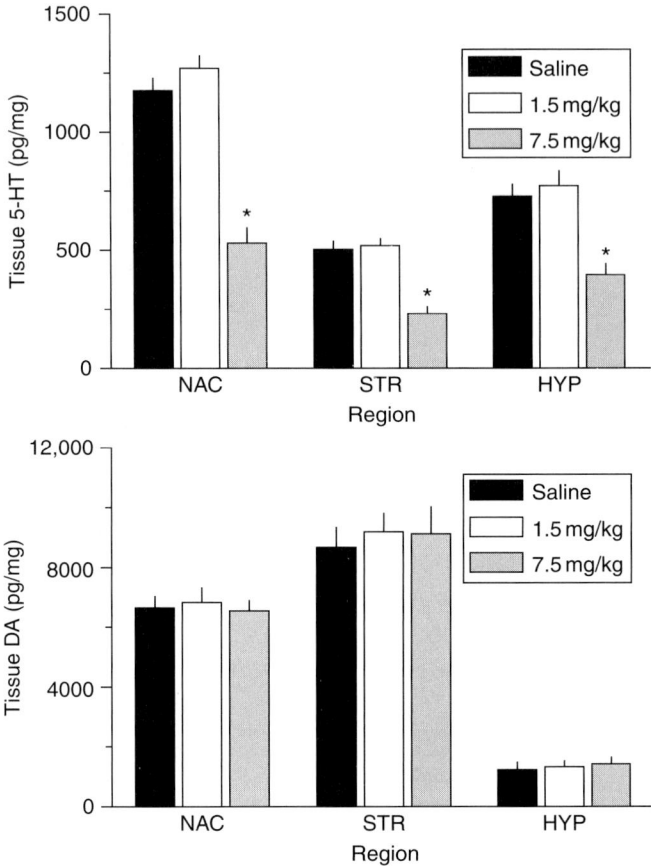

FIG. 7. Effects of MDMA binge administration on tissue levels of 5-HT (top panel) and DA (bottom panel) in microdissected rat brain regions. Male rats received three i.p. injections of saline, 1.5 mg/kg (low dose) or 7.5 mg/kg MDMA (high dose). Rats were decapitated 2 weeks later, and tissue was dissected from n. accumbens (NAC), striatum (STR) and mediobasal hypothalalmus (HYP). Postmortem concentrations of 5-HT and DA were quantified by HPLC-ECD. Data are pg/mg wet weight expressed as mean $\pm$ S.E.M. for $N = 6$ rats/group. *$P < 0.05$ compared to saline control for a specified brain region.

for 5-HT neurons since tissue DA levels are unaffected. The magnitude of 5-HT depletions depicted in Fig. 7 is similar to that observed by others (Battaglia *et al.*, 1987; Schmidt, 1987; Stone *et al.*, 1987).

Given that high-dose MDMA binges cause acute hyperthermia and long-term 5-HT depletions in the same rats, we examined the relationship between body temperature and postmortem tissue monoamines using Pearson correlation analysis. We found that body temperature is negatively correlated with tissue 5-HT in the n. accumbens ($r = -0.8630$, $P < 0.0001$), striatum ($r = -0.8089$, $P < 0.0001$), and hypothalamus ($r = -0.6910$, $P < 0.001$). There are no significant correlations between body temperature and tissue DA. Figure 8 shows representative data from the n. accumbens, where MDMA-induced elevations in core temperature are significantly correlated with depletion of tissue 5-HT but not DA. Collectively, our findings demonstrate that repeated treatments with behaviorally relevant doses of MDMA do not cause acute hyperthermia or long-term 5-HT depletions. In contrast, repeated administration of MDMA at a dose which is fivefold higher than the behavioral dose causes both of these adverse effects. The degree of acute hyperthermia produced by MDMA appears to predict the extent of subsequent 5-HT depletion. Our findings are consistent with those of O'Shea

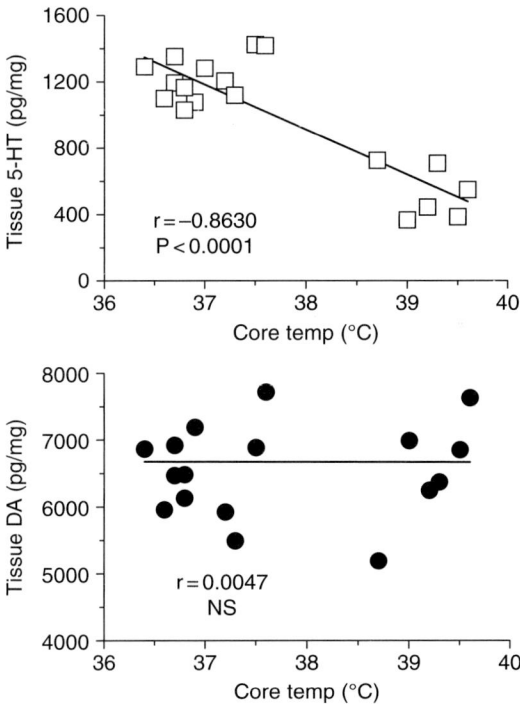

FIG. 8. Correlations between acute body temperature versus postmortem tissue levels of 5-HT (top panel) and DA (bottom panel) in rat n. accumbens measured 2 weeks after MDMA administration. Body temperatures represent average values obtained from 1 to 6 h after MDMA or saline injections, as shown in Fig. 6. Postmortem tissue levels of 5-HT and DA were measured 2 weeks later by HPLC-ECD, as shown in Fig. 7. Data points represent average temperature versus tissue amine concentration from individual rats ($N = 18$ rats), and Pearson correlation coefficients (r) are given. (Modified from Baumann et al., 2008a).

et al. (1998) who reported that high-dose MDMA (10 or 15 mg/kg, i.p.), but not low-dose MDMA (4 mg/kg, i.p.), causes acute hyperthermia and long-term 5-HT depletion in Dark Agouti rats.

V. Functional Consequences of MDMA-Induced 5-HT Depletion

Any definition of neurotoxicity must include the concept that functional impairments accompany neuronal damage (Moser, 2000; Winneke and Lilienthal, 1992). As noted previously, high-dose MDMA causes persistent inactivation of tryptophan hydroxylase which leads to inhibition of 5-HT synthesis and loss of 5-HT (O'Hearn *et al.*, 1988; Stone *et al.*, 1987). Moreover, MDMA-induced reduction in the density of SERT binding sites leads to decreased capacity for 5-HT uptake in nervous tissue (Battaglia *et al.*, 1987; Schmidt, 1987). Regardless of whether these 5-HT deficits reflect neurotoxic damage or long-term adaptations, such changes would be expected to have discernible *in vivo* correlates. Many investigators have examined functional consequences of high-dose MDMA administration, and a comprehensive review of this subject is beyond the scope of this review (reviewed by Cole and Sumnall, 2003; Green *et al.*, 2003). The following discussion will consider long-term effects of MDMA (i.e., >1 week) on *in vivo* indicators of 5-HT function in rats as measured by electrophysiological recording, neuroendocrine secretion, microdialysis sampling, and specific behaviors. A number of key findings are summarized in Table IV. In general, few

TABLE IV

Effects of (±)-MDMA Administration on Functional Indices of 5-HT Transmission in Rats

CNS effect	Dosing regimen	Survival interval	References
No change in 5-HT cell firing	20 mg/kg, s.c., twice daily, 4 days	2 weeks	Gartside *et al.* (1996)
Changes in corticosterone and prolactin secretion	20 mg/kg, s.c., single dose; 20 mg/kg, s.c. twice daily, 4 days	2 weeks 4, 8, and 12 months	Poland *et al.* (1997)
Reductions in evoked 5-HT release *in vivo*	20 mg/kg, s.c., twice daily, 4 days; 10 mg/kg, i.p., twice daily, 4 days	2 weeks, 1 week	Series *et al.* (1994), Shankaran and Gudelsky (1999)
Increased anxiety-like behaviors	5 mg/kg, s.c., 1 or 4 injections, 2 days; 7.5 mg/kg, s.c., twice daily, 3 days	3 months; 2 weeks	McGregor *et al.* (2003), Morley *et al.* (2001),[a] Fone *et al.* (2002)[a]

[a]These investigators noted marked increases in anxiogenic behaviors in the absence of significant MDMA-induced 5-HT depletion in brain.

published studies have been able to relate the magnitude of MDMA-induced 5-HT depletion to the degree of specific functional impairment. Furthermore, MDMA administration rarely causes persistent changes in baseline measures of neural function, and deficits are most readily demonstrated by provocation of the 5-HT system by pharmacological (e.g., drug challenge) or physiological means (e.g., environmental stress).

5-HT projections innervating the rat forebrain have cell bodies residing in the raphe nuclei (Steinbusch, 1981). These neurons exhibit pacemaker-like electrical activity which can be recorded using electrophysiological techniques (Aghajanian et al., 1978; Sprouse et al., 1989). Gartside et al. (1996) used extracellular recording methods to examine 5-HT cell firing in the dorsal raphe of rats previously treated with MDMA. Rats received two daily injections of 20 mg/kg, s.c. MDMA for 4 days and were tested under chloral hydrate anesthesia 2 weeks later. MDMA pretreatment had no effect on the number of classical or burst-firing 5-HT cells encountered during recording. Additionally, 5-HT cell firing rates and action potential characteristics were not different between MDMA- and saline-pretreated groups. These data show that 5-HT neurons and their firing properties are not altered after MDMA administration, and this agrees with immunohistochemical evidence demonstrating MDMA does not destroy 5-HT perikarya. The electrophysiological data from MDMA-pretreated rats differ from the findings reported with 5,7-DHT. In rats treated with i.c.v. 5,7-DHT, the number of classical and burst-firing 5-HT neurons is dramatically decreased in the dorsal raphe, in conjunction with a loss of 5-HT fluorescence (Aghajanian et al., 1978; Hajos and Sharp, 1996). Thus, 5,7-DHT produces reductions in 5-HT cell firing that are attributable to cell death, but MDMA does not.

5-HT neurons projecting from raphe nuclei to the hypothalamus provide stimulatory input for the secretion of adrenocorticotropin (ACTH) and prolactin from the anterior pituitary (Van de Kar, 1991). Accordingly, 5-HT releasers (e.g., fenfluramine) and 5-HT receptor agonists increase plasma levels of these pituitary hormones in rats and humans, while stimulating glucocorticoid secretion from the adrenals of both species (reviewed by Levy et al., 1994). Neuroendocrine challenge experiments have been used to demonstrate changes in serotonergic responsiveness in rats treated with MDMA (Poland, 1990; Poland et al., 1997; Series et al., 1995). In the most comprehensive study, Poland et al. (1997) examined effects of high-dose MDMA on hormone responses elicited by acute fenfluramine challenge. Male Sprague–Dawley rats received single s.c. injections of 20 mg/kg MDMA and were tested 2 weeks later. Prior MDMA exposure did not alter baseline levels of circulating ACTH or prolactin. However, in MDMA-pretreated rats, fenfluramine-induced ACTH secretion was reduced while prolactin secretion was enhanced. The MDMA dosing regimen caused significant depletions of tissue 5-HT in various brain regions, including hypothalamus. In a follow-up time-course study, rats received twice daily s.c. injections of 20 mg/kg MDMA for

4 days, and were challenged with fenfluramine (6 mg/kg, s.c.) at 4, 8, and 12 months thereafter. As observed in the single dose MDMA study, rats exposed to multiple MDMA doses displayed blunted ACTH responses and augmented prolactin responses to fenfluramine. Interestingly, the impaired ACTH response persisted for 12 months in MDMA-pretreated rats, even though tissue levels of 5-HT were not depleted at this time point. The data show that high-dose MDMA can cause functional abnormalities for up to 1 year, and changes in 5-HT responsiveness do not necessarily parallel the extent of recovery from 5-HT depletions in brain.

In our laboratory, we wished to further explore the long-term neuroendocrine consequences of MDMA exposure. Specifically, we wished to compare MDMA-induced hormone responses in rats pretreated with low- versus high-dose MDMA binges. Utilizing the MDMA dosing regimen described previously, male rats received three i.p. injections of 1.5 or 7.5 mg/kg MDMA, one dose every 2 h. Control rats received saline vehicle according to the same schedule. One week after MDMA binges, rats were fitted with indwelling jugular catheters under pentobarbital anesthesia. One week after surgery (i.e., 2 weeks after MDMA or saline), rats were brought into the testing room, and i.v. catheters were connected to extension tubes. Rats received i.v. injections of 1 mg/kg MDMA at time zero, followed by 3 mg/kg MDMA 60 min later. Blood samples were withdrawn via the catheters immediately before and at 15, 30, and 60 min after each dose of MDMA. Elements of the 5-HT syndrome, namely forepaw treading, flattened body posture, and head weaving, were scored during the blood sampling procedure. Plasma levels of prolactin and corticosterone were measured by double-antibody radioimmunoassay methods (Baumann *et al.*, 1998, 2008a).

Figure 9 shows the effects of acute i.v. MDMA on prolactin and corticosterone secretion in rats exposed to MDMA binges. MDMA pretreatment does not alter baseline levels of either hormone. Acute i.v. administration of MDMA elicits dose-related elevations in circulating prolactin and corticosterone, in agreement with the findings of others who administered MDMA to rats via the i.p. route (Nash *et al.*, 1988). Importantly, MDMA pretreatment significantly influences the magnitude of MDMA-induced prolactin $[F_{2,147} = 21.03, P < 0.0001]$ and corticosterone $[F_{2,147} = 12.20, P < 0.001]$ responses. *Post hoc* tests reveal that rats exposed to 7.5 mg/kg MDMA binges display reductions in prolactin and corticosterone secretion evoked by MDMA challenge. Blunted endocrine sensitivity is most severe for prolactin secretion, where responses in the high-dose binge group are about 50–60% lower than control responses. It is noteworthy that MDMA-induced endocrine secretion is similar between the low-dose binge group and saline-pretreated controls. Figure 10 shows the effects of MDMA challenge on 5-HT syndrome in rats previously exposed to MDMA binges. Acute i.v. MDMA produces robust 5-HT syndrome that is significantly affected by MDMA pretreatment $[F_{2,42} = 8.15, P < 0.001]$. Specifically, rats pretreated

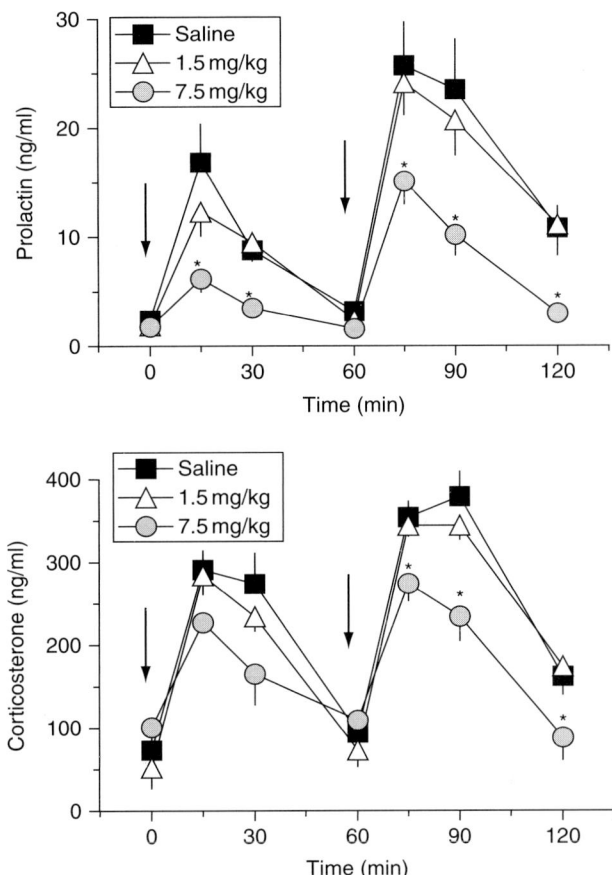

FIG. 9. Effects of i.v. MDMA on plasma concentrations of prolactin (top panel) and corticosterone (bottom panel) in rats pretreated with saline or MDMA binges. Male rats received three i.p. injections of saline, 1.5 mg/kg (low dose) or 7.5 mg/kg MDMA (high dose). Two weeks later, rats were given i.v. challenge injections of 1 mg/kg MDMA at time 0 h, followed by 3 mg/kg at 60 min. Blood samples were withdrawn immediately before and at 15, 30, and 60 min after each drug injection. Plasma hormones were assayed by double-antibody RIA methods. Data are ng/ml expressed as mean ± S.E. M. for $N = 8$ rats/group. *$P < 0.05$ compared to saline-pretreated control at a given time point. (Modified from Baumann *et al.*, 2008a).

with high-dose binges display blunted syndrome scores after 3 mg/kg i.v. MDMA. Similar to the neuroendocrine findings, the behavioral effects of MDMA in the low-dose binge group are comparable to saline-pretreated controls.

Our neuroendocrine and behavioral results are consistent with the development of tolerance to pharmacological effects of MDMA. The present endocrine data do not agree completely with the data of Poland *et al.* (1997); however, our

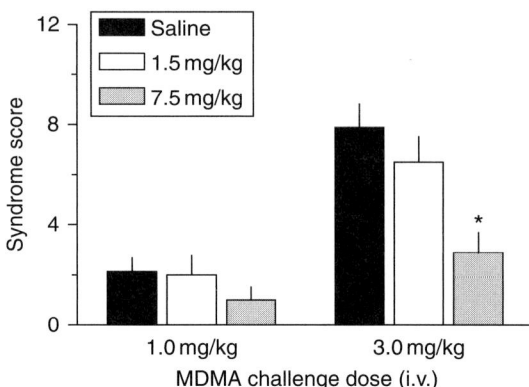

FIG. 10. Effects of i.v. MDMA on 5-HT behavioral syndrome in rats pretreated with saline or MDMA binges. Male rats received three i.p. injections of saline, 1.5 mg/kg (low dose) or 7.5 mg/kg MDMA (high dose). Two weeks later, rats were given i.v. challenge injections of 1 mg/kg MDMA at time 0 h, followed by 3 mg/kg at 60 min. The occurrence of flat-body posture and forepaw treading were scored after each dose of acute i.v. MDMA. Data are summed syndrome scores expressed as mean ± S.E.M. for $N = 8$ rats/group. *$P < 0.05$ compared to saline-pretreated control at the corresponding challenge dose of MDMA. (Modified from Baumann et al., 2008a).

findings do agree with previous data showing blunted hormonal responses to fenfluramine in rats previously treated with high-dose fenfluramine (Baumann et al., 1998). Perhaps more importantly, the data shown in Fig. 9 are strikingly similar to clinical findings in which cortisol and prolactin responses to acute (+)-fenfluramine administration are reduced in human MDMA users (Gerra et al., 1998, 2000; Gouzoulis-Mayfrank et al., 2002). Indeed, Gerra et al. (2000) reported that (+)-fenfluramine-induced prolactin secretion is blunted in abstinent MDMA users for up to 1 year after cessation of drug use. The mechanism(s) underlying reduced sensitivity to (+)-fenfluramine and MDMA are not known, but it is tempting to speculate that MDMA-induced impairments in evoked 5-HT release are involved. While some investigators have cited neuroendocrine changes in human MDMA users as evidence for 5-HT neurotoxicity, Gouzoulis-Mayfrank et al. (2002) provide a compelling argument that endocrine abnormalities in MDMA users could be related to cannabis use rather than MDMA. Further experiments will be required to resolve the precise nature of neuroendocrine changes in human MDMA users.

In vivo microdialysis allows continuous sampling of extracellular fluid from intact brain, and this method has been used to evaluate persistent neurochemical consequences of MDMA exposure (Gartside et al., 1996; Matuszewich et al., 2002; Series et al., 1994; Shankaran and Gudelsky, 1999). For example, Shankaran and Gudelsky (1999) assessed neurochemical effects of acute MDMA challenge in rats that had previously received four i.p. injections of 10 mg/kg MDMA. In vivo microdialysis was performed in the striatum of conscious rats 1 week after high-dose MDMA treatment. Baseline levels of dialysate 5-HT were not altered by prior MDMA exposure even though tissue levels of 5-HT in striatum were depleted by 50%. The ability of MDMA to evoke 5-HT release was severely impaired in MDMA-pretreated rats, while the concurrent DA response was normal. In this same study, effects of MDMA on body temperature and 5-HT syndrome were attenuated in MDMA-pretreated rats, suggesting the development of tolerance. Other investigations using *in vivo* microdialysis methods have shown that 5-HT release in response to physiological or stressful stimuli is impaired in rats pretreated with high-dose MDMA (Gartside *et al.*, 1996; Matuszewich *et al.*, 2002).

Based on our neuroendocrine and behavioral findings, we sought to evaluate whether MDMA tolerance might be reflective of neurochemical changes in the brain. To this end, we performed *in vivo* microdialysis in the n. accumbens and striatum of rats previously exposed to low- and high-dose MDMA binges. For these experiments, rats received three i.p. injections of 1.5 or 7.5 mg/kg MDMA, one dose every 2 h. Control rats received saline vehicle according to the same schedule. One week after MDMA binges, rats were fitted with indwelling jugular catheters and intracranial guide cannulae under pentobarbital anesthesia. Guide cannulae were aimed at the n. accumbens or striatum in separate groups of rats. One week after surgery (i.e., 2 weeks after MDMA or saline), rats were subjected to microdialysis testing. Dialysate samples were collected at 20 min intervals and assayed for 5-HT and DA by HPLC-ECD. After three baseline samples were obtained, rats received i.v. injections of 1 mg/kg MDMA at time zero, followed by 3 mg/kg MDMA 60 min later. Dialysate samples were collected until 1 h after the second dose of MDMA.

Figure 11 illustrates the effects of i.v. MDMA injections on 5-HT and DA release in the n. accumbens of rats previously exposed to MDMA binges. MDMA pretreatment does not affect baseline concentration of either transmitter. Acute i.v. MDMA produces dose-related elevations in dialysate 5-HT and DA, consistent with our prior findings (Baumann *et al.*, 2005, 2007). However, MDMA pretreatment significantly affects dialysate 5-HT responsiveness [$F_{2,161} = 22.95$, $P < 0.0001$], with rats in the 7.5 mg/kg MDMA group displaying reductions in evoked 5-HT release. The magnitude of 5-HT responses in the high-dose MDMA group is about half that of responses in the saline-pretreated group at both challenge doses of MDMA. Rats in the 1.5 mg/kg MDMA group exhibit 5-HT release that is nearly identical to saline-pretreated controls. Interestingly, MDMA pretreatment has no effect on dialysate DA responsiveness. In this case, rats from all pretreatment groups exhibit similar DA release in response to MDMA challenge injections. Microdialysis findings from the striatum gave results analogous to those from n. accumbens. Namely, rats pretreated with high-dose MDMA binges have impairments in MDMA-induced release of 5-HT but not DA.

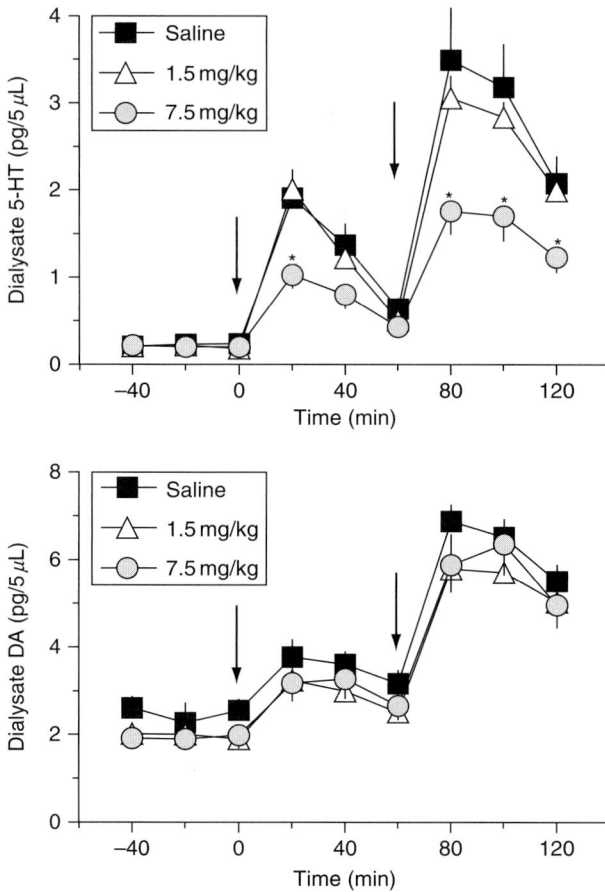

FIG. 11. Effects of i.v. MDMA on dialysate 5-HT (top panel) and DA (bottom panel) in n. accumbens of rats pretreated with saline or MDMA binges. Male rats received three i.p. injections of saline, 1.5 mg/kg (low dose) or 7.5 mg/kg MDMA (high dose). Two weeks later, rats undergoing *in vivo* microdialysis were given i.v. challenge injections of 1 mg/kg MDMA at time 0 h, followed by 3 mg/kg at 60 min. Dialysate samples collected at 20 min intervals were assayed for 5-HT and DA by microbore HPLC-ECD. Data are pg/5 μl expressed as mean ± S.E.M. for $N = 8$ rats/group. *$P < 0.05$ compared to saline-pretreated control at a given time point. (Modified from Baumann *et al.*, 2008a).

Taken together with previous studies, our microdialysis data reveal important long-term consequences of MDMA administration: (1) baseline concentrations of dialysate 5-HT are not altered, despite profound depletions of tissue transmitter, (2) evoked 5-HT release is blunted in response to pharmacological or physiological provocation, (3) impairments in transmitter release are selective for 5-HT since evoked DA release is not affected, and (4) blunted 5-HT release may underlie tolerance to various pharmacological effects of MDMA. The dialysis findings with MDMA resemble those obtained from rats treated with 5-HT neurotoxin, 5,7-dihydroxytryptamine (5,7-DHT) (Hall *et al.*, 1999; Kirby *et al.*, 1995; Romero *et al.*, 1998). In a representative study, Kirby *et al.* (1995) used microdialysis in rat striatum to evaluate the long-term neurochemical effects of i.c.v. 5,7-DHT. These investigators showed that impairments in stimulated 5-HT release and reductions in dialysate 5-HIAA are highly correlated with loss of tissue 5-HT, whereas baseline dialysate 5-HT is not. In fact, depletions of brain tissue 5-HT up to 90% did not affect baseline levels of dialysate 5-HT (Kirby *et al.*, 1995). It seems that adaptive mechanisms serve to maintain normal concentrations of synaptic 5-HT, even under conditions of severe transmitter depletion. A comparable situation exists after lesions of the nigrostriatal DA system where baseline levels of extracellular DA are maintained in the physiological range despite substantial loss of tissue DA (see Zigmond *et al.*, 1990). In the case of high-dose MDMA treatment, it seems feasible that reductions in 5-HT uptake (e.g., less functional SERT protein) and metabolism (e.g., decreased monoamine oxidase activity) compensate for 5-HT depletions in order to keep optimal concentrations of 5-HT bathing nerve cells. On the other hand, deficits in 5-HT release are readily demonstrated in MDMA-pretreated rats when 5-HT systems are taxed by drug challenge or environmental stressors. It is noteworthy that human MDMA users often report tolerance as a major side-effect of chronic drug use (Parrott, 2005; Verheyden *et al.*, 2003). The rat data shown here suggest the possibility that MDMA tolerance in humans may reflect deficits in central 5-HT release mechanisms.

One of the more serious and disturbing clinical findings is that MDMA causes persistent cognitive deficits in some users (Morgan, 2000; Reneman, 2003). Numerous research groups have examined the effects of MDMA treatment on learning and memory processes in rats, yet most studies failed to identify persistent impairments—even when extensive 5-HT depletions were present (Byrne *et al.*, 2000; McNamara *et al.*, 1995; Ricaurte *et al.*, 1993; Robinson *et al.*, 1993; Seiden *et al.*, 1993; Slikker *et al.*, 1989). While a complete review of this literature is not possible, representative findings will be mentioned. In an extensive series of experiments, Seiden *et al.* (1993) evaluated the effects of high-dose MDMA on a battery of tests including open-field behavior, schedule-controlled behavior, one-way avoidance, discriminated two-way avoidance, forced swim, and radial maze performance. Male Sprague–Dawley rats received twice daily s.c. injections of 10–40 mg/kg MDMA for 4 days, and were tested beginning 2 weeks after treatment. Despite large depletions of brain tissue 5-HT, MDMA-pretreated rats exhibited normal behaviors in all paradigms. Likewise, Robinson *et al.* (1993) found that MDMA-induced depletion of cortical 5-HT up to 70% did not alter spatial navigation, skilled forelimb use or foraging behavior in rats. On the other hand, Marston *et al.* (1999) reported that MDMA administration produces persistent deficits in a delayed nonmatch to performance (DNMTP) procedure when long delay intervals are employed (i.e., 30 s). Specifically, saline-pretreated rats exhibited progressive improvement in task performance over successive days of testing, whereas MDMA-pretreated rats did not. The authors theorized that delay-dependent impairments in the DNMTP procedure reflect MDMA-induced deficits in short-term memory, possibly attributable to 5-HT depletion.

With the exception of the findings of Marston *et al.*, the collective behavioral data in rats indicate that MDMA-induced depletions of brain 5-HT have little effect on cognitive processes. There are several potential explanations for this apparent paradox. First, high-dose MDMA administration produces only partial depletion of 5-HT in the range of 40–60% in most brain areas. This level of 5-HT loss may not be sufficient to elicit behavioral alterations, as compensatory adaptations in 5-HT neurons could maintain normal physiological function. Second, MDMA appears to selectively affect fine-diameter fibers arising from the dorsal raphe, and it is possible that these 5-HT circuits may not subserve the behaviors being monitored. Third, the behavioral tests utilized in rat studies might not be sensitive enough to detect subtle changes in learning and memory processes. Finally, the functional reserve capacity in the CNS might be sufficient to compensate for even large depletions of a single transmitter.

While MDMA appears to have few long-term effects on cognition in rats, a growing body of evidence demonstrates that MDMA administration can cause persistent anxiety-like behaviors in this species (Fone *et al.*, 2002; Gurtman *et al.*, 2002; Morley *et al.*, 2001). Morley *et al.* (2001) first reported that MDMA exposure induces long-term anxiety in rats. These investigators gave male Wistar rats one or four i.p. injections of 5 mg/kg MDMA for 2 consecutive days. Subjects were tested 3 months later in a battery of anxiety-related paradigms including elevated plus maze, emergence and social interaction tests. Rats receiving either single or multiple MDMA injections displayed marked increases in anxiogenic behaviors in all three tests when compared to control rats. Because no 5-HT endpoints were examined, it was impossible to relate changes in behavior to changes in 5-HT transmission. In a follow-up study, Gurtman *et al.* (2002) replicated the original findings of Morley *et al.* using rats pretreated with four i. p. injections of 5 mg/kg MDMA for 2 days; persistent anxiogenic effects of MDMA were associated with depletions of 5-HT in the amygdala, hippocampus, and striatum. Interestingly, Fone *et al.* (2002) showed that twice daily injections of 7.5 mg/kg MDMA for 3 days caused impairments in social interaction in adolescent Lister rats, even in the absence of 5-HT depletions or reductions in [³H]-paroxetine-labeled SERT binding sites. These data suggested the possibility that MDMA-induced anxiety does not require 5-HT deficits.

In an attempt to determine potential mechanisms underlying MDMA-induced anxiety, McGregor *et al.* (2003) evaluated effects of the drug on anxiety-related behaviors and a number of postmortem parameters including autoradiography for SERT and 5-HT receptor subtypes. Rats received moderate (5 mg/kg, i.p., 2 days) or high (5 mg/kg, i.p., 4 injections, 2 days) doses of MDMA, and tests were conducted 10 weeks later. This study confirmed that moderate doses of MDMA can cause protracted increases in anxiety-like behaviors without significant 5-HT depletions. Furthermore, the autoradiographic analysis revealed that anxiogenic effects of MDMA may involve long-term reductions in 5-HT$_{2A/2C}$ receptors rather than reductions in SERT binding. Additional work by Bull et $al.$ (2003, 2004) suggests that decreases in the sensitivity of 5-HT$_{2A}$ receptors, but not 5-HT$_{2C}$ receptors, could underlie MDMA-associated anxiety. Clearly, more investigation into this important area of research is warranted.

VI. Reversal of MDMA-Induced 5-HT Depletion by L-5-Hydroxytryptophan

As noted in Section I, accumulating evidence indicates that long-term MDMA abuse is associated with cognitive impairments and mood disturbances that can last for months after cessation of drug intake (Morgan, 2000; Parrott, 2002). There is speculation that psychological problems in MDMA users may result from 5-HT depletions in the CNS (McCann et $al.$, 2008). To the extent that MDMA might produce 5-HT depletion in humans, one feasible approach for treatment would be to restore brain 5-HT by administration of the 5-HT precursor L-5-hydroxytryptophan (5-HTP). To test this idea, male Sprague–Dawley rats received three i.p. injections of 7.5 mg/kg MDMA or saline, one dose every 2 h (Wang et $al.$, 2007). Two weeks later, rats received i.p. injections of saline or 5-HTP plus benserazide (5-HTP-B), each drug at 50 mg/kg. Rats were sacrificed 2 h after treatments. Benserazide is an inhibitor of aromatic acid decarboxylase which prevents peripheral conversion of 5-HTP to 5-HT, thereby facilitating entry of 5-HTP into the CNS. Previous studies show that a similar 5-HTP-B dosing regimen produces large increases in extracellular 5-HT in rat n. accumbens (Halladay et $al.$, 2006). 5-HT concentrations in brain tissue sections were visualized using immunoautoradiography. As reported in Fig. 12, MDMA markedly decreases immunoreactive 5-HT in the caudate n. and hippocampus to about 35% of control. Under the same dosing conditions, SERT binding in these brain regions is reduced by nearly 90%. Importantly, administration of 5-HTP-B to MDMA-pretreated rats significantly increases the 5-HT signal toward normal levels in caudate (85% of control) and hippocampus (66% of control). These experiments suggest that high-dose MDMA treatments leave 5-HT nerve terminals largely intact but emptied of endogenous 5-HT. Administration of the 5-HT precursor 5-HTP appears to "refill" empty nerve terminals, substantially restoring brain 5-HT levels. Our data suggest that precursor loading with 5-HTP might be a clinically useful approach to treatment in abstinent MDMA users who

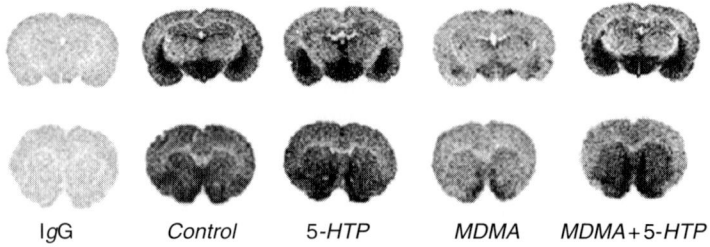

Brain region	Control	5-HTP	MDMA	MDMA+5-HTP
Cortex	100±11.8%	100.5±10.7%	34.2±3.8*%	54.6±7.0†%
Hippocampus	100±10.4%	101.8±10.4%	35.2±3.7*%	65.5±7.8†%
Caudate	100±11.5%	98.6±9.5%	33.6±4.0*%	84.7±11.2†%
Hypothalamus	100±9.4%	100.7±9.6%	22.9±3.1*%	83.6±12.1†%
Amygdala	100±9.9%	99.7±10.4%	25.3±3.3*%	85.7±10.9†%

FIG. 12. Effects of MDMA on brain 5-HT as measured by immunoautoradiography. Male rats received three i.p. injections of saline or 7.5 mg/kg MDMA. Two weeks later, rats were treated with i. p. injections of either saline or 5-HTP/benserazide (5-HTP-B, 50 mg/kg of each drug), and were decapitated 2 h thereafter. 5-HT was measured by immunoautoradiography in postmortem brain sections. Values are mean ± S.D. for $N = 6$ rats/group, and representative autoradiograms are shown. *$P < 0.01$ compared to control; †$P < 0.01$ compared to corresponding MDMA-treated group. Taken from Wang et al., 2007.

experience cognitive and psychiatric morbidity due to persistent 5-HT deficits (Thomasius et al., 2006). Future investigations should test the validity of this hypothesis in carefully controlled clinical studies.

VII. MDMA and Valvular Heart Disease

Recent evidence indicates that MDA, the N-demethylated metabolite of MDMA, displays direct agonist actions at the 5-HT$_{2B}$ receptor subtype (Setola et al., 2003). This finding suggests that illicit MDMA use might increase the risk for VHD. Understanding the possible relationship between MDMA and VHD is best understood in the context of fenfluramine-associated valvulopathy. (±)-Fenfluramine, and its more potent enantiomer, (+)-fenfluramine, were commonly prescribed anorectic medications. These agents were removed from clinical use in

1997 due to the occurrence of VHD in some patients (Connolly and McGoon, 1999). Although a serotonergic mechanism was suspected as a possible cause of the VHD (Connolly et al., 1997), little was known about the pathogenesis of this adverse effect in the late 1990s. In light of the established role of 5-HT as a mitogen (Seuwen et al., 1988), we carried out an investigation to determine if isomers of fenfluramine or its metabolites (i.e., norfenfluramines), might activate a specific mitogenic 5-HT receptor subtype (Rothman et al., 2000). Drugs were tested for activity at various 5-HT receptor sites using high-throughput binding and functional assays. A number of additional test drugs were included in our study to provide both positive and negative controls. "Positive control" drugs included methysergide, its active metabolite methylergonovine (Bredberg et al., 1986) and ergotamine. Methysergide and ergotamine are known to produce left-sided VHD chiefly affecting the mitral valve (Bana et al., 1974; Hendrikx et al., 1996). "Negative control" drugs included phentermine, fluoxetine, and its metabolite norfluoxetine. These drugs interact with monoamine transporters but are not associated with VHD. Our working hypothesis was that "positive control" drugs would share in common the ability to activate a mitogenic 5-HT receptor expressed in heart valves, whereas "negative control" drugs would not.

An initial receptorome screen (Setola and Roth, 2005) led to a detailed evaluation of the binding of these drugs to the 5-HT_2 family of receptors. Our data reveal that isomers of fenfluramine display low affinity for most 5-HT receptors. In contrast, isomers of norfenfluramine possess high affinity ($K_{\mathrm{I}} = 10\text{--}50$ nM) for the 5-HT_{2B} receptor subtype, confirming the findings of others (Fitzgerald et al., 2000; Porter et al., 1999). Moreover, functional studies demonstrate that norfenfluramines are full agonists at the 5-HT_{2B} site. Positive control drugs such as ergotamine, methysergide, and methylergonovine are partial agonists at the 5-HT_{2B} receptor, whereas negative control drugs are not. Taken together, our in vitro findings indicate that 5-HT_{2B} receptors are involved in the valvulopathic effects of fenfluramine. Consistent with this hypothesis, 5-HT_{2B} receptors are richly expressed on mitral and aortic valves (Fitzgerald et al., 2000), where they can stimulate mitogenisis (Lopez-Ilasaca, 1998).

The importance of 5-HT_{2B} receptors in mediating drug-induced VHD is supported by recent data showing carbergoline and pergolide are potent agonists at this receptor site. Both of these medications are known to increase the risk of VHD (for review see Roth, 2007). As noted earlier, Setola et al. have reported that MDMA and MDA are 5-HT_{2B} receptor agonists that elicit prolonged mitogenic responses in human valvular interstitial cells via activation of 5-HT_{2B} receptors (Setola et al., 2003). The data summarized in Table V demonstrate that stereo-isomers of MDA are particularly potent and efficacious agonists at the 5-HT_{2B} site. The in vitro findings predict that MDMA abuse might cause VHD, similar to the effects of other known 5-HT_{2B} agonists. Indeed, a recent clinical investigation reported that MDMA users have markedly increased valvular regurgitation

TABLE V

POTENCY (pEC$_{50}$) AND EFFICACY OF MDMA AND MDA FOR ACTIVATION OF 5-HT$_{2B}$-MEDIATED PI HYDROLYSIS

Test drug	pEC$_{50}$ for 5-HT$_{2B}$-mediated PI hydrolysis (nM)	Efficacy for 5-HT$_{2B}$-mediated PI hydrolysis (relative to 5-HT)
5-HT	1.0	1.0
(±)-MDMA	2000	0.32
(+)-MDMA	900	0.27
(−)-MDMA	6000	0.38
(±)-MDA	190	0.80
(+)-MDA	150	0.76
(−)-MDA	100	0.81
Fenfluramine	400	0.13
Norfenfluramine	60	0.96
Dihydroergotamine	30	0.73
Pergolide	53	1.12
Methysergide	150	0.18
Methylergonivine	0.8	0.40

Data are mean values for $N = 3$ experiments. Data taken from Setola et al. (2003).

compared to normal control subjects (Droogmans et al., 2007). The severity of regurgitation was directly related to the amount of MDMA consumed, indicating a dose-related increase in the risk for VHD. Additional studies will be needed to confirm and extend these important clinical observations.

VIII. Summary

The findings reviewed in this chapter allow a number of conclusions to be drawn with regard to MDMA-induced toxic effects. First, MDMA produces pharmacological actions at the same doses in rats and humans (e.g., 1–2 mg/kg), which argues against the arbitrary use of allometric scaling to adjust doses between these species. Second, high doses of MDMA that produce 5-HT depletions in rats are accompanied by impairments in evoked 5-HT release that are manifest as drug tolerance. The animal data indicate that tolerance in human MDMA users could reflect central 5-HT dysfunction, thereby contributing to dangerous dose escalation. Third, doses of MDMA that fail to cause 5-HT depletions in rats can produce persistent increases in anxiety-like behaviors, suggesting even moderate doses of the drug may pose risks. Finally, the agonist activity of MDA isomers at the 5-HT$_{2B}$ receptor subtype could be involved with the increased risk for VHD in heavy MDMA users. Future investigations are warranted to further elucidate the long-term effects of MDMA exposure on neural and cardiac function in humans.

Acknowledgments

This research was generously supported by the NIDA IRP. The authors are indebted to John Partilla, Chris Dersch, Mario Ayestas, Robert Clark, Fred Franken, and John Rutter for their expert technical assistance during these studies.

References

Aghajanian, G. K., Wang, R. Y., and Baraban, J. (1978). Serotonergic and non-serotonergic neurons of the dorsal raphe: Reciprocal changes in firing induced by peripheral nerve stimulation. *Brain Res.* **153**(1), 169–175.

Bana, D. S., MacNeal, P. S., LeCompte, P. M., Shah, Y., and Graham, J. R. (1974). Cardiac murmurs and endocardial fibrosis associated with methysergide therapy. *Am. Heart J.* **88**, 640–655.

Banken, J. A. (2004). Drug abuse trends among youth in the United States. *Ann. NY Acad. Sci.* **1025**, 465–471.

Bankson, M. G., and Cunningham, K. A. (2001). 3,4-Methylenedioxymethamphetamine (MDMA) as a unique model of serotonin receptor function and serotonin-dopamine interactions. *J. Pharmacol. Exp. Ther.* **297**, 846–852.

Battaglia, G., and De Souza, E. B. (1989). Pharmacologic profile of amphetamine derivatives at various brain recognition sites: Selective effects on serotonergic systems. *NIDA Res. Monogr.* **94**, 240–258.

Battaglia, G., Yeh, S. Y., O'Hearn, E., Molliver, M. E., Kuhar, M. J., and De Souza, E. B. (1987). 3,4-Methylenedioxymethamphetamine and 3,4-methylenedioxyamphetamine destroy serotonin terminals in rat brain: Quantification of neurodegeneration by measurement of [^{3}H]paroxetine-labeled serotonin uptake sites. *J. Pharmacol. Exp. Ther.* **242**, 911–916.

Battaglia, G., Yeh, S. Y., and De Souza, E. B. (1988). MDMA-induced neurotoxicity: Parameters of degeneration and recovery of brain serotonin neurons. *Pharmacol. Biochem. Behav.* **29**, 269–274.

Baumann, M. H., and Rutter, J. J. (2003). Application of *in vivo* microdialysis methods to the study of psychomotor stimulant drugs. *In* Methods in Drug Abuse Research, Cellular and Circuit Analysis, (B. D. Waterhouse, Ed.), pp. 51–86. CRC Press, Boca Raton, FL.

Baumann, M. H., Ayestas, M. A., and Rothman, R. B. (1998). Functional consequences of central serotonin depletion produced by repeated fenfluramine administration in rats. *J. Neurosci.* **18**, 9069–9077.

Baumann, M. H., Ayestas, M. A., Dersch, C. M., and Rothman, R. B. (2001). 1-(*m*-Chlorophenyl) piperazine (mCPP) dissociates *in vivo* serotonin release from long-term serotonin depletion in rat brain. *Neuropsychopharmacology* **24**, 492–501.

Baumann, M. H., Clark, R. D., Budzynski, A. G., Partilla, J. S., Blough, B. E., and Rothman, R. B. (2005). N-substituted piperazines abused by humans mimic the molecular mechanism of 3,4-methylenedioxymethamphetamine (MDMA, or 'Ecstasy'). *Neuropsychopharmacology* **30**(3), 550–560.

Baumann, M. H., Wang, X., and Rothman, R. B. (2007). 3,4-Methylenedioxymethamphetamine (MDMA) neurotoxicity in rats: A reappraisal of past and present findings. *Psychopharmacology (Berl.)* **189**(4), 407–424.

Baumann, M. H., Clark, R. D., Franken, F. H., Rutter, J. J., and Rothman, R. B. (2008a). Tolerance to 3,4-methylenedioxymethamphetamine in rats exposed to single high-dose binges. *Neuroscience* **152** (3), 773–784.

Baumann, M. H., Clark, R. D., and Rothman, R. B. (2008b). Locomotor stimulation produced by 3,4-methylenedioxymethamphetamine (MDMA) is correlated with dialysate levels of serotonin and dopamine in rat brain. *Pharmacol. Biochem. Behav.* **90**(2), 208–217.

Baumgarten, H. G., and Lachenmayer, L. (2004). Serotonin neurotoxins—Past and present. *Neurotox. Res.* **6**(7–8), 589–614.

Benmansour, S., Cecchi, M., Morilak, D. A., Gerhardt, G. A., Javors, M. A., Gould, G. G., and Frazer, A. (1999). Effects of chronic antidepressant treatments on serotonin transporter function, density, and mRNA level. *J. Neurosci.* **19**(23), 10494–10501.

Berger, U. V., Gu, X. F., and Azmitia, E. C. (1992). The substituted amphetamines 3,4-methylene-dioxymethamphetamine, methamphetamine, *p*-chloroamphetamine and fenfluramine induce 5-hydroxytryptamine release via a common mechanism blocked by fluoxetine and cocaine. *Eur. J. Pharmacol.* **215**, 153–160.

Bredberg, U., Eyjolfsdottir, G. S., Paalzow, L., Tfelt-Hansen, P., and Tfelt-Hansen, V. (1986). Pharmacokinetics of methysergide and its metabolite methylergometrine in man. *Eur. J. Clin. Pharmacol.* **30,** 75–77.

Bull, E. J., Hutson, P. H., and Fone, K. C. (2003). Reduced social interaction following 3,4-methyle-nedioxymethamphetamine is not associated with enhanced 5-HT 2C receptor responsivity. *Neuropharmacology* **44**(4), 439–448.

Bull, E. J., Hutson, P. H., and Fone, K. C. (2004). Decreased social behaviour following 3,4-methylenedioxymethamphetamine (MDMA) is accompanied by changes in 5-HT2A receptor responsivity. *Neuropharmacology* **46**(2), 202–210.

Byrne, T., Baker, L. E., and Poling, A. (2000). MDMA and learning: Effects of acute and neurotoxic exposure in the rat. *Pharmacol. Biochem. Behav.* **66**(3), 501–508.

Carlsson, A. (1976). The contribution of drug research to investigating the nature of endogenous depression. *Pharmakopsychiatr. Neuropsychopharmakol.* **9**(1), 2–10.

Charlier, C., Broly, F., Lhermitte, M., Pinto, E., Ansseau, M., and Plomteux, G. (2003). Polymorph-isms in the CYP 2D6 gene: Association with plasma concentrations of fluoxetine and paroxetine. *Ther. Drug Monit.* **25**(6), 738–742.

Chu, T., Kumagai, Y., DiStefano, E. W., and Cho, A. K. (1996). Disposition of methylenedioxy-methamphetamine and three metabolites in the brains of different rat strains and their possible roles in acute serotonin depletion. *Biochem. Pharmacol.* **51**(6), 789–796.

Colado, M. I., O'Shea, E., and Green, A. R. (2004). Acute and long-term effects of MDMA on cerebral dopamine biochemistry and function. *Psychopharmacology (Berl.)* **173**(3–4), 249–263.

Cole, J. C., and Sumnall, H. R. (2003). The pre-clinical behavioural pharmacology of 3,4-methyle-nedioxymethamphetamine (MDMA). *Neurosci. Biobehav. Rev.* **27**(3), 199–217.

Commins, D. L., Vosmer, G., Virus, R. M., Woolverton, W. L., Schuster, C. R., and Seiden, L. S. (1987). Biochemical and histological evidence that methylenedioxymethylamphetamine (MDMA) is toxic to neurons in the rat brain. *J. Pharmacol. Exp. Ther.* **241**(1), 338–345.

Connolly, H. M., and McGoon, M. D. (1999). Obesity drugs and the heart. *Curr. Probl. Cardiol.* **24,** 745–792.

Connolly, H. M., Crary, J. L., McGoon, M. D., Hensrud, D. D., Edwards, B. S., and Schaff, H. V. (1997). Valvular heart disease associated with fenfluramine-phentermine. *N. Engl. J. Med.* **337**(9), 581–588.

Crespi, D., Mennini, T., and Gobbi, M. (1997). Carrier-dependent and Ca(2+)-dependent 5-HT and dopamine release induced by (+)-amphetamine, 3,4-methylendioxymethamphetamine, p-chloro-amphetamine and (+)-fenfluramine. *Br. J. Pharmacol.* **121,** 1735–1743.

Dafters, R. I. (1995). Hyperthermia following MDMA administration in rats: Effects of ambient temperature, water consumption, and chronic dosing. *Physiol. Behav.* **58**(5), 877–882.

Dafters, R. I., and Lynch, E. (1998). Persistent loss of thermoregulation in the rat induced by 3,4-methylenedioxymethamphetamine (MDMA or "Ecstasy") but not by fenfluramine. *Psychopharmacology (Berl.)* **138**(2), 207–212.

de la Torre, R., Farre, M., Ortuno, J., Mas, M., Brenneisen, R., Roset, P. N., Segura, J., and Cami, J. (2000). Non-linear pharmacokinetics of MDMA ('ecstasy') in humans. *Br. J. Clin. Pharmacol.* **49,** 104–109.

de la Torre, R., Farre, M., Roset, P. N., Pizarro, N., Abanades, S., Segura, M., Segura, J., and Cami, J. (2004). Human pharmacology of MDMA: Pharmacokinetics, metabolism, and disposition. *Ther. Drug Monit.* **26**(2), 137–144.

Doblin, R. (2002). A clinical plan for MDMA (Ecstasy) in the treatment of posttraumatic stress disorder (PTSD): Partnering with the FDA. *J. Psychoactive Drugs* **34**(2), 185–194.

Droogmans, S., Cosyns, B., D'Haenen, H., Creeten, E., Weytjens, C., Franken, P. R., Scott, B., Schoors, D., Kemdem, A., Close, L., Vandenbossche, J. L., Bechet, S., *et al.* (2007). Possible association between 3,4-methylenedioxymethamphetamine abuse and valvular heart disease. *Am. J. Cardiol.* **100**(9), 442–445.

Escobedo, I., O'Shea, E., Orio, L., Sanchez, V., Segura, M., de la Torre, R., Farre, M., Green, A. R., and Colado, M. I. (2005). A comparative study on the acute and long-term effects of MDMA and 3,4-dihydroxymethamphetamine (HHMA) on brain monoamine levels after i.p. or striatal administration in mice. *Br. J. Pharmacol.* **144**(2), 231–241.

Finnegan, K. T., Ricaurte, G. A., Ritchie, L. D., Irwin, I., Peroutka, S. J., and Langston, J. W. (1988). Orally administered MDMA causes a long-term depletion of serotonin in rat brain. *Brain Res.* **447,** 141–144.

Fitzgerald, J. L., and Reid, J. J. (1993). Interactions of methylenedioxymethamphetamine with monoamine transmitter release mechanisms in rat brain slices. *Naunyn. Schmiedebergs Arch. Pharmacol.* **347**(3), 313–323.

Fitzgerald, J. L., and Reid, J. J. (1994). Sympathomimetic actions of methylenedioxymethamphetamine in rat and rabbit isolated cardiovascular tissues. *J. Pharm. Pharmacol.* **46,** 826–832.

Fitzgerald, L. W., Burn, T. C., Brown, B. S., Patterson, J. P., Corjay, M. H., Valentine, P. A., Sun, J. H., Link, J. R., Abbaszade, I., Hollis, J. M., Largent, B. L., Hartig, P. R., *et al.* (2000). Possible role of valvular serotonin 5-HT$_{2B}$ receptors in the cardiopathy associated with fenfluramine. *Mol. Pharmacol.* **57,** 75–81.

Fone, K. C., Beckett, S. R., Topham, I. A., Swettenham, J., Ball, M., and Maddocks, L. (2002). Long-term changes in social interaction and reward following repeated MDMA administration to adolescent rats without accompanying serotonergic neurotoxicity. *Psychopharmacology (Berl.)* **159**(4), 437–444.

Forsling, M. L., Fallon, J. K., Shah, D., Tilbrook, G. S., Cowan, D. A., Kicman, A. T., and Hutt, A. J. (2002). The effect of 3,4-methylenedioxymethamphetamine (MDMA, 'Ecstasy') and its metabolites on neurohypophysial hormone release from the isolated rat hypothalamus. *Br. J. Pharmacol.* **135**(3), 649–656.

Frazer, A., and Benmansour, S. (2002). Delayed pharmacological effects of antidepressants. *Mol. Psychiatry* **7**(Suppl. 1), S23–S28.

Gartside, S. E., McQuade, R., and Sharp, T. (1996). Effects of repeated administration of 3,4-methylenedioxymethamphetamine on 5-hydroxytryptamine neuronal activity and release in the rat brain *in vivo*. *J. Pharmacol. Exp. Ther.* **279**(1), 277–283.

Gerra, G., Zaimovic, A., Giucastro, G., Maestri, D., Monica, C., Sartori, R., Caccavari, R., and Delsignore, R. (1998). Serotonergic function after (+/−)3,4-methylene-dioxymethamphetamine ('Ecstasy') in humans. *Int. Clin. Psychopharmacol.* **13**(1), 1–9.

Gerra, G., Zaimovic, A., Ferri, M., Zambelli, U., Timpano, M., Neri, E., Marzocchi, G. F., Delsignore, R., and Brambilla, F. (2000). Long-lasting effects of (+/−)3,4-methylenedioxy-methamphetamine (ecstasy) on serotonin system function in humans. *Biol. Psychiatry* **47**(2), 127–136.

Geyer, M. A. (1996). Serotonergic functions in arousal and motor activity. *Behav. Brain Res.* **73**(1–2), 31–35.

Glennon, R. A., and Higgs, R. (1992). Investigation of MDMA-related agents in rats trained to discriminate MDMA from saline. *Pharmacol. Biochem. Behav.* **43**(3), 759–763.

Gold, L. H., Koob, G. F., and Geyer, M. A. (1988). Stimulant and hallucinogenic behavioral profiles of 3,4-methylenedioxymethamphetamine and N-ethyl-3,4-methylenedioxyamphetamine in rats. *J. Pharmacol. Exp. Ther.* **247**(2), 547–555.

Goodwin, J. S., Larson, G. A., Swant, J., Sen, N., Javitch, J. A., Zahniser, N. R., De Felice, L. J., and Khoshbouei, H. (2008). Amphetamine and methamphetamine differentially affect dopamine transporters *in vitro* and *in vivo*. *J. Biol. Chem.* **284**(5), 2978–2989.

Gouzoulis-Mayfrank, E., Becker, S., Pelz, S., Tuchtenhagen, F., and Daumann, J. (2002). Neuroendo-crine abnormalities in recreational ecstasy (MDMA) users: Is it ecstasy or cannabis? *Biol. Psychiatry* **51**(9), 766–769.

Green, A. R., Mechan, A. O., Elliott, J. M., O'Shea, E., and Colado, M. I. (2003). The pharmacology and clinical pharmacology of 3,4-methylenedioxymethamphetamine (MDMA, "ecstasy"). *Pharmacol. Rev.* **55**(3), 463–508.

Green, A. R., O'Shea, E., and Colado, M. I. (2004). A review of the mechanisms involved in the acute MDMA (Ecstasy)-induced hyperthermic response. *Eur. J. Pharmacol.* **500**(1–3), 3–13.

Gudelsky, G. A., and Nash, J. F. (1996). Carrier-mediated release of serotonin by 3,4-methylenediox-ymethamphetamine: Implications for serotonin-dopamine interactions. *J. Neurochem.* **66,** 243–249.

Gurtman, C. G., Morley, K. C., Li, K. M., Hunt, G. E., and McGregor, I. S. (2002). Increased anxiety in rats after 3,4-methylenedioxymethamphetamine: Association with serotonin depletion. *Eur. J. Pharmacol.* **446**(1–3), 89–96.

Hajos, M., and Sharp, T. (1996). A 5-hydroxytryptamine lesion markedly reduces the incidence of burst-firing dorsal raphe neurones in the rat. *Neurosci. Lett.* **204**(3), 161–164.

Halladay, A. K., Wagner, G. C., Sekowski, A., Rothman, R. B., Baumann, M. H., and Fisher, H. (2006). Alterations in alcohol consumption, withdrawal seizures, and monoamine transmission in rats treated with phentermine and 5-hydroxy-L-tryptophan. *Synapse* **59**(5), 277–289.

Hall, F. S., Devries, A. C., Fong, G. W., Huang, S., and Pert, A. (1999). Effects of 5,7-dihydroxy-tryptamine depletion of tissue serotonin levels on extracellular serotonin in the striatum assessed with *in vivo* microdialysis: Relationship to behavior. *Synapse* **33**(1), 16–25.

Harris, D. S., Baggott, M., Mendelson, J. H., Mendelson, J. E., and Jones, R. T. (2002). Subjective and hormonal effects of 3,4-methylenedioxymethamphetamine (MDMA) in humans. *Psychopharmacology (Berl.)* **162**(4), 396–405.

Hendrikx, M., Van Dorpe, J., Flameng, W., and Daenen, W. (1996). Aortic and mitral valve disease induced by ergotamine therapy for migraine: A case report and review of the literature. *J. Heart Valve Dis.* **5,** 235–237.

Johanson, C. E., Kilbey, M., Gatchalian, K., and Tancer, M. (2005). Discriminative stimulus effects of 3,4-methylenedioxymethamphetamine (MDMA) in humans trained to discriminate among d-amphetamine, meta-chlorophenylpiperazine and placebo. *Drug Alcohol Depend.* **81,** 27–36.

Johnson, M. P., Hoffman, A. J., and Nichols, D. E. (1986). Effects of the enantiomers of MDA, MDMA and related analogues on [3H]serotonin and [3H]dopamine release from superfused rat brain slices. *Eur. J. Pharmacol.* **132**(2–3), 269–276.

Kalia, M., O'Callaghan, J. P., Miller, D. B., and Kramer, M. (2000). Comparative study of fluoxetine, sibutramine, sertraline and dexfenfluramine on the morphology of serotonergic nerve terminals using serotonin immunohistochemistry. *Brain Res.* **858,** 92–105.

Kankaanpaa, A., Meririnne, E., Lillsunde, P., and Seppala, T. (1998). The acute effects of amphetamine derivatives on extracellular serotonin and dopamine levels in rat nucleus accumbens. *Pharmacol. Biochem. Behav.* **59**(4), 1003–1009.

Kirby, L. G., Kreiss, D. S., Singh, A., and Lucki, I. (1995). Effect of destruction of serotonin neurons on basal and fenfluramine-induced serotonin release in striatum. *Synapse* **20**(2), 99–105.

Kish, S. J. (2002). How strong is the evidence that brain serotonin neurons are damaged in human users of ecstasy? *Pharmacol. Biochem. Behav.* **71**(4), 845–855.

Kleiber, M. (1932). Body size and metabolism. *Hilgardia* **6,** 315–353.

Landry, M. J. (2002). MDMA: A review of epidemiologic data. *J. Psychoactive Drugs* **34**(2), 163–169.

Lavelle, A., Honner, V., and Docherty, J. R. (1999). Investigation of the prejunctional alpha2-adrenoceptor mediated actions of MDMA in rat atrium and vas deferens. *Br. J. Pharmacol.* **128**(5), 975–980.

Levy, A. D., Baumann, M. H., and Van de Kar, L. D. (1994). Monoaminergic regulation of neuroendocrine function and its modification by cocaine. *Front. Neuroendocrinol.* **15,** 85–156.

Liechti, M. E., and Vollenweider, F. X. (2001). Which neuroreceptors mediate the subjective effects of MDMA in humans? A summary of mechanistic studies. *Hum. Psychopharmacol.* **16**(8), 589–598.

Liechti, M. E., Baumann, C., Gamma, A., and Vollenweider, F. X. (2000). Acute psychological effects of 3,4-methylenedioxymethamphetamine (MDMA, "Ecstasy") are attenuated by the serotonin uptake inhibitor citalopram. *Neuropsychopharmacology* **22**(5), 513–521.

Lin, J. H. (1998). Applications and limitations of interspecies scaling and in vitro extrapolation in pharmacokinetics. *Drug Metab. Dispos.* **26,** 1202–1212.

Lopez-Ilasaca, M. (1998). Signaling from G-protein-coupled receptors to mitogen-activated protein (MAP)-kinase cascades. *Biochem. Pharmacol.* **56,** 269–277.

Lyles, J., and Cadet, J. L. (2003). Methylenedioxymethamphetamine (MDMA, Ecstasy) neurotoxicity: Cellular and molecular mechanisms. *Brain Res. Brain Res. Rev.* **42**(2), 155–168.

Lyon, R. A., Glennon, R. A., and Titeler, M. (1986). 3,4-Methylenedioxymethamphetamine (MDMA): Stereoselective interactions at brain 5-HT1 and 5-HT2 receptors. *Psychopharmacology (Berl.)* **88**(4), 525–526.

Mahmood, I. (1999). Allometric issues in drug development. *J. Pharm. Sci.* **88,** 1101–1106.

Malberg, J. E., and Seiden, L. S. (1998). Small changes in ambient temperature cause large changes in 3,4- methylenedioxymethamphetamine (MDMA)-induced serotonin neurotoxicity and core body temperature in the rat. *J. Neurosci.* **18,** 5086–5094.

Malpass, A., White, J. M., Irvine, R. J., Somogyi, A. A., and Bochner, F. (1999). Acute toxicity of 3,4-methylenedioxymethamphetamine (MDMA) in Sprague-Dawley and Dark Agouti rats. *Pharmacol. Biochem. Behav.* **64**(1), 29–34.

Marston, H. M., Reid, M. E., Lawrence, J. A., Olverman, H. J., and Butcher, S. P. (1999). Behavioural analysis of the acute and chronic effects of MDMA treatment in the rat. *Psychopharmacology (Berl.)* **144**(1), 67–76.

Mas, M., Farre, M., de la Torre, R., Roset, P. N., Ortuno, J., Segura, J., and Cami, J. (1999). Cardiovascular and neuroendocrine effects and pharmacokinetics of 3, 4- methylenedioxymethamphetamine in humans. *J. Pharmacol. Exp. Ther.* **290,** 136–145.

Matuszewich, L., Filon, M. E., Finn, D. A., and Yamamoto, B. K. (2002). Altered forebrain neurotransmitter responses to immobilization stress following 3,4-methylenedioxymethamphetamine. *Neuroscience* **110**(1), 41–48.

Maurer, H. H., Bickeboeller-Friedrich, J., Kraemer, T., and Peters, F. T. (2000). Toxicokinetics and analytical toxicology of amphetamine-derived designer drugs ('Ecstasy'). *Toxicol. Lett.* **112–113,** 133–142.

McCann, U. D., Szabo, Z., Vranesic, M., Palermo, M., Mathews, W. B., Ravert, H. T., Dannals, R. F., and Ricaurte, G. A. (2008). Positron emission tomographic studies of brain dopamine and serotonin transporters in abstinent (+/−)3,4-methylenedioxymethamphetamine ("ecstasy") users: Relationship to cognitive performance. *Psychopharmacology (Berl.)* **200**(3), 439–450.

McGregor, I. S., Clemens, K. J., Van der Plasse, G., Li, K. M., Hunt, G. E., Chen, F., and Lawrence, A. J. (2003). Increased anxiety 3 months after brief exposure to MDMA ("Ecstasy") in rats: Association with altered 5-HT transporter and receptor density. *Neuropsychopharmacology* **28**(8), 1472–1484.

McNamara, M. G., Kelly, J. P., and Leonard, B. E. (1995). Some behavioural and neurochemical aspects of subacute (+/−)3,4-methylenedioxymethamphetamine administration in rats. *Pharmacol. Biochem. Behav.* **52**(3), 479–484.

Mechan, A. O., Esteban, B., O'Shea, E., Elliott, J. M., Colado, M. I., and Green, A. R. (2002). The pharmacology of the acute hyperthermic response that follows administration of 3,4-methylenedioxymethamphetamine (MDMA, 'ecstasy') to rats. *Br. J. Pharmacol.* **135**(1), 170–180.

Mechan, A., Yuan, J., Hatzidimitriou, G., Irvine, R. J., McCann, U. D., and Ricaurte, G. A. (2006). Pharmacokinetic profile of single and repeated oral doses of MDMA in squirrel monkeys: Relationship to lasting effects on brain serotonin neurons. *Neuropsychopharmacology* **31**(2), 339–350.

Molliver, M. E., Berger, U. V., Mamounas, L. A., Molliver, D. C., OHearn, E., and Wilson, M. A. (1990). Neurotoxicity of MDMA and related compounds: Anatomic studies. *Ann. NY Acad. Sci.* **600,** 649–661(discuss).

Monks, T. J., Jones, D. C., Bai, F., and Lau, S. S. (2004). The role of metabolism in 3,4-(+)-methylenedioxyamphetamine and 3,4-(+)-methylenedioxymethamphetamine (ecstasy) toxicity. *Ther. Drug Monit.* **26**(2), 132–136.

Morgan, M. J. (2000). Ecstasy (MDMA): A review of its possible persistent psychological effects. *Psychopharmacology (Berl.)* **152**(3), 230–248.

Morley, K. C., Gallate, J. E., Hunt, G. E., Mallet, P. E., and McGregor, I. S. (2001). Increased anxiety and impaired memory in rats 3 months after administration of 3,4-methylenedioxymethamphetamine ("ecstasy"). *Eur. J. Pharmacol.* **433**(1), 91–99.

Moser, V. C. (2000). The functional observational battery in adult and developing rats. *Neurotoxicology* **21**(6), 989–996.

Mueller, M., Peters, F. T., Maurer, H. H., McCann, U. D., and Ricaurte, G. A. (2008). Nonlinear pharmacokinetics of (+/−)3,4-methylenedioxymethamphetamine (MDMA, "Ecstasy") and its major metabolites in squirrel monkeys at plasma concentrations of MDMA that develop after typical psychoactive doses. *J. Pharmacol. Exp. Ther.* **327**(1), 38–44.

Nash, J. F., Jr., Meltzer, H. Y., and Gudelsky, G. A. (1988). Elevation of serum prolactin and corticosterone concentrations in the rat after the administration of 3,4-methylenedioxymethamphetamine. *J. Pharmacol. Exp. Ther.* **245**(3), 873–879.

Nichols, D. E., Lloyd, D. H., Hoffman, A. J., Nichols, M. B., and Yim, G. K. (1982). Effects of certain hallucinogenic amphetamine analogues on the release of [3H]serotonin from rat brain synaptosomes. *J. Med. Chem.* **25**(5), 530–535.

Noujaim, S. F., Lucca, E., Munoz, V., Persaud, D., Berenfeld, O., Meijler, F. L., and Jalife, J. (2004). From mouse to whale: A universal scaling relation for the PR Interval of the electrocardiogram of mammals. *Circulation* **110**(18), 2802–2808.

Oberlender, R., and Nichols, D. E. (1988). Drug discrimination studies with MDMA and amphetamine. *Psychopharmacology (Berl.)* **95**(1), 71–76.

O'Cain, P. A., Hletko, S. B., Ogden, B. A., and Varner, K. J. (2000). Cardiovascular and sympathetic responses and reflex changes elicited by MDMA. *Physiol. Behav.* **70**(1–2), 141–148.

O'Hearn, E., Battaglia, G., De Souza, E. B., Kuhar, M. J., and Molliver, M. E. (1988). Methylene-dioxyamphetamine (MDA) and methylenedioxymethamphetamine (MDMA) cause selective ablation of serotonergic axon terminals in forebrain: Immunocytochemical evidence for neuro-toxicity. *J. Neurosci.* **8**(8), 2788–2803.

O'Shea, E., Granados, R., Esteban, B., Colado, M. I., and Green, A. R. (1998). The relationship between the degree of neurodegeneration of rat brain 5-HT nerve terminals and the dose and frequency of administration of MDMA ('ecstasy'). *Neuropharmacology* **37**(7), 919–926.

Parrott, A. C. (2002). Recreational Ecstasy/MDMA, the serotonin syndrome, and serotonergic neurotoxicity. *Pharmacol. Biochem. Behav.* **71**(4), 837–844.

Parrott, A. C. (2004). Is ecstasy MDMA? A review of the proportion of ecstasy tablets containing MDMA, their dosage levels, and the changing perceptions of purity. *Psychopharmacology (Berl.)* **173**(3–4), 234–241.

Parrott, A. C. (2005). Chronic tolerance to recreational MDMA (3,4-methylenedioxymethampheta-mine) or Ecstasy. *J. Psychopharmacol.(Oxford)* **19**(1), 71–83.

Partilla, J. S., Dersch, C. M., Yu, H., Rice, K. C., and Rothman, R. B. (2000). Neurochemical neutralization of amphetamine-type stimulants in rat brain by the indatraline analog (−)-HY038. *Brain Res. Bull.* **53**(6), 821–826.

Pifl, C., Wolf, A., Rebernik, P., Reither, H., and Berger, M. L. (2009). Zinc regulates the dopamine transporter in a membrane potential and chloride dependent manner. *Neuropharmacology* **56**(2), 531–540.

Poland, R. E. (1990). Diminished corticotropin and enhanced prolactin responses to 8-hydroxy-2(di-n-propylamino)tetralin in methylenedioxymethamphetamine pretreated rats. *Neuropharmacology* **29** (11), 1099–1101.

Poland, R. E., Lutchmansingh, P., McCracken, J. T., Zhao, J. P., Brammer, G. L., Grob, C. S., Boone, K. B., and Pechnick, R. N. (1997). Abnormal ACTH and prolactin responses to fenflur-amine in rats exposed to single and multiple doses of MDMA. *Psychopharmacology (Berl.)* **131**(4), 411–419.

Porter, R. H., Benwell, K. R., Lamb, H., Malcolm, C. S., Allen, N. H., Revell, D. F., Adams, D. R., and Sheardown, M. J. (1999). Functional characterization of agonists at recombinant human 5-HT$_{2A}$, 5-HT$_{2B}$ and 5-HT$_{2C}$ receptors in CHO-K1 cells. *Br. J. Pharmacol.* **128,** 13–20.

Ratzenboeck, E., Saria, A., Kriechbaum, N., and Zernig, G. (2001). Reinforcing effects of MDMA ("ecstasy") in drug-naive and cocaine-trained rats. *Pharmacology* **62**(3), 138–144.

Reneman, L. (2003). Designer drugs: How dangerous are they? *J. Neural Transm. Suppl.* **66,** 61–83.

Ricaurte, G. A., Markowska, A. L., Wenk, G. L., Hatzidimitriou, G., Wlos, J., and Olton, D. S. (1993). 3,4-Methylenedioxymethamphetamine, serotonin and memory. *J. Pharmacol. Exp. Ther.* **266,** 1097–1105.

Ricaurte, G. A., Yuan, J., and McCann, U. D. (2000). (+/−)3,4-Methylenedioxymethamphetamine ('Ecstasy')-induced serotonin neurotoxicity: Studies in animals. *Neuropsychobiology* **42**(1), 5–10.

Robinson, T. E., Castaneda, E., and Whishaw, I. Q. (1993). Effects of cortical serotonin depletion induced by 3,4-methylenedioxymethamphetamine (MDMA) on behavior, before and after addi-tional cholinergic blockade. *Neuropsychopharmacology* **8**(1), 77–85.

Romero, L., Jernej, B., Bel, N., Cicin-Sain, L., Cortes, R., and Artigas, F. (1998). Basal and stimulated extracellular serotonin concentration in the brain of rats with altered serotonin uptake. *Synapse* **28**(4), 313–321.

Roth, B. L. (2007). Drugs and valvular heart disease. *N. Engl. J. Med.* **356**(1), 6–9.

Rothman, R. B., and Baumann, M. H. (2006). Therapeutic potential of monoamine transporter substrates. *Curr. Top. Med. Chem.* **6**(17), 1845–1859.

Rothman, R. B., Baumann, M. H., Savage, J. E., Rauser, L., McBride, A., Hufisein, S., and Roth, B. L. (2000). Evidence for possible involvement of 5-HT$_{2B}$ receptors in the cardiac valvulopathy associated with fenfluramine and other serotonergic medications. *Circulation* **102,** 2836–2841.

Rothman, R. B., Baumann, M. H., Dersch, C. M., Romero, D. V., Rice, K. C., Carroll, F. I., and Partilla, J. S. (2001). Amphetamine-type central nervous system stimulants release norepinephrine more potently than they release dopamine and serotonin. *Synapse* **39,** 32–41.

Rudnick, G., and Clark, J. (1993). From synapse to vesicle: The reuptake and storage of biogenic amine neurotransmitters. [Review]. *Biochim. Biophys. Acta* **1144,** 249–263.

Scanzello, C. R., Hatzidimitriou, G., Martello, A. L., Katz, J. L., and Ricaurte, G. A. (1993). Serotonergic recovery after (+/−)3,4-(methylenedioxy) methamphetamine injury: Observations in rats. *J. Pharmacol. Exp. Ther.* **264,** 1484–1491.

Schechter, M. D. (1988). Serotonergic-dopaminergic mediation of 3,4-methylenedioxymethamphetamine (MDMA, "ecstasy"). *Pharmacol. Biochem. Behav.* **31**(4), 817–824.

Schenk, S., Gittings, D., Johnstone, M., and Daniela, E. (2003). Development, maintenance and temporal pattern of self-administration maintained by ecstasy (MDMA) in rats. *Psychopharmacology (Berl)* **169**(1), 21–27.

Schifano, F. (2004). A bitter pill. Overview of ecstasy (MDMA, MDA) related fatalities. *Psychopharmacology (Berl.)* **173**(3–4), 242–248.

Schmidt, C. J. (1987). Neurotoxicity of the psychedelic amphetamine, methylenedioxymethamphetamine. *J. Pharmacol. Exp. Ther.* **240**(1), 1–7.

Schmidt, C. J., Levin, J. A., and Lovenberg, W. (1987). In vitro and in vivo neurochemical effects of methylenedioxymethamphetamine on striatal monoaminergic systems in the rat brain. *Biochem. Pharmacol.* **36**(5), 747–755.

Seiden, L. S., Woolverton, W. L., Lorens, S. A., Williams, J. E., Corwin, R. L., Hata, N., and Olimski, M. (1993). Behavioral consequences of partial monoamine depletion in the CNS after methamphetamine-like drugs: The conflict between pharmacology and toxicology. *NIDA Res. Monogr.* **136,** 34–46(discussion 46–52).

Selken, J., and Nichols, D. E. (2007). Alpha1-adrenergic receptors mediate the locomotor response to systemic administration of (+/−)-3,4-methylenedioxymethamphetamine (MDMA) in rats. *Pharmacol. Biochem. Behav.* **86**(4), 622–630.

Series, H. G., Cowen, P. J., and Sharp, T. (1994). *p*-Chloroamphetamine (PCA), 3,4-methylenedioxymethamphetamine (MDMA) and d-fenfluramine pretreatment attenuates d-fenfluramine-evoked release of 5-HT *in vivo. Psychopharmacology (Berl.)* **116,** 508–514.

Series, H. G., le Masurier, M., Gartside, S. E., Franklin, M., and Sharp, T. (1995). Behavioural and neuroendocrine responses to d-fenfluramine in rats treated with neurotoxic amphetamines. *J. Psychopharmacol.* **9,** 214–222.

Setola, V., and Roth, B. L. (2005). Screening the receptorome reveals molecular targets responsible for drug-induced side effects: Focus on 'fen-phen'. *Expert Opin. Drug Metab. Toxicol.* **1**(3), 377–387.

Setola, V., Hufeisen, S. J., Grande-Allen, K. J., Vesely, I., Glennon, R. A., Blough, B., Rothman, R. B., and Roth, B. L. (2003). 3,4-Methylenedioxymethamphetamine (MDMA, "Ecstasy") induces fenfluramine-like proliferative actions on human cardiac valvular interstitial cells in vitro. *Mol. Pharmacol.* **63**(6), 1223–1229.

Seuwen, K., Magnaldo, I., and Pouyssegur, J. (1988). Serotonin stimulates DNA synthesis in fibroblasts acting through 5-HT1B receptors coupled to a Gi-protein. *Nature* **335,** 254–256.

Shankaran, M., and Gudelsky, G. A. (1999). A neurotoxic regimen of MDMA suppresses behavioral, thermal and neurochemical responses to subsequent MDMA administration. *Psychopharmacology (Berl)* **147**(1), 66–72.

Slikker, W. Jr., Holson, R. R., Ali, S. F., Kolta, M. G., Paule, M. G., Scallet, A. C., McMillan, D. E., Bailey, J. R., Hong, J. S., and Scalzo, F. M. (1989). Behavioral and neurochemical effects of orally administered MDMA in the rodent and nonhuman primate. *Neurotoxicology* **10**(3), 529–542.

Spanos, L. J., and Yamamoto, B. K. (1989). Acute and subchronic effects of methylenedioxymethamphetamine [(+/−)MDMA] on locomotion and serotonin syndrome behavior in the rat. *Pharmacol. Biochem. Behav.* **32,** 835–840[Published erratum appears in *Pharmacol. Biochem. Behav.* 1989 Nov;34 (3):679].

Sprague, J. E., Everman, S. L., and Nichols, D. E. (1998). An integrated hypothesis for the serotonergic axonal loss induced by 3,4-methylenedioxymethamphetamine. *Neurotoxicology* **19,** 427–441.

Sprouse, J. S., Bradberry, C. W., Roth, R. H., and Aghajanian, G. K. (1989). MDMA (3,4-methylenedioxymethamphetamine) inhibits the firing of dorsal raphe neurons in brain slices via release of serotonin. *Eur. J. Pharmacol.* **167**(3), 375–383.

Starr, M. A., Page, M. E., and Waterhouse, B. D. (2008). MDMA (3,4-methylenedioxymethamphetamine)-mediated distortion of somatosensory signal transmission and neurotransmitter efflux in the ventral posterior medial thalamus. *J. Pharmacol. Exp. Ther.* **327**(1), 20–31.

Steinbusch, H. W. (1981). Distribution of serotonin-immunoreactivity in the central nervous system of the rat-cell bodies and terminals. *Neuroscience* **6,** 557–618.

Stone, D. M., Merchant, K. M., Hanson, G. R., and Gibb, J. W. (1987). Immediate and long-term effects of 3,4-methylenedioxymethamphetamine on serotonin pathways in brain of rat. *Neuropharmacology* **26**(12), 1677–1683.

Tancer, M., and Johanson, C. E. (2003). Reinforcing, subjective, and physiological effects of MDMA in humans: A comparison with d-amphetamine and mCPP. *Drug Alcohol Depend.* **72**(1), 33–44.

Thomasius, R., Zaplova, P., Petersen, K., Buchert, R., Andresen, B., Wartberg, L., Nebeling, B., and Schmoldt, A. (2006). Mood, cognition and serotonin transporter availability in current and former ecstasy (MDMA) users: The longitudinal perspective. *J. Psychopharmacol. (Oxford, England)* **20**(2), 211–225.

Van de Kar, L. D. (1991). Neuroendocrine pharmacology of serotonergic (5-HT) neurons. *Annu. Rev. Pharmacol. Toxicol.* **31,** 289–320.

Verheyden, S. L., Henry, J. A., and Curran, H. V. (2003). Acute, sub-acute and long-term subjective consequences of 'ecstasy' (MDMA) consumption in 430 regular users. *Hum. Psychopharmacol.* **18**(7), 507–517.

Vollenweider, F. X., Gamma, A., Liechti, M., and Huber, T. (1998). Psychological and cardiovascular effects and short-term sequelae of MDMA ("ecstasy") in MDMA-naive healthy volunteers. *Neuropsychopharmacology* **19,** 241–251.

Wakonigg, G., Sturm, K., Saria, A., and Zernig, G. (2003). Methylenedioxymethamphetamine (MDMA, 'ecstasy') serves as a robust positive reinforcer in a rat runway procedure. *Pharmacology* **69**(4), 180–182.

Wang, X., Baumann, M. H., Dersch, C. M., and Rothman, R. B. (2007). Restoration of 3,4-methylenedioxymethamphetamine-induced 5-HT depletion by the administration of l-5-hydroxytryptophan. *Neuroscience* **148**(1), 212–220.

West, G. B., Woodruff, W. H., and Brown, J. H. (2002). Allometric scaling of metabolic rate from molecules and mitochondria to cells and mammals. *Proc. Natl. Acad. Sci. USA* **99**(Suppl. 1), 2473–2478.

White, C. R., and Seymour, R. S. (2005). Allometric scaling of mammalian metabolism. *J. Exp. Biol.* **208**(Pt 9), 1611–1619.

Winneke, G., and Lilienthal, H. (1992). Extrapolation from animals to humans: Scientific and regulatory aspects. *Toxicol. Lett.* **64–65**(Spec No: 239–246).

Wu, D., Otton, S. V., Inaba, T., Kalow, W., and Sellers, E. M. (1997). Interactions of amphetamine analogs with human liver CYP2D6. *Biochem. Pharmacol.* **53**(11), 1605–1612.

Yacoubian, G. S. Jr (2003). Tracking ecstasy trends in the United States with data from three national drug surveillance systems. *J. Drug Educ.* **33**(3), 245–258.

Yamamoto, B. K., Nash, J. F., and Gudelsky, G. A. (1995). Modulation of methylenedioxymetham-phetamine-induced striatal dopamine release by the interaction between serotonin and gamma-aminobutyric acid in the substantia nigra. *J. Pharmacol. Exp. Ther.* **273**(3), 1063–1070.

Zigmond, M. J., Abercrombie, E. D., Berger, T. W., Grace, A. A., and Stricker, E. M. (1990). Compensations after lesions of central dopaminergic neurons: Some clinical and basic implications. *Trends Neurosci.* **13**(7), 290–296.

COCAINE-INDUCED BREAKDOWN OF THE BLOOD–BRAIN BARRIER AND NEUROTOXICITY

Hari S. Sharma,* Dafin Muresanu,[†] Aruna Sharma,* and Ranjana Patnaik[‡]

*Laboratory of Cerebrovascular Research & Pain Research Laboratory, Department
of Surgical Sciences, Anesthesiology & Intensive Care Medicine,
University Hospital, Uppsala University, SE-75185 Uppsala, Sweden
[†]Department of Neurology, Medical University of Cluj-Napoca, Cluj-Napoca, Romania
[‡]Department of Biomaterials, School of Biomedical Engineering, Institute of Technology,
Banaras Hindu University, Varanasi 2210005, India

INTERNATIONAL REVIEW OF
NEUROBIOLOGY, VOL. 88
DOI: 10.1016/S0074-7742(09)88011-2

297

Role of cocaine in influencing blood–brain barrier (BBB) function is still unknown. Available evidences suggest that cocaine administration results in acute hyperthermia and alterations in brain serotonin metabolism. Since hyperthermia is capable to induce the breakdown of the BBB either directly or through altered serotonin metabolism, a possibility exists that cocaine may induce neurotoxicity by causing BBB disruption. This hypothesis is discussed in this review largely based on our own laboratory investigations. Our observations in rats demonstrate that cocaine depending on the dose and routes of administration induces profound hyperthermia, increased plasma and brain serotonin levels leading to BBB breakdown and brain edema formation. Furthermore, cocaine was able to enhance cellular stress as seen by upregulation of heat shock protein (HSP 72 kD) expression and resulted in marked neuronal and glial cell damages at the time of the BBB dysfunction. Taken together, these observations are the first to suggest that cocaine-induced BBB disruption is instrumental in precipitating brain pathology. The possible mechanisms of cocaine-induced BBB breakdown and neurotoxicity are discussed.

I. Introduction

Cocaine is a powerful psychostimulant that affects the functions of cardiovascular and the central nervous system (CNS) profoundly depending on its dose and routes of the administration (Brecklin and Bauman, 1999; Kessler *et al.*, 2007; Killam, 1993; Majewska, 1996; Phillips *et al.*, 2009; Sim *et al.*, 2007; Tolat *et al.*, 2000). Repeated use of cocaine leads to dependence and is often associated with severe psychiatric and other behavioral symptoms including death (Pozzi *et al.*, 2008; Shanti and Lucas, 2003; Walsh *et al.*, 2009). One of the important reactions of mild-to-moderate doses of cocaine is induction of profound hyperthermia (Lomax and Daniel, 1990a,b,c) followed by mental disorders such as extreme anxiety, agitation, and hallucination associated with nausea, vomiting, and serious respiratory disturbances (Benowitz, 1992; Daras *et al.*, 1995; De Giorgio *et al.*, 2007; Kleerup *et al.*, 1997; Marzuk *et al.*, 1998; Mott *et al.*, 1994; Tanvetyanon *et al.*, 2001). These symptoms are followed by severe depression, suppression of respiration, convulsion, coma, and eventually death (Castellani *et al.*, 1978a,b; Dandekar *et al.*, 2009; de Wit *et al.*, 2008; Lesse and Collins, 1979; Matsuzaki *et al.*, 1976; Post and Kopanda, 1976). Prolonged use of cocaine induces psychomotor disturbances, suicidal tendency, anorexia, weight loss, and mental deterioration (see Amon *et al.*, 1986; Fowler *et al.*, 1986; Jonas *et al.*, 1987; Rosenberg *et al.*, 1986; Spiehler Reed, 1985; Vicentic and Jones, 2007).

Interestingly, withdrawal of cocaine also results in severe psychiatric and other problems including depression, headaches abdominal pain, chest pain, tachycardia, and collapse (Amon et al., 1986; Denton et al., 2008; Guan et al., 2009; Kosten and Kleber, 1988; Su et al., 2003). Overdose of cocaine is believed to facilitate stroke or transient ischemic attack and results in intracerebral or subarachnoid hemorrhages (Cregler and Mark, 1987; Hardebo and Hindfelt, 1981; Levine and Welch, 1988; Levine et al., 1987; Wojak and Flamm, 1987). In cocaine users, spontaneous cerebral hemorrhage can occur even in normotensive subjects (see Daras et al., 1991; Wojak and Flamm, 1987). However, the details mechanisms of cocaine-induced brain dysfunction or neurotoxicity are still not well understood.

Since cocaine induces profound hyperthermia and increased serotonin turn-over (Hardebo and Hindfelt, 1981), a possibility exists that the psychostimulant may precipitate its neurotoxic effects through modifying the BBB function (see Kiyatkin and Sharma, 2009; Kiyatkin et al., 2007; Sharma and Ali, 2006, 2008; Sharma and Kiyatkin, 2009; Sharma et al., 2007). This idea is supported by our previous works showing that hyperthermia induced by heat stress is able to induce marked BBB breakdown leading to brain edema formation, cerebral hemorrhages, and brain damage in rat models (Sharma, 2005, 2006a,b, 2007a; Sharma and Hoopes, 2003; Sharma et al., 2006a). In human cases, profound hyperthermia beyond 41 °C often results in abdominal pain, respiratory disturbances, nausea, vomiting, loss of consciousness, coma, convulsion, and eventually death (see Argaud et al., 2007; Maron et al., 2009; Pengelly et al., 2007). These observations are in line with the idea that hyperthermia caused by cocaine administration may contribute to most of the CNS disorders either directly or through modification of the BBB function.

A breakdown of the BBB to large molecules, such as proteins could precipitate vasogenic brain edema and induce brain damage (Sharma, 2006a, 2007a, 2008; Sharma et al., 2006a,b, 2007). Thus, a possibility exist that cocaine-induced hyper-thermia could cause BBB breakdown either directly or through alterations in neuro-chemical metabolism, for example, serotonin resulting in neurotoxicity (Mueller et al., 2009; Wallinga et al., 2009). This review is focused on cocaine-induced BBB breakdown and neurotoxicity largely based on our own laboratory investigations. In addition, the role of serotonin the well-known neurochemical mediator of BBB breakdown and edema formation is also examined in cocaine-induced neurotoxicity.

II. Cocaine Facilitates Blood–Brain Barrier Transport to HIV-1 in the Brain

Several observations in the past suggest that cocaine abuse is often associated with vasculitis and stroke (Krendel et al., 1990; Peterson et al., 1991; Treadwell and Robinson, 2007). Thus, there are possibilities that the psychostimulant could be associated with progression of AIDS dementia by facilitating the transport of HIV-1 across the BBB (Dhillon *et al.*, 2007, 2008; Fiala *et al.*, 2005; Kenedi *et al.*, 2008). However, to our knowledge, influence of cocaine on the breakdown of the BBB *in vivo* situation is still not known.

Using cell culture studies, Zhang *et al.* (1998) showed that cocaine in a concentration of 10^{-5} and 10^{-6} M is able to enhance the human brain endothelial cell membrane permeability to HIV-1. This concentration of cocaine also induced apoptosis of brain endothelial cells and monocytes (Fiala *et al.*, 2005; Gan *et al.*, 1999; Zhang *et al.*, 1998). Furthermore that psychostimulant enhanced the secretion of chemokines, for example, interleukin-8, interferon-inducible protein-10, macrophage inflammatory protein-1a, and monocyte chemoattractant protein-1 and cytokine, TNF-α by human monocytes (Dey and Snow, 2007; Irwin *et al.*, 2007; Zhang *et al.*, 1998). It is believed that TNF-α is primarily responsible for enhanced invasion of HIV-1 strains across the endothelial cell membrane (see Sharma, 2009; Sharma and Sharma, 2008; Sharma *et al.*, 2003; Wilson *et al.*, 2009). These observations suggest that cocaine may directly affect the brain endothelial cell membrane permeability *in vitro* situations and could also facilitate tracer transfer across the BBB indirectly through stimulating proinflammatory cytokines and/or chemokines (see Sharma, 2009). However, a direct demonstration of cocaine-induced BBB leakage *in vivo* situation is still lacking and requires further investigation.

III. Cocaine and Cytotoxicity

Several *in vitro* studies demonstrated cytotoxic effects of cocaine in non-neural cells, for example, liver, aorta, adrenal, and kidney probably through generation of free radicals (Barroso-Moguel *et al.*, 1995; Fan *et al.*, 2009; Jover *et al.*, 1993; Lepsch *et al.*, 2009; Powers *et al.*, 1992; Schreiber and Formal, 2007; Welder, 1992; Yao *et al.*, 2009). Based on these results, it appears that cocaine could also induce neurotoxicity in the CNS using identical mechanisms. However, a direct demonstration of cocaine-induced neurotoxicity *in vivo* situations is still not available.

The cytotoxic effects of cocaine are largely mediated through its sympathomimetic actions as well as Ca^{2+}-related mechanisms (Billman, 1995). Thus, cocaine by blocking the reuptake of noradrenaline into the sympathetic nerve terminal leads to its accumulation that could overactivate the physiological receptor activity (del Castillo *et al.*, 2009). Furthermore, cocaine-induced depletion of glutathione with an increase in lipid peroxidation and production of free radicals and other toxic tissue metabolites could also contribute to cellular toxicity (Hahn *et al.*, 2009; Schreiber and Formal, 2007). This is further supported by the fact that flavin containing or cytochrome P450 catalyzed monooxygenase is known to catalyze pathways that generates toxic metabolites of cocaine (Boelsterli and Göldlin, 1991). Available evidences suggest that microsomal oxidation of cocaine results in production of pharmacologically active metabolite, N-hydroxynorcocaine in both animal experiments and human cases that is further catalyzed to potent free radical norcocaine nitroxide damaging cells and tissues (Boyer $et\ al.$, 1988). Thus, it appears that cocaine could have powerful toxic effects on cells and tissue through various mechanisms. It is quite likely that using similar mechanisms cocaine could induce profound neurotoxicity in the CNS $in\ vivo$ situations.

IV. Cocaine Affects Astrocytic Functions

Astrocytes are non-neural cells whose numbers far exceed than neurons in the CNS (see Sharma, 2007b). The glial cells play important roles in neuromodulation and neuroprotection (see Fellin, 2009). Thus, under situation of stress caused by trauma or ischemia they express glial fibrillary acidic protein (GFAP), a cytoskeletal intermediate filament protein that is exclusively expressed in the astrocytes (Bignami, 1984; Bignami and Dahl, 1988). Previous reports from our laboratory showed that acute methamphetamine or morphine exposure in rats results in activation of GFAP in different brain regions (Kiyatkin and Sharma, 2009; Sharma and Ali, 2006, 2008; Sharma and Kiyatkin, 2009; Sharma $et\ al.$, 1992a). However, role of cocaine in influencing astrocytic functions is not well known. Fattore $et\ al.$ (2002) using an $in\ vivo$ exposure of cocaine (20 mg/kg, i.p.) in mice showed massive upregulation of GFAP in dentate gyrus 24 h after single cocaine administration. This effect was lasted until 14 days. Morphological and morphometric analysis by the authors further show that cocaine was able to alter the shape, size, and numbers significantly as compared to saline controls (Fattore $et\ al.$, 2002; see Bowers and Kalivas, 2003). It appears that cocaine-induced neurotoxicity and/or oxidative stress in astrocytes could result in overactivation of GFAP and altered morphology of astrocyte (Kiyatkin and Sharma, 2009; Kiyatkin $et\ al.$, 2007; Korbo, 1999; Sharma and Kiyatkin, 2009; Sharma $et\ al.$, 2007). However, it would be interesting to see whether other brain regions are also affected by acute cocaine toxicity.

V. Cocaine Induces Stress Response and Upregulates Heat Shock Protein Expression

Available evidences suggest that acute cocaine administration is capable to induce stress response by activating hypothalamic–pituitary–adrenal (HPA) axis (Levy $et\ al.$, 1994a; Mendelson $et\ al.$, 2002). It appears that cocaine-induced activation of HPA axis is largely due to augmentation of hypothalamic mono-amine neurotransmission leading to increased circulation of ACTH resulting in subsequent enhancement of corticosterone levels in both animal and human studies (see Blake *et al.*, 1994). The heat shock proteins (HSPs) are a group of cellular stress response proteins that are expressed in almost all cell types following several types of environmental, chemical, and hormonal stressors (see Sharma and Westman, 2003; Sharma *et al.*, 2006c; Welch, 1992; Westman and Sharma, 1998; Westman *et al.*, 2000). Thus, Blake *et al.* (1994) was the first to examine cocaine-induced HSP expression in aorta and adrenal glands in the rat. Their investigations show that cocaine (15 mg/kg, i.p.) induces expression of HSP 70 by activating the transcription and regulatory factors, mRNA and proteins following 24 h after the initial injection (Blake *et al.*, 1994). This increase in HSP expression by cocaine was regulated by ACTH in the adrenal gland, as hypophysectomy completely eliminated this activation. However, HSP expression in aorta could be due to direct effects of cocaine, as this was not affected by hypophysectomy. This suggests that cocaine could induce HSP activation either directly or indirectly through neurohormones in various organs and cell types. It appears that seroto-nin could also influence HSP expression caused by cocaine. Thus, reserpine, a serotonin releaser compound in fact enhanced the cocaine-induced HSP expres-sion in aorta. This suggests that a direct action of cocaine on vascular tissue may also lead to activation of HSP 70 gene (Blake *et al.*, 1994). However, it is not known whether cocaine can influence brain HSP expression by inducing BBB breakdown that could result due to severe neuronal and/or glial stress. Thus, it is quite likely that cocaine could also activate brain HSP expression, a subject that requires additional investigations.

VI. Cocaine Influences Serotonin Metabolism

Acute or chronic use of cocaine results in widespread alterations in serotonin metabolism in the brain (Jaehne *et al.*, 2007; Jean *et al.*, 2007). Experiments carried out in mice and rats showed that single administration of cocaine (20–40 mg/kg) through intraperitoneal or intravenous routes is associated with 200–400% increase in serotonin release in the striatum (Levy *et al.*, 1994b). Alterations in serotonin transporters and increased densities of serotonin uptake sites are known to occur during chronic cocaine administration (Cunningham *et al.*, 1992). These serotonin transporter sites are responsible for synaptic and plasma clearance of serotonin. Thus, binding of cocaine with highest affinity to these serotonin transporters will elevate the synaptic serotonin levels (Pitts and Marwah, 1987). Increased serotonin concentrations in the brain could lead to BBB breakdown (Sharma *et al.*, 1990). Furthermore, release of serotonin in the brain could also induce hyperthermia (Capela *et al.*, 2009; Sharma, 2007a). Taken together, it appears that cocaine could influence hyperthermia probably via elevation of brain serotonin levels that in turn may affect the BBB breakdown either directly or through specific serotonin receptors-mediated mechanisms (Sharma *et al.*, 1990).

VII. Cocaine Affects Cerebral Circulation

Influence of cocaine on the cerebral blood flow (CBF) is still controversial (Copersino *et al.*, 2009). Thus, few studies report decrease in cortical blood flow by cocaine whereas, some studies showed an increase in the regional or subcortical blood flow following cocaine administration in rodents (Ceolin *et al.*, 2007; Schmidt *et al.*, 2006). It appears that cocaine-induced increase in serotonin levels in the brain plays important roles in influencing cerebral circulation (Bonvento *et al.*, 1991). The brain microvasculature is equipped with various serotonin receptors that are activated by exogenous or endogenous serotonin concentrations (Masu *et al.*, 2008). Some serotonin receptors are involved in vasodilatation and some produces vasoconstriction (Bonvento *et al.*, 1991). Thus, the effects of serotonin of the microcirculation depend on the vessel diameter and the type of receptors present (Sharma *et al.*, 1990). It is likely that cocaine can induce vasodilatation and/or vasoconstriction on cerebral microvessels leading to increase in the regional CBF in some areas whereas a decrease can also be seen in another brain regions. However, further studies on cocaine-induced alterations in regional CBF in relation to brain function are needed.

VIII. Cocaine-Induced Hyperthermia

Cocaine abuse is one of the leading causes of death in humans that is largely attributed to cardiovascular emergencies, for example, hypertensive crisis, acute myocardial infarction, and profound hyperthermia (see Crandall *et al.*, 2002). The hyperthermic effects of cocaine are due to increase in heat production caused by hypermetabolic state, for example, agitation and increased locomotor activity (Menaker *et al.*, 2008). Thus, the symptoms of cocaine hyperthermia are quite comparable to that of neuroleptic malignant syndrome and malignant hyperthermia that is characterized by hyperpyrexia, delirium, tachycardia, and hypertension (see Merigian and Roberts, 1987). However, it appears that the impaired heat dissipation could be another important factor in cocaine-induced hyperthermia. This idea is supported by the fact that cocaine-induced hyperthermia is exacerbated at high environmental temperatures (Kiyatkin and Bae, 2008). Thus, it is quite likely that cocaine-induced hyperthermia is crucial for altered brain functions as seen in heat stroke or heat-related illnesses (see Sharma, 2007a).

In human cases, several fatalities have been reported in cocaine users due to hyperthermia (Bauwens *et al.*, 1989). These subjects normally exhibit high fever like conditions (>40 °C body temperature) on arrival to the hospital before death and often associated with diaphoresis, tachycardia, tachypnea, and altered mental states. These symptoms are quite similar to that seen in patients with heat stroke (Shibolet *et al.*, 1976). This indicates that hyperthermia due to cocaine intoxication are largely due to increase in metabolic rate and/or impaired heat loss mechanisms due to peripheral vasoconstriction (Lombard *et al.*, 1988). Severe hyperthermia could directly influence brain microvasculature leading to BBB disruption and subsequently lead to neurotoxicity (Sharma *et al.*, 2007). However, additional studies on cocaine-induced hyperthermia and neurotoxicity *in vivo* situations are needed to establish this point.

IX. Our Investigations on Cocaine-Induced Neurotoxicity

We have undertaken detailed investigations on cocaine-induced neurotoxicity in rodents using *in vivo* models. Our main interest is to examine whether cocaine depending on its dose and routes of administration could affect BBB dysfunction and induce brain damage through serotonin-mediated mechanisms. In addition, whether these effects of cocaine on brain pathology are age related. Furthermore, whether drugs modifying serotonin metabolism could influence cocaine-induced neurotoxicity. In this section, our new findings on cocaine-induced neurotoxicity in a rat model and its modification with age and serotonin 5-HT2 and 5-HT3 receptor modulator agents are discussed.

A. EXPERIMENTAL PROCEDURES

Experiments were carried out on inbred albino Charles Foster male rats (body weight 200–250 g; age 12–14 weeks old) housed at controlled room temperature (21 ± 1 °C) with 12-h light and 12-h dark schedule. The food and tap water were provided *ad libitum* before the experiments.

All experimental procedures and handling of animals described here are performed according to National Institute of Health (NIH) Guidelines and Care of Experimental Animals and approved by Local Institutional Ethics Committee.

1. *Cocaine Administration*

Cocaine (3b-Hydroxy-1aH,5aH-tropane-2b-carboxylic acid methyl ester benzoate hydrochloride) was obtained from TOCRIS Bioscience, UK (Cat No. 2833; Cocaine Hydrochloride) in powder form and stored at laboratory room temperature before experiments. At the time of experiments, cocaine was freshly dissolved in sterile water about 30 min before the onset of experiments and administered in rats through various routes as described below.

2. *Intravenous Administration*

For this purpose, a polythene cannula (PE 10) was surgically implanted aseptically into the right jugular vein of rats about 1 week before the experiments. The venous cannula was flushed everyday with heparinized saline (250 U/100 ml) to maintain its patency (Sharma and Dey, 1981, 1986). The exterior end of the cannula was sealed with heat after flushing. At the time of cocaine administration, the rats were held gently and a syringe filled with cocaine solution was connected with to the cannula and the administration was done according to the require dose and the protocol (see Sharma and Dey, 1986).

3. *Intraperitoneal Administration*

For this purpose, rats were injected with cocaine in a dose of 20–60 mg/kg intraperitoneally according to standard procedure suing sterile syringe (se Sharma, 1987). All doses were adjusted in a volume of 0.3–0.5 ml solution (Sharma, 1987).

4. *Intracerebroventricular Administration*

In separate group of rats, cocaine was administered intracerebroventricularly (i.c.v) using standard procedures as described earlier (Sharma, 1987; Sharma *et al.*, 1995a). For this purpose, a sterilized stainless steel cannula (34 G) was implanted aseptically into the right lateral cerebral ventricle (coordinates +2 mm from bregma on parietal bone and 2.5 mm lateral to the midline, 1.5 mm deep) 8 days before the experiments (see Sharma *et al.*, 1995a). The shaft of cannula was fixed with dental cement (BDH, London, UK). A stainless steel wire stellate was inserted into the cannula in such a way that its tip was protruded 0.5 mm beyond the tip of the cannula (Sharma, 1987; Sharma and Dey, 1986). The cannula was flushed with artificial CSF daily to maintain its patency. At the time of cocaine administration, the steel wire stellate was removed gently and a 38 G needle shaft (London, UK) was inserted gradually into the cannula in identical manner so that its flat tip is protruded 0.5 mm beyond the cannula within the CSF space. A 100 μl (Hamilton, UK) syringe was connected with this cannula through a PE 10 catheter and the desired amount of cocaine was administered using an infusion pump (Harvard apparatus, USA) at the rate of

10 μl/min. The dose of cocaine (20–60 μg) was adjusted as such that each animal received 30 μl of solution through the intracerebroventricular route. At the end of experiment, Evans blue was used to flush the cannula using identical volume and speed to trace the drug passage in to the CSF. In some animals (<5% cases), where the spread of Evans blue was not homogenous within the lateral cerebral ventricle was discarded (see Sharma, 1987).

B. Parameters Measured

The following parameters were measured in the present investigation to study cocaine-induced neurotoxicity in rats.

1. *Physiological Variables*

The mean arterial blood pressure (MABP), blood gases, and body temperature were measured using standard protocol. For measurement of hyperthermia, a thermistor probe (Yellow Springfield, USA) was inserted into the rectum of animals (about 4–5 cm) to record deep visceral body temperature. The thermistor probes were connected to a digital telethermometer (Aplab Electronics, UK) (Sharma, 1987; Sharma and Dey, 1986).

The arterial blood pressure was measured using a cannula implanted into the right common carotid artery (PE 25) aseptically about 1 week before the experiments (Sharma, 1987). At the time of blood pressure measurement, the arterial cannula was connected to a Strain Gauge pressure transducer (Statham P23, USA) and the output was fed to a chart recorder (Electromed, UK). Immediately before connecting the cannula to pressure transducer, a sample of arterial blood was withdrawn for later measurement of blood gases and pH using a Radiometer apparatus (Copenhagen, Denmark).

2. *Blood–Brain Barrier Permeability*

The BBB permeability following cocaine administration was examined using two protein tracers, for example, Evans blue albumin (EBA) and radioiodine ($^{[131]}$I) (Sharma and Dey, 1986). These tracers when administered into systemic circulation bind to serum albumin and thus leakage of these tracers represents extravasation of tracer–protein complex. For this purpose, EBA (2% of 0.3 ml/ 100 g solution, pH 7) and radioiodine (10 μCi/100 g) were administered together through previously implanted jugular vein cannula 5 min before the end of the experiment (Sharma, 1987). After 5 min of tracer injections, the animals were anesthetized with Equithesin (0.3 ml/100 g, i.p.) and the intravascular tracers were washed out with a brief saline rinse through heart at 90 torr = mmHg, perfusion pressure using a peristaltic pump (Harvard Apparatus, USA). After that the brain was dissected out, examined for EBA leakage visually and then dissected into desired brain areas. The brain samples were weighed immediately and counted in a 3-in Gamma counter (Packard, USA). Immediately, before perfusion, about 1 ml of whole blood was withdrawn from the left ventricle after cardiac puncture to determine the whole blood radioactivity (Sharma, 1987). Extravasation of radioiodine was expressed as percentage increase in the brain regions over the whole blood radioactivity.

After measurement of the radioactivity, the Evans blue was extracted from the brain samples in a mixture of sodium sulfate and acetone, and the dye entered into the samples was measured in a spectrophotometer at 620 nm. The concentration of the dye in samples was assessed using standard solution of Evans blue as described earlier (Sharma and Dey, 1986).

3. Brain Edema

In separate group of rats, brain edema was measured using changes in brain water content (Sharma and Cervós-Navarro, 1990). For this purpose, at the end of the experiments, the brains were taken out and placed on the cold saline wetted filter papers and the blood clots, if any along with the large superficial blood vessels were removed and discarded. The brain was then dissected out into several desired regions (see Table III). Each brain samples (sample size 120–300 mg) were weighed immediately and placed in an incubator maintained at 90 °C for 72 h to obtain their dry weights (Sharma and Cervós-Navarro, 1990; Sharma et al., 1991). The brain water content was calculated from the difference between wet and dry weights of the samples (Sharma and Dey, 1987; Sharma et al., 1992a).

4. Cerebral Blood Flow

Cocaine-induced changes in CBF were examined in rats using radiolabeled carbonized microspheres (15 ± 6 μm OD) (Sharma, 1987; Sharma and Dey, 1986; Sharma et al., 1992a). About 10^6 microspheres were administered as a bolus into the left common carotid artery through a polythene cannula (PE 25) implanted 8 days earlier retrogradely toward the heart (Sharma, 1987). In these animals, the peripheral arterial blood from the right femoral artery was withdrawn at the rate of 0.8 ml/min at starting from 30 s before the start of the microsphere administration at every 30 s and continued up to 90 s after the termination of the microspheres injection (Sharma and Dey, 1986). After 120 s of microsphere injections, the brains were removed and dissected into desired regions, weighed immediately and the radioactivity was determined in a 3-in Gamma counter (Packard, USA). Total blood radioactivity in all the serial samples was also determined. The CBF was calculated from the radioactivity present in the timed reference arterial samples and the radioactivity shown in each brain samples as described earlier (Sharma, 1987). The CBF is expressed as ml/g/min (see Sharma, 1987).

5. *Plasma and Brain Serotonin*

Plasma and brain serotonin was measured fluorometrically following cocaine administration in rats according to the standard protocol (Sharma and Dey, 1986). For this purpose, at the end of the experiment, about 1 ml arterial sample was withdrawn under anesthesia and then the brains were taken out and dissected into several brain regions. Each brain samples and plasma were deproteinized and serotonin was extracted in a salt saturated alkaline medium (Sharma and Dey, 1981). The serotonin fluorescence was developed using ninhydrin at 60 °C and the fluorescence was measured in samples cooled at room temperature in an Aminco Bowman Spectrophotofluorometer at excitation 460 and emission 590 nM wavelengths. Standard serotonin solution was used to calculate concentrations of the amine in plasma and blood samples (Sharma and Dey, 1986).

6. *Histopathology*

Cellular changes in brains after cocaine treatment was examined using standard histological protocol on paraffin section after staining with Hematoxylin & Eosin (Sharma and Cervós-Navarro, 1990; Sharma *et al.*, 1991). For this purpose, after the end of the experiments, the animals were perfused transcardially (at 90-torr perfusion pressure) with 4% paraformaldehyde solution (pH 7.4) preceded with a brief saline rinse. After perfusion, the brains were removed and placed in same fixative for 1 week at 4 °C. After that, coronal sections from the rat brain passing through hippocampus were cut (4–5 mm thick) and embedded in paraffin. About 3-μm thick sections were cut and stained with Hematoxylin & Eosin (Sharma and Cervós-Navarro, 1990). These sections were examined under a bright field inverted light microscope (Carl Zeiss) and the digital images were recorded and analyzed by at least two independent workers in a blinded fashion.

7. *Immunohistochemistry*

Cocaine-induced changes in astrocytic function and stress protein responses were examined using immunohistochemistry of GFAP and heat shock protein 72 kD (HSP), respectively, in several brain regions (Cervós-Navarro *et al.*, 1998; Sharma *et al.*, 1992b, 1995b). The HSP expression was studied on 40-μm thick Vibratome sections obtained from specific brain areas (see Table IV) and the GFAP immunostaining was investigated on 3-μm thick paraffin sections as described earlier (Sharma *et al.*, 1992b, 1995b). In brief, free-floating Vibratome sections were incubated with monoclonal HSP antibodies (1:5000) at 4 °C under constant agitation for overnight. The immunoreaction was developed using peroxidase–antiperoxidase reaction (see Sharma *et al.*, 1995b). For GFAP immunoreactivity, paraffin sections were incubated with monoclonal GFAP antibodies (1:2000) at 4 °C for overnight and the immune complexes were developed using ABC techniques (Sharma *et al.*, 1992b). The immunoreaction for HSP and GFAP were examined under a bright field inverted light microscope and digital images were recorded and processed according to standard procedures (see Sharma and Sjöquist, 2002).

C. STATISTICAL ANALYSES OF THE DATA OBTAINED

Quantitative data were analyzed to evaluate statistical significance using ANOVA followed by Dunnet's test for multiple group comparison from one control group. Semiquantitative data were evaluated using nonparametric Chi-square test. A p-value < 0.05 was considered significant.

X. Novel Observations on Cocaine-Induced Brain Pathology

Our observations clearly show that cocaine is capable to induce marked BBB breakdown, edema formation and increase in brain serotonin levels leading to profound brain dysfunctions and brain pathology (see below). This effect of cocaine, however, depends on the dose and route of its administration.

A. COCAINE INDUCES HYPERTHERMIA

Administration of cocaine either through intraperitoneal, intravenous, or intracerebroventricular routes induced marked hyperthermia within 4 h (see Table I). The most pronounced hyperthermia was seen when cocaine was administered through intracerebroventricular route as compared to other modes of administration (Table I). Accordingly, cocaine given intravenously resulted in rise of body temperature by more than 4 °C (41.24 ± 0.38 °C) from the control groups (36.8 ± 0.80 °C) if administered in high doses (30 mg/kg). The lower doses were much less effective in inducing hyperthermia (Table I). Almost identical hyperthermia was seen when cocaine was administered in a dose of 60 mg/kg intraperitoneally (41.54 ± 0.43 °C) at 4 h. Lower doses of cocaine (40 or 20 mg/kg, i.p.) were not that effective (see Table I). On the other hand, intracerebroventricular administration of cocaine (60 μg) enhanced the body temperature by 5 °C within 4-h period (Table I). Lower doses of cocaine given through intracerebroventricular route produced only mild-to-moderate degrees of hyperthermia (40 μg: 40.39 ± 0.21 °C and 20 μg: 39.23 ± 0.23 °C). This suggests that intracerebroventricular administration of cocaine results in the most pronounced hyperthermia as compared to its systemic administration. Direct exposure of cocaine to the CNS could influence the thermoregulatory mechanisms more

TABLE I
Effect of Cocaine on BBB Permeability, Cerebral Blood Flow, and Physiological Variables After 4 h of Administration in the Rat

Expt. type	n	BBB breakdown Rect T (°C)	BBB breakdown EBA mg (%)	BBB breakdown [131] Iodine (%)	Blood flow CBF (ml/g/min)	Physiological variables Arterial (pH)	PaO2 (torr)	PaCO2 (torr)	MABP (torr)
Saline control	6	36.8 ± 0.80	0.24 ± 0.04	0.34 ± 0.04	1.02 ± 0.05	7.38 ± 0.04	80.78 ± 0.23	33.56 ± 0.28	110 ± 6
Intravenous cocaine									
10 mg/kg	8	38.24 ± 0.80**	0.56 ± 0.06*	0.89 ± 0.12**	1.21 ± 0.08*	7.36 ± 0.08	81.34 ± 0.07	33.24 ± 0.18	128 ± 8
20 mg/kg	6	39.04 ± 0.34**	0.89 ± 0.04**	0.96 ± 0.10**	1.28 ± 0.06*	7.35 ± 0.06	81.56 ± 0.21	33.14 ± 0.12	134 ± 6
30 mg/kg	8	41.24 ± 0.38**	1.14 ± 0.10**	1.89 ± 0.14**	1.56 ± 0.04**	7.32 ± 0.08	81.89 ± 0.32	32.24 ± 0.16	156 ± 8
Intraperitoneal cocaine									
20 mg/kg	6	38.48 ± 0.44**	0.48 ± 0.08*	0.78 ± 0.08**	1.08 ± 0.07	7.37 ± 0.08	81.54 ± 0.15	33.18 ± 0.17	130 ± 7
40 mg/kg	8	39.89 ± 0.76**	0.54 ± 0.07*	0.86 ± 0.10**	1.16 ± 0.06*	7.35 ± 0.06	81.78 ± 0.21	33.04 ± 0.12	140 ± 6
60 mg/kg	8	41.54 ± 0.43**	0.90 ± 0.10**	1.21 ± 0.08**	1.28 ± 0.08**	7.32 ± 0.08	81.89 ± 0.21	32.87 ± 0.21	156 ± 7
Intracerebroventricular cocaine									
20 µg	6	39.23 ± 0.23**	0.58 ± 0.08**	0.88 ± 0.10**	1.28 ± 0.08**	7.36 ± 0.07	81.55 ± 0.12	33.28 ± 0.08	138 ± 8
40 µg	6	40.39 ± 0.21**	0.89 ± 0.06**	1.24 ± 0.13**	1.38 ± 0.12**	7.33 ± 0.06	81.76 ± 0.22	33.18 ± 0.08	147 ± 8
60 µg	8	41.86 ± 0.23**	1.21 ± 0.15**	1.98 ± 0.21**	1.47 ± 0.13**	7.28 ± 0.08	81.89 ± 0.14	32.87 ± 0.12	167 ± 8

Values are mean ± SD; BBB, blood–brain barrier; EBA, Evans Blue albumin; CBF, cerebral blood flow. *$p < 0.05$, **$p < 0.01$, ANOVA followed by Dunnet's test (data from H.S. Sharma et al., unpublished observations).

effectively than its peripheral administration (Füzesi *et al.*, 2008). Increased behavioral activity after intracerebroventricular administration of cocaine as compared to its intravenous or intraperitoneal injection is in line with this idea (results not shown).

B. COCAINE ALTERS CARDIOVASCULAR RESPONSES

Cocaine depending on dose and route of its administration enhanced markedly the MABP in rats after 4-h period (Table I). This effect of cocaine was most pronounced when the compound was administered into the cerebral ventricles as compared to its intravenous or intraperitoneal injections (Table I). A rise in MABP by 45 torr can be achieved by 30 mg/kg intravenous doses or 60 mg/kg intraperitoneal administration of cocaine. Whereas, more than 55-torr rise in MABP was achieved when cocaine was administered 60 μg into the lateral cerebral ventricle (Table I). These observations suggest that cocaine could affect central cardiovascular regulatory mechanism to induce hypertension (Erzouki *et al.*, 1995).

This idea is further strengthened by the fact that cocaine administration in any doses by either route did not influence the blood gases or the arterial pH markedly (Table I). Thus, a slight reduction in arterial pH was seen after cocaine administration in high doses after its administration through the intracerebroventricular route (Table I). A dose-dependent cocaine-induced mild increase in PaO_2 was also seen irrespective of the routes of its administration (Table I). Whereas, a dose-dependent slight reduction in $PaCO_2$ by cocaine was observed that was most pronounced after its administration into the brain ventricles followed by intraperitoneal and intravenous injections (Table I). This suggests that peripheral changes in the microvasculature, arterial pH, or blood gases are not directly involved in cocaine-induced hypertension (cf. Gillis *et al.*, 1995).

C. COCAINE ENHANCES BBB PERMEABILITY TO PROTEINS

In our investigations, cocaine was able to induce a dose-dependent breakdown of the BBB to protein tracers, EBA and radioiodine, 4 h after its administration given either through intravenous, intracerebroventricular, or intraperitoneal routes (Table I). This effect was most pronounced when the cocaine was injected into the cerebral ventricles followed by its intravenous administration (Table I). Intraperitoneal administration also resulted in a considerable increase in the BBB permeability to proteins. A higher increase in the BBB permeability to radioiodine as compared to EBA indicates that cocaine could induce a size-dependent increase of tracers across the brain microvessels (see Table I,

Sharma and Dey, 1984, 1987). Size-dependent entry is mostly related to an increase in membrane permeability of the cerebral endothelial cells rather than widening of its tight junctions (Sharma et al., 1995a). This observation suggests that cocaine either directly or through stimulation of other neurochemicals and/ or its receptors located on the endothelial cell membrane could lead to increased transport of tracer substances across the BBB (see Sharma and Ali, 2006, 2008; Sharma and Dey, 1986; Sharma et al., 1995a).

To further examine the sensitivity of cocaine on cerebral microvessels in different brain regions leading to alterations in the regional BBB permeability, we examined the leakage of radioiodine in several brain areas (see Fig. 1). Our observations show that the selective vulnerability of different brain regions was markedly affected by cocaine dependent on its route of the administration (Fig. 1). Thus, intravenous administration of cocaine (30 mg/kg) resulted in large extravasation radioiodine tracer in occipital cortex and caudate nucleus followed by frontal cortex, cerebellar cortex, hypothalamus, hippocampus, thalamus, parietal cortex, cingulate cortex, colliculi, temporal cortex, brainstem, and spinal cord (Fig. 1). Whereas, intraperitoneal administration of cocaine (60 mg/kg) showed a much less accumulation of radioiodine in the occipital cortex, cerebellum, and hippocampus although the remaining brain areas showed more or less similar degree of radiotracer leakage as seen by its intravenous injection (30 mg/kg) (Fig. 1). The magnitude of radiotracer leakage was highest in almost all the brain regions following intracerebroventricular injection of cocaine (60 μg) as compared to its intravenous or intraperitoneal injections. The most marked increase in the BBB disruption to radioiodine by i.c.v. cocaine was seen in caudate nucleus followed by hippocampus, cerebellum, thalamus, colliculi, frontal cortex, occipital cortex, hypothalamus, parietal cortex, brain stem, cingulate cortex, and the spinal cord (Fig. 1).

These observations demonstrate that the routes of administration play important roles in cocaine-induced BBB dysfunction. Thus, it appears that cocaine-induced BBB disruptions may partly be dependent on the concentration of the drug reaching to different brain areas. Alternatively, different brain regions may have different sensitivity to cocaine based on its bioavailability. Obviously, the bioavailability may depend on the route and dose of the cocaine administration. Thus, intracerebroventricular administration of cocaine could easily reach to the hippocampus, caudate nucleus, cerebellum, and other deep brain regions in high concentrations when given into the cerebral ventricles. These changes in cocaine-induced BBB permeability may not be related to its different effects on the luminal (given through intravenous route) and abluminal (administered through i.c.v. route) sides of the endothelial cell membranes. There are reasons to believe that both abluminal and luminal side of the endothelial cell membrane are equally sensitive to cocaine and other neurochemical mediators, for example, serotonin or histamine (Sharma and Kiyatkin,

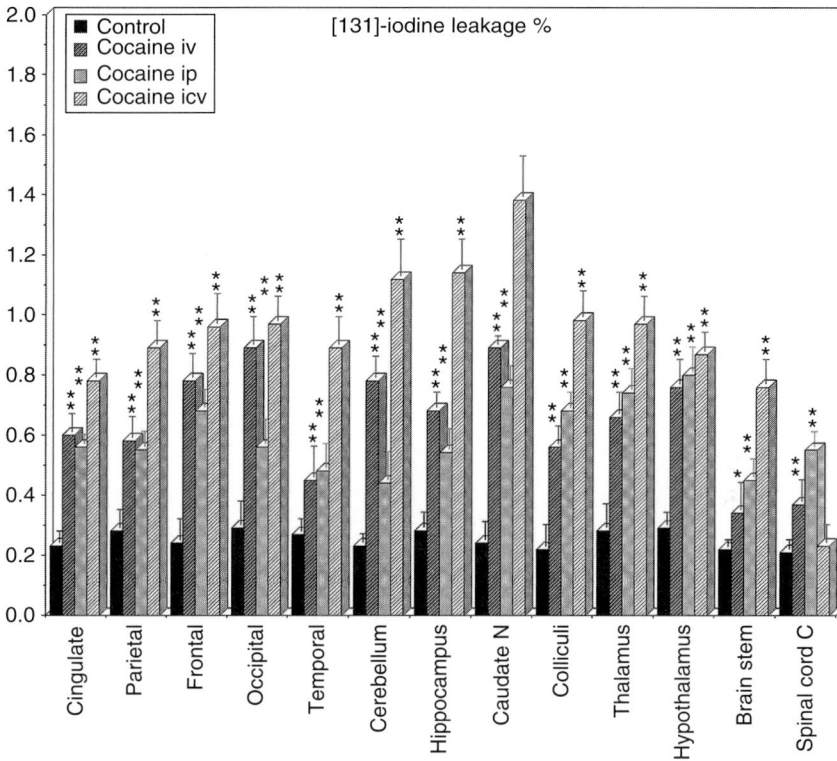

FIG. 1. Regional blood–brain barrier permeability in saline- (control) or cocaine-treated animals. Cocaine was administered either intravenously (30 mg/kg, i.v.), intraperitoneally (60 mg/kg, i.p.), or intracerebroventricularly (60 μg, i.c.v.) and the regional BBB permeability to radioiodine was measured 4 h after the drug administration. **$p < 0.01$, *$p < 0.05$, ANOVA followed by Dunnet's test for multiple group comparison from one saline-treated group. For details see text (H.S. Sharma *et al.*, unpublished observations).

2009; Sharma *et al.*, 1990, 1992b, 2007). However, further studies are needed to explore the regional or luminal versus abluminal sensitivity of cocaine to cerebral microvessels in order to clarify these points.

D. Cocaine Induces Brain Edema Formation

Extravasation of serum proteins into the brain microenvironment leads to vasogenic edema formation (Sharma *et al.*, 1998). Thus, it is quite likely that cocaine-induced leakage of protein tracers could result in vasogenic brain edema formation (Sharma and Cervós-Navarro, 1990; Sharma *et al.*, 1991). Analysis of regional brain water content showed a close correspondence between regional BBB permeability to radiotracer and accumulation of brain water (see Fig. 2). Thus, intracerebroventricular administration of cocaine (60 µg) induced most marked edema formation in the hippocampus, caudate nucleus, colliculi, cerebellum, thalamus, cingulate cortex, parietal cortex, frontal cortex, followed by occipital cortex, temporal cortex, and brain stem (Fig. 2). On the other hand, intravenous cocaine resulted in brain edema formation that was most prominent in hippocampus, caudate nucleus followed by colliculi, thalamus, hypothalamus, cerebellum, cerebral cortex, and brain stem (Fig. 2). Intraperitoneal injection of cocaine-induced brain edema was very similar to that seen following intravenous

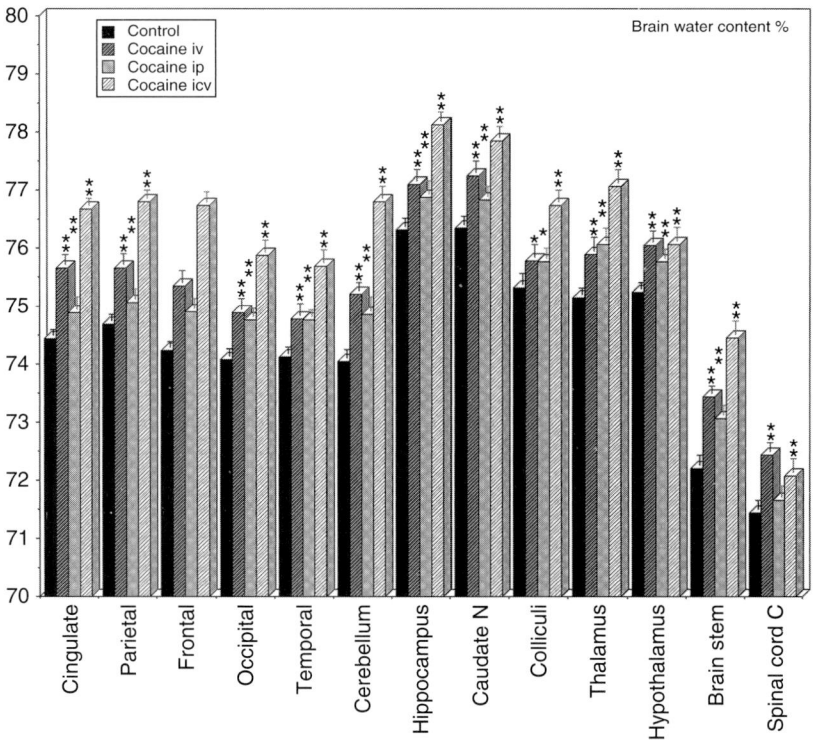

FIG. 2. Regional brain edema formation in saline- (control) or cocaine-treated animals. Cocaine was administered either intravenously (30 mg/kg, i.v.), intraperitoneally (60 mg/kg, i.p.), or intracerebroventricularly (60 µg, i.c.v.) and the regional brain water content was measured 4 h after the drug administration. **p < 0.01, *p < 0.05, ANOVA followed by Dunnet's test for multiple group comparison from one saline-treated group. For details see text (H.S. Sharma *et al.*, unpublished observations).

cocaine infusion (Fig. 2). This suggests that cocaine is capable to induce brain edema formation by causing leakage of proteins into the brain microenvironment.

E. COCAINE ALTERS SEROTONIN METABOLISM

Measurement of serotonin in plasma and different brain regions showed a profound increase in the amine content following cocaine administration (Table II). This effect was most pronounced when cocaine was administered through intravenous or intracerebroventricular routes. However, this increase in brain or plasma serotonin caused by cocaine was markedly different following different routes of the drug administration. Thus, in some brain areas, for example, hippocampus, hypothalamus, and cerebellum intraperitoneal administration of cocaine (60 mg/kg) resulted in a higher increase in serotonin content as compared to its intravenous injection (30 mg/kg; Table II). On the other hand, a higher level of serotonin was observed in the cortex, thalamus, and spinal cord following intravenous cocaine injection (Table II). Whereas, intracerebroventricular administration of cocaine showed the highest increase in serotonin levels in the cortex, hippocampus, and cerebellum (Table II). These observations suggest that cocaine depending on its route of administration could influence the regional changes in the brain serotonin levels within 4 h. These regional variations in serotonin levels may also affect the regional BBB permeability and regional CBFs (Sharma and Dey, 1986; Sharma et al., 1990). Since serotonin is a well-known mediator of BBB disruption and brain edema formation, it is likely that cocaine-induced alterations in regional serotonin level could play important roles in drug-induced neurotoxicity.

F. COCAINE INFLUENCES REGIONAL CEREBRAL BLOOD FLOW

Cocaine either given through intravenous, intraperitoneal, or intracerebro-ventricular routes resulted in marked alterations in the regional CBF (see Fig. 3). It appears that routes of administration of cocaine have a significant effect on the redistribution of the CBF in different brain regions (Fig. 3). Thus, intravenous cocaine administration (30 mg/kg) significantly enhanced the regional CBF in most of the cortical regions, for example, cingulate, parietal, frontal, and occipital cortices. Whereas, intraperitoneal cocaine administration (50 mg/kg) resulted in a significant decrease in blood flow to these regions and was most pronounced in the frontal and the occipital cortices (Fig. 3). In temporal cortex, intravenous cocaine reduced the blood flow but intraperitoneal cocaine resulted in marked elevation in the blood flow in this brain region (Fig. 3). On the other hand, intravenous cocaine slightly but significantly reduced the blood flow in cerebellar

TABLE II

EFFECTS OF COCAINE ADMINISTRATION ON PLASMA AND BRAIN SEROTONIN LEVELS IN THE RAT

Expt. type	n	Serotonin Plasma (µg/ml)	Brain serotonin (µg/g) Cortex	Hippocampus	Thalamus	Hypothalamus	Cerebellum	Brain stem	Spinal cord
Saline control	6	0.30 ± 0.04	0.67 ± 0.23	0.56 ± 0.21	0.78 ± 0.32	0.89 ± 0.32	0.45 ± 0.21	0.44 ± 0.23	0.65 ± 0.21
Intravenous cocaine									
30 mg/kg	8	0.98 ± 0.10**	1.23 ± 0.14**	0.89 ± 0.21**	0.98 ± 0.10*	0.45 ± 0.14**	0.89 ± 0.19**	0.46 ± 0.12	0.89 ± 0.11*
Intraperitoneal cocaine									
60 mg/kg	8	1.04 ± 0.08**	0.98 ± 0.21**	0.94 ± 0.12*	0.78 ± 0.12	0.78 ± 0.23**	0.94 ± 0.14**	0.48 ± 0.12	0.78 ± 0.21
Intracerebroventricular cocaine									
60 µg	8	0.67 ± 0.08**	1.56 ± 0.21**	1.23 ± 0.21**	0.74 ± 0.08	0.34 ± 0.08**	1.05 ± 0.21**	0.38 ± 0.21	0.56 ± 0.12

Values are mean ± SD; *$p < 0.05$, **$p < 0.01$, ANOVA followed by Dunnet's test (data from H.S. Sharma et al., unpublished observations).

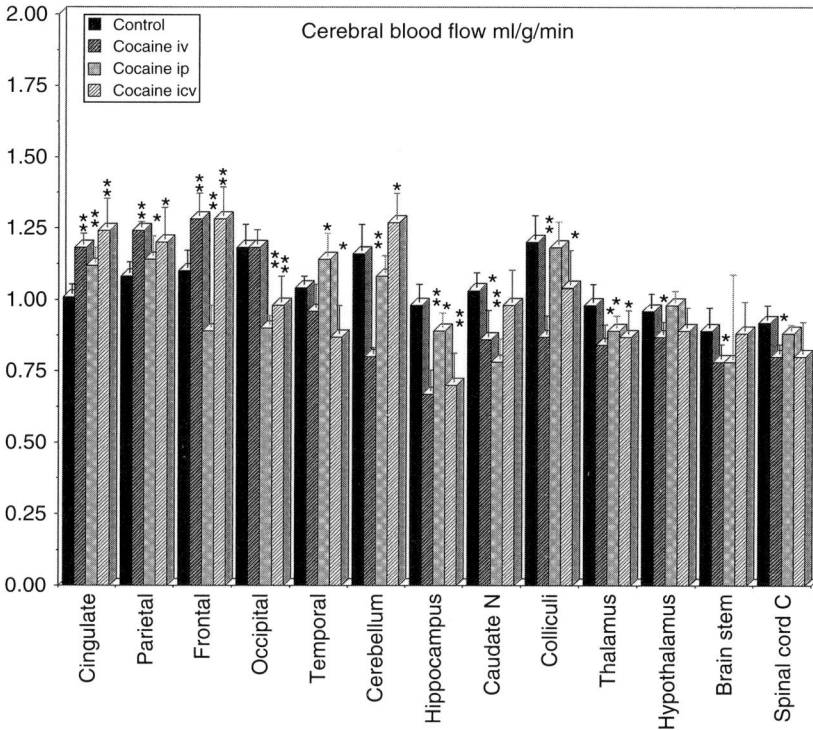

FIG. 3. Regional cerebral blood flow (CBF) changes in saline- (control) or cocaine-treated animals. Cocaine was administered either intravenously (30 mg/kg, i.v.), intraperitoneally (60 mg/kg, i.p.), or intracerebroventricularly (60 μg, i.c.v.) and the regional CBF was measured 4 h after the drug administration using radioiodine-labeled carbonized microspheres. **$p < 0.01$, *$p < 0.05$, ANOVA followed by Dunnet's test for multiple group comparison from one saline-treated group. For details see text (H.S. Sharma *et al.*, unpublished observations).

cortex; however, this reduction in blood flow was accentuated by intraperitoneal cocaine administration (Fig. 3). Interestingly, intracerebroventricularly administered cocaine (50 μg) enhanced the blood flow profoundly in the cingulate, parietal, frontal, and cerebellar cortices, a significant reduction in the blood flow was seen in the occipital and temporal cortices in this group (Fig. 3).

In the subcortical areas, for example, the hippocampus, caudate nucleus, and thalamus, cocaine induced mild-to-moderate reductions in the blood flow irrespective of its routes of its administration (Fig. 3). The colliculi, hypothalamus, brain stem, and the spinal cord showed only minor changes in the regional blood flow after cocaine administration (Fig. 3). Thus, in colliculi and hypothalamus, slight reductions in the blood flow were seen by intravenous or intracerebroventricular cocaine administration, whereas intraperitoneal injection did not alter the blood flow significantly in anyone of these brain areas (Fig. 3). In the brain stem and the spinal cord, only intravenous cocaine resulted in slight reduction in blood flow but the intracerebroventricular or intraperitoneal cocaine was ineffective in inducing any blood flow alterations in these brain regions (Fig. 3). These observations suggest that cocaine has a very sensitive and specific effect on the regional brain microvasculature resulting in precise alterations in the local CBF. It is likely that cocaine-induced alterations in the regional serotonin metabolism could play important roles (Sharma *et al.*, 1990). Since distribution of serotonin receptors could be different in different bran regions and the regional changes in serotonin levels also differ accordto the routes of cocaine administration, it is quite likely that increased serotonin levels could affect various serotonin receptors located in different brain regions. This will lead to either an increase or decrease in the blood flows depending on the nature and type of the receptors activated in different brain areas (Bonvento *et al.*, 1991). Alternatively, cocaine-induced regional changes in the BBB could also affect in part the access of serotonin to various 5-HT receptors within the brain microvessels. However, this is a new subject that requires additional investigations.

G. Cocaine Induces Neuronal Damage

Marked neuronal changes were seen after cocaine administration in various brain regions at the time of the BBB breakdown (Table III; Fig. 4). These neuronal changes were most pronounced by cocaine after its administration through intracerebroventricular (50 μg) route followed by intraperitoneal (50 mg) or intravenous (30 mg) administration (Table III). The most marked neuronal changes were seen in the subcortical structures, for example, hippocampus and caudate nucleus. Cerebral and cerebellar cortices also showed profound neuronal damages after cocaine that were most marked by intracerebroventricular cocaine as compared to its systemic administration (Table III). Brain stem and spinal cord showed mild-to-moderate neuronal damages.

A representative example of hippocampal neuronal damage following cocaine administration is shown in Fig. 4. As evident from the figure, the dentate gyrus neurons showed marked damage and distortion following intracerebroventricular cocaine followed by its intravenous and intraperitoneal injections (Fig. 4). In cocaine-treated rats, damage of neurons such as, condensed cytoplasm, perineuronal vacuolation, and edema are most prominent. Most damaged neurons were dark in appearance and a clear nucleus is hard to be seen (Fig. 4). In some distorted neurons where cell nucleus is still visible, eccentric nucleoli with altered shape are common findings. A general sponginess and edema is clearly seen in the background and neurons located in these areas often show much distortion and

TABLE III

Semiquantitative Analysis of Cocaine Administration on Neuronal and Glial Cell Changes in the Rat

Brain regions	Saline control Neuronal reaction (n = 6)	Saline control Glial reaction (n = 6)	Cocaine i.v. (30 mg/kg) Neuronal reaction (n = 8)	Cocaine i.v. (30 mg/kg) Glial reaction (n = 8)	Cocaine i.p. (60 mg/kg) Neuronal reaction (n = 6)	Cocaine i.p. (60 mg/kg) Glial reaction (n = 6)	Cocaine i.c.v. (50 µg) Neuronal reaction (n = 8)	Cocaine i.c.v. (50 µg) Glial reaction (n = 8)
1. Cingulate cortex	0	1 ± 1	8 ± 2*	9 ± 3*	6 ± 2*	8 ± 2*	12 ± 5*	14 ± 7*
2. Parietal cortex	0	0	12 ± 6*	10 ± 5*	8 ± 4*	10 ± 4*	16 ± 6*	18 ± 8*
3. Hippocampus CA-3	0	2 ± 2	18 ± 6**	22 ± 7**	12 ± 5*	14 ± 8*	22 ± 6**	28 ± 9**
4. Thalamus dorsal N	1 ± 1	2 ± 2	14 ± 8**	24 ± 7**	16 ± 6**	23 ± 8**	28 ± 6**	32 ± 6**
5. Hypothalamus ventral	0	0	12 ± 6**	16 ± 6**	20 ± 8**	26 ± 9**	34 ± 12**	38 ± 14**
6. Cerebellum vermis	1 ± 1	2 ± 3	18 ± 4**	28 ± 7**	28 ± 7**	34 ± 9**	46 ± 18**	54 ± 20**
7. Brain stem medulla	1 ± 2	4 ± 2	9 ± 4*	12 ± 6*	8 ± 6*	10 ± 6*	18 ± 9**	23 ± 12**
8. Spinal cord cervical	0	2 ± 2	12 ± 6*	14 ± 8*	10 ± 6*	12 ± 8*	12 ± 8*	16 ± 9*

Values are mean $\pm$ SD; *$p < 0.05$, **$p < 0.01$, Chi-square test from control group. Cells were counted in blinded fashion by at least two independent observers in each animal and the median values were used for calculation. Neuronal distortion, shrinkage and/or swelling, dark nucleus with condensed cytoplasm, and altered shape are used as criteria for neuronal reaction or damage. Intense expression of glial fibrillary acidic protein (GFAP) immunoreactivity with distorted shape of astrocytes and their presence in edematous region are considered as glial reaction (data from H.S. Sharma et al., unpublished observations).

Cocaine neurotoxicity

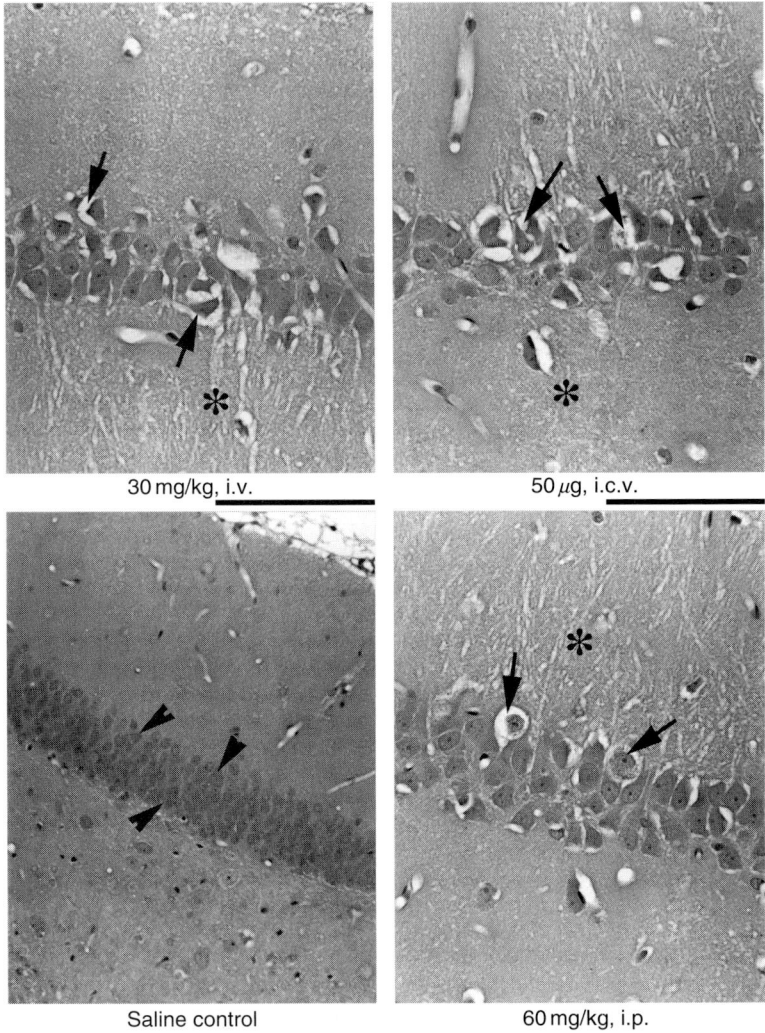

FIG. 4. Representative example of neuronal damage in the hippocampus in saline- (control) or cocaine-treated animals. Cocaine was administered either intravenously (30 mg/kg, i.v.), intraperitoneally (60 mg/kg, i.p.), or intracerebroventricularly (50 μg, i.c.v.) and the neuronal damage was examined after 4 h survival on 3-μm thick paraffin sections passing through hippocampus and stained with Hematoxylin & Eosin (H&E) using standard procedures. Damaged neurons (arrows) are frequent in cocaine-treated animals as compared to saline treatment where no obvious neuronal damage can be seen (arrowheads). Perineuronal edema seen as wide space around the neurons is most prominent in rats that received 50 μg of cocaine by intracerebroventricular route (i.c.v.) followed by its intravenous (30 mg, i.v.) administration. Intraperitoneal cocaine (60 mg) also showed considerable neuronal damage (arrows). A general sponginess and edema (*) is quite frequent in cocaine-treated rats as compared to saline treatment. Bar = 50 μm. For details see text (H.S. Sharma *et al.*, unpublished observations). (See Color Insert.)

damage. Thus, most neurons were either swollen or shrunken and perineuronal vacuolation is much frequent in the neuropil (see Fig. 4).

These observations are the first to show that cocaine could induce neurotoxicity in the CNS *in vivo* situations. This effect of cocaine on neuronal damage depends on the specific regions of the brain. This suggests that cocaine can induce selective damage to the nerve sells when administered through intracerebroventricular or systemic routes. It appears that cocaine-induced hyperthermia and breakdown of the BBB could be crucial factors in inducing neuronal damages (Sharma and Sjöquist, 2002; Sharma *et al.*, 1998). When the BBB is altered to proteins, vasogenic edema formation will occur (Sharma and Cervós-Navarro, 1990; Sharma *et al.*, 1991, 1995). Spread of edema fluid within the brain microenvironment associated with other neurotoxic elements, for example, serotonin could affect neuronal structure and function depending on the specific 5-HT receptor types located in those brain areas, either on the neurons, glial cells, or on the microvessels (Sharma, 2009). Since hyperthermia could affect both the increase in serotonin and BBB permeability (Sharma and Hoopes, 2003; Sharma *et al.*, 1998), the magnitude and intensity of cocaine-induced changes in these parameters could markedly induce brain pathology.

H. Cocaine Activates Astrocytic Activity

Administration of cocaine results in massive upregulation of GFAP activity in different brain regions depending on its route of administration. In general, intracerebroventricular injection of cocaine showed most pronounced upregulation of GFAP in various brain regions followed by its systemic applications (Table IV; Fig. 5). The pattern of astrocytic activation closely corresponded to neuronal damage in different brain areas (Table III). Thus, hippocampus, caudate nucleus, hypothalamus, cerebral, and cerebellar cortices showed pronounced astrocytic activation followed by brain stem and spinal cord. A representative example of GFAP activation by cocaine is shown in Fig. 5. It appears that most of the perineuronal and perivascular astrocytes showed pronounced activation by cocaine administration. Dark and dense GFAP-immunoreactive astrocytes seen around neuronal dendrites and microvessels are in line with this idea.

Breakdown of the BBB and spread of edema fluid in various brain regions could result in astrocytic activation in such a short period of cocaine exposure (see Cervós-Navarro *et al.*, 1998; Sharma *et al.*, 1992a). In addition, hyperthermia also could influence GFAP activation deepening on the local or regional changes in the brain temperature (see Kiyatkin and Sharma, 2009; Kiyatkin *et al.*, 2007; Sharma and Kiyatkin, 2009; Sharma *et al.*, 1992a). In addition, local changes in serotonin metabolism may also activate astrocytic reactions (Cervós-Navarro

TABLE IV

EFFECTS OF COCAINE ADMINISTRATION ON HSP 72 KD EXPRESSION IN THE RAT BRAIN

HSP-immunoreactive cells (nr)

Expt. type	n	Cingulate cortex	Parietal cortex	Hippocampus CA-3	Thalamus dorsal N	Hypothalamus ventral	Cerebellum vermis	Brain stem medulla	Spinal cord cervical
Saline control	6	2 ± 2	1 ± 1	2 ± 2	0	0	1 ± 2	0	0
Intravenous cocaine									
30 mg/kg	8	34 ± 8**	48 ± 12**	56 ± 18**	34 ± 8**	38 ± 10**	58 ± 11**	18 ± 5**	16 ± 4**
Intraperitoneal cocaine									
60 mg/kg	6	18 ± 7**	22 ± 6**	32 ± 13**	24 ± 8**	22 ± 9**	38 ± 12**	8 ± 4*	8 ± 5*
Intracerebroventricular cocaine									
60 µg	8	38 ± 12**	43 ± 10**	48 ± 8**	38 ± 6**	30 ± 12**	54 ± 8**	6 ± 3*	6 ± 2*

Values are mean ± SD; *$p < 0.05$, **$p < 0.01$, Chi-square test from control group. Cells were counted in blinded fashion by at least two independent observers in each animal and the median values were used for calculation. HSP-positive cells that could be either neuron or glial cell were counted on Vibratome sections obtained from the designated area in each animal (data from H.S. Sharma et al., unpublished observations).

Cocaine neurotoxicity

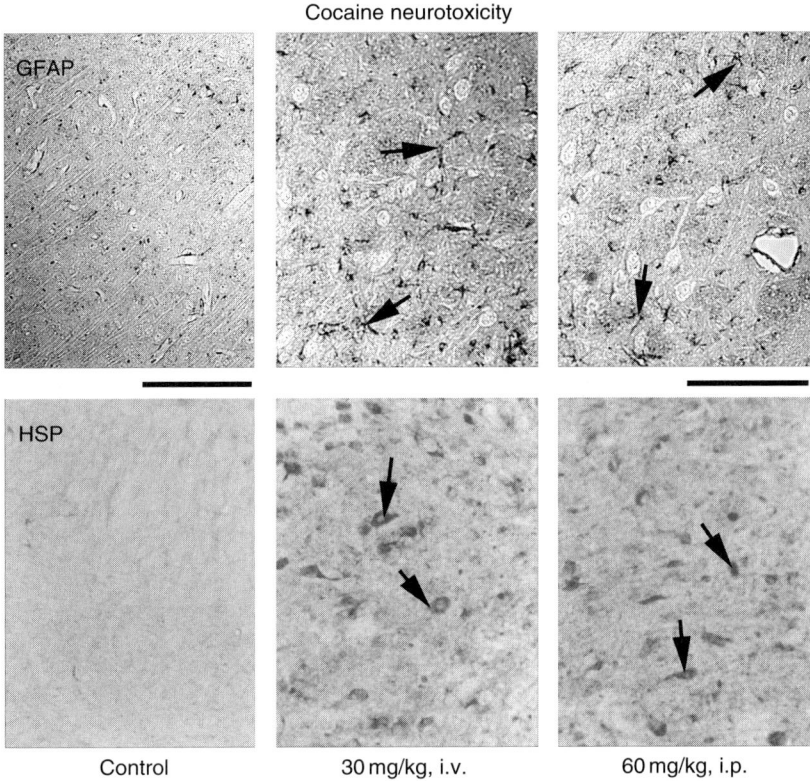

FIG. 5. Representative examples showing immunostaining of glial fibrillary acidic protein (GFAP) and heat shock protein (HSP 72 kD) in saline- (control) or cocaine-treated animals. Cocaine was administered either intravenously (30 mg/kg, i.v.) or intraperitoneally (60 mg/kg, i.p.) and the GFAP and HSP expression was examined 4 h after the drug administration. GFAP immunostaining on 3-μm thick paraffin sections showed marked activation of astrocytes in cocaine-treated rats as compared to saline treatment. HSP expression was examined on 40-μm thick Vibratome sections that show over-expression of the stress protein in cocaine-treated animals as compared to saline-treated control rat. In general, the HSP and GFAP expression were more pronounced in the rat that received cocaine through intravenous route as compared to its intraperitoneal administration. For details see text (H.S. Sharma et al., unpublished observations). (See Color Insert.)

et al., 1998). Blockade of serotonin synthesis reduces GFAP activation following brain injury (Sharma et al., 1993). Thus, it appears that cocaine-induced serotonin alteration, BBB breakdown, and brain edema formation all could contribute to GFAP activation. However, exact mechanism of cocaine-induced astrocytic activation requires further studies.

I. Cocaine Upregulates HSP Expression

Cocaine administration in rats induced marked increase in HSP 72 kD immunoreactivity in almost all brain regions examined (see Table IV; Fig. 5). This increase in HSP expression was most pronounced following intravenous cocaine administration followed by its intracerebroventricular and intraperitoneal injections (Table IV). Hippocampus and cerebellum showed the most pronounced HSP overexpression followed by parietal cortex, hypothalamus, and cingulate cortex (Table IV). Brain stem and spinal cord showed only mild upregulation of HSP by cocaine (Table IV). A representative example of HSP expression in cingulate cortex following intravenous and intraperitoneal administration of cocaine is shown in Fig. 5. The HSP expression was seen largely in the cell cytoplasm of the most neurons and in some cases, the cell nucleus was also stained (Fig. 5). In a given area of the brain only selective neurons showed HSP expression and several neurons in the vicinity are HSP negative (Fig. 5). This suggests that cocaine-induced stress reactions are very specific and precise. The possible reason for such a selective increase in HSP expression in the brain by cocaine is still not known.

On the other hand, intraperitoneal administration of cocaine showed much less HSP-positive neurons located at different places even in the identical brain areas. Whereas, intracerebroventricular cocaine exhibited much more widespread immunostaining of HSP with greater intensity in these brain areas (results not shown). This suggests that the cocaine-induced cellular stress response could be one of the crucial factors in influencing the magnitude and intensity of HSP expression. Obviously, the magnitude of cellular stress response may be related with the concatenation of cocaine in the brain. It is uncertain, whether direct exposure of cocaine to brain cells could also be responsible for dense and intense HSP reaction. Alternatively, cocaine-induced hyperthermia, breakdown of the BBB function, and spread of edema fluid may also contribute to HSP induction in different brain regions.

XI. Modulation of Serotonin 5-HT2 and 5-HT3 Receptors Influences Cocaine-Induced Neurotoxicity

Since cocaine administration increases serotonin concentrations in plasma and brain, it appears that serotonin could play major roles in cocaine-induced BBB breakdown and brain pathology. Serotonin-induced BBB breakdown is mediated through 5-HT2 receptors and thus blockade of 5-HT2 receptor with ritanserin or ketanserin attenuated BBB disruption in animal models of trauma and hyperthermia (see Sharma, 2009; Sharma and Hoopes, 2003). On the basis of these observations, we pretreated animals with either ketanserin or ritanserin (2 mg/kg, s.c.) 30 min before cocaine administration. Pretreatment with 5-HT2 receptor blockers markedly attenuated cocaine-induced BBB breakdown and brain pathology in rats. This indicates that 5-HT2 receptors play important roles in cocaine-induced BBB disruption and brain damage. Obviously, a reduction in BBB leakage is crucial in attenuating brain pathology (Sharma *et al.*, 1998).

Furthermore, there are evidences that blockade of 5-HT3 receptors are able to reduce anxiety and drug addiction induced changes in the brain function (Sharma *et al.*, 2007). Thus, we also examined the effects of a potent 5-HT3 receptor antagonist, ondansetron on cocaine-induced brain dysfunction. For this purpose, ondansetron (3 mg/kg, s.c.) was administered 30 min before cocaine administration and brain pathology and BBB dysfunction were examined. Our observations show that blockade of 5-HT3 receptors significantly attenuated cocaine-induced brain pathology and BBB leakage. However, this effect was significantly less pronounced in cocaine-treated animals as compared to the blockade of 5-HT2 receptors. Taken together, these observations suggest that both 5-HT2 and 5-HT3 receptors participate in cocaine-induced neurotoxicity.

XII. Cocaine-Induced BBB Dysfunction and Neurotoxicity Is Age Related

Preliminary data from our laboratory suggests that the effects of cocaine on brain pathology could vary according to the age of the animals. Thus, aged rats when given identical doses of cocaine through intravenous, intraperitoneal, or intracerebroventricular routes exhibited much more neuronal damage than the young rats (results not shown). In old animals exacerbation of BBB breakdown, HSP expression and astrocytic activation induced by cocaine was also most pronounced than the young rats (results not shown). This indicates that cocaine in identical doses could induce greater brain damage and other adverse effects in old age groups. The possible mechanisms of a higher sensitivity to cocaine on the BBB dysfunction in aged rats are currently not well understood. A lack of neuronal plasticity and aging-induced alterations in brain function could have contributed to this effect. However, additional studies are needed to identify the cellular and molecular mechanisms of cocaine-induced exacerbation of brain pathology in the aged animals. It remains to be seen whether the effects of serotonin modulating drugs that reduce the cocaine-induced brain pathology in young rats are also able to attenuate brain dysfunction in aged animals. This is a subject that is being investigated currently in our laboratory.

XIII. General Conclusion and Future Perspectives

In conclusion, our observations demonstrate that cocaine as a powerful psychostimulant if given in moderate or high amounts in the systemic or cerebral circulation comparable to that accumulated in the blood and brains of chronic cocaine abusers results in profound BBB breakdown and brain edema formation. In addition, hypertension associated with increased brain serotonin levels and altered regional blood flow together with BBB breakdown will further allow several toxic substances from the blood to enter into the brain fluid compartments. These events associated with cocaine use are crucial to induce brain pathology. It is also likely that acute effects of cocaine could open the BBB permeability enabling several unwanted substances to enter into the brain leading to long-term consequences on brain function. There are reasons to believe the cellular and molecular changes in the brain caused by cocaine administration could precipitate in an early and/or slowly developing neurodegenerative brain diseases. This idea is well reflected by our findings on cocaine-induced alterations in astrocytic activation along with overexpression of HSP activity. It may be that cocaine-induced hyperthermia and increase in brain serotonin levels could further contribute to marked cellular and molecular dysfunction in addition to their effects on BBB breakdown and/or brain pathology.

Our observations further point out that the route of cocaine administration is an important factor in inducing brain pathology. We used three different routes of administration of cocaine using three different doses. Using brain pathology as the endpoint observations, our findings suggest that 30 mg/kg intravenous dose is quite comparable to that of 60 mg/kg cocaine through intraperitoneal route. Furthermore, only 50-μg intracerebroventricular cocaine is capable to induce brain pathology that is quite similar to that seen during 30 mg/kg intravenous or 60 mg/kg intraperitoneal cocaine. This idea is further supported by our observations that lower doses of cocaine given to either routes were not that effective on causing brain pathology. However, the regional changes of the BBB function, brain edema, cerebral blood flow, serotonin levels, and glial or cellular stress responses following cocaine administration were quite different in different brain regions depending on the route of its administration. This suggests that apart from comparable doses, the routes of cocaine administration play important roles in precipitating brain pathology.

Although we did not examine the reversibility of cocaine-induced brain pathology or BBB dysfunction, the magnitude and intensity of brain damage seen in cocaine-treated animals suggest that most of these changes could be irreversible in nature and may trigger further extensive neurodegenerative changes over time. However, it remains to be seen whether repeated use of cocaine in small doses, or long-term survival following cocaine administration could result in clear neurodegenerative changes after several weeks or months of the initial drug injection.

Moreover, it is still not known whether cocaine use could also increase the vulnerability of the brain to other neurological disease, such as stroke, neuroinflammation, and neuropsychiatric disease. These aspects of long-term effects of cocaine on brain degeneration and its susceptibility to additional stress situation, for example, hyperthermia, ischemia, stroke, bacterial, viral infections, and/or nanoparticles exposure in exacerbating brain pathology are currently being investigated in our laboratory.

Acknowledgments

Research reported in this investigation is supported by Grants from Swedish Medical Research Council Grant no. 2710 (HSS): Alexander von Humboldt Foundation (HSS), Bonn, Germany; The University Grants Commission (HSS), New Delhi, India; Indian Council of Medical Research (HSS), New Delhi, India; and Astra-Zeneca Mölndal (HSS), Sweden. We sincerely thank the reviewers for their constructive suggestions for further improvement of this review with regard to elaboratation of serotoninergic mechanisms of cocaine actions. Technical assistance of Kärstin Flink, Kerstin Rystedt (Uppsala), Franziska Drum, Angela Jan (Berlin), Shiv Mandir Singh, and Rakesh K. Gupta (Varanasi) are highly appreciated.

References

Amon, C. A., Tate, L. G., Wright, R. K., and Matusiak, W. (1986). Sudden death due to ingestion of cocaine. *J. Anal. Toxicol.* **10**(5), 217–218.

Argaud, L., Ferry, T., Le, Q. H., Marfisi, A., Ciorba, D., Achache, P., Ducluzeau, R., and Robert, D. (2007). Short- and long-term outcomes in heatstroke following the 2003 heat wave in Lyon, France. *Arch. Intern. Med.* **167**(20), 2177–2183.

Barroso-Moguel, R., Mendez-Armenta, M., and Villeda-Hernandez, J. (1995). Experimental nephropathy by chronic administration of cocaine in rats. *Toxicology* **98**(1–3), 41–46.

Bauwens, J. E., Boggs, J. M., and Hartwell, P. S. (1989). Fatal hyperthermia associated with cocaine use. *West J. Med.* **150**(2), 210–212.

Benowitz, N. L. (1992). How toxic is cocaine? *Ciba Found. Symp.* **166,** 125–143 discussion 143–148; (Review).

Bignami, A. (1984). Glial fibrillary acidic (GFA) protein in Müller glia. Immunofluorescence study of the goldfish retina. *Brain Res.* **300**(1), 175–178.

Bignami, A., and Dahl, D. (1988). Expression of brain-specific hyaluronectin (BHN), a hyaluronate-binding protein, in dog postnatal development. *Exp. Neurol.* **99**(1), 107–117.

Billman, G. E. (1995). Cocaine: A review of its toxic actions on cardiac function. *Crit. Rev. Toxicol.* **25**(2), 113–132 (Review).

Blake, M. J., Buckley, A. R., Buckley, D. J., LaVoi, K. P., and Bartlett, T. (1994). Neural and endocrine mechanisms of cocaine-induced 70-kDa heat shock protein expression in aorta and adrenal gland. *J. Pharmacol. Exp. Ther.* **268**(1), 522–529.

Boelsterli, U. A., and Göldlin, C. (1991). Biomechanisms of cocaine-induced hepatocyte injury mediated by the formation of reactive metabolites. *Arch. Toxicol.* **65**(5), 351–360 (Review).

Bonvento, G., MacKenzie, E. T., and Edvinsson, L. (1991). Serotonergic innervation of the cerebral vasculature: Relevance to migraine and ischaemia. *Brain Res. Brain Res. Rev.* **16**(3), 257–263 (Review).

Boyer, C. S., Ross, D., and Petersen, D. R. (1988). Sex and strain differences in the hepatotoxic response to acute cocaine administration in the mouse. *J. Biochem. Toxicol.* **3**, 295–307.

Bowers, M. S., and Kalivas, P. W. (2003). Forebrain astroglial plasticity is induced following withdrawal from repeated cocaine administration. *Eur. J. Neurosci.* **17**(6), 1273–1278.

Brecklin, C. S., and Bauman, J. L. (1999). Cardiovascular effects of cocaine: Focus on hypertension. *J. Clin. Hypertens. (Greenwich)* **1**(3), 212–217.

Capela, J. P., Carmo, H., Remião, F., Bastos, M. L., Meisel, A., and Carvalho, F. (2009). Molecular and cellular mechanisms of ecstasy-induced neurotoxicity: An overview. *Mol. Neurobiol.* **39**(3), 210–271 (Review).

Castellani, S. A., Ellinwood, E. H. Jr., and Kilbey, M. M. (1978a). Tolerance to cocaine-induced convulsions in the cat. *Eur. J. Pharmacol.* **47**(1), 57–61.

Castellani, S., Ellinwood, E. H. Jr, and Kilbey, M. M. (1978b). Behavioral analysis of chronic cocaine intoxication in the cat. *Biol. Psychiatry* **13**(2), 203–215.

Ceolin, L., Schwarz, A. J., Gozzi, A., Reese, T., and Bifone, A. (2007). Effects of cocaine on blood flow and oxygen metabolism in the rat brain: Implications for phMRI. *Magn. Reson. Imaging* **25**(6), 795–800.

Cervós-Navarro, J., Sharma, H. S., Westman, J., and Bongcam-Rudloff, E. (1998). Glial reactions in the central nervous system following heat stress. *Prog. Brain Res.* **115**, 241–274 (Review).

Copersino, M. L., Heming, R. I., Better, W., Cadet, J. L., and Gorelick, D. A. (2009). EEG and cerebral blood flow velocity abnormalities in chronic cocaine users. *Clin. EEG Neurosci.* **40**(1), 39–42.

Crandall, C. G., Vongpatanasin, W., and Victor, R. G. (2002). Mechanism of cocaine-induced hyperthermia in humans. *Ann. Intern. Med.* **136**(11), 785–791.

Cregler, L. L., and Mark, H. (1987). Relation of stroke to cocaine abuse. *N. Y. State J. Med.* **87**(2), 128–129.

Cunningham, K. A., Paris, J. M., and Goeders, N. E. (1992). Chronic cocaine enhances serotonin autoregulation and serotonin uptake binding. *Synapse* **11**(2), 112–123.

Dandekar, M. P., Singru, P. S., Kokare, D. M., and Subhedar, N. K. (2009). Cocaine- and amphetamine-regulated transcript peptide plays a role in the manifestation of depression: Social isolation and olfactory bulbectomy models reveal unifying principles. *Neuropsychopharmacology* **34**(5), 1288–1300.

Daras, M., Tuchman, A. J., and Marks, S. (1991). Central nervous system infarction related to cocaine abuse. *Stroke* **22**(10), 1320–1325.

Daras, M., Kakkouras, L., Tuchman, A. J., and Koppel, B. S. (1995). Rhabdomyolysis and hyperthermia after cocaine abuse: A variant of the neuroleptic malignant syndrome? *Acta Neurol. Scand.* **92**(2), 161–165.

De Giorgio, F., Vetrugno, G., Fucci, N., Rainio, J., Tartaglione, T., Di Lazzaro, V., and Carbone, A. (2007). Fatal stroke in a young cocaine drug addict: Chemical hair analysis and cervical artery examination twenty months after death. *Folia Neuropathol.* **45**(3), 149–152.

de Wit, M., Gennings, C., Zilberberg, M., Burnham, E. L., Moss, M., and Balster, R. L. (2008). Drug withdrawal, cocaine and sedative use disorders increase the need for mechanical ventilation in medical patients. *Addiction* **103**(9), 1500–1508.

del Castillo, C., Morales, L., Alguacil, L. F., Salas, E., Garrido, E., Alonso, E., and Pérez-García, C. (2009). Proteomic analysis of the nucleus accumbens of rats with different vulnerability to cocaine addiction. *Neuropharmacology* **57**(1), 41–48.

Denton, J. S., Donoghue, E. R., McReynolds, J., and Kalelkar, M. B. (2008). An epidemic of illicit fentanyl deaths in Cook County, Illinois: September 2005 through April 2007. *J. Forensic Sci.* **53**(2), 452–454.

Dey, S., and Snow, D. M. (2007). Cocaine exposure *in vitro* induces apoptosis in fetal locus coeruleus neurons through TNF-alpha-mediated induction of Bax and phosphorylated c-Jun NH(2)-terminal kinase. *J. Neurochem.* **103**(2), 542–556.

Dhillon, N. K., Williams, R., Peng, F., Tsai, Y. J., Dhillon, S., Nicolay, B., Gadgil, M., Kumar, A., and Buch, S. J. (2007). Cocaine-mediated enhancement of virus replication in macrophages: Implications for human immunodeficiency virus-associated dementia. *J. Neurovirol.* **13**(6), 483–495.

Dhillon, N. K., Peng, F., Bokhari, S., Callen, S., Shin, S. H., Zhu, X., Kim, K. J., and Buch, S. J. (2008). Cocaine-mediated alteration in tight junction protein expression and modulation of CCL2/CCR2 axis across the blood–brain barrier: Implications for HIV-dementia. *J. Neuroimmune Pharmacol.* **3**(1), 52–56.

Erzouki, H. K., Allen, A. C., Newman, A. H., Goldberg, S. R., and Schindler, C. W. (1995). Effects of cocaine, cocaine metabolites and cocaine pyrolysis products on the hindbrain cardiac and respiratory centers of the rabbit. *Life Sci.* **57**(20), 1861–1868.

Fan, L., Sawbridge, D., George, V., Teng, L., Bailey, A., Kitchen, I., and Li, J. M. (2009). Chronic cocaine-induced cardiac oxidative stress and mitogen-activated protein kinase activation: The role of Nox2 oxidase. *J. Pharmacol. Exp. Ther.* **328**(1), 99–106.

Fattore, L., Puddu, M. C., Picciau, S., Cappai, A., Fratta, W., Serra, G. P., and Spiga, S. (2002). Astroglial *in vivo* response to cocaine in mouse dentate gyrus: A quantitative and qualitative analysis by confocal microscopy. *Neuroscience* **110**(1), 1–6.

Fellin, T. (2009). Communication between neurons and astrocytes: Relevance to the modulation of synaptic and network activity. *J. Neurochem.* **108**(3), 533–544 (Review).

Fiala, M., Eshleman, A. J., Cashman, J., Lin, J., Lossinsky, A. S., Suarez, V., Yang, W., Zhang, J., Popik, W., Singer, E., Chiappelli, F., Carro, E., *et al.* (2005). Cocaine increases human immunodeficiency virus type 1 neuroinvasion through remodeling brain microvascular endothelial cells. *J. Neurovirol.* **11**(3), 281–291.

Fowler, R. C., Rich, C. L., and Young, D. (1986). San Diego Suicide Study. II. Substance abuse in young cases. *Arch. Gen. Psychiatry* **43**(10), 962–965.

Füzesi, T., Sánchez, E., Wittmann, G., Singru, P. S., Fekete, C., and Lechan, R. M. (2008). Regulation of cocaine- and amphetamine-regulated transcript-synthesising neurons of the hypothalamic paraventricular nucleus by endotoxin; implications for lipopolysaccharide-induced regulation of energy homeostasis. *J. Neuroendocrinol.* **20**(9), 1058–1066.

Gan, X., Zhang, L., Berger, O., Stins, M. F., Way, D., Taub, D. D., Chang, S. L., Kim, K. S., House, S. D., Weinand, M., Witte, M., Graves, M. C., *et al.* (1999). Cocaine enhances brain endothelial adhesion molecules and leukocyte migration. *Clin. Immunol.* **91**(1), 68–76.

Gillis, R. A., Hernandez, Y. M., Erzouki, H. K., Raczkowski, V. F., Mandal, A. K., Kuhn, F. E., and Dretchen, K. L. (1995). Sympathetic nervous system mediated cardiovascular effects of cocaine are primarily due to a peripheral site of action of the drug. *Drug Alcohol Depend.* **37**(3), 217–230.

Guan, X., Zhang, R., Xu, Y., and Li, S. (2009). Cocaine withdrawal enhances long-term potentiation in rat hippocampus via changing the activity of corticotropin-releasing factor receptor subtype 2. *Neuroscience* **161**(3), 665–670.

Hahn, J., Hopf, F. W., and Bonci, A. (2009). Chronic cocaine enhances corticotropin-releasing factor-dependent potentiation of excitatory transmission in ventral tegmental area dopamine neurons. *J. Neurosci.* **29**(20), 6535–6544.

Hardebo, J. E., and Hindfelt, B. (1981). The effect of temperature elevation on the cerebrovascular response to noradrenaline and 5-hydroxytryptamine. *Acta Physiol. Scand.* **112**(4), 413–416.

Irwin, M. R., Olmos, L., Wang, M., Valladares, E. M., Motivala, S. J., Fong, T., Newton, T., Butch, A., Olmstead, R., and Cole, S. W. (2007). Cocaine dependence and acute cocaine induce decreases of monocyte proinflammatory cytokine expression across the diurnal period: Autonomic mechanisms. *J. Pharmacol. Exp. Ther.* **320**(2), 507–515.

Jaehne, E. J., Salem, A., and Irvine, R. J. (2007). Pharmacological and behavioral determinants of cocaine, methamphetamine, 3,4-methylenedioxymethamphetamine, and para-methoxyamphetamine-induced hyperthermia. *Psychopharmacology (Berlin)* **194**(1), 41–52.

Jean, A., Conductier, G., Manrique, C., Bouras, C., Berta, P., Hen, R., Charnay, Y., Bockaert, J., and Compan, V. (2007). Anorexia induced by activation of serotonin 5-HT4 receptors is mediated by increases in CART in the nucleus accumbens. *Proc. Natl. Acad. Sci. USA* **104**(41), 16335–16340.

Jonas, J. M., Gold, M. S., Sweeney, D., and Pottash, A. L. (1987). Eating disorders and cocaine abuse: A survey of 259 cocaine abusers. *J. Clin. Psychiatry* **48**(2), 47–50.

Jover, R., Ponsoda, X., Gómez-Lechón, J., and Castell, J. V. (1993). Cocaine hepatotoxicity: Two different toxicity mechanisms for phenobarbital-induced and non-induced rat hepatocytes. *Biochem. Pharmacol.* **46**(11), 1967–1974.

Kenedi, C. A., Joynt, K. E., and Goforth, H. W. (2008). Comorbid HIV encephalopathy and cocaine use as a risk factor for new-onset seizure disorders. *CNS Spectr.* **13**(3), 230–234.

Kessler, F. H., Woody, G., Portela, L. V., Tort, A. B., De Boni, R., Peuker, A. C., Genro, V., von Diemen, L., de Souza, D. O., and Pechansky, F. (2007). Brain injury markers (S100B and NSE) in chronic cocaine dependents. *Rev. Bras. Psiquiatr.* **29**(2), 134–139.

Killam, A. L. (1993). Cardiovascular and thrombosis pathology associated with cocaine use. *Hematol. Oncol. Clin. North Am.* **7**(6), 1143–1151 (Review).

Kiyatkin, E. A., and Bae, D. (2008). Behavioral and brain temperature responses to salient environmental stimuli and intravenous cocaine in rats: Effects of diazepam. *Psychopharmacology (Berlin)* **196**(3), 343–356.

Kiyatkin, E. A., and Sharma, H. S. (2009). Permeability of the blood–brain barrier depends on brain temperature. *Neuroscience* **161**(3), 926–939.

Kiyatkin, E. A., Brown, P. L., and Sharma, H. S. (2007). Brain edema and breakdown of the blood–brain barrier during methamphetamine intoxication: Critical role of brain hyperthermia. *Eur. J. Neurosci.* **26**(5), 1242–1253.

Kleerup, E. C., Wong, M., Marques-Magallanes, J. A., Goldman, M. D., and Tashkin, D. P. (1997). Acute effects of intravenous cocaine on pulmonary artery pressure and cardiac index in habitual crack smokers. *Chest* **111**(1), 30–35.

Korbo, L. (1999). Glial cell loss in the hippocampus of alcoholics. *Alcohol Clin. Exp. Res.* **23**(1), 164–168.

Kosten, T. R., and Kleber, H. D. (1988). Rapid death during cocaine abuse: A variant of the neuroleptic malignant syndrome? *Am. J. Drug Alcohol Abuse* **14**(3), 335–346 (Review).

Krendel, D. A., Ditter, S. M., Frankel, M. R., and Ross, W. K. (1990). Biopsy-proven cerebral vasculitis associated with cocaine abuse. *Neurology* **40**(7), 1092–1094.

Lepsch, L. B., Munhoz, C. D., Kawamoto, E. M., Yshii, L. M., Lima, L. S., Curi-Boaventura, M. F., Salgado, T. M., Curi, R., Planeta, C. S., and Scavone, C. (2009). Cocaine induces cell death and activates the transcription nuclear factor kappa-b in pc12 cells. *Mol. Brain* **2**(1), 3.

Lesse, H., and Collins, J. P. (1979). Effects of cocaine on propagation of limbic seizure activity. *Pharmacol. Biochem. Behav.* **11**(6), 689–694.

Levine, S. R., and Welch, K. M. (1988). Cocaine and stroke. *Stroke* **19**(6), 779–783 (Review).

Levine, S. R., Washington, J. M., Jefferson, M. F., Kieran, S. N., Moen, M., Feit, H., and Welch, K. M. (1987). "Crack" cocaine-associated stroke. *Neurology* **37**(12), 1849–1853.

Levy, A. D., Rittenhouse, P. A., Li, Q., Yracheta, J., Kunimoto, K., and Van de Kar, L. D. (1994a). Influence of repeated cocaine exposure on the endocrine and behavioral responses to stress in rats. *Psychopharmacology (Berlin)* **113**(3–4), 547–554.

Levy, A. D., Baumann, M. H., and Van de Kar, L. D. (1994b). Monoaminergic regulation of neuroendocrine function and its modification by cocaine. *Front. Neuroendocrinol.* **15**(2), 85–156 (Review).

Lomax, P., and Daniel, K. A. (1990a). Cocaine and body temperature in the rat: Effects of exercise and age. *Pharmacology* **41**(6), 309–315.

Lomax, P., and Daniel, K. A. (1990b). Cocaine and body temperature in the rat: Effects of ambient temperature. *Pharmacology* **40**(2), 103–109.

Lomax, P., and Daniel, K. A. (1990c). Cocaine and body temperature in the rat: Effect of exercise. *Pharmacol. Biochem. Behav.* **36**(4), 889–892.

Lombard, J., Wong, B., and Young, J. H. (1988). Acute renal failure due to rhabdomyolysis associated with cocaine toxicity. *West J. Med.* **148**(4), 466–468 (no abstract available).

Majewska, M. D. (1996). Cocaine addiction as a neurological disorder: Implications for treatment. *NIDA Res. Monogr.* **163**, 1–26 (Review).

Maron, B. J., Doerer, J. J., Haas, T. S., Tierney, D. M., and Mueller, F. O. (2009). Sudden deaths in young competitive athletes: Analysis of 1866 deaths in the United States, 1980–2006. *Circulation* **119**(8), 1085–1092.

Marzuk, P. M., Tardiff, K., Leon, A. C., Hirsch, C. S., Portera, L., Iqbal, M. I., Nock, M. K., and Hartwell, N. (1998). Ambient temperature and mortality from unintentional cocaine overdose. *JAMA* **279**(22), 1795–1800.

Masu, K., Saino, T., Kuroda, T., Matsuura, M., Russa, A. D., Ishikita, N., and Satoh, Y. (2008). Regional differences in 5-HT receptors in cerebral and testicular arterioles of the rat as revealed by Ca^{2+} imaging of real-time confocal microscopy: Variances by artery size and organ specificity. *Arch. Histol. Cytol.* **71**(5), 291–302.

Matsuzaki, M., Spingler, P. J., Misra, A. L., and Mule, S. J. (1976). Cocaine: Tolerance to its convulsant and cardiorespiratory stimulating effects in the monkey. *Life Sci.* **19**(2), 193–203.

Menaker, J., Farcy, D. A., Boswell, S. A., Stein, D. M., Dutton, R. P., Hess, J. R., and Scalea, T. M. (2008). Cocaine-induced agitated delirium with associated hyperthermia: A case report. *J. Emerg. Med.* DOI:10.1016/j.jemermed.2008.02.055.

Mendelson, J. H., Mello, N. K., Sholar, M. B., Siegel, A. J., Mutschler, N., and Halpern, J. (2002). Temporal concordance of cocaine effects on mood states and neuroendocrine hormones. *Psychoneuroendocrinology* **27**(1–2), 71–82.

Merigian, K. S., and Roberts, J. R. (1987). Cocaine intoxication: Hyperpyrexia, rhabdomyolysis and acute renal failure. *J. Toxicol. Clin. Toxicol.* **25**(1–2), 135–148.

Mott, S. H., Packer, R. J., and Soldin, S. J. (1994). Neurologic manifestations of cocaine exposure in childhood. *Pediatrics* **93**(4), 557–560.

Mueller, M. A., Yuan, J., Neudorffer, A., Peters, F., Maurer, H., McCann, U., Felim, A. F., Largeron, M., and Ricaurte, G. A. (2009). Further studies on the role of metabolites in MDMA-induced serotonergic neurotoxicity. *Drug Metab. Dispos.* DOI: 10.1124/dmd.109.028340.

Pengelly, L. D., Campbell, M. E., Cheng, C. S., Fu, C., Gingrich, S. E., and Macfarlane, R. (2007). Anatomy of heat waves and mortality in Toronto: Lessons for public health protection. *Can. J. Public Health* **98**(5), 364–368.

Peterson, P. L., Roszler, M., Jacobs, I., and Wilner, H. I. (1991). Neurovascular complications of cocaine abuse. *J. Neuropsychiatry Clin. Neurosci.* **3**(2), 143–149.

Phillips, K., Luk, A., Soor, G. S., Abraham, J. R., Leong, S., and Butany, J. (2009). Cocaine cardiotoxicity: A review of the pathophysiology, pathology, and treatment options. *Am. J. Cardiovasc. Drugs* **9**(3), 177–196 DOI:10.2165/00129784-200909030-00005 (Review).

Pitts, D. K., and Marwah, J. (1987). Cocaine modulation of central monoaminergic neurotransmission. *Pharmacol. Biochem. Behav.* **26**(2), 453–461.

Post, R. M., and Kopanda, R. T. (1976). Cocaine, kindling, and psychosis. *Am. J. Psychiatry* **133**(6), 627–634 (Review).

Powers, J. F., Alroy, J., and Shuster, L. (1992). Hepatic morphologic and biochemical changes induced by subacute cocaine administration in mice. *Toxicol. Pathol.* **20**(1), 61–70.

Pozzi, M., Roccatagliata, D., and Sterzi, R. (2008). Drug abuse and intracranial hemorrhage. *Neurol. Sci.* **29**(Suppl. 2), S269–S270 (Review).

Rosenberg, J., Pentel, P., Pond, S., Benowitz, N., and Olson, K. (1986). Hyperthermia associated with drug intoxication. *Crit. Care Med.* **14**(11), 964–969.

Schmidt, K. F., Febo, M., Shen, Q., Luo, F., Sicard, K. M., Ferris, C. F., Stein, E. A., and Duong, T. Q. (2006). Hemodynamic and metabolic changes induced by cocaine in anesthetized rat observed with multimodal functional MRI. *Psychopharmacology (Berlin)* **185**(4), 479–486.

Schreiber, A. L., and Formal, C. S. (2007). Spinal cord infarction secondary to cocaine use. *Am. J. Phys. Med. Rehabil.* **86**(2), 158–160.

Shanti, C. M., and Lucas, C. E. (2003). Cocaine and the critical care challenge. *Crit. Care Med.* **31**(6), 1851–1859 (Review).

Sharma, H. S. (1987). Effect of captopril (a converting enzyme inhibitor) on blood–brain barrier permeability and cerebral blood flow in normotensive rats. *Neuropharmacology* **26**(1), 85–92.

Sharma, H. S. (2005). Heat-related deaths are largely due to brain damage. *Indian J. Med. Res.* **121**(5), 621–623.

Sharma, H. S. (2006a). Hyperthermia induced brain oedema: Current status and future perspectives. *Indian J. Med. Res.* **123**(5), 629–652 (Review).

Sharma, H. S. (2006b). Hyperthermia influences excitatory and inhibitory amino acid neurotransmitters in the central nervous system. An experimental study in the rat using behavioural, biochemical, pharmacological, and morphological approaches. *J. Neural Transm.* **113**(4), 497–519.

Sharma, H. S., Lundstedt, T., Boman, A., Lek, P., Seifert, E., Wiklund, L., and Ali, S. F. (2006c). A potent serotonin-modulating compound AP-267 attenuates morphine withdrawal-induced blood-brain barrier dysfunction in rats. *Ann. NY Acad. Sci.* Aug **1074,** 482–496.

Sharma, H. S. (2007a). Methods to produce hyperthermia-induced brain dysfunction. *Prog. Brain Res.* **162**, 173–199 (Review).

Sharma, H. S. (2007b). Neurodegeneration and neuroregeneration: Recent advancements and future perspectives. *Curr. Pharm. Des.* **13**(18), 1825–1827.

Sharma, H. S. (2008). New perspectives for the treatment options in spinal cord injury. *Expert Opin. Pharmacother.* **9**(16), 2773–2800 (Review).

Sharma, H. S. (2009). Blood–central nervous system barriers: The gateway to neurodegeneration, neuroprotection and neuroregeneration. *In* "Handbook of Neurochemistry and Molecular Neurobiology: Brain and Spinal Cord Trauma" (A. Lajtha, N. Banik, and S. K. Ray, Eds.), **11**, pp. 363–457. Springer-Verlag, Berlin Chapter 17.

Sharma, H. S., and Ali, S. F. (2006). Alterations in blood–brain barrier function by morphine and methamphetamine. *Ann. N. Y. Acad. Sci.* **1074,** 198–224.

Sharma, H. S., and Ali, S. F. (2008). Acute administration of 3,4-methylenedioxymethamphetamine induces profound hyperthermia, blood–brain barrier disruption, brain edema formation, and cell injury. *Ann. N. Y. Acad. Sci.* **1139,** 242–258.

Sharma, H. S., and Cervós-Navarro, J. (1990). Brain oedema and cellular changes induced by acute heat stress in young rats. *Acta Neurochir. Suppl. (Wien)* **51,** 383–386.

Sharma, H. S., and Dey, P. K. (1981). Impairment of blood–brain barrier (BBB) in rat by immobilization stress: role of serotonin (5-HT). *Indian J. Physiol. Pharmacol.* **25**(2), 111–122.

Sharma, H. S., and Dey, P. K. (1984). Role of 5-HT on increased permeability of blood–brain barrier under heat stress. *Indian J. Physiol. Pharmacol.* **28**(4), 259–267.

Sharma, H. S., and Dey, P. K. (1986). Influence of long-term immobilization stress on regional blood–brain barrier permeability, cerebral blood flow and 5-HT level in conscious normotensive young rats. *J. Neurol. Sci.* **72**(1), 61–76.

Sharma, H. S., and Dey, P. K. (1987). Influence of long-term acute heat exposure on regional blood-brain barrier permeability, cerebral blood flow and 5-HT level in conscious normotensive young rats. *Brain Res.* **424**(1), 153–162.

Sharma, H. S., and Hoopes, P. J. (2003). Hyperthermia induced pathophysiology of the central nervous system. *Int. J. Hyperthermia* **19**(3), 325–354 (Review).

Sharma, H. S., and Kiyatkin, E. A. (2009). Rapid morphological brain abnormalities during acute methamphetamine intoxication in the rat: An experimental study using light and electron microscopy. *J. Chem. Neuroanat.* **37**(1), 18–32.

Sharma, H. S., and Sharma, A. (2008). Antibodies as promising novel neuroprotective agents in the central nervous system injuries. *Central Nerv. Syst. Agents Med. Chem.* **8**(3), 143–169.

Sharma, H. S., and Sjöquist, P. O. (2002). A new antioxidant compound H-290/51 modulates glutamate and GABA immunoreactivity in the rat spinal cord following trauma. *Amino Acids* **23**(1–3), 261–272.

Sharma, H. S., and Westman, J. (2003). Depletion of endogenous serotonin synthesis with *p*-CPA attenuates upregulation of constitutive isoform of heme oxygenase-2 expression, edema formation and cell injury following a focal trauma to the rat spinal cord. *Acta Neurochir. Suppl.* **86**, 389–394.

Sharma, H. S., Olsson, Y., and Dey, P. K. (1990). Changes in blood–brain barrier and cerebral blood flow following elevation of circulating serotonin level in anesthetized rats. *Brain Res.* **517**(1–2), 215–223.

Sharma, H. S., Cervós-Navarro, J., and Dey, P. K. (1991). Acute heat exposure causes cellular alteration in cerebral cortex of young rats. *Neuroreport* **2**(3), 155–158.

Sharma, H. S., Nyberg, F., Cervos-Navarro, J., and Dey, P. K. (1992a). Histamine modulates heat stress-induced changes in blood–brain barrier permeability, cerebral blood flow, brain oedema and serotonin levels: An experimental study in conscious young rats. *Neuroscience* **50**(2), 445–454.

Sharma, H. S., Zimmer, C., Westman, J., and Cervós-Navarro, J. (1992b). Acute systemic heat stress increases glial fibrillary acidic protein immunoreactivity in brain: Experimental observations in conscious normotensive young rats. *Neuroscience* **48**(4), 889–901.

Sharma, H. S., Olsson, Y., and Cervós-Navarro, J. (1993). *p*-Chlorophenylalanine, a serotonin synthesis inhibitor, reduces the response of glial fibrillary acidic protein induced by trauma to the spinal cord. An immunohistochemical investigation in the rat. *Acta Neuropathol* **86**(5), 422–427.

Sharma, H. S., Westman, J., Navarro, J. C., Dey, P. K., and Nyberg, F. (1995a). Probable involvement of serotonin in the increased permeability of the blood–brain barrier by forced swimming. An experimental study using Evans blue and ^{131}I-sodium tracers in the rat. *Behav. Brain Res.* **72**(1–2), 189–196.

Sharma, H. S., Olsson, Y., and Westman, J. (1995b). A serotonin synthesis inhibitor, *p*-chlorophenylalanine reduces the heat shock protein response following trauma to the spinal cord: An immunohistochemical and ultrastructural study in the rat. *Neurosci. Res.* **21**(3), 241–249.

Sharma, H. S., Westman, J., and Nyberg, F. (1998). Pathophysiology of brain edema and cell changes following hyperthermic brain injury. *Prog. Brain Res.* **115**, 351–412 (Review).

Sharma, H. S., Winkler, T., Stålberg, E., Gordh, T., Alm, P., and Westman, J. (2003). Topical application of TNF-alpha antiserum attenuates spinal cord trauma induced edema formation, microvascular permeability disturbances and cell injury in the rat. *Acta Neurochir. Suppl.* **86**, 407–413.

Sharma, H. S., Duncan, J. A., and Johanson, C. E. (2006a). Whole-body hyperthermia in the rat disrupts the blood–cerebrospinal fluid barrier and induces brain edema. *Acta Neurochir. Suppl.* **96**, 426–431.

Sharma, H. S., Lundstedt, T., Boman, A., Lek, P., Seifert, E., Wiklund, L., and Ali, S. F. (2006b). A potent serotonin-modulating compound AP-267 attenuates morphine withdrawal-induced blood–brain barrier dysfunction in rats. *Ann. N. Y. Acad. Sci.* **1074,** 482–496.

Sharma, H. S., Sjöquist, P. O., and Ali, S. F. (2007). Drugs of abuse-induced hyperthermia, blood–brain barrier dysfunction and neurotoxicity: Neuroprotective effects of a new antioxidant compound H-290/51. *Curr. Pharm. Des.* **13**(18), 1903–1923 (Review).

Shibolet, S., Lancaster, M. C., and Danon, Y. (1976). Heat stroke: A review. *Aviat. Space Environ. Med.* **47**(3), 280–301 (Review).

Sim, M. E., Lyoo, I. K., Streeter, C. C., Covell, J., Sarid-Segal, O., Ciraulo, D. A., Kim, M. J., Kaufman, M. J., Yurgelun-Todd, D. A., and Renshaw, P. F. (2007). Cerebellar gray matter volume correlates with duration of cocaine use in cocaine-dependent subjects. *Neuropsychopharmacology* **32**(10), 2229–2237.

Spiehler, V. R., and Reed, D. (1985). Brain concentrations of cocaine and benzoylecgonine in fatal cases. *J. Forensic Sci.* **30**(4), 1003–1011.

Su, J., Li, J., Li, W., Altura, B. T., and Altura, B. M. (2003). Cocaine induces apoptosis in cerebral vascular muscle cells: Potential roles in strokes and brain damage. *Eur. J. Pharmacol.* **482**(1–3), 61–66.

Tanvetyanon, T., Dissin, J., and Selcer, U. M. (2001). Hyperthermia and chronic pancerebellar syndrome after cocaine abuse. *Arch. Intern. Med.* **161**(4), 608–610.

Tolat, R. D., O' Dell, M. W., Golamco-Estrella, S. P., and Avella, H. (2000). Cocaine-associated stroke: Three cases and rehabilitation considerations. *Brain Inj.* **14**(4), 383–391.

Treadwell, S. D., and Robinson, T. G. (2007). Cocaine use and stroke. *Postgrad. Med. J.* **83**(980), 389–394 (Review).

Vicentic, A., and Jones, D. C. (2007). The CART (cocaine- and amphetamine-regulated transcript) system in appetite and drug addiction. *J. Pharmacol. Exp. Ther.* **320**(2), 499–506 (Review).

Wallinga, A. E., ten Voorde, A. M., de Boer, S. F., Koolhaas, J. M., and Buwalda, B. (2009). MDMA-induced serotonergic neurotoxicity enhances aggressiveness in low- but not high-aggressive rats. *Eur. J. Pharmacol.* **618**(1–3), 22–27.

Walsh, S. L., Stoops, W. W., Moody, D. E., Lin, S. N., and Bigelow, G. E. (2009). Repeated dosing with oral cocaine in humans: Assessment of direct effects, withdrawal, and pharmacokinetics. *Exp. Clin. Psychopharmacol.* **17**(4), 205–216.

Welch, W. J. (1992). Mammalian stress response: Cell physiology, structure/function of stress proteins, and implications for medicine and disease. *Physiol. Rev.* **72**(4), 1063–1081 (Review).

Welder, A. A. (1992). A primary culture system of adult rat heart cells for the evaluation of cocaine toxicity. *Toxicology* **72**(2), 175–187.

Westman, J., and Sharma, H. S. (1998). Heat shock protein response in the central nervous system following hyperthermia. *Prog. Brain Res.* **115**, 207–239 (Review).

Westman, J., Drieu, K., and Sharma, H. S. (2000). Antioxidant compounds EGB-761 and BN-520 21 attenuate heat shock protein (HSP 72 kD) response, edema and cell changes following hyperthermic brain injury. An experimental study using immunohistochemistry in the rat. *Amino Acids* **19**(1), 339–350.

Wilson, C. M., Gaber, M. W., Sabek, O. M., Zawaski, J. A., and Merchant, T. E. (2009). Radiation-induced astrogliosis and blood–brain barrier damage can be abrogated using anti-TNF treatment. *Int. J. Radiat. Oncol. Biol. Phys.* **74**(3), 934–941.

Wojak, J. C., and Flamm, E. S. (1987). Intracranial hemorrhage and cocaine use. *Stroke* **18**(4), 712–715.

Yao, H., Allen, J. E., Zhu, X., Callen, S., and Buch, S. (2009). Cocaine and human immunodeficiency virus type 1 gp120 mediate neurotoxicity through overlapping signaling pathways. *J. Neurovirol.* **15**(2), 164–175.

Zhang, L., Looney, D., Taub, D., Chang, S. L., Way, D., Witte, M. H., Graves, M. C., and Fiala, M. (1998). Cocaine opens the blood–brain barrier to HIV-1 invasion. *J. Neurovirol.* **4**(6), 619–626.

CANNABINOID RECEPTORS IN BRAIN: PHARMACOGENETICS, NEUROPHARMACOLOGY, NEUROTOXICOLOGY, AND POTENTIAL THERAPEUTIC APPLICATIONS

Emmanuel S. Onaivi

Department of Biology, William Paterson University, Wayne, New Jersey 07470, USA

Much progress has been achieved in cannabinoid research. A major break-through in marijuana-cannabinoid research has been the discovery of a previously unknown but elaborate endogenous endocannabinoid system (ECS), complete with endocannabinoids and enzymes for their biosynthesis and degradation with genes encoding two distinct cannabinoid (CB1 and CB2) receptors (CBRs) that are activated by endocannabinoids, cannabinoids, and marijuana use. Physical and genetic localization of the CBR genes *CNR1* and *CNR2* have been mapped to chromosome 6 and 1, respectively. A number of variations in CBR genes have been associated with human disorders including osteoporosis, attention deficit hyperactivity disorder (ADHD), posttraumatic stress disorder (PTSD), drug dependency, obesity, and depression. Other family of lipid receptors including vanilloid (VR1) and lysophosphatidic acid (LPA) receptors appear to be related to the CBRs at the phylogenetic level. The ubiquitous abundance and differential distribution of the ECS in the human body and brain along with the coupling to many signal transduction pathways may explain the effects in most biological system and the myriad behavioral effects associated with smoking marijuana. The neuropharmacological and neuroprotective features of phytocannabinoids

INTERNATIONAL REVIEW OF
NEUROBIOLOGY, VOL. 88
DOI: 10.1016/S0074-7742(09)88012-4

335

and endocannabinoid associated neurogenesis have revealed roles for the use of cannabinoids in neurodegenerative pathologies with less neurotoxicity. The remarkable progress in understanding the biological actions of marijuana and cannabinoids have provided much richer results than previously appreciated cannabinoid genomics and raised a number of critical issues on the molecular mechanisms of cannabinoid induced behavioral and biochemical alterations. These advances will allow specific therapeutic targeting of the different components of the ECS in health and disease. This review focuses on these recent advances in cannabinoid genomics and the surprising new fundamental roles that the ECS plays in the retrograde signaling associated with cannabinoid inhibition of neurotransmitter release to the genetic basis of the effects of marijuana use and pharmacotherpeutic applications and limitations. Much evidence is provided for the complex *CNR1* and *CNR2* gene structures and their associated regulatory elements. Thus, understanding the ECS in the human body and brain will contribute to elucidating this natural regulatory mechanism in health and disease.

I. Introduction

This review presents the remarkable new understanding that the cellular, biochemical, and behavioral responses to marijuana, which remains one of the most widely used and abused drugs in the world, are coded in our genes. The discovery that specific gene codes for cannabinoid receptors (CBRs) that are activated by marijuana use, and that the human body makes its own marijuana-like substances—endocannabinoids (Onaivi *et al.*, 1996, 2006b)—that also activates CBRs have provided surprising new knowledge about cannabinoid genomic and proteomic profiles. These remarkable advances in understanding the biological actions of marijuana, cannabinoids, and endocannabinoids, is unraveling the genetic basis of marijuana use and the implication in human health and disease. In the era of personalized medicine, we are eager to understand how to utilize cannabinoid pharmacotherapeutic agents based on our new knowledge of the genetic basis of cannabinoid and endocannabinoid action. We know that the two well-characterized cannabinoid CB1 and CB2 receptors are encoded by *CNR1* and *CNR2* genes that have been mapped to human chromosome 6 and 1, respectively. A number of variations in CBR genes have been associated with human disorders including osteoporosis (Karsak *et al.*, 2005; Sipe *et al.*, 2005), ADHD (Lu *et al.*, 2008), posttraumatic stress disorder (PTSD) (Lu *et al.*, 2008), drug dependency (Onaivi *et al.*, 2006b), obesity (Cota *et al.*, 2003; Jesudason and Wittert, 2008), and depression (Onaivi *et al.*, 2006a; Serra and Fratta, 2007).

Thus, because of the ubiquitous distribution and role of the endocannabinoid system in the regulation of a variety of normal human physiology, drugs that are targeted to different aspects of this system are already benefiting cancer subjects and those with AIDs and metabolic syndromes (Jesudason and Wittert, 2008). In the coming era of personalized medicine, genetic variants and haplotypes in *CNR1* and *CNR2* genes associated with obesity or addiction phenotypes may help identify specific targets in conditions of endocannabinoid dysfunction.

Our previous investigations had defined a number of features of the *CNR1* gene's structure, regulation, and variation (Zhang *et al.*, 2004), but many of these features still remain poorly defined. Nevertheless, we and others have now demonstrated and reported that variants of the *CNR1* gene are associated with a number of disorders and substance abuse vulnerability in diverse ethnic groups including European–American, African–American, and Japanese subjects (Zhang *et al.*, 2004). Most strikingly, variants of *CNR* genes co-occur with other genetic variations and share biological susceptibility that underlies comorbidity in many neuropsychiatric disturbances (Palomo *et al.*, 2007). Thus, emerging evidence indicates that the endocannabinoid system exerts a powerful modulatory action on retrograde signaling associated with inhibition of synaptic transmission (Lovinger, 2008; Onaivi *et al.*, 2006b). Interestingly, a role for variations in *CNR1* gene has been associated with striatal responses to happy, but not to disgust, faces (Chakrabarti *et al.*, 2006) with implication that functional variation of *CNR* genotypes may be associated with disturbances of the brain involving emotional and social stimuli, such as autism (Chakrabarti *et al.*, 2006) and depression (Domschke *et al.*, 2008; Onaivi *et al.*, 2008a). This review focuses on these recent advances in cannabinoid genomics and the surprising new fundamental roles that the endogenous endocannabinoid system (ECS) plays in the genetic basis of the effects of marijuana use. The powerful influence of cannabinoid-induced retrograde signaling modulates GABAergic and glutamatergic systems indicate that the main excitatory and inhibitory systems are in part under the influence of the endocannabinoid system. Additional evidence is provided for the complex *CNR1* and *CNR2* gene structures and their associated regulatory elements. Therefore, understanding the ECS in the human body and brain will contribute to elucidating this natural regulatory mechanism in health and disease.

II. Cannabinoid Genomics

Most of the physiological effects of smoking marijuana are probably mediated via CBRs, actions at other non-CBRs, and nonreceptor mechanism(s). Thus, the genetic basis of marijuana use and perhaps dependence may be associated at least in part to their effects on these G-protein-coupled CBRs. The presently known

CBRs and their gene transcripts can now be analyzed in human blood samples (Onaivi *et al.*, 1999), fetal brain (Mato *et al.*, 2003; Wang *et al.*, 2003), placenta (Park *et al.*, 2003), uterus during pregnancy (Dennedy *et al.*, 2004), Alzheimer's disease brains (Benito *et al.*, 2003), prefrontal cortex of depressed suicide victims (Hungund *et al.*, 2004), cognitive impairment in multiple sclerosis (Woolmore *et al.*, 2008), substance abuse and dependence (Herman *et al.*, 2006; Schmidt *et al.*, 2002; Zhang *et al.*, 2004; Zuo *et al.*, 2007), and obesity phenotypes (Russo *et al.*, 2007). The human CB1-R cDNA was isolated by Gerard *et al.* (1991) from a human brain stem cDNA library using a 600 bp DNA probe and polymerase chain reaction. The deduced amino acid sequences of the rat and human receptors showed that they encode protein residues of 473 and 472 amino acids with 97.3% homology. These proteins share the seven hydrophobic transmembrane domains and residues common among the family of G-protein receptors (Gerard *et al.*, 1991; Matsuda *et al.*, 1990; Shire *et al.*, 1995) (Table I). The human (Fig. 1) and rat CB1-Rs also share pharmacological characteristics including the inhibition of adenylate cyclase activity via Gi/o in a stereoselective and pertusis sensitive manner following activation by cannabinoids. The CB1-Rs alter potassium channel conductance and decrease calcium channel conductance. The expression of the CB1-Rs is particularly dense in presynaptic terminals and endocannabinoids act as retrograde messengers at many synapses in the CNS. Although, the transcripts for both CB1 and CB2 receptors (Figs. 1 and 2) were reported to be present in the human placenta, the identification of CB1-Rs and fatty acid amide hydrolase in the human placenta may be targets for the effects marijuana use during pregnancy (Park *et al.*, 2003). Most of the studies, examining the expression and role of CBRs in the reproductive system have used mice where both CB1 and CB2 receptor gene transcripts were identified in mouse preimplantation embryos (Battista *et al.*, 2008). There is a close link between the endocannabinoid system and sex hormones in male and female reproductive system (Battista *et al.*, 2008). From mouse model to humans, the CBRs are a placental and uterine site of action of endocannabinoids and marijuana use during pregnancy and parturition. The identification of CBRs in the uterus, placenta, and the high densities of CB1-Rs, functionally coupled to G-protein in prenatal developmental stages throughout the human brain, suggest the involvement of the cannabinoid system in neural development (Mato *et al.*, 2003). It has also been suggested that the high expression of CB1 mRNA in the human fetal limbic structures may render such brain structures more vulnerable to prenatal cannabis exposure (Wang *et al.*, 2003). In adult diseased brains, an upregulation of CB1-Rs and agonist-stimulated [^{35}S] GTPγS binding in the prefrontal cortex of diseased suicide victims was reported in comparison to matched controls (Hungund *et al.*, 2004). This and other observations of the involvement of the ECS in a variety of CNS disorders renders different component of the ECS possible targets in the treatment of a variety of CNS mental and neurological disturbances. A trinucleotide repeat (AAT)

TABLE I
SUBTYPES OF CANNABINOID RECEPTORS

Hydrophobic transmembrane domains	CB1 Receptors	CB2 Receptors
Amino acids	472 AA	360 AA
Chr Location	6q14–15	1p36.11
Gene name	*CNR1*	*CNR2*
Endogenous ligand	2-AG	2-AG
CNS distribution	Yes	Yes
Peripheral distribution	Yes	Yes
Subtypes[a]	CB1 and CB1A-CB1n	CB2A and CB2B

[a]See text for isoforms and variants of CB1 and CB2 receptors.

Alternative splicing forms of cannabinoid receptor 1 gene

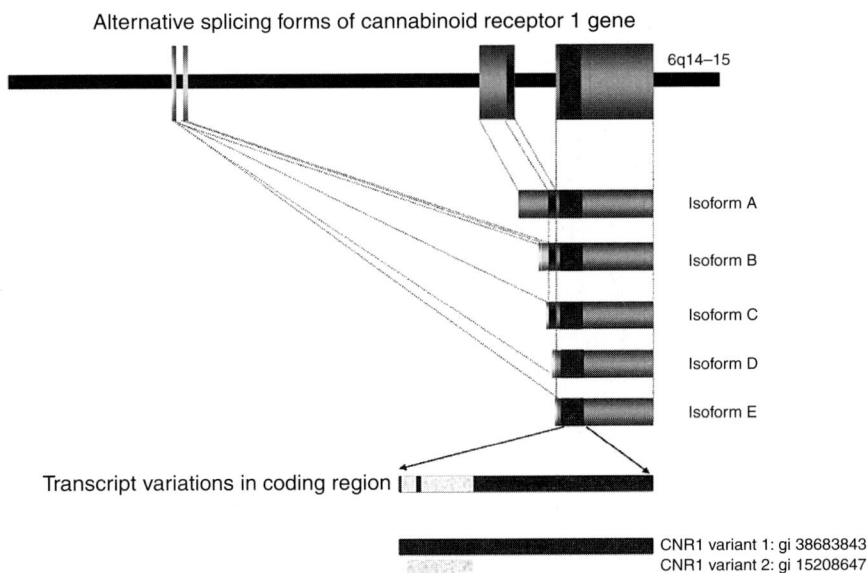

FIG. 1. Structure of CB1 *CNR* gene. Genetic structure of the CB1-R *CNR* gene. The gene is mapped on two finished genomic sequences: The complete mRNA sequence and transcript size is known. The sequences consist of exons, one of them include the whole coding region of the gene. A number of ESTs are identified and surprisingly some of them are between the exons of the published mRNA. A number of polymorphisms have been identified in the CB1 *CNR* gene and discussed. (See Color Insert.)

polymorphism of the *CNR1* gene has been associated with appetite disorder involving the binging/purging type of anorexia nervosa, which is a severe and disabling psychiatric disorder characterized by profound weight loss and body image disturbance (Siegfried *et al.*, 2004).

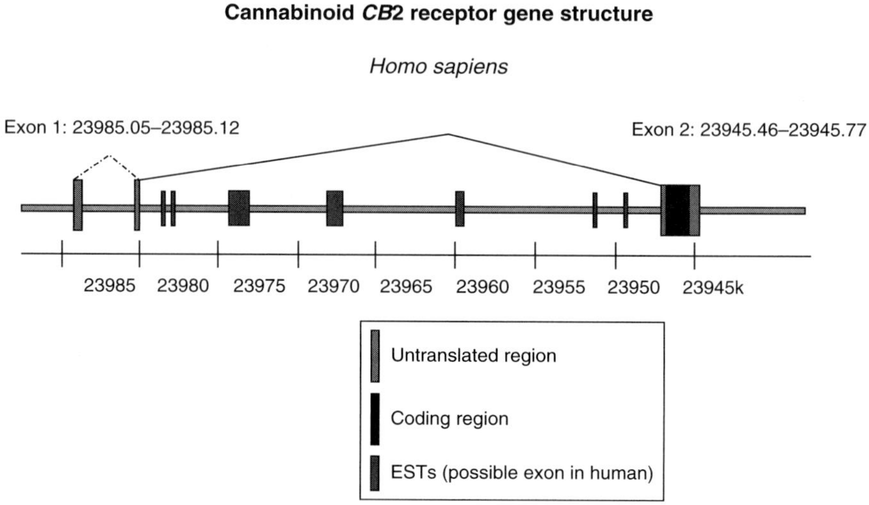

Cannabinoid *CB2* receptor gene structure

Homo sapiens

Exon 1: 23985.05–23985.12   Exon 2: 23945.46–23945.77

23985   23980   23975   23970   23965   23960   23955   23950   23945k

Untranslated region

Coding region

ESTs (possible exon in human)

FIG. 2. Structure of CB2 *CNR* gene. Genetic structure of the CB2-R *CNR* gene. Some variants have been identified and some known polymorphisms in the CB2 *CNR* gene are discussed. (See Color Insert.)

The assembly of the human CB1-Rs and the impact of its long N-terminal tail has been examined by Andersson *et al.* (2003). They found that the long N-tail of CB1-Rs that lack a typical signal sequence and cannot be efficiently translocated across endoplasmic reticulum membrane might not be relevant to ligand binding or activation of the receptor (Andersson *et al.*, 2003). With the sheer abundance of CBRs, the functional significance of the existence of a CB1-R with long N-terminal and a shorter N-terminal tail in CB1A receptor are not immediately clear, but may be related to rapid formation of different tonically active states allowing for sequestration of G-proteins, making them unavailable to couple to other receptors (Vasquez and Lewis, 1999). One of the major advantages for the physiological actions of these CBRs may be associated with the quick response needed for the sequestration of G-proteins and the retrograde signaling of endo-cannabinoids on presynaptic CB1-Rs to inhibit neurotransmitter release. The second cannabinoid CB2-R clone was isolated from myeloid cells by PCR and degenerate primers using cDNA template from human promyelocytic leukaemic line HL60 (Munro *et al.*, 1993). Similar hydrophobic domains 1, 2, 5, 6, and 7, in which 50% or greater of the amino acids were identical between CB1-Rs and CB2-Rs. The extracellular domain also contained sequence motifs that were common to both clones (Matsuda, 1997). The protein encoded by the *CNR2* gene shows 44% identity with the human *CNR1* gene (Table I). A number of functional and expression studies have been performed with the *CNR2* gene and the results indicate that the CB2-R is the predominant CBR in the immune system, where it is expressed in B and T cells (Gurwitz and Kloog, 1998; Munro et al., 1993; Schatz et al., 1997). A number of laboratories were not able to detect the presence of CB2-Rs in healthy brains, but there had been demonstration of CB2-R expression in rat microglial cells and other brain associated cells during inflammation (Onaivi et al., 2006b). We and others have now identified and reported the presence of CB2-Rs in brain neuronal and glial process (Beltramo et al., 2006; Gong et al., 2006; Onaivi et al., 2006b, Van Sickle et al., 2005). To further improve understanding of the role of CB2-Rs in the brain, we hypothesized that genetic variants of CNR2 gene might be associated with depression in a human population and that alteration in CNR2 gene expression may be involved in the effects of abused substances in rodents. Our data reveals that CB2-Rs are expressed in brain and plays a role in depression and substance abuse (Onaivi et al., 2008b). The relevance of CB2-Rs, if any to problem and compulsive marijuana use is unknown. However, a specific enzyme activity by a phosphodiesterase of the phospholipase D-type responsible for generating endolipids like anandamide has been identified, cloned from mouse, rat and human and characterized, with the expression of the mRNA, and proteins widely distributed in murine brain, kidney, and testis (Okamoto et al., 2004). The direct contribution of this enzyme and its influence in the effects of abused substances and marijuana dependence has not been determined, but the levels of N-acylethanolamines are known to increase in a variety of animal models of tissue degeneration (Okamoto et al., 2004). Furthermore, cells and tissues involved in neuroinflammation express functional CBRs with an upregulation of components of the endocannabinoid signaling system in the inflammatory response (Walter and Stella, 2004). As the cannabinoid signaling machinery is involved in immune function, it is yet unknown if neuroinflammatory processes contribute to continued marijuana use and dependence, once initiated. Other, as yet unknown, components of the cannabinoid system have been shown to be involved in neuroinflammatory processes that are sensitive to the nonpsychotropic CBRs whose molecular identity is currently unknown. This additional evidence about cannabinoid regulation of neuroinflammation and immunomodulation has been provided by 2-AG stimulation of microglial cell migration that is blocked by nonpsychotropic CBR antagonists, cannabinoid, and abnormal cannabidiol (Walter and Stella, 2004). It appears therefore that the vital role played by cannabinoids system in the neuroinflammation and immunomodulation may be exploited in the development of immunopharmactherapy against drugs of abuse by targeting some component of the ECS. As there is accumulating evidence indicating a central role for the endocannabinoid system in the regulation of the rewarding effects of abused substances, an endocannabinoid hypothesis of drug reward and addiction has been postulated (Onaivi et al., 2008). In humans, the genetic liability following fetal exposure to marijuana during pregnancy might be a predisposing vulnerability factor to use marijuana in adulthood. Earlier studies have reported on the capability of marijuana use to induce genotoxity (see Li and Lin, 1998, for a review). Many other studies have shown that the allele frequency distributions of many genetic variants at *CNR1* gene differ by population and sex (Zhang *et al.*, 2004; Zuo *et al.*, 2007).

III. Cannabinoid Neuropharmacology, Neurogenesis, Neurotoxicology, and Neuroprotective Properties

The neuropharmacological and neuroprotective features of phytocannabinoids (de Lago and Fernandez-Ruiz, 2007) and endocannabinoid associated neurogenesis (Goncalves *et al.*, 2008) have revealed roles for the use of cannabinoids in neurodegenerative pathologies with less neurotoxicity. A role for CB2-Rs and not for CB1-Rs has been reported in the regulation of adult subventricular zone neurogenesis in an age-dependent manner (Goncalves *et al.*, 2008). Acute toxicity studies show marijuana use or acute administration of THC, the main psychoactive ingredient of cannabis does not lead to death (Beaulieu, 2005). In deed unlike the opiate receptors, CBRs appear to be sparsely distributed in areas of the brain that lead to respiratory depression in opiate overdose causing death. It has been demonstrated that in traumatic brain injury, neuroprotection is exerted by the release of 2-AG that may serve as a molecular regulator of pathophysiological events in reducing brain damage (Mechoulam and Shohami, 2007). There is also growing evidence that the endocannabinoids are lipid mediators that are involved in the control of neuron survival (Galve-Roperh *et al.*, 2008). Therefore, different mechanisms have been associated with CBRs and their role in neuroprotection (Van Der Stelt and Di Marzo, 2005).

IV. Chromosomal Mapping of the *CNR* Genes

The two well-documented human CBR *CNR* genes and of many other mammalian species have been mapped to their respective chromosomes. Human *CNR1* and *CNR2* genes have been mapped to chromosome 6q14-q15 and 1p34-p35, respectively. Using genetic linkage mapping and chromosomal *in situ* hybridization the genomic location of the human *CNR* gene was determined, thus confirming the linkage analysis and defining a precise alignment of the genetic and cytogenetic maps (Hoehe *et al.*, 1991). These investigators found that the location of the human CB1 *CNR* gene is very near the gene encoding the alpha subunit of chorionic gonadotropin (CGA). After we cloned and sequenced the mouse CB1 *CNR* gene (Chakrabarti *et al.*, 1995), we determined that the mouse CB1 and CB2 *CNR* genes are located in proximal chromosome 4 (Stubbs *et al.*, 1996). This location is within a region to which other homologs of human 6q genes are located. The CB1 *CNR* gene, GABRR1, GABRR2, and Cga are linked together both in the mouse chromosome 4 and on human chromosome 6q (Fig. 3). The results of the chromosomal location of the human *CNR* gene (Hoehe *et al.*, 1991) and the mouse (Stubbs *et al.*, 1996) *CNR* genes add a new marker to this region of the mouse–human homology, and confirm the close linkage of *CNR* genes in both species Fig. 3. Further, the location of the rat CB1 *CNR* gene in the rat genome has been determined and fits the rodent–human homology as the CB1 *CNR* genes are highly conserved in the mammalian species. The genomic sequences (using complete or draft) of a part of chromosome 6p covers more than 200kb around the CB1/CNR gene and 90 Kb around the human *CNR2* gene. The program GENESCAN was used to estimate four fairly possible gene sequences including CB1 *CNR* gene in this region. These estimated sequences showed similarity to known protein sequences, revealed by BLAST. As the neurobiological effects of marijuana and other cannabinoids suggest the involvement of the *CNR* genes in mental and neurological disturbances, the mapping of the genes will undoubtedly enhance our understanding of the linkage and possible cannabinoid genetic abnormalities. The chromosomal location and genomic structure of human and mouse fatty acid amide hydrolase (FAAH) genes

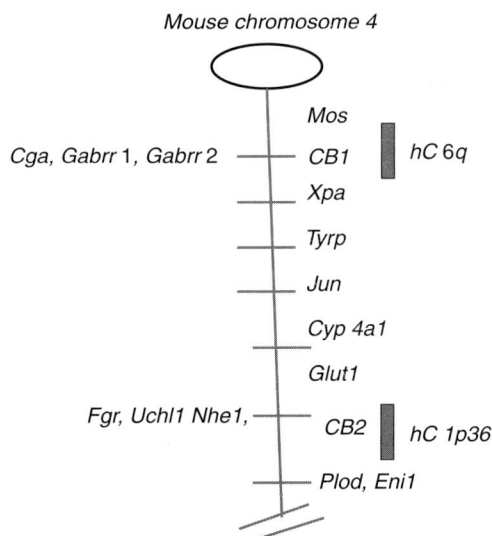

FIG. 3. Chromosomal localization of CB1-R and CB2-R *CNR* genes. The chromosomal locations of CB1-R and CB2-R CNR genes are on mouse chromosome 4 and human chromosome 6 and 1P36, respectively.

have been mapped to chromosomes 1p34-p35 and 4 (Wan *et al.*, 1998). The localization of FAAH and CB2 *CNR* genes are in same chromosomal regions in the mouse and human and again adds a new marker to this region of the mouse–human homology, and confirms the close linkage of FAAH and *CNR* genes in both species.

V. Polymorphic Structure of *CNR* Genes

Although two CBR G-protein coupled receptor (GPCR) subtypes have been cloned and well characterized, studies now indicate the functional existence of splice variations of CB1- and CB2-Rs. New information about *CNR* genes and their allelic variants in humans and rodents is adding to our understanding of vulnerabilities to addictions and other neuropsychiatric disorders and the genetic basis of marijuana use and addiction in vulnerable individuals. Some of the complexities are however due to our lack of understanding of the regulation of the CBR genes and the molecular identity of other subtypes (if any). The known and cloned CBRs are designated as *CNR1* and *CNR2* or CB1-R and CB2-R. They belong to the large super family of receptors that couple to guanine–nucleotide–binding proteins and that thread through cell membranes seven times (heptahelical receptors). The CB1-R is predominantly expressed in brain and spinal cord, and thus often referred to as the brain CBR. A spliced variant of human CB1-R, cDNA, CB1A, was isolated and characterized (Rinaldi-Carmona *et al.*, 1996; Shire *et al.*, 1995). It appears that the CB1A resulted from the excision of an intron at the 5′-extremity of the coding region of the human receptor mRNA (Shire *et al.*, 1995). This subtype of CBRs were stably expressed in Chinese hamster ovary cell lines and the truncated and modified CB1 isoform CB1A was shown to exhibit all the pharmacological properties of CB1-R to a slightly attenuated extent (Rinaldi-Carmona *et al.*, 1996). A study of the distribution of the human CB1A mRNA by reverse transcriptase PCR showed the presence of minor quantities of this isozyme together with CB1-R throughout the brain and in all the peripheral tissues examined (Shire *et al.*, 1995). While the presence of CB1A has not been detected in rodents, evidence for the existence of unknown CBR subtypes in mouse brain has been described (Onaivi *et al.*, 2006b). The CB2-R is, at times, referred to as the peripheral CBR because of its largely peripheral expression in immune cells. Results of new studies indicate that CB2-Rs are expressed in the neuron and glial cells in the brain particularly during pathological events (Benito *et al.*, 2003) and in mast cells (Samson *et al.*, 2003). CB1 and CB2 receptor gene products are expressed in relative abundance in specific tissues and cell types. CB1-Rs are considered to be the most abundant GPCRs in the mammalian brain (Matyas *et al.*, 2006; Onaivi *et al.*, 2002a). These CB1-Rs are highly expressed in brain areas such as the cortex, limbic system, hippocampus, cerebellum, and several nuclei of the basal ganglia; areas associated with emotionality, cognition, motor, and executive function. CB2-Rs on the other hand were known to be expressed in peripheral and immune tissues and have traditionally been referred to as peripheral CB2-Rs, because a number of previous studies did not detect the presence of CB2-Rs in the mammalian brain (Brusco *et al.*, 2008a). Although glial expression of CB2-Rs was to be expected and have been reported by many investigators (Brusco *et al.*, 2008b), the functional expression of neuronal of CB2-Rs had been controversial. We and others have now identified and reported the presence of CB2-Rs in brain neuronal cells (Ashton *et al.*, 2005; Gong *et al.*, 2006; Jhaveri *et al.*, 2008; Onaivi *et al.*, 2006a; Van Sickle *et al.*, 2005) and this warrants a re-evaluation of the role of CB2-Rs in the brain. With novel and precise cannabinoid probes, our results indicate the expression of brain CB2-Rs in a mouse model of depression and in the effects of abused substances (Onaivi *et al.*, 2008a).

This review discusses the current state of description of the genes encoding CBRs and examines the genetic basis of marijuana use; and also defines some of the limitations of current knowledge. The molecular biology and pharmacology of the CBRs is still less well understood than that of many GPCRs. Only scanty information describes how these CBR *CNR* genes are regulated. A moderate store of data describes some of the complex signal transduction pathways engaged by CBR activation (Onaivi *et al.*, 2002b). However, many topics including the ways in which the abundant CB1-Rs could alter activities mediated through other co expressed GPCRs by sequestering G-proteins and other means are still been investigated. With regards to these GPCR-CBRs, we also need to be aware of the possible ligand gated ion channels influenced by cannabinoids and interactions at the vanilloid receptors. We need to bear in mind data suggesting that cannabinoids can exert receptor-independent effects on biological (Hillard *et al.*, 1985; Makriyannis *et al.*, 1989) and enzyme systems such as protein kinase C. Despite these caveats, it is impressive to consider the large amount of data about CBRs amassed over the last decade, which has enhanced our knowledge of the physiological factors that triggers and contributes to the cycle of marijuana use.

There is increasing information and knowledge available at the molecular level about *CNR* gene structure, regulation and polymorphisms. Different human *CNR* gene polymorphisms have been reported (Table II). These polymorphic sequences are either single nucleotide polymorphisms (SNPs) or larger changes in the number of repeats. We have identified polymorphisms in nontranslated region that affect splicing and promoter activity. As described in Table II, some of the SNPs are silent to translation, while others lead to amino acid substitutions in the CBR protein. Many of the currently known genetic polymorphisms of the endocannabinoid system have been linked with a number of conditions including substance abuse, mental disorders and energy metabolism and have been

TABLE II

GENETIC POLYMORPHISMS OF CANNABINOID RECEPTOR GENES (*CNR* GENES)

CNR genes	Polymorphism	Linkage or association
CB1	Two allele DNA polymorphism	Associated with *CNR1* gene (Caenazzo *et al.*, 1991)
	1359 G/A CNR1 variant	Associated with alcohol, dependence (Gadzicki *et al.*, 1999; Schmidt *et al.*, 2002)
	1359 G/A CNR1 variant	Not associated with Tourette syndrome (Gadzicki *et al.*, 2004)
	1359 G/A CNR1 variant	Not associated with Alcohol Withdrawal tremens (Preuss *et al.*, 2003)
	1359 G/A CNR1 variant	Associated with weight loss (Aberle *et al.*, 2007, 2008)
	3813 A/G and 4895 A/G variant	Associated with obesity in men (Russo *et al.*, 2007)
	CNR1 SNPs	No association in obesity in German children (Aberle *et al.*, 2007; Muller *et al.*, 2007)
	CNR1 SNPs	Associated with obesity and BMI (Benzinou *et al.*, 2008; Gazzerro *et al.*, 2007; Jaeger *et al.*, 2008; Peeters *et al.*, 2007)
	CNR1, FAAH, DRD2 gene	Associated with comorbidity of alcoholism and antisocial behavior (Hoenicka *et al.*, 2007).
	(AAT)n repeat of CNR1 gene	Conflicting associations with drug dependence (Ballon *et al.*, 2006; Comings, 1996; Comings *et al.*, 1997; Covault *et al.*, 2001; Heller *et al.*, 2001; Jesudason and Wittert, 2008).
	CNR1 variants, SNPs, 'TAG' haplotype	Associated with polysubstance abuse (Zhang *et al.*, 2004; Zuo *et al.*, 2007)
	CNR1 SNPs	Not associated with polysubstance abuse (Herman *et al.*, 2006)
	CNR1 SNPs	Associated with cannabis dependence (Agrawal *et al.*, 2008, 2009)
	CBR haplotype	Associated with fewer cannabis dependence symptoms in adolescents (Hopfer *et al.*, 2003, 2006, 2007).
	CNR1 SNPs	Associated with alcohol and nicotine dependence (Chen *et al.*, 2008; Hutchison *et al.*, 2008)
	CNR1 SNPs	No association with anorexia nervosa (Muller *et al.*, 2008)
	CNR1 (AAT)n repeats	Associated with restricting and binging/purging anorexia nervosa (Siegfried *et al.*, 2004)
	CNR1 (AAT)n repeats	Associated with depression in Parkinson's disease (Barrero *et al.*, 2005)
	CNR1 SNPs	Associated to striatal responses to facial exp. (Chakrabarti *et al.*, 2006)
	(AAT)n repeats	Association with ADHD in alcoholics (Lu *et al.*, 2008; Ponce *et al.*, 2003)

(*continued*)

TABLE II *(continued)*

CNR genes	Polymorphism	Linkage or association
	CNR1 SNP haplotype	Risk factor for ADHD and PTSD (Lu *et al.*, 2008)
	1359 G/A *CNR1* variant	Associated with schizophrenia (Leroy *et al.*, 2001)
	(AAT)n repeats	Not associated with schizophrenia (Li *et al.*, 2000; Tsai *et al.*, 2000) and mood disorders (Tsai *et al.*, 2001)
	(AAT)n repeats	Associated with schizophrenia (Martinez-Gras *et al.*, 2006)
	(AAT)n repeats	Associated with hebephrenic schizophrenia (Chavarria-Siles *et al.*, 2008; Ujike *et al.*, 2002)
	CNR1 variants	Associated with depression and anxiety (Domschke *et al.*, 2008)
	CNR1 variants and (AAT)n repeats	Associated with impulsivity (Ehlers *et al.*, 2007a,b)
	1359 G/A *CNR1* tag SNP	Associated with antipsychotic response but not schizophrenia (Hamdani *et al.*, 2008)
	CNR1 SNPs	No association with cognitive impairment in MS (Woolmore *et al.*, 2008)
CB2	*CNR2* SNPs and haplotypes	Associated with human Osteoporosis (Karsak *et al.*, 2005)
	CNR2 SNPs	Not associated with myocardial infarction or cardiovascular risk factors (Reinhard *et al.*, 2008)
	CNR2 SNPs	Associated with bone mass (Yamada *et al.*, 2007)
	CNR2 (Q63R) SNP	Risk factor for autoimmune disorders (Sipe *et al.*, 2005)
	CNR2 (Q63R) but not (H316Y)	Associated with alcoholism and depression (Ishiguro *et al.*, 2007)

reviewed (Norrod and Puffenbarger, 2007). A silent mutation of a substitution from G to A, at nucleotide position 1359 in codon 453 (Thr) that turned out to be a common polymorphism in the German population was reported (Gadzicki *et al.*, 1999). In this study, allelic frequencies of 1359(G/A) in genomic DNA samples from German Gilles de la Tourette syndrome (GTS) patients and controls were determined by screening the coding exon of the CB1 *CNR* gene using PCR single-stranded conformation polymorphism (PCR-SSCP) analysis (Gadzicki *et al.*, 2004). This was accomplished by the use of a PCR based assay by artificial creation of a MSP1 restriction site in amplified wild-type DNA (G-allele), which is destroyed by A-allele (Gadzicki *et al.*, 2004). They found no significant differences of the allelic distributions between GTS patients and controls within the coding region of the CB1 *CNR* gene. In our studies the frequencies of this polymorphism are significantly different between Caucasian, African–American, and Japanese population (Zhang *et al.*, 2004). There is also a HindIII restriction fragment length polymorphism (RFLP) located in an intron approximately 14 kb in 5' region of the initiation codon of the CB1 *CNR* gene. Caenazzo *et al.* (1991) genotyped 96 unrelated Caucasians using hybridization of human DNA digested with HindIII and identified a two allele with bands at 5.5 (A1) and 3.3 kb (A2). The frequencies of these alleles were 0.23 and 0.77, respectively. Another polymorphism is a triplet repeat marker for CB1 *CNR* gene. This is a simple sequence repeat polymorphism (SSRP) consisting of nine alleles containing (AAT) 12–20 repeat sequences that was identified by Dawson (1991). This polymorphism has been used in linkage and association studies of the CB1 *CNR* gene with mental illness and drug abuse in different population. This CB1 *CNR* gene triplet repeat marker was used to test for linkage with schizophrenia using 23 multiplex schizophrenia pedigrees (Dawson, 1991) and association with heroin abuse in a Chinese population (Li *et al.*, 2000) and also for association with intravenous (IV) drug use in Caucasians (Comings *et al.*, 1997). There was no linkage and association of the marker with schizophrenia indicating that the *CNR1* gene is not a polymorphism of major aetiological effect for schizophrenia but *CNR1* gene might be a susceptibility locus in certain individuals with schizophrenia, particularly those whose symptoms are apparently precipitated or exacerbated by cannabis use (Dawson, 1991). Comings *et al.* (1997), hypothesized those genetic variants of CB1 *CNR* gene might be associated with susceptibility to alcohol or drug dependence and analyzed the triplet repeat marker in the CB1 *CNR* gene. They found a significant association of the CB1 *CNR* gene with a number of different types of drug dependence (cocaine, amphetamine, and cannabis), and intravenous drug use but no significant association with variables related to alcohol abuse/dependence in non-Hispanic Caucasians (Table II). In addition, this group also reported that a significant association of the triplet repeat marker in the CB1 *CNR* gene alleles with the P300 event related potential that has have been implicated in substance abuse (Johnson *et al.*, 1997). Li *et al.* (2000) attempted to replicate the finding of Comings *et al.* (1997), in a sample of Chinese heroin addicts and did not find any evidence that the CB1 *CNR* gene AAT repeat polymorphism confers susceptibility to heroin abuse. CB1 *CNR* gene is located in human chromosome 6q14-q15 and it is interesting that previous reports showed evidence for suggestive linkage to schizophrenia with chromosome 6q markers (Martinez *et al.*, 1999) and also suggestive evidence exist for a schizophrenia susceptibility locus on chromosome 6q (Cao *et al.*, 1997). Although there was no linkage and association of the CB1 *CNR* triplet marker with schizophrenia, it remains to be determined if linkage and association to schizophrenia might exist with other unknown polymorphisms that might exist in the CB1 *CNR* gene structure that is currently poorly characterized. Three other variants have been reported in the CB1 *CNR* gene of an epilepsy patient (Kathmann *et al.*, 2000). This was obtained from PCR assay with cDNA from hippocampal tissue taken from patients undergoing neurosurgery for intractable epilepsy. They detected four mutations in the coding region of the CB1 *CNR* gene, with the first three mutations yielding amino acid substitutions.

Our previous studies analyzed CB1 *CNR* gene structure, regulation, and expression in the mouse and human models to determine genotypic and/or haplotypic associations of CB1 *CNR* gene with addictions and other neuropsychiatric disturbances. The human Chr 6, 91.8~96.1 cM CB1/*CNR1* locus encodes at least six exons which account for 24–28 kb of sequence (Zhang *et al.*, 2004). Examination of CB1 *CNR* gene sequence variations in distinct populations has revealed a G/A SNP in CB1 5′ flanking sequences. The initial values for linkage disequilibrium between these markers and genotypic frequencies of the markers in drug abusing and control populations were calculated. We tested for associations between some SNPs and the risk for polysubstance use disorders in Japanese, European–American, and African–American populations and reported the association between some SNPs in *CNR1* gene and vulnerability to polysubstance use disorder (Zhang *et al.*, 2004). Some other studies have reported similar findings while others have been equivocal (Table II). While polymorphisms in *CNR2* gene have not been well characterized in linkage or association studies in comparison to *CNR1* gene, *CNR2* gene polymorphism has been linked to autoimmune disorders including osteoporosis (Karsak *et al.*, 2005; Sipe *et al.*, 2005) and with bone mineral density in Japanese population (Yamada *et al.*, 2007). We have reported the association between Q63R polymorphism of the CB2 *CNR2* gene and alcoholism (Ishiguro *et al.*, 2007) and depression (Onaivi, 2006) in Japanese population. Another area of research is on the polymorphisms associated with the metabolizing enzymes of the endocannabinods which is not addressed in the review. Thus, polymorphisms at the CBR *CNR* genes may be associated with the diverse actions of marijuana use, and may therefore appear to be a major factor which when triggered by environmental, age, and metabolic factors could lead to the continuous circle of marijuana dependence and use. Cannabinoid research and the use of cannabis products continue to attract significant attention. The current significant advances in molecular biology and technology that increased scientific knowledge in cannabinoid research will certainly contribute to a better policy on the medical use of marijuana. For example, studies with CBR antagonists have contributed to resolving the long-standing debate about addiction to marijuana. Specifically, the controversial question of physical dependence on psychoactive cannabinoids has now been addressed, using the antagonist, rimonabant, also known as SR 141716A [*N*-(piperidin-1-yl)5-(4-chlorophenyl)-(2,4-dichlorophenyl)-4-methyl-1H-pyrazole-3-carboxamidehydrochloride, to precipitate withdrawal reactions in rats injected with increasing doses of Δ^9-THC. Aceto *et al.* (1996) reported a precipitated withdrawal syndrome that was absent in control animals, providing evidence that Δ^9-THC could produce physical dependence. In general it had been claimed that the psycho activity and euphoria induced by cannabinoids limit their use in the clinic for numerous therapeutic applications for which they are currently being evaluated. With the availability of these genes, gene products and other CBR research tools, it is speculated that the properties of these genes and regulation will be intensely studied as to reveal how the psycho activity can be dissociated from the therapeutic properties of marijuana and cannabinoids. Or could it be that certain therapeutic actions of marijuana and cannabinoids cannot be separated from their psycho activity?

VI. Genes Encoding Endocannabinoid Transporter(s) as Pharmacotherapeutic Targets

Although there is evidence from functional studies for the existence of some form of cannabinoid transporter(s), their identity, sequence information, and biological characteristics at the molecular level are unknown. This is therefore "a difficult issue to handle," as concluded by the review on anandamide transport (Glaser et al., 2005). Without a molecular identity and with the different experimental approaches in assay conditions that support or refute the existence of a transporter for anandamide, the cloning and identification of genes encoding these endocannabinoid transporter(s) will certainly resolve the issue. While there is ample scientific evidence to support the concept that anandamide transport across membranes is protein mediated, definitive evidence awaits its molecular characterization. It has also been suggested that endocytosis via a caveolae/lipid raft mediates anandamide transport. However, the differential uptake of the different endocannabinoids, for example anandamide and 2-AG in different cell types may indicate the possibility of different cannabinoid transporters for the different endocannabinoids (Glaser et al., 2005). If there is a specific 2-AG membrane transport, only a few studies have addressed the transport of 2-AG. The scanty reports based upon saturation kinetics have proposed that 2-AG enters the cell by a specific 2-AG transporter, via the anandamide transporter or by simple diffusion. It was therefore suggested that the mechanism of 2-AG transport appears to be rate-limited diffusion through the membrane (Hermann et al., 2006). Further research will show whether the additional endocannabinoids been discovered, like noladin ether will be metabolized and taken up by similar metabolic and uptake inhibitors. This will not be unprecedented since the monoamines have different transporters for dopamine, serotonin and norepinephrine. Nevertheless, endocannabinoids may be a distinct class of agonist since they are hydrophobic and neutral, exhibiting similar biophysical properties to some anesthetics that diffuse freely through the cell membrane (Hermann et al., 2006). Regardless of the elusive existence of transporters for endocannabinoids, some studies have provided preclinical pharmacotherapeutic evidence that targeting the anandamide transport system using putative anandamide transport inhibitors like AM 404, may be an approach in reducing alcohol intake and in some emotional disturbances like anxiety and depression (Bortolato et al., 2006; Cippitelli et al., 2007). On the other hand, some studies have provided evidence against the presence of an anandamide transporter (Glaser *et al.*, 2003). It will be premature to dismiss the existence of other endocannabinoid transporters, including 2-AG transporter, but of course the molecular identification of any endocannabinoid transporter will validate them as pharmacotherapeutic targets. Thus the mechanisms of action of endocannabinoid transporters, if they exist remain to be elucidated and whether or not 2-AG transporters and other endocannabinoid transporters exist.

VII. Variations in Cannabinoid Receptor Genes: Functional and Pharmacotherapeutic Implications

There is accumulating physiological evidence for the existence of other subtypes CBRs different from the currently known CB1 and CB2 receptors. The vanilloid receptor 1 (VRI), the site at which capsaicin in hot chili peppers acts, is a site that anandamide is a full agonist. As anandamide is a partial agonist at the CBRs, some have suggested that vanilloid (VR1) receptor be classified as a CBR subtype—may be CB3. In fact the endocannabinoid that is a full agonist at the CBRs is 2-arachidonyl glycerol (2-AG) (Gonsiorek *et al.*, 2000; Sugiura *et al.*, 1999, 2006). The GPR55 has been suggested as a CBR that increases intracellular calcium and inhibits mM current (Lauckner *et al.*, 2008). However, using a strategy for defining CBR functional fingerprints from mutagenesis and molecular recognition literature data, it was noted that hGPR55 does not appear to share similar fingerprint with the hCB1R and hCB2R (Petitet *et al.*, 2006). While this could not be considered as a proof to exclude GPR55 from the CBR family, the data from other studies strongly suggests that GPR55 is a specific functional receptor for lysophosphatidylinositol receptor (Henstridge *et al.*, 2008; Oka *et al.*, 2007). It has also been reported that 2-arachidonoyl-sn-glycero-3-phosphoinositol, a molecular species of lysophosphatidylinositol, is a possible natural ligand for GPR55 (Oka *et al.*, 2008). Thus far, it appears that GPR55 is quite distinct from other GPCRs and represents an intriguing and unique therapeutic target whose functional receptor requires further validation and characterization (Henstridge *et al.*, 2008).

We have reported that the human CB1R have a number of splice variants, which may in part account for the myriad behavioral effects of smoking marijuana (Fig. 1 and Table II). Some effects may include actions at CB2-Rs that have received much less attention than CB1-Rs. However, we and others have now identified and characterized glial endothelial and neuronal CB2-Rs in the brain. Nonetheless, many features of the CB2R *CNR2* gene structure, regulation and variation remain poorly characterized compared to the CB1-Rs. In humans the *CNR2* gene is reported to consist of a single translated exon flanked by 5' and 3' untranslated regions and a single untranslated exon (Sipe *et al.*, 2005) (Fig. 2). Most regions of the *CNR2* gene are highly conserved, but the human has glutamine at position 63 (Karsak *et al.*, 2005, 2007; Sipe *et al.*, 2005) and another SNP H316Y has been reported and linked to autoimmune disorders (Karsak *et al.*, 2007; Sipe *et al.*, 2005). There has been little or no data on the role of CB2-Rs in neuropsychiatric disorders. However, in neurological disorders associated with inflammation, the expression of CB2-Rs has been reported in limited populations of microglial including plaque-associated glia in Alzheimer's disease brains (Nunez *et al.*, 2004; Pazos *et al.*, 2004). Indeed our studies provide the first evidence for a role of CB2-Rs in depression and substance abuse (Ishiguro *et al.*, 2007; Onaivi, 2006; Onaivi *et al.*, 2006a, 2008a,b). We and others have identified splice variants of the human CB1-Rs and CB2-Rs but have thus far been poorly characterized for functional specificity apart from the broad roles associated with CB1-R and CB2-R subtypes. Alternative splicing of RNAs appears to be more common than previously thought in people, and can generate a variety of proteins, with most genes producing at least two variants. The characterization of CBR variants will add validity to the functional evidence for the existence of multiple CBR subtypes. It has been demonstrated *in vitro* that amino-terminal processing of the hCB1-R may involve rapid N-terminal truncation in the cytoplasma prior to translocation to the endoplasmic reticulum membrane. It was suggested that such a truncation process might be a way to create a novel type of CB1-R isoforms but exactly how the truncated CB1-R may be formed and how the processing is regulated remains to be determined (Nordstrom and Andersson, 2006), In comparison to the monoaminergic system, the application of modern techniques to cannabinoid research is new. For example molecular cloning has revealed the presence of serotonin (5-hydroxytryptamine; 5-HT) receptor subtypes, which can be subdivided in seven subfamilies (Gerhardt and van Heerikhuizen, 1997), and 15 serotonin (5-HT) receptor subtypes and growing. Like 5-HT receptors that include ligand-gated ion channels, it has been suggested that the VR1 receptor should be reclassified as a CBR subtype since the endocannabinoid, anandamide is a full agonist at VR1 receptor, but acts as a partial agonist at CB1-s. New knowledge on cannabinoid transcriptional and posttranslational modifications, such as alternate splicing and perhaps RNA editing may indicate formation of multiple proteins could unravel specific mechanisms associated with numerous behavioral and physiological effects of marijuana use. The cloning and sequencing of CB1-R *CNR1* gene from 62 species has also been reported by Murphy *et al.* (2001) and awaits full characterization. As predicted here the identification and characterization of these putative CBR isozymes and different elements of the endocannabinoid system may reveal novel targets for medication development. However, the limitless signaling capabilities and the endless complexity of the cannabinoid system require continuous intensive investigation.

VIII. Genetic Basis of Cannabis use and Dependence

The question of physical and psychological addiction to marijuana is no longer a debate as the significant progress in marijuana-cannabinoid research have applied modern techniques using new molecular probes that were previously unavailable to resolve the issue of addiction to marijuana. There is substantial evidence in favor of heritable factors influencing the liability to cannabis use behaviors (Agrawal *et al.*, 2008). Furthermore, previous twin studies indicate that there is a role of genetic influences on early stages of cannabis use, such as lifetime history of use, early onset use and frequency of use, but no conclusive replicable genomic locations explains these phenotypes for cannabis use behaviors in all populations (Agrawal *et al.*, 2009). And of course environmental factors and the availability of cannabis is undoubtedly a contributing factor. Other studies also suggest the presence of reliable and empirically valid withdrawal syndrome for marijuana use and that genetic factors and heritability to cannabis use account for a significant portion of variance in use, abuse, and dependence (Haughey *et al.*, 2008). Thus, it has been demonstrated that marijuana withdrawal after abstinence and cue-elicited craving in humans are associated with polymorphisms in *CNR1* and fatty acid hydrolase (*FAAH*) genes (Haughey *et al.*, 2008). In humans the above study indicates that abstinence from marijuana use precipitates withdrawal in daily smokers, making craving and withdrawal variables as possible cannabis dependence endophenotypes (Haughey *et al.*, 2008). Marijuana dependency syndrome is similar to, than different from the more recognized substance dependence syndromes (Budney, 2006). In addition, animal models allowing the evaluation of genetic and neural correlates of marijuana-cannabinoid dependence and abuse potential have now been developed. While the question of marijuana-cannabinoid dependence had been controversial, the discovery of rimonabant, the CB1-R antagonist, precipitated withdrawal in rodent models indicative of cannabinoid dependence (Cook *et al.*, 1998). Similar questions are no longer posed about cigarette and opiate dependence. As this review focuses on the molecular and genetic basis of the biological effects of marijuana and the cannabinoid system, we cannot dismiss the larger question and current debate about the consequences of medicinal and recreational use of marijuana. Thus, is there evidence for the existence of a diagnosable cannabis specific withdrawal syndrome in human users? This issue has been addressed by the review of the published literature into cannabis withdrawal symptoms in human users (Smith, 2002). Long-term marijuana use can lead to addiction for some people; that is, they use the drug compulsively even though it often interferes with family, school, work, and recreational activities. Along with craving, withdrawal symptoms can make it hard for long-term marijuana smokers to stop using the drug. People trying to quit report irritability, difficulty sleeping, and anxiety.

They also display increased aggression on psychological tests, peaking approximately 1 week after they last used the drug. Tolerance, which varies with the effects of marijuana and cannabinoids, has been attributable to pharmacodynamic changes that may be associated with CBR downregulation and/or receptor desensitization (Grotenhermen, 2004). Although withdrawal symptoms from chronic dosing with high doses of THC in humans are usually mild, the risk for physical and psychic dependence is low compared to opiates, tobacco, alcohol, and benzodiazepines (Grotenhermen, 2004). Converging evidence indicate that gene–environment interaction are contributing factors to marijuana problem use in adolescence (Agrawal and Lynskey, 2008; Young *et al.*, 2006) as gender and early onset of use is a well-established factor for the progression from use to abuse and dependence (Bucholz *et al.*, 2000). After the discovery of endocannabinoid system in the brain, it is now known that it plays a crucial role in regulating the effects of addictive substances, including psychostimulants like nicotine, cocaine, opioids, alcohol, and benzodiazepines. While it is obvious that the initial targets of abused drugs are different, evidence indicates a converging common liability of the endocannabinoid system underlying the dependence to abused drugs. Thus, new findings and hypothesis are investigating the ECS as a common neurobiological mechanism underlying drug addiction (Onaivi, 2008). A number of studies have now shown that CBR genes are not only associated with polysubstance abuse but with other individual drug dependences and with psychiatric disorders (Chen *et al.*, 2008). Therefore, the powerful neuromodulatory action of the ECS on synaptic transmission and interactions with the effects of addictive substances may be an important natural regulatory mechanism for drug reward and therefore a target for the treatment of addictive disorders.

In considering the genetic basis of marijuana use and dependence, it is important to take into account that cannabis and cannabinoid medicinal preparations may no longer be exclusively considered recreational drugs, as they have a number of therapeutic applications and thus it has become of enormous interest to examine their mechanism(s) of action. Data from family, twin, adoption and multivariate genetic studies have provided evidence that cannabis use, abuse, and dependence can be attributed to polygenetic and environmental factors (Agrawal and Lynskey, 2006). We now know that marijuana or cannabis use has dependence liability because of the physical and psychological effects that leads to the classical relapse cycle of abused substances. Our new understanding from linkage and association studies continues to provide substantial evidence for the heritability of cannabis use, abuse, and dependence. There is therefore a role of the endocannabinoid system in the sequelae of environment–genetic interaction for marijuana use and dependence. Emerging evidence comes from accumulating knowledge on cannabinoid genetics; animal modeling of key features of marijuana-cannabinoid dependence that can be evaluated in electrophysiological, pharmacological and linkage and association, and immunohistochemical studies.

For example, behavioral cannabinoid withdrawal syndrome and gene transcriptions along with hormonal alterations have been described in the mouse model following cessation of treatment from CP-55,940, a cannabinoid agonist (Oliva *et al.*, 2003). It is important to note that from mice to human subjects the gene alterations are not limited to *CNR* genes only. The polygenic factors have been defined to mean that many genes with varying effect sizes are responsible for the genetic variation associated with marijuana use and dependency (Agrawal and Lynskey, 2006). Thus, there is accumulating genomic data from the identification of candidate genes and variants, and changes in gene transcription associated with the genetic basis of cannabinoid actions. Such genetic predisposition may include genes that regulate drug sensitivity and predisposition, influence sensation-seeking, or influence general problem behavior (Agrawal and Lynskey, 2006). It appears that a pharmacogenetic approach in drug addiction has potential for improving treatment outcome as gene variants that affect pharmacodynamic and pharmacokinetic factors are identified. It appears gene mutations might guide pharmacotherapeutic agent of choice for treatment of drug addiction (Haile *et al.*, 2008).

While, the etiology of marijuana dependence is a complex interaction of psychological, environmental, and biological factors associated with genetic vulnerability, the study of the contribution of genetic and environmental factors in cannabis dependence is further complicated by polysubstance use. Therefore overlapping gene–environment interactions may contribute to comorbid use, abuse, and dependence on marijuana and other drugs. Comorbidity and genetic loading studies indicate that adults using marijuana on a daily basis have a high frequency of co-abuse of other substances, like nicotine and alcohol. Therefore, an individual who becomes dependent on marijuana will exhibit tolerance, withdrawal reactions, craving and relapse, preventing cessation of use that characterizes the addiction cycle. Comorbidity with other psychiatric disturbances such as depression, panic attacks, anxiety, and antisocial personality disorders (Comings, 1996) have also been reported. Many studies have therefore examined the genetic and environmental contributions to risk of cannabis dependence. The notion of combined use of multiple substances with dependence potential sharing a common genetic influence has been an enduring issue in addiction research (Crabbe, 2002). However, each abused substance appears to be linked to independent genetic influences that have made it difficult to identify addiction specific gene(s), or if an addictive personality gene(s) exists. The questions exist about marijuana dependence and what genes are involved in the continuous use, withdrawal syndrome, and relapse to cessation of use? Our current knowledge shows that specific CBRs, which are encoded by CBR *CNR* genes, are activated by smoking marijuana or by the administration of cannabinoids. The myriad behavioral, physiological, and modulation of programs of gene expression following activation/inhibition of the CBRs is the subject of intense investigation.

While there is increasing evidence that genetic factors are involved in marijuana dependence through studies of familial transmission of marijuana use, abuse, and dependence (e.g., Hopfer *et al.*, 2003), new gene–environment interplay, and linkage association studies, genome scans, and molecular genetic analysis are providing insights into novel phenotypes and endophenotypes of marijuana use, dependency, and comorbidity. Therefore, risk for cannabinoid dependence is likely to be the result of a number of genes, each contributing a small fraction of the overall risk, with allelic variants contributing to different aspects of marijuana dependence. One commonly used approach to identify genetic influences on addictions is the candidate gene approach, which directly tests the effects of genetic variants of a potentially contributing gene in an association study (Kwon and Goate, 2000). The age of onset of marijuana dependence and smoking may also be influenced by genetic risk factors, because it may be unlikely for a 70-year-old man or woman who never initiated smoking marijuana to become dependent on marijuana. As with tobacco and alcohol dependence, application of oligogenic linkage analysis and determination of endophenotypes or trait markers (Tyndale, 2003), may improve the ability to identify specific genetic variants contributing to marijuana dependence. Perhaps as with genetic association studies of alcoholism, similar problems may be encountered with the candidate gene approach in marijuana dependence because of the complexity of addictions (Buckland, 2001).

We have demonstrated that alternative *CNR1* transcripts exist in the CNS that is associated with poly substance abuse in humans (Zhang *et al.*, 2004). Other studies have reported a combined drug dependence phenotype to regions on chromosome 3, 9, 11, and 20. It appears that the genome-wide scan for loci influencing cannabis dependence symptoms on chromosome 3q21 suggest that the region may contain a quantitative trait loci linking cannabis dependence and other substance use disorders (Hopfer *et al.*, 2007). Chromosomes 1 and 6 that habors the CB-R genes can be assumed from our data are genomic regions of considerable interest not only in polysubstance abuse but as possible targets for future mapping in depression and other neuropsychiatric disorders. The CB1-R *CNR1* gene, which is a target of marijuana smoke, is amongst the addiction "hot spots" in the human chromosomes. It is also a logical candidate gene for vulnerability toward developing symptoms of cannabis dependence. The central role of the cannabinoid system in the effects of other abused substances as demonstrated by CB1 receptor mutant mice and from association studies make CB1-R locus, a strong candidate with variants that might contribute to individual differences in drug abuse vulnerability (Zhang *et al.*, 2004). The retrograde action of endocannabinoids on the inhibition of classical neurotransmitter release (e.g., on glutamate, GABA, serotonin, dopamine neurotransmitters) may be differentially modified by genetic variants of the CB1-Rs, which are the main molecular targets for the endocannabinoids and marijuana-cannabinoids. Until recently, many features of the CB1 *CNR1* gene's structure, expression, regulatory regions, polymorphisms, haplotypes, and association with polysubstance abuse were poorly defined. Some studies have examined CB1-R *CNR* genomic variation and association studies have used known markers and increasingly CB1-R associated SNPs have appeared (NCBI:http://www.ncbi.nlm.nih.gov). For example two such transcripts variants derived from the 3′ end, encoding a longer isoform and short isoform, missing a segment near the 5′ end results in a frameshift with different N-terminus compared to the long isoform was reported. Our recent studies have now demonstrated the existence of novel exons, splice variants and candidate promoter region sequences that confer reporter gene expression in cells that express CB1-Rs (Zhang *et al.*, 2004). The results from the association studies show common human CB1 polymorphisms that reveal patterns of linkage disequilibrium in Caucasian and African–American individuals; while a 5′ CB1/ *CNR1* "TAG" haplotype displays significant allelic frequency differences between substance abusers and controls in Caucasian, African–American, and Japanese samples. A common CBR haplotype associated with fewer cannabis dependence symptoms in adolescents who have experimented with cannabis has also been reported (Hopfer *et al.*, 2006). This new work of improved definition of the CB1/ CNR1 locus and its variants therefore adds to the interesting features of regulating the neural circuits important for the reinforcing effects of most abused substances with a critical role in addiction vulnerability (Zhang *et al.*, 2004). The involvement of CB1-R variants in other phenotypes may have significant roles in cannabinoid pleiotropy as the effects of activation of the CB1-R have been implicated in a number of behavioral functions. The emerging structure of the genomic CB1-R *CNR1* gene is complex with multiple exonic sequences, splice variations, polymorphisms, and regulatory regions that is the central feature of the endocannabinoid physiological control system. Certainly, understanding the molecular structure and regulation of genes involved with the endocannabinoid system might be a first step in characterizing the genetic basis of smoking marijuana. Unfortunately, our knowledge of how genes affect complex traits is currently poorly understood.

IX. Functional Implication of Cannabinoid Pharmacotherapeutics

The use of cannabis for both recreational and medicinal purposes dates back for thousands of years. In recent times there has been increased attention that marijuana should be legalized for medicinal use in AIDS, cancer, obesity, multiple sclerosis, and other medical conditions where patients might benefit from the pharmacological effects of cannabis. Synthetic cannabinoids such as dronabinol, marinol, and nabilone already have an established use as antiemetics in nausea and vomiting associated with cancer chemotherapy. The reported beneficial effects in cancer and AIDS patients might reflect improved weight gain, owing to the well-documented antiemetic and appetite stimulating effects of cannabinoids. This might be a major advantage for cancer patients undergoing rigorous chemotherapy, or advanced AIDS patients. The potential draw back in the use of cannabis for medicine has been the psychoactivity induced by smoking marijuana and the potential for abuse and dependence. As presented in this review, there have been major advances in understanding the genetic and thus the biological mechanisms whereby cannabis and cannabinoids interacts with the brain in producing psychoactive and potentially pharmacotherapeutic effects (Onaivi et al., 2006b). The discovery of specific genes encoding CBRs activated by smoking marijuana, and the finding of endocannabinoids, which also the receptors, have transformed cannabinoid research into main science with significant implications for human health and disease. Thus the various components and elements of the endocannabinoid system, including cannabinoid genes, receptors, their endogenous ligands and metabolizing, and synthetic enzymes are potential pharmacogenetic and pharmacotherapeutic targets. We have exhaustively documented our current knowledge and functioning of the naturally occurring marijuana—like substances in human biology (Onaivi et al., 2002b, 2006b). We now know that the existence of the endocannabinoid system, directly impacts human development, health, and therefore is a target in disease conditions. The ubiquitous distribution of this system throughout the entire human physiology makes the components of the endocannabinoid system a target for therapeutic intervention during cell development, cancer, immune disorders, in reproduction and digestion, CNS function and dysfunction, as in mental illness, and neurodegenerative diseases (Onaivi et al., 2002a, 2006b). In addition to CBR—independent action of antitumor effects, evidence now exists that phytocannabinoids and endocannabinoids inhibit the proliferation of some tumors and cancer cells by mechanisms involving inhibition of angiogenesis and the activity of K-ras oncogene products. It has been shown that cannabinoids and endocannabinoids have some role in cancer cell growth and could be potential adjuncts as targets in the treatment of cancer.

Interestingly, although cannabis is widely used as a recreational drug in humans, only a few studies have revealed an appetite stimulant potential of cannabinoids in animals. However, evidence for the aversive effects of Δ^9-THC, WIN 55,212-2, and CP 55,940 on appetite is more readily obtained in a variety of tests. The blockade of CB1-R by rimonabant impaired the perception of the appetitive value of positive reinforcers (cocaine, morphine, and food) and reduced the motivation for sucrose, beer and alcohol consumption, indicating that positive incentive and/or motivational processes could be under a permissive control of CBR-related mechanisms. With the implication of the ECS in appetite regulation and critical roles in reward pathways, it appears that blockade of the CB1-Rs could suppress the powerful cravings that drive the cycle of addiction. Taking together, with the appearance of the endocannabinoid system in development, its presence in breast milk, and its activation by drug and alcohol use, including smoking marijuana, may provide a target to control cravings for food and chemical dependency. Rimonabant (Accomplia) was approved in Europe in 2006 for use in obesity and other metabolic indications but was suspended in 2008 because of suicides and depression as side effects in vulnerable individuals. It must be noted that in many obese patients rimonanbant was most effective as an antiobesity medication and the other complications including diabetes. The abundance of ECS in the human body and brain in comparison to other G-protein-coupled receptors support the critical role of this system in most physiological and biochemical functions that affects behavioral changes after smoking marijuana. Furthermore, there are misconceptions that people addicted to drugs lacked willpower and were morally weak, or that abused drugs release dopamine in the brain's reward centers to produce pleasure and euphoria leading to addiction in vulnerable individuals (Salamone et al., 2005; Spanagel and Weiss, 1999). We now know that drug addiction is a brain disease and experiments have been conducted to test the endocannabinoid hypothesis of drug reward and addiction and whether or not CBRs ligands might be useful as drug abuse pharmacotherapeutic targets (Onaivi, 2008). Our data have established a frame work for an endocannabinoid hypothesis of drug reward and addiction and as a target for the development of cannabinoid therapeutic agents for substance abuse disorders. Thus, several cannabinoid-mediated effects are of interest for therapeutic applications. However, whether or not the antagonism of the CBRs will be effective in treating individuals with problematic and uncontrollable marijuana smoking addiction and who wishes to quit remains to be demonstrated. Furthermore, endocannabinoids and other endogenous fatty acid ethanolamides with their roles in sleep and inflammation are emerging as a new important biological signaling molecules and targets for therapeutic intervention. Other targets for potential development of therapeutic agents are cannabinoid uptake inhibitors, if they exist. When the physiological role of CBRs and endocannabinoids are established, it is likely that other therapeutic targets may be uncovered. Based on what we currently know about the anatomical distribution of CBRs and endocannabinoids and from marijuana users, disorders associated with memory and motor coordination may benefit from novel cannabinoids. Similarly, CBRs and endocannabinoids can be targeted in immune disorders and blood pressure regulation. This is to be expected as many novel drugs are based on chemical modifications of transmitters. Therefore, it seems reasonable to expect that, as with other receptor transmitter systems, excess or lack of CBRs or endogenous ligands may be the cause for disorders in the CNS or those associated with the immune and other peripheral organs.

X. Conclusions and Future Directions

The recent significant progress on marijuana cannabinoid research has lead to new understanding about the biological effects and the unique therapeutic possibilities of targeting the endocannabinoid system which are activated by cannabinoids, endocannabinoids, and marijuana use. In just a few years since the discovery of endocannabinoids that serve as cannabinoids *in vivo*, these lipid signaling molecules and endolipids have been shown to participate in a broad array of physiological and pathological processes. However, the impact of "lipidomics" and these new lipid signaling pathways in modulating behavioral neurobiology and aspects of drug addiction is of current interest. It is now clear that marijuana use is addicting in vulnerable population. Just like most medications that are not benign, benzodiazepines, opiates, and marijuana have side effects that accompany their therapeutic use. Unlike, opiates the intensity of the craving and withdrawal reactions appears to be mild and vary from individual to individual with marijuana dependence. Thus we can look forward to a bright future of discoveries concerning the role of endocannabinoids in brain function, immune function, reproductive function, and emotional behavior, as well as the discovery of potential medicines to improve the health of many who suffer from various disorders. Marijuana (cannabinoid) research appears to have a solid scientific background for significant contribution in understanding human biology and in the development of cannabinoid based therapeutics. The only medical uses for which there has been rigorous scientific evidence are in the treatment of sickness associated with cancer chemotherapy and to counter the loss of appetite and wasting syndrome in AIDS (Iversen, 2000). There is however scientific evidence for the potential therapeutic use of cannabinoids for the treatment of psychomotor disorders, including spasticity, multiple sclerosis, epilepsy, disk prolapse, ADHDs, and pain is no longer a matter of speculation. It is a matter of time when new medications for the pharmacological manipulation of the link between endocannabinoid and other neurotransmitter systems can be developed as targets for the numerous indications of marijuana and endocannabinoid ligands. In the absence of clinical data on the abuse liability of endocannabinoids, the preclinical data indicate that like marijuana, the endocannabinoids may have minimal abuse liability in humans. The safety profile of THC, the active ingredient of cannabis, is good. It has low toxicity in both the short and long term. As concluded by Iversen (2000), marijuana is not a completely benign substance. It is a powerful drug with a variety of effects. However, except for the harms associated with smoking, the adverse effects of marijuana use are within the range of the side effects tolerated for other medications. Unraveling the genetics of the endocannabinoid system that is a target for smoking marijuana is dependent on further understanding of the complex *CNR1*, *CNR2*, and *CNRn* gene structure and their associated regulatory elements that are still poorly understood. It is concluded, that with cannabis use coded in our genes, research on the use of cannabis as medicine with a clear focus on scientific outcome as discussed in this review, rather than on paranoia or repression of cannabis use is the way forward. The perceived dangers of cannabis use no longer outweigh its potential beneficial effects to certain groups of chronically ill patients.

Acknowledgments

I acknowledge support from Center for Research, Dean of Science and Health, Student Workers Fund, and the Provost office at William Paterson University for the assigned release time. I am indebted to Dr. Adriana Patricia Tagliaferro for her patience during the preparation of this chapter and her generous support. I am also associated as a guest researcher with NIDA-NIH and indebted to collaborators including, Drs. Ishiguro, Liu, and Uhl, and others across the globe, present and past students, and postdoctoral fellows.

References

Aberle, J., Fedderwitz, I., Klages, N., George, E., and Beil, F. U. (2007). Genetic variation in two proteins of the endocannabinoid system and their influence on body mass index and metabolism under low fat diet. *Horm. Metab. Res.* **39,** 395–397.

Aberle, J., Flitsch, J., Beck, N. A., Mann, O., Busch, P., Peitsmeier, P., and Beil, F. U. (2008). Genetic variation may influence obesity only under conditions of diet: Analysis of three candidate genes. *Mol. Genet. Metab.* **95,** 188–191.

Aceto, M. D., Scates, S. M., Lowe, J. A., and Martin, B. R. (1996). Dependence on delta 9-THC: Studies on precipitated and abrupt withdrawal. *J. Pharmacol. Exp. Ther.* **278,** 1290–1295.

Agrawal, A., and Lynskey, M. T. (2006). The genetic epidemiology of cannabis use, abuse and dependence. *Addiction* **101,** 801–812.

Agrawal, A., and Lynskey, M. T. (2008). Are there genetic influences on addiction: Evidence from family, adoption and twin studies? *Addiction* **103,** 1069–1081.

Agrawal, A., Morley, K.I, Hansell, N. K., Pergadia, M. L., Montgomery, G. W., Statham, D. J., Todd, R. D., Madden, P. A., Heath, A. C., Whitfield, J., Martin, N. G., Lynskey, M. T., *et al.* (2008). Autosomal linkage analysis for cannabis use behaviors in Australian adults. *Drug Alchol Depend.* **98,** 185–190.

Agrawal, A., Wetherill, L., Dick, D. M., Xuei, X., Hinrichs, A., Hesselbrock, V., Kramer, J., Nurnberger, J. I. Jr, Schuckit, M., Bierut, L. J., Edenberg, H. J., Foroud, T., *et al.* (2009). Evidence for association between polymorphisms in the cannabinoid receptor 1 (CNR1) gene and cannabis dependence. *Am. J. Med. Genet. Part B* **150,** 736–740.

Andersson, H., D'Antona, A. M., Kendall, D. A., Von Heijne, G., and Chin, C. N. (2003). Membrane assembly of the cannabinoid receptor 1: Impact of a long N-terminal tail. *Mol. Pharmacol.* **64,** 570–577.

Ashton, J. C., Friberg, D., Darlington, C. L., and Smith, P. F. (2005). Expression of the cannabinoid CB2 receptor in the rat cerebellum: An immunohistochemical study. *Neurosci. Lett.* **396,** 113–116.

Ballon, N., Leroy, S., Roy, C., Bourdel, M. C., Charles-Nicolas, A., Krebs, M. O., and Poirier, M. F. (2006). (AAT)n repeat in the cannabinoid receptor gene (CNR1): Association with cocaine addiction in an African-Caribbean population. *Pharmacogenomics J.* **6,** 126–130.

Barrero, F. J., Ampuero, I., Morales, B., Vives, F., De Dios Luna Del Castillo, J., Hoenicka, J., and García Yébenes, J. (2005). Depression in Parkinson's disease is related to a genetic polymorphism of the cannabinoid receptor gene (CNR1). *Pharmcogenomics J.* **5,** 135–141.

Battista, N., Pasquariello, N., Di Tommaso, M., and Maccarrone, M. (2008). Interplay between endocannabinoids, steroids and cytokines in the control of human reproduction. *J Neuroendocrinol.* **20,** 82–89.

Beaulieu, P. (2005). Toxic effects of cannabis and cannabinoids: Animal data. *Pain Res. Manag.* **10,** 23A–26A.

Beltramo, M., Bernardini, N., Bertorelli, R., Campanella, M., Nicolussi, E., Fredduzzi, S., and Reggiani, A. (2006). CB2 receptor-mediated antihyperalgesia: Possible direct involvement of neural mechanisms. *Eur. J. Neurosci.* **23,** 1530–1538.

Benito, C., Numez, E., Tolon, R. M., Carrier, E. J., Rábano, A., Hillard, C. J., and Romero, J. (2003). Cannabinoid CB2 receptors and fatty acid amide hydrolase are selectively overexpressed in neuritic plaque-associated glia in Alzheimer's disease brains. *J. Neurosci.* **23,** 11136–11141.

Benzinou, M., Chevre, J.C, Ward, K. J., Lecoeur, C., Dina, C., Lobbens, S., Durand, E., Delplanque, J., Horber, F. F., Heude, B., Balkau, B., Borch-Johnsen, K., *et al.* (2008). Endocannabinoid receptor 1 gene variations increase risk for obesity and modulate body mass index in European populations. *Hum. Mol. Genet.* **17,** 1916–1921.

Bortolato, M., Camolongo, P., Mangieri, R. A., Scattoni, M. L., Frau, R., Trezza, V., La Rana, G., Russo, R., Calignano, A., Gessa, G. A., Cuomo, V., Piomelli, D., *et al.* (2006). Anxiolytic-like properties of the anandamide transport inhibitor AM404. *Neuropsychopharmacology* **31,** 2652–2659.

Brusco, A., Tagliaferro, P., Saez, T., and Onaivi, E. S. (2008a). Ultrastructural localization of neuronal brain CB2 cannabinoid receptors. *Ann. NY Acad. Sci.* **1139,** 450–457.

Brusco, A., Tagliaferro, P., Saez, T., and Onaivi, E. S. (2008b). Postsynaptic localization of CB2 cannabinoid receptors in the rat hippocampus. *Synapse* **62,** 944–949.

Bucholz, K. K., Heath, A. C., and Madden, P. A. (2000). Transitions in drinking in adolescent female: Evidence from the Missouri adolescent female twin study. *Alcohol Clin. Exp. Res.* **24,** 914–923.

Buckland, P. R. (2001). Genetic associations studies of alcoholism: Problems with the candidate gene approach. *Alcohol Alcoholism* **36,** 99–102.

Budney, A. J. (2006). Are specific dependence criteria necessary for different substances: How can research on cannabis inform this issue? *Addiction* **101,** 125–133.

Caenazzo, L., Hoehe, M. R., Hsieh, W.-T., Berrettini, W. H., Bonner, T. I., and Gershon, E. S. (1991). HindIII identifies a two allele DNA polymorphism of the human cannabinoid receptor gene (CNR). *Nucleic Acids Res.* **19,** 4798.

Cao, Q., Martinez, M., Zhang, J., Sanders, A. R., Badner, J.A, Cravchik, A., Markey, C,J, Beshah, E., Guroff, J. J., Maxwell, M. E., Kazuba, D. M., Whiten, R., *et al.* (1997). Suggestive evidence for a schizophrenia susceptibility locus on chromosome 6q and a confirmation in an independent series of pedigrees. *Genomics* **43,** 1–8.

Chakrabarti, A., Onaivi, E. S., and Chaudhuri, G. (1995). Cloning and sequencing of a cDNA encoding the mouse brain-type cannabinoid receptor protein. *DNA Sequence* **5,** 385–388.

Chakrabarti, B., Kent, L., Suckling, J., Bullimore, E., and Baron-Cohen, S. (2006). Variations in the human cannabinoid receptor (*CNR1*) gene modulate striatal responses to happy faces. *Eur. J. Neurosci.* **23,** 1944–1948.

Chavarria-Siles, I., Contrras-Rojas, J., Hare, E., Walss-Bass, C., Quezada, P., Dassori, A., Contreras, S., Medina, R., Ramírez, M., Salazar, R., Raventos, H., Escamilla, M. A., *et al.* (2008).

Cannabinoid receptor 1 gene (CNR1) and susceptibility to a quantitative phenotype for hebephrenic schizophrenia. *Am. J. Med. Genet.* **147,** 279–284.

Chen, X., Williamson, V. S., An, S. S., Hettema, J. M., Aggen, S. H., Neale, M. C., and Kendler, K. S. (2008). Cannabinoid receptor 1 gene association with nicotine dependence. *Arch. Gen. Psychiatr.* **65,** 816–824.

Cippitelli, A., Bilbao, A., Gorriti, M. A., Navarro, M., Massi, M., Piomelli, D., Ciccocioppo, R., and de Fonseca, R. (2007). The anandamide transport inhibitor AM404 reduces ethanol seldadminiatration. *Eur. J. Neurosci.* **26,** 476–486.

Comings, D. E. (1996). Genetic factors in drug abuse and dependence. *NIDA Res. Monogr.* **159,** 16–48.

Comings, D. E., Muhleman, D., Gade, R., Johnson, P., Verde, R., Saucier, G., and MacMurray, J. (1997). Cannabinoid receptor gene (CNR1): Association with IV drug use. *Mol. Psychiatr.* **2,** 161–168.

Cook, S. A., Lowe, J. A., and Martin, B. R. (1998). CB1 receptor antagonist precipitates withdrawal in mice exposed to Delta9-tetrahydrocannabinol. *J. Pharmacol. Exp Ther.* **285,** 1150–1156.

Cota, D., Marsicano, G., Lutz, B., Vicennati, V., Stalla, G. K., Pasquali, R., and Pagotto, U. (2003). Endogenous cannabinoid system as a modulator of food intake. *Int. J. Obes.* **27,** 289–301.

Covault, J., Gelernter, J., and Kranzler, H. (2001). Association study of cannabinoid receptor gene (CNR1) alleles and drug dependence. *Mol. Psychiatr.* **6,** 501–502.

Crabbe, J. C. (2002). Genetic contributions to addictions. *Ann. Rev. Psychol.* **53,** 435–462.

Dawson, E. (1991). Identification of a polymorphic triplet marker for the brain cannabinoid receptor gene: Use in linkage and association studies of schizophrenia. *Psychiatr. Genet.* **5,** S50.

de Lago, E., and Fernandez-Ruiz, J. (2007). Cannabinoids and neuroprotection in motor-related disorders. *CNS Neuro Disord. Drug Targets* **6,** 377–387.

Dennedy, M. C., Friel, A. M., Houlihan, D. D., Broderick, V. M., Smith, T., and Morrison, J. J. (2004). Cannabinoids and human uterus during pregnancy. *Am. J. Obstet. Gynecol.* **190,** 2–9.

Domschke, K., Dannlowski, U., Ohrmann, P., Lawford, B., Bauer, J., Kugel, H., Heindel, W., Young, R., Morris, P., Arolt, V., Deckert, J., Suslow, T., *et al.* (2008). Cannabinoid receptor 1 (*CNR1*) gene: Impact on anti depressant treatment response and emotion processing in major depression. *Eur. Neuropsychopharmacol.* **18,** 751–759.

Ehlers, C. L., Slutske, W. S., Gilder, D. A., and Lau, P. (2007a). Age of first marijuana use and the occurrence of marijuana use disorders in Southwest California Indians. *Pharmacol. Biochem. Behav.* **86,** 290–296.

Ehlers, C. L., Slutske, W. S., Lind, P. A., and Wilhelmsen, K. C. (2007b). Association between single nucleotide polymorphisms in the cannabinoid receptor gene (CNR1) and impulsivity in South west California Indians. *Twin Res. Hum. Genet.* **10,** 805–811.

Gadzicki, D., Muller-Vahl, K., and Stuhrmann, M. (1999). A frequent polymorphism in the coding exon of the human cannabinoid receptor (CNR1) gene. *Mol. Cell. Probes* **13,** 321–323.

Gadzicki, D., Müller-Vahl, K. R., Heller, D., Ossege, S., Nöthen, M. M., Hebebrand, J., and Stuhrmann, M. (2004). Tourette syndrome is not caused by mutations in the central cannabinoid receptor (CNR1) gene. *Am. J. Med. Genet. B Neuropsychiatr. Genet.* **127,** 97–103.

Galve-Roperh, I., Aguado, T., Palazuelos, J., and Guzman, M. (2008). Mechanisms of control of neuron survival by the endocannabinoid system. *Curr. Pharm. Des.* **14,** 2279–2288.

Gazzero, P., Caruso, M.G, Notarnicola, M., Misciagna, G., Guerra, V., Laezza, C., and Bifulco, M. (2007). Association between cannabinoid type-1 receptor polymorphism and body mass index in a southern Italian population. *Int. J. Obes.* **31,** 908–912.

Gerard, C. M., Mollereau, C., Vassart, G., and Parmentier, M. (1991). Molecular cloning of a human cannabinoid receptor which is also expressed in the testis. *Biochem. J.* **279,** 129–134.

Gerhardt, C. C., and van Heerikhuizen, H. (1997). Functional characteristics of heterologously expressed 5-HT receptors. *Eur. J. Pharmacol.* **334,** 1–23.

Glaser, S. T., Abumrad, N. A., Fatade, F., Kaczocha, M., Studholme, K. M., and Deutsch, D. G. (2003). Evidence against the presence of an anandamide transporter. *PNAS* **100,** 4269–4274.

Glaser, S. T., Kaczocha, M., and Deutsch, D. G. (2005). Anandamide transport: A critical review. *Life Sci.* **77,** 1584–1604.

Goncalves, M. B., Suetterlin, P., Yip, P., Molina-Holgado, F., Walker, D. J., Oudin, M. J., Zentar, M. P., Pollard, S., Yanez-Munoz, R. J., Williams, G., Walsh, F. S., Pangalos, M. N., *et al.* (2008). A diacylglycerol lipase-CB2 cannabinoid pathway regulates adult subventricular zone neurogenesis in an age-dependent manner. *Mol. Cell. Neurosci.* **38,** 526–536.

Gong, J. P., Onaivi, E. S., Ishiguro, H., Liu, Q. R., Tagliaferro, P. A., Brusco, A., and Uhl, G. R. (2006). Cannabinoid CB2 receptors: Immunohistochemical localization in rat brain. *Brain Res.* **1071,** 10–23.

Gonsiorek, W., Lunn, C., Fan, X., Narula, S., Lundell, D., and Hipkin, R. W. (2000). Endocannabinoid 2-Arachidonyl glycerol is a full agonist through human type 2 cannabinoid receptor: Antagonism by anandamide. *Mol Pharmacol.* **57,** 1045–1050.

Grotenhermen, F. (2004). Pharmacology of cannabinoids. *Neuroendocrinol. Lett.* **25,** 14–23.

Gurwitz, D., and Kloog, Y. (1998). Do endogenous cannabinoids contribute to HIV-mediated immune failure? *Mol. Med. Today* **4,** 196–200.

Haile, C. N., Kosten, T. A., and Kosten, T. R. (2008). Pharmagenetic treatments for drug addiction: Alcohol and opiates. *Am. J. Drug Alcohol Abuse* **34,** 355–381.

Hamdani, N., Tabeze, J- P., Ramoz, N., Ades, J., Hamon, M., Sarfati, Y., Boni, C., and Gorwood, P. (2008). The *CNR1* gene as a pharmacogenetic factor for antipsychotics rather than a susceptibility gene for schizophrenia. *Eur. Neuropsychopharmacol.* **18,** 34–40.

Haughey, H. M., Marshall, E., Schacht, J. P., Louis, A., and Hutchison, K. E. (2008). Marijuana withdrawal and craving: Influence of the cannabinoid 1 (*CNR1*) and fatty acid amide hydrolase (*FAAH*) genes. *Addiction* **103,** 1678–1686.

Heller, D., Schneider, U., Seifert, J., Cimander, K. F., and Stuhrmann, M. (2001). The cannabinoid receptor gene (CNR1 is not affected in German i.v. drug users. *Addict. Biol.* **6,** 183–187.

Henstridge, C. M., Balenga, N. A. B., Ford, L. A., Ross, R. A., Waldhoer, M., and Irving, A. J. (2008). The GPR55 ligand L-α-lysophosphatidylinositol promotes RhoA-dependent Ca^{2+} signaling and NFAT activation. *FASEB J.* **23,** 183–193.

Herman, A.I, Kranzler, H. R., Cubells, J. F., Gelernter, J., and Covault, J. (2006). Association study of the CNR1 gene exon 3 alternative promoter region polymorphisms and substance abuse. *Am. J. Med. Genet. B Neuropsychiatr. Genet.* **141B,** 499–503.

Hermann, A., Kaczocha, M., and Deutsch, D. G. (2006). 2-Arachidonylglycerol (2-AG) membrane transport: History and outlook. *AAPS J.* **8,** E409–E412.

Hillard, C. J., Harris, R. A., and Bloom, A. S. (1985). Effects of the cannabinoids on physical properties of brain membranes and phospholipid vesicles: Fluorescence studies. *J. Pharmacol. Exp. Ther.* **232,** 579–588.

Hoehe, M. R., Caenazzo, L., Martinez, M. M., Hsieh, W. T., Modi, W. S., Gershon, E. S., and Bonner, T. I. (1991). Genetic and physical mapping of the human cannabinoid receptor gene to chromosome 6q14-q15. *New Biologist* **3,** 880–885.

Hoenicka, J., Ponce, G., Jimenez-Arriero, M. A., Ampuero, I., Rodríguez-Jiménez, R., Rubio, G., Aragüés, M., Ramos, J. A., and Palomo, T. (2007). Association in alcoholic patients between psychopathic traits and the additive effect of allelic forms of CNR1 and FAAH endocannabinoid genes, and the 3′ region of the DRD2 gene. *Neurotox Res.* **11,** 51–60.

Hopfer, C. J., Stallings, M. C., Hewith, J. K., and Crowley, T. J. (2003). Family transmission of marijuana use, abuse, and dependence. *J. Am. Acad. Child Adolesc. Psychiatr.* **42,** 834–841.

Hopfer, C. J., Young, S. E., Purcell, S., Crowley, T. J., Stallings, M. C., Corley, R. P., Rhee, S. H., Smolen, A., Krauter, K., Hewitt, J. K., and Ehringer, M. A. (2006). Cannabis receptor haplotype associated with fewer cannabis dependence symptoms in adolescents. *Am. J. Med. Genet. B Neuropsychiatr. Genet.* **141,** 895–901.

Hopfer, C. J., Lessem, J. M., Hartman, C. A., Stallings, M. C., Cherny, S. S., Corley, R. P., Hewitt, J. K., Krauter, K. S., Mikulich-Gilbertson, S. K., Rhee, S. H., Smolen, A., Young, S. E., *et al.* (2007). A genome-wide scan influencing adolescent cannabis dependence symptoms: Evidence for linkage on chromosomes 3 and 9. *Drug Alchol Depend.* **89,** 34–41.

Hungund, B. L., Vinod, K. Y., Kassir, S. A., Basavarajappa, B. S., Yalamanchili, R., Cooper, T. B., Mann, J. J., and Arango, V. (2004). Upregulation of CB1 receptors and agonist-stimulated [35] GTPgammaS binding in the prefrontal cortex of depressed suicide victims. *Mol. Psychiatr.* **9,** 184–190.

Hutchison, K. E., Haughley, H., Niculescu, M., Schacht, J., Kaiser, A., Stitzel, J., Horton, W. J., and Filbey, F. (2008). The incentive salience of alcohol: Translating the effects of genetic variant in CNR1. *Arch. Gen. Psychiatr.* **65,** 841–850.

Ishiguro, H., Iwasaki, S., Teasenfitz, L., Higuchi, S., Horiuchi, Y., Saito, T., Arinami, T., and Onaivi, E. S. (2007). Involvement of cannabinoid CB2 receptor in alcohol preference in mice and alcoholism in humans. *Pharmacogenomics J.* **7,** 380–385.

Iversen, L. L. (2000). The Science of Marijuana Oxford University Press, Inc., New York.

Jaeger, J. P., Mattevi, V. S., Callegri-Jacques, S. M., and Hutz, M. H. (2008). Cannabinoid type-1 receptor gene polymorphisms are associated with central obesity in a Southern Brazilian population. *Dis. Markers* **25,** 67–74.

Jesudason, D., and Wittert, G. (2008). Endocannabinoid system in food intake and metabolic regulation. *Curr. Opin. Lipidol.* **19,** 344–348.

Jhaveri, M. D., Elmes, S. J., Richardson, D., Barrett, D. A., Kendall, D. A., Mason, R., and Chapman, V. (2008). Evidence for a novel functional role of cannabinoid CB(2) receptors in the thalamus of neuropathic rats. *Eur. J. Neurosci.* **27,** 1722–1730.

Johnson, J. P., Muhleman, D., MacMurray, J., Gade, R., Verde, R., Ask, M., Kelley, J., and Comings, D. E. (1997). Association between the cannabinoid receptor gene (CNR1) and the P300 event-related potential. *Mol. Psychiatr.* **2,** 169–171.

Karsak, M., Cohen-Solal, M., Freudenberg, J., Ostertag, A., Morieux, C., Kornak, U., Essig, J., Erxlebe, E., Bab, I., Kubisch, C., de Vernejoul, M. C., and Zimmer, A. (2005). Cannabinoid receptor type 2 gene is associated with human osteoporosis. *Hum. Mol. Genet.* **14,** 3389–3396.

Karsak, M., Gaffal, E., Date, R., Wang-Eckhardt, L., Rehnelt, J., Petrosino, S., Starowicz, K., Steuder, R., Schlicker, E., Cravatt, B., Mechoulam, R., Buettner, R., *et al.* (2007). Attenuation of allergic contact dermatitis through the endocannabinoid system. *Science.* **316,** 1494–1497.

Kathmann, M., Haug, K., Heils, A., Nothen, M. M., and Schlicker, E. (2000). Exchange of three amino acids in the cannabinoid CB1 receptor (CNR1) of an epilepsy patient. Symposium on the Cannabinoids, Burlington, Vermont, ICRS Abst, p. 18.

Kwon, J. M., and Goate, A. M. (2000). The candidate gene approach. *Alcohol Res. Health.* **24,** 164–168.

Lauckner, J. E., Jensen, J. B., Chen, H- Y., Lu, H. C., Hille, B., and Mackie, K. (2008). GPR55 is a cannabinoid receptor that increases intracellular calcium and inhibits M current. *PNAS* **105,** 2699–2704.

Leroy, S., Griffon, N., Bourdel, M. C., Olié, J. P., Poirier, M. F., and Krebs, M. O. (2001). Schizophrenia and the cannabinoid receptor type 1 (CB1): Association study using a single-base polymorphism in coding exon 1. *Am. J. Med. Genet.* **105,** 749–752.

Li, J.-H., and Lin, L.-F. (1998). Genetic toxicology of abused drugs: A brief review. *Mutagenesis.* **13,** 557–565.

Li, T., Liu, X., Zhu, Z. H., Zhao, J., Hu, X., Ball, D. M., Sham, P. C., and Collier, D. A. (2000). No association between (AAT)n repeats in the cannabinoid receptor gene (CNR1) and heroin abuse in a Chinese population. *Mol. Psychiatr.* **5,** 128–130.

Lovinger, D. M. (2008). Presynaptic modulation by endocannabinoids. *Handb. Exp. Pharmacol.* **184,** 435–477.

Lu, A. T., Ogdie, M. N., Järvelin, M. R., Moilanen, I. K., Loo, S. K., McCracken, J. T., McGough, J. J., Yang, M. H., Peltonen, L., Nelson, S. F., Cantor, R. M., Smalley, S. L., *et al.* (2008). Association of the cannabinoid receptor gene (CNR1) with ADHD and post-traumatic stress disorder. *Am. J. Med. Genet. B Neuropsychiatr. Genet.* **147,** 1488–1489.

Makriyannis, A., Banijamali, A., Jarrell, H. C., and Yang, D. P. (1989). The orientation of (-)-delta-9-tetrahydrocannabinol in DPPC bilayers as determined by solid state 2H-NMR. *Biochim. Biophys. Acta.* **986,** 141–145.

Martinez, M., Goldin, L. R., Cao, Q., Zhang, J., Sanders, A. R., Nancarrow, D. J., Taylor, J. M., Levinson, D. F., Kirby, A., Crowe, R. R., Andreasen, N.C, Black, D. W., *et al.* (1999). Follow-up study on a susceptibility locus for schizophrenia on chromosome 6q. *Am. J. Med. Genet.* **88,** 337–343.

Martinez-Gras, I., Hoenicka, J., Ponce, G., Rodríguez-Jiménez, R., Jiménez-Arrierom, M. A., Pérez-Hernandez, E., Ampuero, I., Ramos-Atance, J. A., Palomo, T., and Rubio, G. (2006). (AAT)n repeat in the cannabinoid receptor gene, CNR1: Association with schizophrenia in a Spanish population. *Eur. Arch. Psychiatr. Clin. Neurosci.* **256,** 437–441.

Mato, S., Olmo, E. D., and Pazos, A. (2003). Ontogenetic development of cannabinoid receptor expression and signal transduction functionality in the human brain. *Euro. J. Neurosci.* **17,** 1747–1754.

Matsuda, L. A. (1997). Molecular aspects of cannabinoid receptors. *Crit. Rev. Neurobiol.* **11,** 143–166.

Matsuda, L. A., Lolait, S. J., Brownstein, M. J., Young, A. C., and Bonner, T. I. (1990). Structure of a cannabinoid receptor and functional expression of the cloned cDNA. *Nature* **346,** 561–564.

Matyas, F., Yanovsky, Y., Mackie, K., Kelsch, W., Misgeld, U., and Freund, T. F. (2006). Subcellular localization of type 1 cannabinoid receptors in the rat basal ganglia. *Neuroscience* **137,** 337–361.

Mechoulam, R., and Shohami, E. (2007). Endocannabinoids and traumatic brain injury. *Mol. Neurobiol.* **36,** 68–74.

Muller, T.D, Reichwald, K., Wermter, A- K., Brönner, G., Nguyen, T. T., Friedel, S., Koberwitz, K., Engeli, S., Lichtner, P., Meitinger, T., Schäfer, H., and Hebebrand, J. (2007). No evidence for an involvement of variants in the cannabinoid receptor gene (CNR1 in obesity in German children and adolescents. *Mol. Genet. Metabol.* **90,** 429–434.

Muller, T. D., Reichwald, K., Bronner, G., Kirschner, J., Nguyen, T. T., Scherag, A., Herzog, W., Herpertz-Dahlmann, B., Lichtner, P., Meitinger, T., Platzer, M., and Schafer, H. (2008). Lack of association of genetic variants in genes of the endocannabinoid system with anorexia nervosa. *Child Adolesc Psychiatr. Health.* **2,** 33–40.

Munro, S., Thomas, K. L., and Abu-Shaar, M. (1993). Molecular characterization of a peripheral cannabinoid receptor. *Nature* **365,** 61–65.

Murphy, W. J., Elzirik, E., Johnson, W. E., Zhang, Y. P., Ryder, O. A., and O'Brien, S. J. (2001). Molecular phylogenetics and origins of placental mammals. *Nature* **409,** 614–618.

Nordstrom, R., and Andersson, H. (2006). Amino-terminal processing of the human cannabinoid receptor 1. *J. Receptor Signal Transduction* **26,** 259–267.

Norrod, A. G., and Puffenbarger, R. B. (2007). Genetic polymorphisms of the endocannabinoid system. *Chem. Biodivers.* **4,** 1926–1932.

Nunez, E., Benito, C., Pazos, M. R., Barbachano, A., Fajardo, O., González, S., Tolón, R. M., and Romero, J. (2004). Cannabinoid CB2 receptors are expressed by perivascular microglia cells in the human brain: An immunohistochemical study. *Synapse* **53,** 208–213.

Oka, S., Nakajima, K., Yamashita, A., Kishimoto, S., and Sugiura, T. (2007). Identification of GPR55 as a lysophosphatidyinositol receptor. *Biochem. Biophys. Res. Com.* **362,** 928–934.

Oka, S., Toshida, T., Maruyama, K., Nakajima, K., Yamashita, A., and Sugiura, T. (2008). 2-Arachidonoyl-sn-glycero-3-phosphoinositol: A Possible Natural Ligand for GPR55. *J. Biochem.* **145,** 13–20.

Okamoto, Y., Morishita, J., Tsuboi, K., Tonai, T., and Ueda, N. (2004). Molecular characterization of a phospholipase D generating anandamide and its congeners. *J. Biol. Chem.* **279,** 5298–5305.

Oliva, J. M., Ortiz, S., Palomo, T., and Manzanares, J. (2003). Behavioral and gene transcription alterations induced by spontaneous cannabinoid withdrawal in mice. *J. Neurochem.* **85,** 94–104.

Onaivi, E. S. (2006). Neuropsychobiological evidence for the functional presence and expression of cannabinoid CB2 receptors in the brain. *Neuropsychobiology* **54,** 231–246.

Onaivi, E. S. (2008). An endocannabinoid hypothesis of drug reward and drug addiction. *Ann. NY Acad. Sci.* **1139,** 412–421.

Onaivi, E. S., Chakrabarti, A., and Chaudhuri, G. (1996). Cannabinoid receptor genes. *Prog. Neurobiol.* **48,** 275–305.

Onaivi, E. S., Chaudhuri, G., Abaci, A.S, Parker, M., Manier, D. H., Martin, P. R., and Hubbard, J. R. (1999). Expression of cannabinoid receptors and their gene transcripts in human blood cells. *Prog. Neuropsychopharmacol. Biol. Psychiatr.* **23,** 1063–1077.

Onaivi, E. S., Ishiguro, H., Lin, Z., Akinshola, B. E., Zhang, P- W., and Uhl, G. R. (2002a). Cannabinoid receptor genetics and behavior. *In* Biology of Marijuana: From Gene to behavior, (E. S. Onaivi, Ed.), pp. 1–44. CRC Press, Taylor and Francis Group, Boca Raton.

Onaivi, E. S., Leonard, C. M., Ishiguro, H., Zhang, P. W., Lin, Z., Akinshola, B. E., and Uhl, G. R. (2002b). Endocannabinoids and cannabinoid receptor genetics. *Prog. Neurobiol.* **66,** 307–344.

Onaivi, E. S., Ishiguro, H., Gong, J. P., Patel, S., Perchuk, A., Meozzi, P. A., Myers, L., Mora, Z., Tagliaferro, P., Gardner, E., Brusco, A., Akinshola, B. E., *et al.* (2006a). Discovery of the presence and functional expression of cannabinoid CB2 receptors in brain. *Ann. NY Acad. Sci.* **1074,** 514–536.

Onaivi, E. S., Ishiguro, H., Zhang, P- W., Lin, Z., Akinshola, B. E., Leonard, C. M., Chirwa, S. S., Gong, J.-P., and Uhl, G. R. (2006b). Endocannabinoid receptor genetics and marijuana use. *In* Endocannabinoids: The Brain and Body's Marijuana and Beyond, (E. S. Onaivi, T. Sugiura, and V. DiMarzo, Eds.), pp. 57–118. CRC Press, Taylor and Francis Group, Boca Raton.

Onaivi, E. S., Ishiguro, H., Gong, J.P, Patel, S., Meozzi, P. A., Myers, L., Perchuk, A., Mora, Z., Tagliaferro, P. A., Gardner, E., Brusco, A., Akinshola, B. E., *et al.* (2008a). Brain neuronal CB2 cannabinoid receptors in drug abuse and depression: From mice to human subjects. *PLoS ONE* **3,** e1640.

Onaivi, E. S., Ishiguro, H., Gong, J. P., Patel, S., Meozzi, P. A., Myers, L., Perchuk, A., Mora, Z., Tagliaferro, P. A., Gardner, E., Brusco, A., Akinshola, B. F., *et al.* (2008b). Functional expression of brain neuronal CB2 cannabinoid receptors are involved in the effects of drugs of abuse and in depression. *Ann. NY Acad. Sci.* **1139,** 434–449.

Palomo, T., Kostrzewa, R. M., Beninger, R. J., and Archer, T. (2007). Genetic variation and spared biological susceptibility underlying comorbidity in neuropsychiatry. *Neurotox. Res.* **12,** 29–42.

Park, B., Gibbons, H. M., Mitchell, M. D., and Glass, M. (2003). Identification of the CB1 cannabinoid receptor and fatty acid amide hydrolase (FAAH) in the human placenta. *Placenta* **24,** 990–995.

Pazos, M. R., Nunez, E., Benito, C., Tolón, R. M., and Romero, J. (2004). Role of the endocannabinoid system in Alzheimers's disease: New perspectives. *Life Sci.* **75,** 1907–1915.

Peeters, A., Beckers, S., Mertens, I., Van Hul, W., and Van Gaal, L. (2007). The G1422A variant of the cannabinoid receptor gene (CNR1) is associated with abdominal adiposity in obese men. *Endocrine* **31,** 138–141.

Petitet, F., Donlan, M., and Michel, A. (2006). GPR55 as a new cannabinoid recptor: Still a long way to prove it. *Chem. Biol. Drug Des.* **67,** 252–253.

Ponce, G., Hoenicka, J., Rubio, G., Ampuero, I., Jiménez-Arriero, M. A., Rodríguez-Jiménez, R., Palomo, T., and Ramos, J. A. (2003). Association between cannabinoid receptor gene (CNR1) and childhood attention deficit/hyperactivity disorder in Spanish male alcoholic patients. *Mol. Psychiatr.* **8,** 466–467.

Preuss, U. W., Koller, G., Zill, P., Bondy, B., and Soyka, M. (2003). Alcoholism related phenotypes and genetic variants of the CB1 receptor. *Eur. Arch. Psychiatr. Clin. Neurosci.* **253,** 275–280.

Reinhard, W., Stark, K., Neureuther, K., Sedlacek, K., Fischer, M., Baessler, A., Weber, S., Kaess, B., Wiedmann, S., Erdmann, J., Lieb, W., Jeron, A., *et al.* (2008). Common polymorphisms in the cannabinoid CB2 gene (CNR2) are not associated with myocardial infarction and cardiovascular risk factors. *Int. J. Mol. Med.* **22,** 165–174.

Rinaldi-Carmona, M., Calandra, B., Shire, D., Bouaboula, M., Oustric, D., Barth, F., Casellas, P., Ferrara, P., and Le Fur, G. (1996). Characterization of two cloned CB1 cannabinoid receptor isoforms. *J. Pharmacol. Exp. Ther.* **278,** 871–878.

Russo, P., Strazzullo, P., Cappuccio, F.P., Tregouet, D. A., Lauria, F., Loguercio, M., Barba, G., Versiero, M., and Siani, A. (2007). Genetic variations at the endocannabinoid type 1 receptor gene (CNR1) are associated with obesity phenotype in men. *J. Clin. Endocrinol. Metabol.* **92,** 2382–2386.

Salamone, J. D., Correa, M., Mingote, S. M., and Weber, S. M. (2005). Beyond the reward hypothesis: Alternative functions of nucleus accumbens dopamine. *Curr. Opin. Pharmacol.* **5,** 34–41.

Samson, M- T., Small-Howard, A., Shimoda, L. M., Koblan-Huberson, M., Stokes, A. J., and Turner, H. (2003). Differential roles of CB1 and CB2 cannabinoid receptors in Mast cells. *J. Immunol.* **170,** 4953–4962.

Schatz, A. R., Lee, M., Condie, R. B., Pulaski, J. T., and Kaminski, N. E. (1997). Cannabinoid receptors CB1 and CB2: A characterization of expression and adenylate cyclase modulation within the immune system. *Toxicol. Appl. Pharmacol.* **142,** 278–287.

Schmidt, L. G., Samochowiec, J., Finckh, U., Fiszer-Piosik, E., Horodnicki, J., Wendel, B., Rommelspacher, H., and Hoehe, M. R. (2002). Association of a CB1 cannabinoid receptor gene (CNR1) polymorphism with severe dependence. *Drug Alcohol Depend.* **65,** 221–224.

Serra, G., and Fratta, W. (2007). A possible role for the endocannabinoid system in the neurobiology of depression. *Clin. Practice Epidemiol. Mental Health* **3.25,** 1–11.

Shire, D., Carillon, C., Kaghad, M., Calandra, B., Rinaldi-Carmona, M., Le Fur, G., Caput, D., and Ferrara, P. (1995). An amino terminal variant of the central cannabinoid receptor resulting from alternative splicing. *J. Biol. Chem.* **270,** 3726–3731.

Siegfried, Z., Kanyas, K., Latzer, Y., Karni, O., Bloch, M., Lerer, B., and Berry, E. M. (2004). Association study of cannabinoid receptor gene (CNR1) alleles and anorexia nervosa: Differences between restricting and binging/purging subtypes. *Am. J. Med. Genet. B Neuropsychiatr. Genet.* **125,** 126–130.

Sipe, J. C., Arbour, N., Gerber, A., and Beutler, E. (2005). Reduced endocannabinoid immune modulation by a common cannabinoid 2 (CB2) receptor gene polymorphism: Possible risk for autoimmune disorders. *J. Leukoc. Biol.* **78,** 231–238.

Smith, N. T. (2002). A review of the published literature into cannabis withdrawal symptoms in human users. *Addiction* **97,** 621–632.

Spanagel, R., and Weiss, F. (1999). The dopamine hypothesis of reward: Past and current status. *Trend. Neurosci.* **22,** 521–527.

Stubbs, L., Chittenden, L., Chakrabarti, A., and Onaivi, E. S. (1996). The mouse cannabinoid receptor gene is located in proximal chromosome 4. *Mammal. Genome.* **7,** 165–166.

Sugiura, T., Kodaka, T., Nakane, S., Miyashita, T., Kondo, S., Suhara, Y., Takayama, H., Waku, K., Seki, C., Baba, N., and Ishima, Y. (1999). Evidence that cannabinoid CB1 receptor is a 2-arachidonyl glycerol receptor. *J. Biol. Chem.* **274,** 2794–2801.

Sugiura, T., Kishimoto, S., Oka, S., and Gokoh, M. (2006). Biochemistry, pharmacology and physiology of 2-arachidonyl glycerol, an endogenous cannabinoid receptor ligand. *Prog. Lip. Res.* **45,** 405–446.

Tsai, S. J., Wang, Y.C, and Hong, C. J. (2000). Association study of a cannabinoid receptor gene (CNR1) polymorphism and schizophrenia. *Psychiatr. Genet.* **10,** 149–151.

Tsai, S. J., Wang, Y.C, and Hong, C. J. (2001). Association study between cannabinoid receptor gene (CNR1) and pathogenesis and psychotic symptoms of mood disorders. *Am. J. Med. Genet.* **105,** 219–221.

Tyndale, R. F. (2003). Genetics of alcohol and tobacco use in humans. *Ann. Med.* **35,** 94–121.

Ujike, H., Takaki, M., Nakata, K., Tanaka, Y., Takeda, T., Kodama, M., Fujiwara, Y., Sakai, A., and Kuroda, S. (2002). CNR1, central cannabinoid receptor gene, associated with susceptibility to hebephrenic schizophrenia. *Mol. Psychiatr.* **7,** 515–518.

Van Der Stelt, V., and Di Marzo, V. (2005). Cannabinoid receptors and their role in neuroprotection. *Neuromolecular Med.* **7,** 37–50.

Van Sickle, M. D., Duncan, M., Kingsley, P. J., Mouihate, A., Urbani, P., Mackie, K., Stella, N., Makriyannis, A., Piomelli, D., Davison, J. S., Marnett, L. J., Di Marzo, V., *et al.* (2005). Identification and functional characterization of brain stem cannabinoid CB2 receptors. *Science* **27,** 539–545.

Vasquez, C., and Lewis, D. L. (1999). The CB1 cannabinoid receptor can sequester G-proteins, making them unavailable to couple to other receptors. *J. Neurosci.* **19,** 9271–9280.

Walter, L., and Stella, N. (2004). Cannabinoids and neuroinflammation. *Br. J. Pharmacol.* **141,** 775–785.

Wan, M., Cravatt, B. F., Ring, H. Z., Zhang, X., and Francke, U. (1998). Conserved chromosomal location and genomic structure of human and mouse fatty-acid amide hydrolase genes and evaluation of clasper as a candidate neurological mutation. *Genomics* **54,** 408–414.

Wang, X., Dow-Edwards, D., Keller, E., and Hurd, Y. L. (2003). Preferential limbic expression of the cannabinoid receptor mRNA in the human fetal brain. *Neurosci* **118,** 681–694.

Woolmore, J. A., Stone, M. J., Holley, S., Jenkinson, P., Ike, A., Jones, P., Fryer, A., Strange, R., Stephens, R., Langdon, D., and Hawkins, C. (2008). Polymorphism of the cannabinoid 1 recpetor gene and cognitive impairment in multiple sclerosis. *Mult. Scler.* **14,** 177–182.

Yamada, Y., Ando, F., and Shimkata, H. (2007). Association of candidate gene polymorphisms with bone mineral density in community-dwelling Japanese women and men. *Int. J. Mol. Med.* **19,** 791–801.

Young, S. E., Rhee, S. H., Stallings, M. C., Corley, R. P., and Hewitt, J. K. (2006). Genetic and environmental vulnerabilities underlying adolescent substance use and problem use: General or specific? *Behavior. Genet.* **36,** 603–615.

Zhang, P. W., Ishiguro, H., Ohtsuki, T., Hess, J., Carillo, F., Walther, D., Onaivi, E. S., Arinami, T., and Uhl, G. R. (2004). Human cannabinoid receptor 1:5' exons, candidate regulatory regions, polymorphisms, haplotypes and association with polysubstance abuse. *Mol. Psychiatr.* **9,** 916–931.

Zuo, L., Kranzler, H. R., Luo, Z., Covault, J., and Gelernter, J. (2007). CNR1 variation modulates risk for drug and alcohol dependence. *Biol. Psychiatr.* **62,** 616–626.

INTERMITTENT DOPAMINERGIC STIMULATION CAUSES BEHAVIORAL SENSITIZATION IN THE ADDICTED BRAIN AND PARKINSONISM

Francesco Fornai,*,† Francesca Biagioni,† Federica Fulceri,* Luigi Murri,‡ Stefano Ruggieri,† and Antonio Paparelli*

*Department of Human Morphology and Applied Biology, University of Pisa, Pisa, Italy
†Laboratory of Neurobiology of Movement Disorders I.R.C.C.S.,
I.N.M. Neuromed (IS), Pozzilli, Isernia, Italy
‡Department of Neurosciences, Clinical Neurology, University of Pisa, Pisa, Italy

The gold standard therapy for Parkinson's disease (PD) consists in chronic administration of pulses of the dopamine (DA) precursor L-dihydroxyphenylalanine (L-DOPA). Although the main brain area which is DA-deficient is the dorsal striatum (more the putamen than the caudate nucleus), other DA-innervated brain regions (i.e., the ventral striatum and other limbic areas) are affected by systemic administration of L-DOPA. While such a therapy produces an increase in synaptic and nonsynaptic DA, which replace the neurotransmitter deficiency, peaks of extracellular DA in the course of disease progression produce abnormal involuntary movements related to behavioral sensitization.

Methamphetamine (METH), a widely abused drug, is known to produce behavioral sensitization, related to DA release (more in the ventral than dorsal striatum as well as other limbic regions).

INTERNATIONAL REVIEW OF
NEUROBIOLOGY, VOL. 88
DOI: 10.1016/S0074-7742(09)88013-6

The present review discusses the overlapping between these treatments, based on pulses of DA stimulation with an emphasis on the class of DA receptors; signal transduction pathways; rearranged expression of neurotransmitters, cotransmitters, and their receptors coupled with ultrastructural changes. In fact, all these levels of synaptic plasticity show a surprising homology following these treatments, posing the mechanisms of behavioral sensitization during DA-replacement therapy in PD very close to the neurobiological mechanisms operating during METH abuse. In line with this view is the growing evidence of addictive behaviors in PD patients during the course of DA-replacement therapy.

I. General Overview

Investigation in the neurobiology of dopamine (DA)-related diseases leads to unexpected similarities between distinct pathological conditions which mutually help to unravel the fine mechanisms leading to motor and nonmotor abnormalities. This seems to be the case of the effects induced by drugs of abuse like methamphetamine (METH) and pulses of DA-replacement therapy in Parkinson's disease (PD). In fact, both looking at patients and experimental models, these conditions are characterized by the onset of behavioral sensitization related to abnormal stimulation of DA receptors (Nishikawa *et al.*, 1983). This occurs either for METH (Chen *et al.*, 2007; Zhang *et al.*, 2006a,b), or following chronic pulses of L-dihydroxyphenylalanine (L-DOPA) (Bejjani *et al.*, 2000; Chase, 1998a; Nutt, 2007; Taylor *et al.*, 2005). In both conditions, the stimulation of DA receptors is chronic, but intermittent (Hamamura *et al.*, 1991; Kuribara, 1994; Yamada *et al.*, 1988). Recent data suggest that, when stimulation of DA receptors occurs following such variable intensity, the transduction pathways activated by D_1 dopamine receptor (D_1R) are persistently modified. In fact, chronic pulses of L-DOPA in a DA-denervated striatum alter the D_1R signaling (Fiorentini *et al.*, 2006; St-Hilaire *et al.*, 2005) both in parkinsonian rats and primates (Aubert *et al.*, 2005; Guigoni *et al.*, 2005), as well as humans (Corvol *et al.*, 2006). This is in line with an elegant model proposed by Gerfen (2003) on the role of abnormal metabolic pathways, placed downstream to D_1R, which are recruited when a DA-depleted striatum is challenged by peaks of extracellular DA. This consists in the activation of abnormal transduction mechanisms (Corvol *et al.*, 2004).

On the other hand, since the pioneer studies on the neurochemistry of METH, the D_1R was considered as a major determinant for the postsynaptic effects induced by this drug (Sonsalla *et al.*, 1986). At present, the D_1R is the major candidate to mediate the onset of METH-induced sensitization (Kuribara,

1995a,b; Yoshida *et al.*, 1995). Accordingly, phenotypic changes produced by METH on postsynaptic neurons depend on the activation of D_1R (Horner *et al.*, 2006). In turn, a specific pattern in the genetic polymorphism of the D_1R locus is associated with a higher frequency of METH-induced sensitization and abuse (Liu *et al.*, 2006).

Thus, following an altered DA transmission, a cascade of events take place, since both chronic L-DOPA and METH administration change the phosphorylation of the same specific proteins as well as the striatal expression of the same specific genes. These alterations occur in parallel (i.e., both compounds either decrease or increase the same molecules). The aim of the present review is to analyze all these common features to draw an integrated perspective of the phenomena underlying the consequence of pulses of DA stimulation in different pathological context. At the same time, we will focus on the distinct features which are based on a different neuropathological background at onset (i.e., a DA-denervated striatum in PD, versus a normal striatal DA content, at least at the beginning, during METH abuse). Plastic changes leading to abnormal activity of DA-dependent neural substrates will be discussed in relation with common and distinct behavioral effects and structural alterations.

II. DA Release Produced by METH in the Normal Striatum

METH abuse has reached epidemic proportions throughout the United States during the past decade (Lineberry and Bostwick, 2006). METH is a powerful DA/NA releaser in the brain (Rothman *et al.*, 2001; Yui *et al.*, 2004) and, repetitive administration of this compound, leads to behavioral sensitization which can be reproduced quite constantly along different animal species. The mechanism of action is very specific, since it produces a disruption of the physiological DA vesicular storage both producing an altered proton gradient through the DA storing vesicles (Cubells *et al.*, 1994), and impairing the activity of the high-affinity vesicular DA uptake (vesicular monoamine transporter type 2, VMAT-2). In particular, METH produces a direct inhibition of the VMAT-2 (Ary and Komiskey, 1980; Brown *et al.*, 2000), it also redistributes VMAT-2 from vesicles to cytoplasm (Sandoval *et al.*, 2002, 2003) and it promotes a VMAT-2-mediated DA efflux from the vesicles to the cytoplasm (Volz *et al.*, 2007; for review on these specific effects of METH on VMAT-2, see Fleckenstein *et al.*, 2007; Volz *et al.*, 2007). Moreover, METH reverts the direction of the membrane DA transporter (DAT, Schmidt and Gibb, 1985), thus promoting the efflux of DA from the axon terminals in the extracellular space (Raiteri *et al.*, 1979). In addition, the METH-induced increase of extracellular DA is further potentiated by the competitive inhibition of the most important enzyme for intracellular DA metabolism: monoamine oxidase

(MAO) (Suzuki *et al.*, 1980). Altogether, these molecular effects converge to produce a rapid and massive increase in extracellular DA levels following METH administration (Fig. 1).

This occurs throughout the brain, and mostly in those areas which possess a rich density of DA axon terminals. Thus, the dorsal and ventral striatum are strongly affected although specific limbic regions and the frontal cortex are

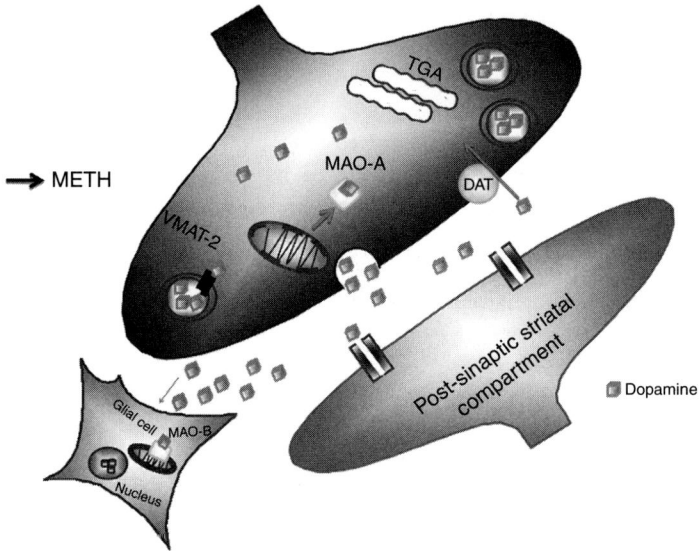

Fig. 1. The target of methamphetamine in the DA terminals. When methamphetamine is administered systemically at doses above 1 mg/kg, three main molecular targets are affected: (A) the vesicular monoamine transporter type 2 (VMAT-2), which is responsible for storing DA within the synaptic vesicles; (B) the dopamine transporter (DAT), which takes up extracellular DA within the DA terminals; and (C) monoamine oxidase type A (MAO-A), which is the main intracellular enzyme responsible for DA metabolism. Methamphetamine impairs the activity of these proteins at several levels. In fact, it directly inhibits VMAT-2 and reverts its direction, thus facilitating the outflow of DA from the vesicles into the axoplasm; VMAT-2 is also redistributed by methamphetamine in nonvesicles compartments (i.e., GERL system), this decreases the number of vesicles able to take up cytosolic DA. The competitive inhibition of monoamine oxidase occludes the physiological metabolism of cytosolic DA, thus producing very high amounts of DA within the terminals. Due to the reversal of the direction of DAT, DA is extruded from the axon into the extracellular space where it can no longer be taken up by DAT, since it is inhibited on the external side of the membrane. All together, these mechanisms contribute to generating high extracellular DA levels after methamphetamine administration. Such high DA amounts are not related to the physiological action potential of the DA neuron and diffuse far from the synaptic sites. In summary, during methamphetamine administration DA terminals are present, but only from an anatomical point of view since they are functionally impaired both as storage sites and metabolic compartments. This results in extracellular DA concentrations varying from peaks to very low levels and renders the DA stimulation extremely variable (pulsatile). (See Color Insert.)

involved as well (Abekawa *et al.*, 1994; Nishijima *et al.*, 1996; Piccini *et al.*, 2003; Stephans and Yamamoto, 1995; Uehara *et al.*, 2004). When multiple doses of METH are administered at different time intervals, a process named behavioral sensitization occurs. This is substantiated by an increased behavioral response to the same dose of the drug. Occurrence of a sensitized behavioral response is joined with specific biochemical correlates. For instance, repeated METH administrations produce an enhancement of striatal DA release (Kazahaya *et al.*, 1989) related with enhanced motor activity. Although the molecular basis of this latter effect rely on persistent changes at presynaptic level, it is rather the postsynaptic site which plays a leading role in sustaining behavioral sensitization (Suzuki *et al.*, 1997). Recently, Frankel *et al.* (2007) demonstrated that a very brief exposure to a subthreshold dose of METH is sufficient to induce a sensitized response to a further administration (a phenomenon which appear to us to be very close to the so-called "priming," a potentiated response occurring with a single dose of DA agonist in a DA-denervated striatum). Three main features (i.e., time, space, and concentration) are likely to provide a strong contribution to the onset of DA-dependent behavioral sensitization at postsynaptic level following METH administration. 1. *The temporal pattern* of DA release becomes almost independent from the physiological action potential, while it is largely grounded on the presynaptic effects of the drug. 2. *The spatial pattern* of DA release is altered by the redistribution of the storage sites along the DA terminals as a consequence of the interaction with VMAT-2 and DAT. Since METH redistributes VMAT-2 within the cytoplasm, it leaves the vesicles deficient of VMAT-2 activity (Sandoval *et al.*, 2002, 2003). This effect alters the pooling of DA stores in the space related with synaptic activity. Moreover, the DAT-mediated DA efflux creates nonphysiological sites of DA release, thus contributing to the altered spatial pattern which characterizes METH-induced DA stimulation. This is further altered by the inhibitory effect on MAO activity, since the affinity of METH for MAO-A is 10-fold higher compared with MAO-B, this produces a marked alteration in the amount and the sites of DA metabolism. In fact, MAO-A are present within the DA terminals, while MAO-B are mainly placed in the glial cells (Agid *et al.*, 1973; Fornai *et al.*, 1999; Gesi *et al.*, 2001; Tipton *et al.*, 2004; Youdim, 1974; Youdim *et al.*, 2006). Thus, occurrence of high intracellular DA levels cannot be metabolized by intracellular MAO-A and the increase of extracellular DA, close to DA terminals, tend to diffuse massively to extrasynaptic sites, being the DA uptake reverted and the MAO-A inhibited. Altogether, these effects are supposed to magnify the physiological volume transmission of DA (Fuxe *et al.*, 2003) causing a consistent diffusion of the neurotransmitter throughout the neural trim surrounding the DA terminals toward unusual extrasynaptic sites. This is supposed to enhance an abnormal spatial pattern of DA stimulation. 3. *The concentration pattern* is markedly altered by METH which leads to striatal extracellular DA levels reaching 10-fold the baseline values as shown by brain dialysis (Battaglia *et al.*, 2002a,b;

Lazzeri *et al.*, 2007). On the other hand, the high extracellular DA concentration which is obtained acutely during METH intake is followed by a drop in extracellular DA levels below the physiological baseline values within few hours. Thus, extracellular DA concentration varies toward excess and deficiency, depending on the time window being considered during METH abuse.

Altogether, the molecular mechanisms of METH-induced DA release lead to marked peaks of extracellular DA according to a nonphysiological sort of discontinuous temporal pattern extending to large areas going beyond the physiological space dedicated to DA synaptic transmission. Thus, although METH-induced DA release occurs in a normally DA-innervated striatum, the mechanisms of action of METH generate DA extracellular levels which are quite independent by the anatomy of DA terminals and their physiological action potentials. To a certain extent, we may assume that under the action of METH, striatal DA levels become quite independent from the physiological integrity of the mesostriatal pathway.

III. Increase of Extracellular DA Produced by Supplementing Exogenous L-DOPA in a DA-Denervated Striatum

Such a time and space aberrancy of DA release, joined with altered amount of extracellular DA (peaks followed by drops in synaptic and extrasynaptic DA concentrations) induced by METH, closely mimics the effects of L-DOPA administration in a DA-denervated striatum (i.e., DA-replacement therapy in advanced PD). In fact, while the occurrence of a physiological integrity of mesostriatal DA axons prevents L-DOPA to produce abnormal extracellular DA levels, the loss of DA terminals erases the main physiological DA storage system. Thus, while following METH there is a functional loss of DA storage, due to the impairment of DAT and VMAT-2 to selectively take up and store otherwise diffusible DA, in PD there is the anatomical loss of DAT and VMAT-2, since they are placed, fair selectively, within degenerated DA axons. The consequence is quite similar, since other compartments are able to convert L-DOPA into DA (serotonin neurons, glia, striatal intrinsic neurons) leading to diffusible DA in the absence of effective storage mechanisms. These phenomena, joined with a comparable loss of MAO-A (which is competitively inhibited by METH, while it disappears along with the DA terminals in PD), make exogenous L-DOPA a powerful trigger to generate high levels of extracellular DA followed by a drop in striatal extracellular concentration. This occurs for a limited time window being L-DOPA administered according to an intermittent schedule. Thus, even in this case, excess of extracellular DA are followed by absence of detectable concentrations of the neurotransmitter. This produce: 1. *an altered temporal pattern* of stimulation of DA receptors; 2. *an altered spatial distribution* of DA stimulation; 3. *an altered concentration* of the neurotransmitter which is independent from action potentials in the destroyed DA terminals. Interestingly, when a DA-substitution therapy is delivered into a normally innervated striatum, behavioral sensitization is hardly occur, while the lack of integrity of DA terminals seems to be essential for inducing the priming phenomenon, behavioral sensitization as well as L-DOPA-induced dyskinesia. In line with this, the enhancement of volume transmission was recently demonstrated in a DA-denervated striatum and it was related to biochemical changes occurring during the onset of dyskinesia (Fuxe *et al.*, 2003, 2005, 2006; Rivera *et al.*, 2006). In fact, the occurrence of L-DOPA-induced dyskinesia in a DA-denervated striatum is grounded on the progressive disappearance of DA terminals while it is triggered by intermittent DA-substitution therapy (Blin *et al.*, 1988). Interestingly, the ability to reduce L-DOPA-induced dyskinesia through a continuous DA stimulation is paralleled by the therapeutic potential of long-acting DAT blockers to cure METH addiction both at experimental and clinical level (Baumann *et al.*, 2002a,b). Surprisingly, continuous stimulation with DA agonists also provides therapeutic effects in METH -induced neurotoxicity as well as in experimental parkinsonism by counteracting the neurodegenerative process (Battaglia *et al.*, 2002a,b; Fornai *et al.*, 2001) (Fig. 2).

IV. The Postsynaptic Site: Occurrence of Sensitization Converge on Common Phosphorylation Processes

Going back from the clinical data to animal behavior, the occurrence of sensitization following pulses of DA-replacement therapy is well demonstrated in different animal species, where the occurrence of sensitization is defined by an increased motor response following the same dose of L-DOPA in rodents (Juncos *et al.*, 1989) and primates (Calon *et al.*, 1995). In a recent paper, this concept was reinforced by the direct demonstration of a sensitization process in parkinsonian patients taking L-DOPA (Evans *et al.*, 2006). This concept has been originally well defined by the Chase's group (Juncos *et al.*, 1989), they were able to address· the dependence of behavioral sensitization to L-DOPA to an intermittent administration of the drug, while no sensitized response occurred following continuous stimulation. Soon after, the behavioral syndrome produced by the first intermittent DA stimulation in a parkinsonian striatum was named "priming": a persistent alteration of the postsynaptic response was postulated to underlay this form of

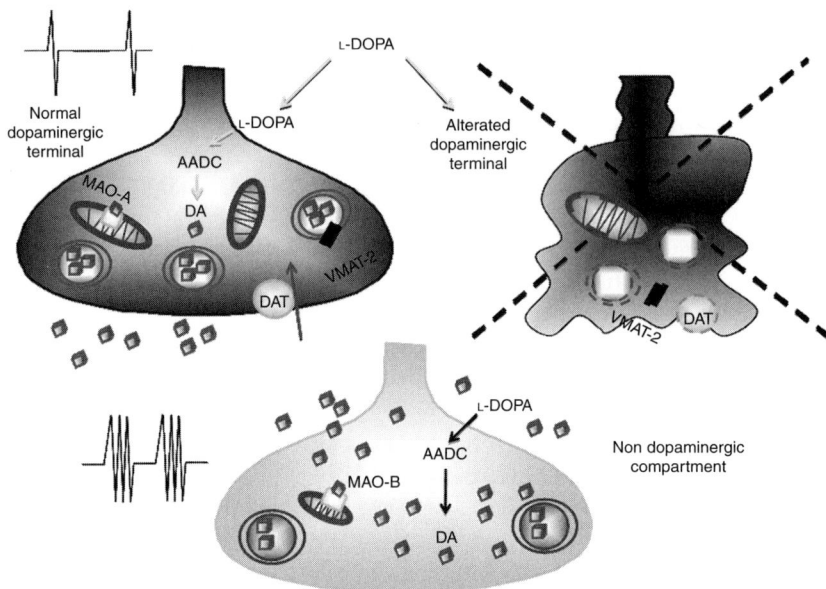

FIG. 2. Effects of L-DOPA administration in a DA-denervated striatum. Following the subtotal loss of DA terminals, which occurs in advanced PD, the physiological compartments responsible for DA storage, metabolism, and release (see also Fig. 1) are degenerated. The absence of these compartments causes exogenous L-DOPA to be metabolized by extra-DA sites that do not possess the ability to store DA and, hence, take up extracellular amounts. Thus, as in the case of methamphetamine (Fig. 1), the high extracellular DA levels produced diffuse well beyond the synaptic space, thereby enhancing the physiological volume transmission of this catecholamine. In this way, peaks of extracellular DA are followed by drops in extracellular DA levels, generating an abnormal stimulation of DA receptors that is particularly deleterious for D_1R receptor signaling. (See Color Insert.)

nonconditioned behavioral sensitization (Carey, 1991; Morelli *et al.*, 1993). Evidence for postsynaptic changes consisting of altered signal transduction was further provided (Barone *et al.*, 1994). In particular, Di Chiara and coworkers demonstrated that the onset of behavioral sensitization was accompanied by the phosphorylation of dopamine and adenosine 3¢–5¢-monophosphate-regulated phosphoprotein (DARPP-32), a DA- and cyclic AMP-regulated phosphoprotein functionally linked to D_1R receptors in striatum (Barone *et al.*, 1994).

These findings are strikingly similar to what occurs in chronic METH abusers, in which clinical symptoms of sensitizations are evident upon chronic use (Ujike and Sato, 2004). In humans, the sensitizing effects of METH are considered as a major determinant to the relapse of psychosis (Yui *et al.*, 1999). Also in this case, the occurrence of sensitization is well documented in animal models from rodent

(Nishikawa *et al.*, 1983) to primates (Kashiwabara, 1983) and the same occurs for the striatal biochemical alterations, which, again consist ofincreased phosphorylation processes (pCREB; Muratake *et al.*, 1998), occurring during METH-induced behavioral sensitization (Mc Daid *et al.*, 2006).

V. The Postsynaptic Site: Common Gene Expression

A. METHAMPHETAMINE

The natural target of pulses of extracellular DA is the medium size GABA neurons in which, following DA stimulation, an altered expression of immediate early genes like c-fos and zif/268 takes place (Horner and Keefe, 2006). These genes regulate a variety of cell activities and are characterized by a wide spectrum of effects and undergo an early synthesis. Among these genes, special attention has been paid recently to delta fos B which is increased following METH (Ujike, 2001) and it is currently proposed as a molecular switch in triggering the processes of physiological motivation, learning, and addiction to a variety of drugs and behaviors (Green *et al.*, 2006; Hyman *et al.*, 2006; Olausson *et al.*, 2006; Tamminga and Nestler, 2006; Zachariou *et al.*, 2006). However, METH addiction includes also the increase of genes which regulate the neuronal phenotype such as those coding for NMDA (Yamamoto *et al.*, 2006) or AMPA (Palmer *et al.*, 2005) glutamate receptor subunits, GABA A (Zhang *et al.*, 2006a,b) or GABA B (Palmer *et al.*, 2005) receptor subunits as well as the GABA synthesizing enzyme GAD 67 (Zhang *et al.*, 2006a,b). Also the genes coding for specific cotransmitter peptides and their receptors are altered following METH: such as the preproenkefaline (Tien *et al.*, 2007), preprodynorphin (Horner and Keefe, 2006), or the opioid receptors (Tien *et al.*, 2007). Most of these alterations accompany and sustain METH-induced sensitization and trigger the onset of addiction, as further confirmed by a correlation between genetic variants of the opioid system and the onset of METH addiction in humans (Nomura *et al.*, 2006). Again, changes produced in the cannabinoid system by METH administration are supposed to underlay its sensitizing properties, craving and relapse and to accompany the onset of dyskinesia as well as long-lasting behavioral changes (Anggadiredja *et al.*, 2004; Fattore *et al.*, 2007; Landa *et al.*, 2006; Vinklerova *et al.*, 2002). Thus, it is evident that the process of METH-induced sensitization/addiction leads to a real switch in the phenotype of those neuronal networks in the dorsal and ventral striatum as well as other areas which is expected to modify profoundly the physiology of the brain.

B. DA REPLACEMENT THERAPY IN PARKINSONISM

These phenotype changes are surprisingly similar to those occurring in a parkinsonian striatum following chronic pulses of DA-replacement therapy (St-Hilaire *et al.*, 2005).

Substitution therapy with L-DOPA produces a change in the expression of immediate early genes like c-fos, c-jun, and zif (Canales and Graybiel, 2000; Gerfen, 2000; Svenningsson *et al.*, 2000). Among these regulatory genes there is also delta fos B, which is considered as a molecular switch to trigger drug addiction and it is markedly increased by L-DOPA in the DA-denervated striatum (Stefanova *et al.*, 2004), where it correlates with the onset and severity of L-DOPA-induced dyskinesia. Moreover, L-DOPA alters gene expression-related neuronal phenotype for striatal glutamatergic transmission (Chase and Oh, 2000; Oh and Chase, 2002), both affecting NMDA (Oh *et al.*, 1999) and non-NMDA (Marin *et al.*, 2000) glutamate receptor subunits. This mechanism has been implicated in motor sensitization leading to dyskinesia as demonstrated by blocking NMDA (Blanchet *et al.*, 1997) or AMPA (Konitsiotis *et al.*, 2000) glutamate receptors to reduce L-DOPA-induced dyskinesia. These findings extended the role of gluta-mate antagonists from preventing the pathogenesis of PD to the treatment of drug-induced motor complications (Fornai *et al.*, 1996, 1997; Zuddas *et al.*, 1992). Again, during L-DOPA-induced dyskinesia there is a switch in the expression of genes belonging to the opioid system (Samadi *et al.*, 2006) including peptides like enkephalins (Henry *et al.*, 1999) and dynorphins (Van de Witte *et al.*, 1998) and the opioid receptors (Brooks, 2000; Chen *et al.*, 2005). This is in line with the antidyskinetic effects of opioid agonists during L-DOPA administration (Samadi *et al.*, 2004). The dyskinetic effects of pulses of L-DOPA are also related to modifications in GABA A (Katz *et al.*, 2005) or GABA B (Calon *et al.*, 2003) receptors subunits or the GABA synthesizing enzymes (GAD 65 and 67) with a specific correlation between changes in the expression of GAD 67 and L-DOPA-induced behavioral abnormalities (Stephenson *et al.*, 2005). In particular, an altered expression of GAD 67 is found to correlate with the onset of DA-induced priming (Carta *et al.*, 2003). Extending the extraordinary level of overlapping is the recent comprehensive demonstration of an altered transmission in the endo-cannabinoid system in the DA-depleted striatum (Lee *et al.*, 2006) which is in line both with the effects of cannabinoids on L-DOPA-induced dyskinesia (Van der Stelt *et al.*, 2005) and, in turn, the role of L-DOPA in regulating the expression of proteins belonging to the endocannabinoid system (Ferrer *et al.*, 2003; Maccarrone *et al.*, 2003).

All the effects pose a solid evidence on common neurobiological changes, which discloses a surprising similarity between the METH addicted brain and the parkinsonian brain chronically challenged with pulses of DA-substitution therapy.

This neurobiological background needs to be kept in mind when analyzing overlapping clinical syndromes in Section IX.

VI. The Common Trigger at the Postsynaptic Site: Inappropriate Activation of D_1 Receptors

Interestingly, the regulation of these genes and proteins mostly occurs downstream to the activation of DA receptors, thus making up a sort of causal chain of events, mainly placed at postsynaptic level, within medium size GABA neurons of the dorsal and ventral striatum. This phenotypic arrangement accompanies the onset of abnormal involuntary movements called dyskinesia in PD and addiction/stereotipies/automatism/hyperkinesia during METH sensitization. The common link consists in variations of extracellular DA concentration (peaks and drops of DA levels) occurring for a short period of time, thus producing a nonhomogeneous temporal and spatial pattern of DA receptor stimulation (Nishikawa et al., 1983). This occurs either for METH (Chen et al., 2007; Zhang et al., 2006a,b), or following chronic administration of L-DOPA (Bejjani et al., 2000; Taylor et al., 2005). In both conditions the sensitization process depends on the chronic, though intermittent stimulation of DA receptors (Hamamura et al., 1991; Kuribara, 1994; Yamada et al., 1988). In particular, recent data suggest the sensitization process as a mechanism triggered by an altered signaling of D_1R. In fact, chronic L-DOPA administration produces an altered D_1R signaling (Fiorentini et al., 2006; St-Hilaire et al., 2005) both in parkinsonian rats and primates (Aubert et al., 2005; Guigoni et al., 2005), in line with the elegant model proposed by Gerfen (2003). This alteration occurs through the recruitment of abnormal transduction pathways (Corvol et al., 2004).

On the other hand, the D_1R is the major candidate to trigger the sensitization process induced by METH. In fact, multiple high doses of METH produce phenotypic changes which depend on the activation of D_1R (Horner et al., 2006), thus confirming the evidence provided more than a decade ago suggesting the pivotal role of D_1R in triggering sensitization to METH (Kuribara, 1995a,b; Yoshida et al., 1995).This phenomenon is associated with addiction in line with a linkage between METH abuse and polymorphism of D_1R (Liu et al., 2006). Thus, following an altered DA transmission, inappropriate stimulation of D_1R, represents the key mechanism to trigger a cascade of events both during chronic L-DOPA and METH administration ranging from biochemical changes to gene expression (Cai et al., 2006), and sustaining a pathological behavior.

Of course, important discrepancies exist: first of all, during METH intake a normally innervated striatum undergo abnormal DA stimulation, while in parkinsonism this occurs in a DA-denervated striatum. This difference was emphasized in Sections II and III and justifies crucial differences between METH abuse and DA-replacement therapy. Nonetheless, in both cases the physiological activation of D_1R is lost and peaks and drops of extracellular DA concentration occurs. This abnormality is likely to sustain most of the biochemical and behavioral consequences.

In fact, when a DA-replacement therapy in PD is delivered through low, continuous levels of DA agonists, dyskinesia are reduced as shown by a recent elegant paper by the Lees' group (Katzenschlager *et al.*, 2005). This study carried out in humans affected by PD demonstrates both a marked reduction of dyskinesia during continuous subcutaneous apomorphine (a mixed DA agonist) therapy, and later on reduced dyskinesias following a challenge with a pulse of L-DOPA (Katzenschlager *et al.*, 2005). This latter finding was obtained also by using continuous infusion of another DA agonist, lisuride (Stocchi *et al.*, 2002). These data support the concept that reducing pulses of DA stimulation by delivering continuous DA receptor stimulation may reverse, at least partially, the sensitization process which mediates the development of drug-induced dyskinesias in PD (Chase, 1998b; Katzenschlager *et al.*, 2005; Nutt, 2007; Stocchi and Olanow, 2004). Interestingly, as the dyskinetic syndrome is attenuated by continuous low levels of DA agonists, also the molecular changes described above tend to withdraw, thus confirming the causal link between altered gene expression and abnormal behavior following repeated peaks of DA stimulation. In fact, the presence of continuous DA stimulation prevents both behavioral abnormalities produced by intermittent DA-replacement therapy (L-DOPA) in PD (Bibbiani *et al.*, 2005) and behavioural effects induced by METH (Miller *et al.*, 2001). Interestingly, Baumann *et al.* (2002b) demonstrated that a long-acting blocker of DA uptake, which produces continuous low level of DA stimulation and prevents peaks induced by METH may be useful to treat METH addiction. Mc Daid *et al.* (2007) substantiating the role of D_1R just found the suppression of METH-induced sensitization by administering the selective D_1R antagonist SCH 23390. Altogether these data confirm the hypothesis that peaks and drops of DA levels produce behavioral sensitization through the abnormal stimulation of D_1R. In line with this, the blockade of D_1R, but not of other DA receptors, abolishes the induction of both dynorphins- and enkefalins-related genes following METH exposure (Wang and McGinty, 1996).

VII. Convergence on Common Pathways: From Activation of D1 Receptors to Gene-Dependent Plastic Changes

In the previous section, we evidenced similar biochemical and molecular effects produced by pulses of L-DOPA in PD and METH intake/administration. In this section, we will overview whether a common chain of events can be drawn following L-DOPA and METH to justify the coherency of effects induced by chronic exposure to these compounds. A major standpoint both in an early normal and a DA-denervated striatum (during METH and DA-replacement therapy, respectively) consists in the abnormal stimulation of DA receptors (Hamamura et al., 1991; Kuribara, 1994; Yamada et al., 1988) due to abnormal activity of D_1R (Gerfen, 2003; Kuribara, 1995a,b; Yoshida et al., 1995). Such an altered activation of D_1R produces a permanent switch in the signal transduction pathways (Gerfen, 2003). This is the first event involved in the altered sensitivity to further DA stimulation and can be recognized as an early step during the sensitization process. Thus, further alterations depend on second messengers and downstream mechanisms triggered by abnormal D_1R activity rather than a simple change in the receptor number and/or affinity. As stated above, a classic example of altered signaling following chronic L-DOPA administration to parkinsonian monkeys consists in increased DA- and cAMP-regulated phosphoprotein of 32 kDa (DARPP-32). However, also cyclin-dependent kinase 5 (Cdk5) is altered depending on the activation of D_1R. These changes are related with the onset of dyskinesia (Aubert et al., 2005), while the number of D_1R or D_2R is not and no correlation exists for D_2R signaling pathways (Aubert et al., 2005).

The very same proteins are associated with METH-induced sensitization as shown for Cdk5 (Chen and Chen, 2005) and DARPP-32 (Fukui et al., 2003); also in this case, Cdk5 and DARPP-32 are modified depending on the selective activation of a D_1R-dependent pathway. Thus, recent data add to the evidence provided in Section IV showing that phosphorylation of transduction proteins in behavioral sensitization is a complex phenomenon, being far to be fully elucidated. Both CREB and DARPP-32 phosphorylation depend on the activation of D_1R as shown by Hotte et al. (2006), who found this effect in relation to plastic changes associated with memory retrieval performance. Interestingly, Edwards et al. (2002) measured the protein kinase A (PKA)-mediated phosphorylation of DARPP-32 and NMDA glutamate receptor subunit NR1 and they found both processes to be dependent on the selective activation of D_1R. On the other hand, Nishi et al. (2005) found that activation of both NMDA and AMPA glutamate receptors promotes the phosphorylation of DARPP-32 in the striatum, thus posing the hypothesis of a vicious circle of reciprocal enhancement between glutamate receptor activation and D_1R activity. In fact, these phosphorylation processes reciprocally increase the responsivity of the signaling cascade. This hypothesis might explain why D_1R antagonists and both NMDA and non-NMDA glutamate antagonists reduce L-DOPA-induced dyskinesia. In their elegant work, Nishi et al. (2005) found that glutamate activates at least five different signaling cascades (including the signaling dependent by the activation of metabotropic glutamate receptors) with different time dependencies, resulting in a complex regulation of protein kinase and protein phosphatase activities. This finding produces the first evidence for a cross talk between glutamate and

D_1R converging on phosphorylation pathways (mostly, DARRP-32). This might explain why strengthening of synaptic connections in medium spiny neurons of the striatum occurs when glutamatergic input (from cortex) and dopaminergic input (from substantia nigra) are received simultaneously. In fact, both pathways converge in medium spiny neurons to strengthen synapses acting synergistically on the same signaling molecules. Recent evidence shows a dimerization of D_1R and glutamate receptor subunit to form receptor hybrids which move out from the synaptic area during L-DOPA-induced dyskinesia (Fiorentini *et al.*, 2006; Missale *et al.*, 2006). These data add further evidence on the synergism between D_1R and glutamate receptor during synaptic plasticity induced by pulses of L-DOPA posing the D_1R /NMDA dimers as novel molecular target to treat dyskinesia (Missale *et al.*, 2006). A major consequence of such interaction involves the phosphoprotein DARPP-32, which is highly expressed in striatal medium spiny neurons. DARPP-32 is regulated by several neurotransmitters through a complex network of intracellular signaling pathways involving cAMP (increased through DA stimulation) and calcium (increased through glutamate stimulation) (Lindskog *et al.*, 2006). The convergence of such altered signaling during dyskinesia in PD and drugs of abuse is also evident by the effects on Cdk-5 (Chergui *et al.*, 2004). This kinase is recruited both by D_1R and glutamate receptor in the striatum and plays an essential role in motor- and reward-related behaviors. Cyclin-dependent kinase 5 (Cdk5) negatively regulates postsynaptic signaling of DA in the striatum. In addition to a postsynaptic role, Cdk5 negatively regulates DA and glutamate release in the striatum. Moreover, Chergui *et al.* (2004) found that inhibition of CDK5 increases the activity and phosphorylation of NMDA receptors, while these effects are reduced by a D_1R antagonist. The role of D_1R in modifying glutamate activity goes beyond the signaling pathway bound to NMDA receptors and involves the phosphorylation of AMPA receptors leading to its altered activity as well as the expression of AMPA receptors on the cell membrane (Gao *et al.*, 2006; Swayze *et al.*, 2004). Such an effect is operating both during METH, and DA administration (Snyder *et al.*, 2000). Phosphorylation of cytoplasmic proteins following METH leads to altered synthesis of genes coding for NMDA (Yamamoto *et al.*, 2006) or AMPA (Palmer *et al.*, 2005) glutamate receptor subunits, GABA A (Zhang *et al.*, 2006a,b) or GABA B (Palmer *et al.*, 2005) receptor subunits, as well as the GABA synthesizing enzyme GAD 67 (Zhang *et al.*, 2006a,b). These genes are also affected by L-DOPA administration through a process which is grounded on the same signal transduction pathways. The occurrence of these alterations makes the striatal neurons different phenotypic entity and underlies the sensitization process which is expressed as an altered behavior following repeated administration of METH or L-DOPA.

VIII. Convergence on Ultrastructural Changes and Site Specificity

According to very recent data, the overlapping between DA-replacement therapy in PD and METH administration extends to the ultrastructural changes produced by pulses of DA stimulation. In fact, following chronic METH administration in the nigrostriatal system of rodents (Fornai *et al.*, 2004) and humans (Quan *et al.*, 2005), ubiquitinated structures occur which are reminiscent of those occurring in PD. This finding is matched by the presence of ubiquitin-positive subcellular structures occurring in striatal neurons following intermittent DA-replacement therapy in a parkinsonian striatum (Fulceri *et al.*, 2007). Not surprisingly, when primary cultured striatal neurons are exposed to peaks of exogenous DA, these ubiquitin-positive subcellular aggregates can be easily reproduced (Lazzeri *et al.*, 2007). According to the evidence provided, these ubiquitin-positive structures are reduced by D_1R antagonists and can be generated by selective D_1R agonists, while D_2R ligands do not produce any noticeable effect (Lazzeri *et al.*, 2007) (Fig. 3). At present, it remains unclear the relationship between these

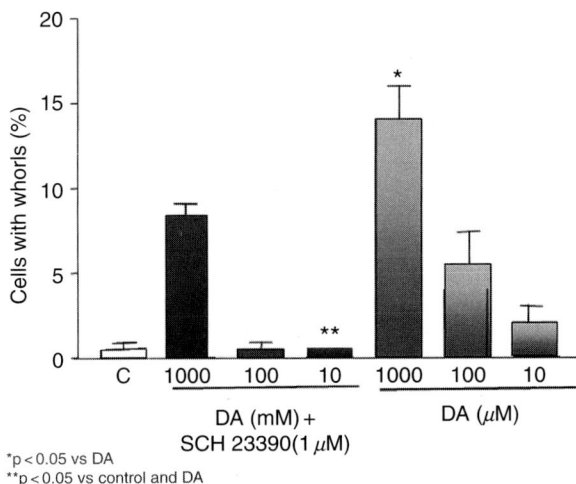

Fig. 3. Effects of a D1R antagonist on ubiquitinated bodies within striatal neurons. Following DA administration to primary striatal cell cultures, a dose-dependent formation of ubiquitin-positive neuronal bodies occurs. Such an effect is mediated by the activation of D_1R as shown by the preventive effects of the selective D_1R receptor antagonist SCH-23390. This effect confirms the pivotal role of the receptor in mediating the postsynaptic effects of DA peaks on postsynaptic striatal neurons (see also Lazzeri *et al.*, 2007).

morphological findings and abnormal motor behavior. However, according to recent studies, the ubiquitin proteasome system is responsible for synaptic changes leading to increased synaptic plasticity and long-term potentiation, which are reminiscent of the striatal mechanisms leading to abnormal motor patterns (Patrick, 2006; Yi and Ehlers, 2005). Ubiquitination in PD relates to the presence of inclusion bodies (Shults, 2006); however, increased ubiquitinated bodies within striatal neurons are also described following pulses of extracellular DA. Thus, the occurrence of striatal ubiquitination may reflect either a consequence of an ongoing toxicity or being simply related to the presence of plastic changes.

While METH may lead to a damage of nigrostriatal DA axons in rodents and primates including humans, thus predisposing to PD (McCann *et al.*, 1998), this effect may recover (Volkow *et al.*, 2001) and even when persisting, DA levels are more reduced in the caudate (39% of controls) compared with the putamen nucleus (50% of controls) (Moszczynska *et al.*, 2004), thus showing a trend which is opposite to the pattern of striatal DA loss which occurs in PD (more in the putamen, less in the caudate). Such a site specificity is critical and needs to be considered carefully when focussing on the divergent effects of METH and L-DOPA therapy in PD (this concept will be further developed in relation with the ventral striatum in Section IX). Nonetheless, according to Moszczynska *et al.* (2004), the near-total DA loss in the caudate could explain reports of cognitive disturbances, sometimes disabling, in some drug users, and suggests that treatment with L-DOPA during drug rehabilitation for METH abusers might be helpful. However, chronic pulses of L-DOPA in some PD patients may lead to rewarding effects and to the onset of craving and abnormal behavior leading to a syndrome recently defined by Lees and coworkers as hedonistic homeostatic dysregulation (Giovannoni *et al.*, 2000; Bearn *et al.*, 2004). This consists of an abuse habit which leads to manic and psychotic episodes, which are similar to those occurring as a consequence of METH abuse and represents the major behavioral effect of METH-induced sensitization. Thus, such an overlapping may be very dangerous.

IX. Behavioral Sensitization: METH-Induced Addiction and L-DOPA-Induced Dyskinesia: Two Faces of a Coin

The major difference we already analyzed between these forms of behavioral sensitization consists in the integrity of the mesostriatal pathway (at least at the beginning of METH addiction) compared with the severe loss of such a system (during a DA-replacement therapy in PD). We discussed these points at Sections II, III, and VIII, showing also that an analogy can be found (due to the functional abolishment of the physiology of DA terminals produced by the mechanisms of action of METH, and the occurrence of mesostriatal DA denervation following chronic METH abuse. In light of what reported in the previous section, the occurrence of mesostriatal denervation in METH abusers is slighter and sometime reversible compared with PD (Volkow *et al.*, 2001) and occurs more in the caudate compared with the putamen nucleus, while the latter is more affected in PD (Moszczynska *et al.*, 2004). Again, the functional studies demonstrate that during METH intake, the DA release is more pronounced in the ventral (limbic) striatum, while the main target of the DA-replacement therapy in PD is the dorsal striatum. In keeping with these anatomical discrepancies, the behavioural alterations expressed by the sensitization processes are quite different. In fact, the brain circuitry claimed to be involved in METH-induced sensitization is primarily represented by the limbic system and, mostly the ventral striatum, while in the case of L-DOPA-induced dyskinesia is the dorsal striatum which plays a major role. This is evidenced by distinct behavioral features: while the major behavioral expression associated with METH-induced sensitization consists in addiction, in the case of L-DOPA-induced behavioral sensitization, dyskinesia represents the main abnormal behavior. Nonetheless, the literature in the last few years provided intriguing evidence for an overlapping consisting in the onset of movement disorders in METH abusers, and abuse habits in PD patients treated with pulses of DA-replacement therapy. Thus, in PD the occurrence of behavioral sensitization is not limited to dyskinesia, but may extend to abnormal addictive behaviors such as gambling or a psychomotor activation (punding), which overlap the effects of METH addiction. Interestingly, the occurrence of punding in PD patients treated with pulses of L-DOPA is associated with the severity of L-DOPA-induced dyskinesia (Silveira-Moriyama *et al.*, 2006), thus suggesting that a common neurobiological substrate underlies two distinct behavioral pattern of sensitization by acting on different brain circuits (ventral and dorsal striatum, respectively) responsible for these phenomena. Again, as shown by the Lees group (Giovannoni *et al.*, 2000; Bearn *et al.*, 2004), some PD patients use DA drugs following a pattern which fulfills the diagnostic core criteria for substance dependence. A phenomenon designated by the authors as hedonistic homeostatic dysregulation syndrome. As mentioned above such a behavior (also known as DA-dependent syndrome) resembles what occurs during METH abuse. Interestingly, a recent work carried out by Evans *et al.* (2006) analyzed the functional anatomy of DA release in a group pf PD patients carrying such psychopathological syndrome following replacement therapy with L-DOPA. These patients were analyzed for DA release using the displacement of radioactive raclopride. Lending substance to the site specificity for different kinds of behavioral sensitization, these patients had a powerful DA release in the ventral striatum following exogenous L-DOPA (Evans *et al.*, 2006). This contrasts with the preferential DA release in the dorsal striatum induced by L-DOPA in PD patients, while it provides a clear anatomical and clinical example for the overlapping between these forms of behavioral sensitization. In fact, according to the authors, the exaggerated DA release in the ventral striatum represents a sensitized response related to psychomotor activation induced by drug intake (Evans *et al.*, 2006). Such interesting data provide a direct demonstration in humans that the ventral striatum is involved in the behavioral sensitization expressed as addictive behavior by L-DOPA, in line with the role of the ventral striatum in the addiction produced by repeated METH administration.

One major behavioral consequence of METH-induced sensitization is the occurrence of psychosis (Dore and Sweeting, 2006; Ujike and Sato, 2004; Yui *et al.*, 1999. In particular, Ujike and Sato (2004) found noticeable evidence of behavioral sensitization in METH abusers undergoing progressive psychopathology. These authors, by profiting of the pure/sole METH abuse occurring as the most frequent abuse phenomenon in the last 50 years in Japan, observed the sequential alteration of psychiatric symptoms solely induced by METH. The results reveal three core features: (a) Progressive qualitative alteration in mental symptoms from a nonpsychotic, to a prepsychotic, to a severely psychotic state. (b) Enhanced vulnerability to relapse of psychosis; and (c) very long duration of the vulnerability to relapse. This indicates that sensitization to METH develops during abuse and plays a key role in the susceptibility to the onset of psychosis. Such an evidence is confirmed through several studies showing a tight relationship between METH-induced sensitization, and the onset/relapse of psychosis even at long time interval following METH withdrawal (Akiyama *et al.*, 1994; Nishikawa *et al.*, 1983; Sato *et al.*, 1983).

In line with this, a typical side effect of DA-substitution therapy in PD is the onset of psychotic symptoms (Calne *et al.*, 1969; Jenkins and Groh, 1970). During L-DOPA administration, the onset of psychotic episodes is related with the onset of dyskinesia (Friedman and Sienkiewicz, 1991), this association is often time-locked suggesting a common neurobiological process. Namely, Borek and Friedman (2005) recently described two anedoctical cases of hedonistic homeostatic dysregulation sustained by L-DOPA addiction in parkinsonian patient. In one patient, the intake of L-DOPA was not triggered by the alleviation of motor symptoms but rather by the psychiatric effects. Interestingly, this patient developed L-DOPA-induced psychosis as a consequence of compulsive intake of high doses of the drug which led, at the same time, to both psychosis and dyskinesia. Kaiser *et al.* (2003) found that the onset of L-DOPA induced dyskinesia and psychosis are both related to the same polymorphism in the gene coding for a major target of METH which is the DAT (see Section II). Psychotic symptoms induced by METH and L-DOPA are characterized by a high frequency of paranoid behaviors.

On the other hand, METH abusers may show the tendency to develop dyskinesia (Sperling and Horowitz, 1994), and METH-induced hyperkinesias are related with METH-induced psychosis (Batki and Harris, 2004).

In line with this, the occurrence of METH-induced stereotipies as well as L-DOPA-induced stereotypic behavior in hemiparkinsonian animals is well established.

The overlapping of behavioral expression of drug-induced sensitization is grounded on the systemic effects produced by these drugs. In fact, although the dorsal striatum is the main target of chronic administration of L-DOPA, such DA precursor reaches all the brain, just like the systemic intake of METH. We discussed the occurrence of a sensitization process based on the abnormal pulses of extracellular DA due to the mechanism of action of METH and the inappropriate handling of exogenous L-DOPA by DA-denervated structures. In fact, it is likely that the occurrence of addiction-like features following L-DOPA administration is bound to the progression of catecholamine denervation to the ventral striatum or other brain regions, which are progressively altered during the pathological course of the disease (Del Tredici et al., 2002). A major point concerning the gold standard of L-DOPA as a replacement therapy in PD consists in its ability to be converted not only to DA but also to NA. This concept is now emerging (Mercuri and Bernardi, 2005; Gesi et al., 2000) in view of the clear evidence of a dramatic NA loss which occurs in PD (Braak et al., 2000, 2001; Gesi et al., 2000). Thus, the sensitization process may also be related to the abnormal NA activity induced by L-DOPA in a NA-denervated brain. In a recent paper, Fulceri et al. (2007) demonstrated that L-DOPA-induced abnormal involuntary movements are dramatically enhanced when a severe NA depletion is associated with nigrostriatal DA loss. On the other hand, the mechanism of action of METH extends to NA terminals where the effects described at Section II for DA are replicated within the NA axons. It is noteworthy that METH release NA more potently than DA (Rothman et al., 2001). Lending further implications to these findings, recent literature emphasized the role of increased NA release in producing sensitization following METH administration (Yui et al., 2004).

Acknowledgment

We thank Dr. Giorgio Consoli for his support and helpful discussions.

References

Abekawa, T., Ohmori, T., and Koyama, T. (1994). Effects of repeated administration of a high dose of methamphetamine on dopamine and glutamate release in rat striatum and nucleus accumbens. Brain Res. **643,** 276–281.

Agid, Y., Javoy, F., and Youdim, M. B. (1973). Monoamine oxidase and aldehyde dehydrogenase activity in the striatum of rats after 6-hydroxydopamine lesion of the nigrostriatal pathway. *Br. J. Pharmacol.* **48,** 175–178.

Akiyama, K., Kanzaki, A., Tsuchida, K., and Ujike, H. (1994). Methamphetamine-induced behavioral sensitisation and its implications for relapse of schizophrenia. *Schizophr. Res.* **12,** 251–257.

Anggadiredja, K., Nakamichi, M., Hiranita, T., Tanaka, H., Shoyama, Y., Watanabe, S., and Yamamoto, T. (2004). Endocannabinoid system modulates relapse to methamphetamine seeking: possible mediation by the arachidonic acid cascade. *Neuropsychopharmacology* **29,** 1470–1478.

Ary, T. E., and Komiskey, H. L. (1980). Basis of phencyclidine's ability to decrease the synaptosomal accumulation of 3H-catecholamines. *Eur. J. Pharmacol.* **61,** 401–405.

Aubert, I., Guigoni, C., Hakansson, K., Li, Q., Dovero, S., Barthe, N., Bioulac, B. H., Gross, C. E., Fisone, G., Bloch, B., and Bezard, E. (2005). Increased D1 dopamine receptor signaling in levodopa-induced dyskinesia. *Ann. Neurol.* **57,** 17–26.

Barone, P., Morelli, M., Popoli, M., Cicarelli, G., Campanella, G., and Di Chiara, G. (1994). Behavioural sensitisation in 6-hydroxydopamine lesioned rats involves the dopamine signal transduction: changes in DARPP-32 phosphorylation. *Neuroscience* **61,** 867–873.

Batki, S. L., and Harris, D. S. (2004). Quantitative drug levels in stimulant psychosis: Relationship to symptom severity, catecholamines and hyperkinesia. *Am. J. Addict.* **13,** 461–470.

Battaglia, G., Busceti, C. L., Cuomo, L., Giorgi, F. S., Orzi, F., De Blasi, A., Nicoletti, F., Ruggieri, S., and Fornai, F. (2002a). Continuous subcutaneous infusion of apomorphine rescues nigro-striatal dopaminergic terminals following MPTP injection in mice. *Neuropharmacology* **42,** 367–373.

Battaglia, G., Gesi, M., Lenzi, P., Busceti, C. L., Soldani, P., Orzi, F., Rampello, L., Nicoletti, F., Ruggieri, S., and Fornai, F. (2002b). Morphological and biochemical evidence that apomorphine rescues striatal dopamine terminals and prevents methamphetamine toxicity. *Ann. N.Y. Acad. Sci.* **965,** 254–266.

Baumann, M. H., Ayestas, M. A., Sharpe, L. G., Lewis, D. B., Rice, K. C., and Rothman, R. B. (2002a). Persistent antagonism of methamphetamine-induced dopamine release in rats pre treated with GBR12909 decanoate. *J. Pharmacol. Exp. Ther.* **301,** 1190–1197.

Baumann, M. H., Phillips, J. M., Ayestas, M. A., Ali, S. F., Rice, K. C., and Rothman, R. B. (2002b). Preclinical evaluation of GBR12909 decanoate as a long-acting medication for methamphetamine dependence. *Ann. N.Y. Acad. Sci.* **965,** 92–108.

Bearn, J., Evans, A., Kelleher, M., Turner, K., and Lees, A. (2004). Recognition of a dopamine replacement therapy dependence syndrome in Parkinson's disease: A pilot study. *Drug Alcohol Depend.* **76,** 305–310.

Bejjani, B. P., Arnulf, I., Demeret, S., Damier, P., Bonnet, A. M., Houeto, J. L., and Agid, Y. (2000). Levodopa-induced dyskinesias in Parkinson's disease: Is sensitisation reversible? *Ann. Neurol.* **47,** 655–658.

Bibbiani, F., Costantini, L. C., Patel, R., and Chase, T. N. (2005). Continuous dopaminergic stimulation reduces risk of motor complications in parkinsonian primates. *Exp. Neurol.* **192,** 73–78.

Blanchet, P. J., Papa, S. M., Metman, L. V., Mouradian, M. M., and Chase, T. N. (1997). Modulation of levodopa-induced motor response complications by NMDA antagonists in Parkinson's disease. *Neurosci. Biobehav. Rev.* **21,** 447–453.

Blin, J., Bonnet, A. M., and Agid, Y. (1988). Does levodopa aggravate Parkinson's disease? *Neurology* **38,** 1410–1416.

Borek, L. L., and Friedman, J. H. (2005). Levodopa addiction in idiopathic Parkinson disease. *Neurology* **65,** 1508.

Braak, H., Rub, U., Sandmann-Keil, D., Gai, W. P., de Vos, R. A., Jansen Steur, E. N., Arai, K., and Braak, E. (2000). Parkinson's disease: Affection of brain stem nuclei controlling premotor and motor neurons of the somatomotor system. *Acta Neuropathol.* **99,** 489–495.

Braak, E., Sandmann-Keil, D., Rub, U., Gai, W. P., de Vos, R. A., Steur, E. N., Arai, K., and Braak, H. (2001). Alpha-synuclein immunopositive Parkinson's disease-related inclusion bodies in lower brain stem nuclei. *Acta Neuropathol.* **101,** 195–201.

Brooks, D. J. (2000). PET studies and motor complications in Parkinson's disease. *Trends Neurosci.* **23,** S101–S108.

Brown, J. M., Hanson, G. R., and Fleckenstein, A. E. (2000). Methamphetamine rapidly decreases vesicular dopamine uptake. *J. Neurochem.* **74,** 2221–2223.

Cai, N. S., McCoy, M. T., Ladenheim, B., Lyles, J., Ali, S. F., and Cadet, J. L. (2006). Serial analysis of gene expression in the rat striatum following methamphetamine administration. *Ann. N.Y. Acad. Sci.* **1074,** 13–30.

Calne, D. B., Stern, G. M., Laurence, D. R., Sharkey, J., and Armitage, P. (1969). L-DOPA in postencephalitic parkinsonism. *Lancet* **1,** 744746.

Calon, F., Goulet, M., Blanchet, P. J., Martel, J. C., Piercey, M. F., Bedard, P. J., and Di Paolo, T. (1995). Levodopa or D2 agonist induced dyskinesia in MPTP monkeys: correlation with changes in dopamine and GABAA receptors in the striatopallidal complex. *Brain Res.* **680,** 43–52.

Calon, F., Morissette, M., Rajput, A. H., Hornykiewicz, O., Bedard, P. J., and Di Paolo, T. (2003). Changes of GABA receptors and dopamine turnover in the postmortem brains of parkinsonians with levodopa-induced motor complications. *Mov. Disord.* **18,** 241–253.

Canales, J. J., and Graybiel, A. M. (2000). Patterns of gene expression and behavior induced by chronic dopamine treatments. *Ann. Neurol.* **47,** S53–S59.

Carey, R. J. (1991). Chronic L-DOPA treatment in the unilateral 6-OHDA rat: Evidence for behavioral sensitisation and biochemical tolerance. *Brain Res.* **568,** 205–214.

Carta, A. R., Fenu, S., Pala, P., Tronci, E., and Morelli, M. (2003). Selective modifications in GAD67 mRNA levels in striatonigral and striatopallidal pathways correlate to dopamine agonist priming in 6-hydroxydopamine-lesioned rats. *Eur. J. Neurosci.* **18,** 2563–2572.

Chase, T. N. (1998a). Levodopa therapy: Consequences of the nonphysiologic replacement of dopamine. *Neurology* **50,** S17–S25.

Chase, T. N. (1998b). The significance of continuous dopaminergic stimulation in the treatment of Parkinson's disease. *Drugs* **55,** S1–S9.

Chase, T. N., and Oh, J. D. (2000). Striatal dopamine- and glutamate-mediated dysregulation in experimental parkinsonism. *Trends Neurosci.* **23,** S86–S91.

Chen, P. C., and Chen, J. C. (2005). Enhanced Cdk5 activity and p35 translocation in the ventral striatum of acute and chronic methamphetamine-treated rats. *Neuropsychopharmacology* **30,** 538–549.

Chen, L., Togasaki, D. M., Langston, J. W., Di Monte, D. A., and Quik, M. (2005). Enhanced striatal opioid receptor-mediated G-protein activation in L-DOPA-treated dyskinetic monkeys. *Neuroscience* **132,** 409–420.

Chen, P. C., Lao, C. L., and Chen, J. C. (2007). Dual alteration of limbic dopamine D(1) receptor-mediated signalling and the Akt/GSK3 pathway in dopamine D(3) receptor mutants during the development of methamphetamine sensitisation. *J. Neurochem.* **100,** 225–241.

Chergui, K., Svenningsson, P., and Greengard, P. (2004). Cyclin-dependent kinase 5 regulates dopaminergic and glutamatergic transmission in the striatum. *Proc. Natl. Acad. Sci. USA* **101,** 2191–2196.

Corvol, J. C., Muriel, M. P., Valjent, E., Feger, J., Hanoun, N., Girault, J. A., Hirsch, E. C., and Herve, D. (2004). Persistent increase in olfactory type G-protein alpha subunit levels may underlie D1 receptor functional hypersensitivity in Parkinson disease. *J. Neurosci.* **24,** 7007–7014.

Corvol, J. C., Girault, J. A., and Herve, D. (2006). Role and regulation of dopamine D1 receptors in the striatum: implications for the genesis of dyskinesia in Parkinson's disease. *Rev. Neurol.* **162,** 691–702.

Cubells, J. F., Rayport, S., Rajendran, G., and Sulzer, D. (1994). Methamphetamine neurotoxicity involves vacuolation of endocytic organelles and dopamine-dependent intracellular oxidative stress. *J. Neurosci.* **14,** 2260–2271.

Del Tredici, K., Rub, U., De Vos, R. A., Bohl, J. R., and Braak, H. (2002). Where does parkinson disease pathology begin in the brain? *J. Neuropathol. Exp. Neurol.* **61,** 413–426.

Dore, G., and Sweeting, M. (2006). Drug-induced psychosis associated with crystalline methamphetamine. *Australas. Psychiatry* **14,** 86–89.

Edwards, S., Simmons, D. L., Galindo, D. G., Doherty, J. M., Scott, A. M., Hughes, P. D., and Wilcox, R. E. (2002). Antagonistic effects of dopaminergic signaling and ethanol on protein kinase A-mediated phosphorylation of DARPP-32 and the NR1 subunit of the NMDA receptor. *Alcohol Clin. Exp. Res.* **26,** 173–180.

Evans, A. H., Pavese, N., Lawrence, A. D., Tai, Y. F., Appel, S., Doder, M., Brooks, D. J., Lees, A. J., and Piccini, P. (2006). Compulsive drug use linked to sensitized ventral striatal dopamine transmission. *Ann. Neurol.* **59,** 852–858.

Fattore, L., Spano, M. S., Deiana, S., Melis, V., Cossu, G., Fadda, P., and Fratta, W. (2007). An endocannabinoid mechanism in relapse to drug seeking: a review of animal studies and clinical perspectives. *Brain Res. Rev.* **53,** 1–16.

Ferrer, B., Asbrock, N., Kathuria, S., Piomelli, D., and Giuffrida, A. (2003). Effects of levodopa on endocannabinoid levels in rat basal ganglia: Implications for the treatment of levodopa-induced dyskinesias. *Eur. J. Neurosci.* **18,** 1607–1614.

Fiorentini, C., Pizzetti, M. C., Busi, C., Bontempi, S., Collo, G., Spano, P., and Missale, C. (2006). Loss of synaptic D1 dopamine/N-methyl-D-aspartate glutamate receptor complexes in L-DOPA-induced dyskinesia in the rat. *Mol. Pharmacol.* **69,** 805–812.

Fleckenstein, A. E., Volz, T. J., Riddle, E. L., Gibb, J. W., and Hanson, G. R. (2007). New insights into the mechanism of action of amphetamines. *Annu. Rev. Pharmacol. Toxicol.* **47,** 681–698.

Fornai, F., Vaglini, F., Maggio, R., Bonuccelli, U., and Corsini, G. U. (1996). Excitatory amino acids and MPTP toxicity. *Adv. Neurol.* **69,** 167–176.

Fornai, F., Vaglini, F., Maggio, R., Bonuccelli, U., and Corsini, G. U. (1997). Species differences in the role of excitatory amino acids in experimental parkinsonism. *Neurosci. Biobehav. Rev.* **21,** 401–415.

Fornai, F., Chen, K., Giorgi, F. S., Gesi, M., Alessandri, M. G., and Shih, J. C. (1999). Striatal dopamine metabolism in monoamine oxidase B-deficient mice: A brain dialysis study. *J. Neurochem.* **73,** 2434–2440.

Fornai, F., Battaglia, G., Gesi, M., Orzi, F., Nicoletti, F., and Ruggieri, S. (2001). Dose-dependent protective effects of apomorphine against methamphetamine-induced nigrostriatal damage. *Brain Res.* **898,** 27–35.

Fornai, F., Lenzi, P., Gesi, M., Soldani, P., Ferrucci, M., Lazzeri, G., Capobianco, L., Battaglia, G., De Blasi, A., Nicoletti, F., and Paparelli, A. (2004). Methamphetamine produces neuronal inclusions in the nigrostriatal system and in PC12 cells. *J. Neurochem.* **88,** 114–123.

Frankel, P. S., Hoonakker, A. J., Danaceau, J. P., and Hanson, G. R. (2007). Mechanism of an exaggerated locomotor response to a low-dose challenge of methamphetamine. *Pharmacol. Biochem. Behav.* **86,** 511–515.

Friedman, A., and Sienkiewicz, J. (1991). Psychotic complications of long-term levodopa treatment of Parkinson's disease. *Acta Neurol. Scand.* **84,** 111–113.

Fukui, R., Svenningsson, P., Matsuishi, T., Higashi, H., Nairn, A. C., Greengard, P., and Nishi, A. (2003). Effect of methylphenidate on dopamine/DARPP signalling in adult, but not young, mice. *J. Neurochem.* **87,** 1391–1401.

Fulceri, F., Biagioni, F., Ferrucci, M., Lazzeri, G., Bartalucci, A., Galli, V., Ruggieri, S., Paparelli, A., and Fornai, F. (2007). Abnormal involuntary movements (AIMs) following pulsatile dopaminergic stimulation: Severe deterioration and morphological correlates following the loss of locus coeruleus neurons. *Brain Res.* **1135,** 219–229.

Fuxe, K., Jacobsen, K. X., Hoistad, M., Tinner, B., Jansson, A., Staines, W. A., and Agnati, L. F. (2003). The dopamine D1 receptor-rich main and paracapsular intercalated nerve cell groups of the rat amygdala: Relationship to the dopamine innervation. *Neuroscience* **119,** 733–746.

Fuxe, K., Rivera, A., Jacobsen, K. X., Hoistad, M., Leo, G., Horvath, T. L., Staines, W., De la Calle, A., and Agnati, L. F. (2005). Dynamics of volume transmission in the brain. Focus on catecholamine and opioid peptide communication and the role of uncoupling protein 2. *J. Neural. Transm.* **112,** 65–76.

Fuxe, K., Manger, P., Genedani, S., and Agnati, L. (2006). The nigrostriatal DA pathway and Parkinson's disease. *J. Neural. Transm. Suppl.* **70,** 71–83.

Gao, C., Sun, X., and Wolf, M. E. (2006). Activation of D1 dopamine receptors increases surface expression of AMPA receptors and facilitates their synaptic incorporation in cultured hippocampal neurons. *J. Neurochem.* **98,** 1664–1677.

Gerfen, C. R. (2000). Dopamine-mediated gene regulation in models of Parkinson's disease. *Ann. Neurol.* **47,** S42–S52.

Gerfen, C. R. (2003). D1 dopamine receptor supersensitivity in the dopamine-depleted striatum animal model of Parkinson's disease. *Neuroscientist* **9,** 455–462.

Gesi, M., Soldani, P., Giorgi, F. S., Santinami, A., Bonaccorsi, I., and Fornai, F. (2000). The role of the locus coeruleus in the development of Parkinson's disease. *Neurosci. Biobehav. Rev.* **24,** 655–668.

Gesi, M., Santinami, A., Ruffoli, R., Conti, G., and Fornai, F. (2001). Novel aspects of dopamine oxidative metabolism (confounding outcomes take place of certainties). *Pharmacol. Toxicol.* **89,** 217–224.

Giovannoni, G., O'Sullivan, J. D., Turner, K., Manson, A. J., and Lees, A. J. (2000). Hedonistic homeostatic dysregulation in patients with Parkinson's disease on dopamine replacement therapies. *J. Neurol. Neurosurg. Psychiatry* **68,** 423–428.

Green, T. A., Alibhai, I. N., Hommel, J. D., DiLeone, R. J., Kumar, A., Theobald, D. E., Neve, R. L., and Nestler, E. J. (2006). Induction of inducible cAMP early repressor expression in nucleus accumbens by stress or amphetamine increases behavioral responses to emotional stimuli. *J. Neurosci.* **26,** 8235–8242.

Guigoni, C., Aubert, I., Li, Q., Gurevich, V. V., Benovic, J. L., Ferry, S., Mach, U., Stark, H., Leriche, L., Hakansson, K., Bioulac, B. H., Gross, C. E., *et al.* (2005). Pathogenesis of levodopa-induced dyskinesia: Focus on D1 and D3 dopamine receptors. *Parkinsonism Relat. Disord.* **11,** S25–S29.

Hamamura, T., Akiyama, K., Akimoto, K., Kashihara, K., Okumura, K., Ujike, H., and Otsuki, S. (1991). Co-administration of either a selective D1 or D2 dopamine antagonist with methamphetamine prevents methamphetamine-induced behavioral sensitization and neurochemical change, studied by in vivo intracerebral dialysis. *Brain Res.* **546,** 40–46.

Henry, B., Crossman, A. R., and Brotchie, J. M. (1999). Effect of repeated L-DOPA, bromocriptine, or lisuride administration on preproenkephalin-A and preproenkephalin-B mRNA levels in the striatum of the 6-hydroxydopamine-lesioned rat. *Exp. Neurol.* **155,** 204–220.

Horner, K. A., and Keefe, K. A. (2006). Regulation of psychostimulant-induced preprodynorphin, c-fos and zif/268 messenger RNA expression in the rat dorsal striatum by mu opioid receptor blockade. *Eur. J. Pharmacol.* **532,** 61–73.

Horner, K. A., Westwood, S. C., Hanson, G. R., and Keefe, K. A. (2006). Multiple high doses of methamphetamine increase the number of preproneuropeptide Y mRNA-expressing neurons in the striatum of rat via a dopamine D1 receptor-dependent mechanism. *J. Pharmacol. Exp. Ther.* **319,** 414–421.

Hotte, M., Thuault, S., Lachaise, F., Dineley, K. T., Hemmings, H. C., Nairn, A. C., and Jay, T. M. (2006). D1 receptor modulation of memory retrieval performance is associated with changes in pCREB and pDARPP-32 in rat prefrontal cortex. *Behav. Brain Res.* **171,** 127–133.

Hyman, S. E., Malenka, R. C., and Nestler, E. J. (2006). Neural mechanisms of addiction: The role of reward-related learning and memory. *Ann. Rev. Neurosci.* **29,** 565–598.

Jenkins, R. B., and Groh, R. H. (1970). Mental symptoms in Parkinsonian patients treated with L-DOPA. *Lancet* **2,** 177–179.

Juncos, J. L., Engber, T. M., Raisman, R., Susel, Z., Thibaut, F., Ploska, A., Agid, Y., and Chase, T. N. (1989). Continuous and intermittent levodopa differentially affect basal ganglia function. *Ann. Neurol.* **25,** 473–478.

Kaiser, R., Hofer, A., Grapengiesser, A., Gasser, T., Kupsch, A., Roots, I., and Brockmoller, J. (2003). L-DOPA-induced adverse effects in PD and dopamine transporter gene polymorphism. *Neurology* **60,** 1750–1755.

Kashiwabara, K. (1983). A long-term qualitative behavioral change following chronic methamphetamine administration in Japanese monkeys (*Macaca fuscata*). *Yakubutsu Seishin Kodo* **3,** 137–148.

Katz, J., Nielsen, K. M., and Soghomonian, J. J. (2005). Comparative effects of acute or chronic administration of levodopa to 6-hydroxydopamine-lesioned rats on the expression of glutamic acid decarboxylase in the neostriatum and GABAA receptors subunits in the substantia nigra, pars reticulata. *Neuroscience* **132,** 833–842.

Katzenschlager, R., Hughes, A., Evans, A., Manson, A. J., Hoffman, M., Swinn, L., Watt, H., Bhatia, K., Quinn, N., and Lees, A. J. (2005). Continuous subcutaneous apomorphine therapy improves dyskinesias in Parkinson's disease: A prospective study using single-dose challenges. *Mov. Disord.* **20,** 151–157.

Kazahaya, Y., Akimoto, K., and Otsuki, S. (1989). Subchronic methamphetamine treatment enhances methamphetamine- or cocaine-induced dopamine efflux *in vivo. Biol. Psychiatry* **25,** 903–912.

Konitsiotis, S., Blanchet, P. J., Verhagen, L., Lamers, E., and Chase, T. N. (2000). AMPA receptor blockade improves levodopa-induced dyskinesia in MPTP monkeys. *Neurology* **54,** 1589–1595.

Kuribara, H. (1994). Early post-treatment with haloperidol retards induction of methamphetamine sensitization in mice. *Eur. J. Pharmacol.* **256,** 295–299.

Kuribara, H. (1995a). Inhibition of methamphetamine sensitization by post-methamphetamine treatment with SCH 23390 or haloperidol. *Psychopharmacology* **119,** 34–38.

Kuribara, H. (1995b). Dopamine D1 receptor antagonist SCH 23390 retards methamphetamine sensitization in both combined administration and early posttreatment schedules in mice. *Pharmacol. Biochem. Behav.* **52,** 759–763.

Landa, L., Sulcova, A., and Slais, K. (2006). Involvement of cannabinoid CB1 and CB2 receptor activity in the development of behavioural sensitization to methamphetamine effects in mice. *NeuroEndocrinol. Lett.* **27,** 63–69.

Lazzeri, G., Lenzi, P., Busceti, C. L., Ferrucci, M., Falleni, A., Bruno, V., Paparelli, A., and Fornai, F. (2007). Mechanisms involved in the formation of dopamine-induced intracellular bodies within striatal neurons. *J. Neurochem.* **101,** 1414–1427.

Lee, J., Di Marzo, V., and Brotchie, J. M. (2006). A role for vanilloid receptor 1 (TRPV1) and endocannabinoid signalling in the regulation of spontaneous and L-DOPA induced locomotion in normal and reserpine-treated rats. *Neuropharmacology* **51,** 557–565.

Lindskog, M., Kim, M., Wikstrom, M. A., Blackwell, K. T., and Kotaleski, J. H. (2006). Transient calcium and dopamine increase PKA activity and DARPP-32 phosphorylation. *PLoS Comput. Biol.* **2,** 119.

Lineberry, T. W., and Bostwick, J. M. (2006). Methamphetamine abuse: A perfect storm of complications. *Mayo Clin. Proc.* **81,** 77–84.

Liu, H. C., Chen, C. K., Leu, S. J., Wu, H. T., and Lin, S. K. (2006). Association between dopamine receptor D1 A-48G polymorphism and methamphetamine abuse. *Psychiatry Clin. Neurosci.* **60,** 226–231.

Maccarrone, M., Gubellini, P., Bari, M., Picconi, B., Battista, N., Centone, D., Bernardi, G., Finazzi-Agro, A., and Calabresi, P. (2003). Levodopa treatment reverses endocannabinoid system abnormalities in experimental parkinsonism. *J. Neurochem.* **85,** 1018–1025.

Marin, C., Jimenez, A., Bonastre, M., Chase, T. N., and Tolosa, E. (2000). Non-NMDA receptor-mediated mechanisms are involved in levodopa-induced motor response alterations in Parkinsonian rats. *Synapse* **36,** 267–274.

McCann, U. D., Wong, D. F., Yokoi, F., Villemagne, V., Dannals, R. F., and Ricaurte, G. A. (1998). Reduced striatal dopamine transporter density in abstinent methamphetamine and methcathinone users: Evidence from positron emission tomography studies with [11C]WIN-35,428. *J. Neurosci.* **18,** 8417–8422.

Mc Daid, J., Graham, M. P., and Napier, T. C. (2006). Methamphetamine-induced sensitization differentially alters pCREB and DeltaFosB throughout the limbic circuit of the mammalian brain. *Mol. Pharmacol.* **70,** 2064–2074.

Mc Daid, J., Tedford, C. E., Mackie, A. R., Dallimore, J. E., Mickiewicz, A. L., Shen, F., Angle, J. M., and Napier, T. C. (2007). Nullifying drug-induced sensitization: Behavioral and electrophysiological evaluations of dopaminergic and serotonergic ligands in methamphetamine-sensitized rats. *Drug Alcohol Depend.* **86,** 55–66.

Mercuri, N. B., and Bernardi, G. (2005). The "magic" of L-DOPA: Why is it the gold standard Parkinson's disease therapy? *Trends Pharmacol. Sci.* **26,** 341–344.

Miller, D. K., Crooks, P. A., Teng, L., Witkin, J. M., Munzar, P., Goldberg, S. R., Acri, J. B., and Dwoskin, L. P. (2001). Lobeline inhibits the neurochemical and behavioral effects of amphetamine. *J. Pharmacol. Exp. Ther.* **296,** 1023–1034.

Missale, C., Fiorentini, C., Busi, C., Collo, G., and Spano, P. F. (2006). The NMDA/D1 receptor complex as a new target in drug development. *Curr. Top. Med. Chem.* **6,** 801–808.

Morelli, M., Pontieri, F. E., Linfante, I., Orzi, F., and Di Chiara, G. (1993). Local cerebral glucose utilization after D1 receptor stimulation in 6-OHDA lesioned rats: Effect of sensitization (priming) with a dopaminergic agonist. *Synapse.* **13,** 264–269.

Moszczynska, A., Fitzmaurice, P., Ang, L., Kalasinsky, K. S., Schmunk, G. A., Peretti, F. J., Aiken, S. S., Wickham, D. J., and Kish, S. J. (2004). Why is parkinsonism not a feature of human methamphetamine users? *Brain* **127,** 363–370.

Muratake, T., Toyooka, K., Hayashi, S., Ichikawa, T., Kumanishi, T., and Takahashi, Y. (1998). Immunohistochemical changes of the transcription regulatory factors in rat striatum after methamphetamine administration. *Ann. N.Y. Acad. Sci.* **844,** 21–26.

Nishi, A., Watanabe, Y., Higashi, H., Tanaka, M., Nairn, A. C., and Greengard, P. (2005). Glutamate regulation of DARPP-32 phosphorylation in neostriatal neurons involves activation of multiple signaling cascades. *Proc. Natl. Acad. Sci. USA* **102,** 1199–1204.

Nishijima, K., Kashiwa, A., Hashimoto, A., Iwama, H., Umino, A., and Nishikawa, T. (1996). Differential effects of phencyclidine and methamphetamine on dopamine metabolism in rat frontal cortex and striatum as revealed by *in vivo* dialysis. *Synapse* **22,** 304–312.

Nishikawa, T., Mataga, N., Takashima, M., and Toru, M. (1983). Behavioral sensitization and relative hyperresponsiveness of striatal and limbic dopaminergic neurons after repeated methamphetamine treatment. *Eur. J. Pharmacol.* **88,** 195–203.

Nomura, A., Ujike, H., Tanaka, Y., Otani, K., Morita, Y., Kishimoto, M., Morio, A., Harano, M., Inada, T., Yamada, M., Komiyama, T., Sekine, Y., *et al.* (2006). Genetic variant of prodynorphin gene is risk factor for methamphetamine dependence. *Neurosci. Lett.* **400,** 158–162.

Nutt, J. G. (2007). Continuous dopaminergic stimulation: Is it the answer to the motor complications of Levodopa? *Mov. Disord.* **22,** 1–9.

Oh, J. D., and Chase, T. N. (2002). Glutamate-mediated striatal dysregulation and the pathogenesis of motor response complications in Parkinson's disease. *Amino Acids* **23,** 133–139.

Oh, J. D., Vaughan, C. L., and Chase, T. N. (1999). Effect of dopamine denervation and dopamine agonist administration on serine phosphorylation of striatal NMDA receptor subunits. *Brain Res.* **821,** 433–442.

Olausson, P., Jentsch, J. D., Tronson, N., Neve, R. L., Nestler, E. J., and Taylor, J. R. (2006). DeltaFosB in the nucleus accumbens regulates food-reinforced instrumental behavior and motivation. *J. Neurosci.* **26,** 9196–9204.

Palmer, A. A., Verbitsky, M., Suresh, R., Kamens, H. M., Reed, C. L., Li, N., Burkhart-Kasch, S., McKinnon, C. S., Belknap, J. K., Gilliam, T. C., and Phillips, T. J. (2005). Gene expression differences in mice divergently selected for methamphetamine sensitivity. *Mamm. Genome* **6,** 291–305.

Patrick, G. N. (2006). Synapse formation and plasticity: Recent insights from the perspective of the ubiquitin proteasome system. *Curr. Opin. Neurobiol.* **16,** 90–94.

Piccini, P., Pavese, N., and Brooks, D. J. (2003). Endogenous dopamine release after pharmacological challenges in Parkinson's disease. *Ann. Neurol.* **53,** 647–653.

Quan, L., Ishikawa, T., Michiue, T., Li, D. R., Zhao, D., Origani, S., Zhu, B. L., and Maeda, H. (2005). Ubiquitin-immunoreactive structures in the midbrain of methamphetamine abusers. *Leg. Med.* **7,** 144–150.

Raiteri, M., Cerrito, F., Cervoni, A. M., and Levi, G. (1979). Dopamine can be released by two mechanisms differentially affected by the dopamine transport inhibitor nomifensine. *J. Pharmacol. Exp. Ther.* **208,** 195–202.

Rivera, A., Agnati, L. F., Horvath, T. L., Valderrama, J. J., De La Calle, A., and Fuxe, K. (2006). Uncoupling protein 2/3 immunoreactivity and the ascending dopaminergic and noradrenergic neuronal systems: Relevance for volume transmission. *Neuroscience* **137,** 1447–1461.

Rothman, R. B., Baumann, M. H., Dersch, C. M., Romero, D. V., Rice, K. C., Carroll, F. I., and Partilla, J. S. (2001). Amphetamine-type central nervous system stimulants release norepinephrine more potently than they release dopamine and serotonin. *Synapse* **39,** 32–41.

Samadi, P., Gregoire, L., and Bedard, P. J. (2004). The opioid agonist morphine decreases the dyskinetic response to dopaminergic agents in parkinsonian monkeys. *Neurobiol. Dis.* **16,** 46–53.

Samadi, P., Bedard, P. J., and Rouillard, C. (2006). Opioids and motor complications in Parkinson's disease. *Trends Pharmacol. Sci.* **27,** 512–517.

Sandoval, V., Riddle, E. L., Hanson, G. R., and Fleckenstein, A. E. (2002). Methylphenidate redistributes vesicular monoamine transporter-2: role of dopamine receptors. *J. Neurosci.* **22,** 8705–8710.

Sandoval, V., Riddle, E. L., Hanson, G. R., and Fleckenstein, A. E. (2003). Methylphenidate alters vesicular monoamine transport and prevents methamphetamine-induced dopaminergic deficits. *J. Pharmacol. Exp. Ther.* **304,** 1181–1187.

Sato, M., Chen, C. C., Akiyama, K., and Otsuki, S. (1983). Acute exacerbation of paranoid psychotic state after long-term abstinence in patients with previous methamphetamine psychosis. *Biol. Psychiatry* **18,** 429–440.

Schmidt, C. J., and Gibb, J. W. (1985). Role of the dopamine uptake carrier in the neurochemical response to methamphetamine: effects of amfonelic acid. *Eur. J. Pharmacol.* **109,** 73–80.

Shults, C. W. (2006). Lewy bodies. *Proc. Natl. Acad. Sci. USA* **103,** 1661–1668.

Silveira-Moriyama, L., Evans, A. H., Katzenschlager, R., and Lees, A. J. (2006). Punding and dyskinesias. *Mov. Disord.* **21,** 2214–2217.

Snyder, G. L., Allen, P. B., Fienberg, A. A., Valle, C. G., Huganir, R. L., Nairn, A. C., and Greengard, P. (2000). Regulation of phosphorylation of the GluR1 AMPA receptor in the neostriatum by dopamine and psychostimulants *in vivo. J. Neurosci.* **20,** 4480–4488.

Sonsalla, P. K., Gibb, J. W., and Hanson, G. R. (1986). Roles of D1 and D2 dopamine receptor subtypes in mediating the methamphetamine-induced changes in monoamine systems. *J. Pharmacol. Exp. Ther.* **238,** 932–937.

Sperling, L. S., and Horowitz, J. L. (1994). Methamphetamine-induced choreoathetosis and rhabdomyolysis. *Ann. Intern. Med.* **121,** 986.

Stefanova, N., Lundblad, M., Tison, F., Poewe, W., Cenci, M. A., and Wenning, G. K. (2004). Effects of pulsatile L-DOPA treatment in the double lesion rat model of striatonigral degeneration (multiple system atrophy). *Neurobiol. Dis.* **15,** 630–639.

Stephans, S. E., and Yamamoto, B. Y. (1995). Effect of repeated methamphetamine administrations on dopamine and glutamate efflux in rat prefrontal cortex. *Brain Res.* **700,** 99–106.

Stephenson, D. T., Li, Q., Simmons, C., Connell, M. A., Meglasson, M. D., Merchant, K., and Emborg, M. E. (2005). Expression of GAD65 and GAD67 immunoreactivity in MPTP-treated monkeys with or without L-DOPA administration. *Neurobiol. Dis.* **20,** 347–359.

St-Hilaire, M., Landry, E., Levesque, D., and Rouillard, C. (2005). Denervation and repeated L-DOPA induce complex regulatory changes in neurochemical phenotypes of striatal neurons: implication of a dopamine D1-dependent mechanism. *Neurobiol. Dis.* **20,** 450–460.

Stocchi, F., and Olanow, C. W. (2004). Continuous dopaminergic stimulation in early and advanced Parkinson's disease. *Neurology* **62,** S56–S63.

Stocchi, F., Ruggirei, S., Vacca, L., and Olanow, C. W. (2002). Prospective randomized trial of lisuride infusion versus oral levodopa in patients with Parkinson's disease. *Brain* **125,** 2058–2066.

Suzuki, O., Hattori, H., Asano, M., Oya, M., and Katsumata, Y. (1980). Inhibition of monoamine oxidase by d-methamphetamine. *Biochem. Pharmacol.* **29,** 2071–2073.

Suzuki, H., Shishido, T., Watanabe, Y., Abe, H., Shiragata, M., Honda, K., Horikoshi, R., and Niwa, S. (1997). Changes of behavior and monoamine metabolites in the rat brain after repeated methamphetamine administration: Effects of duration of repeated administration. *Prog. Neuropsychopharmacol. Biol. Psychiatry* **21,** 359–369.

Svenningsson, P., Gunne, L., and Andren, P. E. (2000). L-DOPA produces strong induction of c-fos messenger RNA in dopamine-denervated cortical and striatal areas of the common marmoset. *Neuroscience* **99,** 457–468.

Swayze, R. D., Lise, M. F., Levinson, J. N., Phillips, A., and El-Husseini, A. (2004). Modulation of dopamine mediated phosphorylation of AMPA receptors by PSD-95 and AKAP79/150. *Neuropharmacology* **47,** 764–778.

Tamminga, C. A., and Nestler, E. J. (2006). Pathological gambling: Focusing on the addiction, not the activity. *Am. J. Psychiatry* **163,** 180–181.

Taylor, J. L., Bishop, C., and Walker, P. D. (2005). Dopamine D1 and D2 receptor contributions to L-DOPA-induced dyskinesia in the dopamine-depleted rat. *Pharmacol. Biochem. Behav.* **81,** 887–893.

Tien, L. T., Ho, I. K., Loh, H. H., and Ma, T. (2007). Role of mu -opioid receptor in modulation of preproenkephalin mRNA expression and opioid and dopamine receptor binding in methamphetamine-sensitized mice. *J. Neurosci. Res.* **85,** 673–680.

Tipton, K. F., Boyce, S., O'Sullivan, J., Davey, G. P., and Healy, J. (2004). Monoamine oxidases: Certainties and uncertainties. *Curr. Med. Chem.* **11,** 1965–1982.

Uehara, T., Sumiyoshi, T., Itoh, H., and Kurachi, M. (2004). Inhibition of dopamine synthesis with alpha-methyl-*p*-tyrosine abolishes the enhancement of methamphetamine-induced extracellular dopamine levels in the amygdala of rats with excitotoxic lesions of the entorhinal cortex. *Neurosci. Lett.* **356,** 21–24.

Ujike, H. (2001). Advanced findings on the molecular mechanisms for behavioral sensitization to psychostimulant. *Nippon Yakurigaku Zasshi* **117,** 5–12.

Ujike, H., and Sato, M. (2004). Clinical features of sensitization to methamphetamine observed in patients with methamphetamine dependence and psychosis. *Ann. N.Y. Acad. Sci.* **1025,** 279–287.

Van de Witte, S. V., Drukarch, B., Stoof, J. C., and Voorn, P. (1998). Priming with L-DOPA differently affects dynorphin and substance P mRNA levels in the striatum of 6-hydroxydopamine-lesioned rats after challenge with dopamine D1-receptor agonist. *Brain Res. Mol. Brain Res.* **61,** 219–223.

Van der Stelt, M., Fox, S. H., Hill, M., Crossman, A. R., Petrosino, S., Di Marzo, V., and Brotchie, J. M. (2005). A role for endocannabinoids in the generation of parkinsonism and levodopa-induced dyskinesia in MPTP-lesioned non-human primate models of Parkinson's disease. *FASEB J.* **19,** 1140–1142.

Vinklerova, J., Novakova, J., and Sulcova, A. (2002). Inhibition of methamphetamine self-administration in rats by cannabinoid receptor antagonist AM 251. *J. Psychopharmacol.* **16,** 139–143.

Volkow, N. D., Chang, L., Wang, G. J., Fowler, J. S., Franceschi, D., Sedler, M., Gatley, S. J., Miller, E., Hitzemann, R., Ding, Y. S., and Logan, J. (2001). Loss of dopamine transporters in methamphetamine abusers recovers with protracted abstinence. *J. Neurosci.* **21,** 9414–9418.

Volz, T. J., Fleckenstein, A. E., and Hanson, G. R. (2007). Methamphetamine-induced alterations in monoamine transport: Implications for neurotoxicity, neuroprotection and treatment. *Addiction.* **102,** 44–48.

Wang, J. Q., and McGinty, J. F. (1996). D1 and D2 receptor regulation of preproenkephalin and preprodynorphin mRNA in rat striatum following acute injection of amphetamine or methamphetamine. *Synapse* **22,** 114–122.

Yamada, S., Kojima, H., Yokoo, H., Tsutsumi, T., Takamuki, K., Anraku, S., Nishi, S., and Inanaga, K. (1988). Enhancement of dopamine release from striatal slices of rats that were subchronically treated with methamphetamine. *Biol. Psychiatry* **24,** 399–408.

Yamamoto, H., Imai, K., Kamegaya, E., Takamatsu, Y., Irago, M., Hagino, Y., Kasai, S., Shimada, K., Yamamoto, T., Sora, I., Koga, H., and Ikeda, K. (2006). Repeated methamphetamine administration alters expression of the NMDA receptor channel epsilon2 subunit and kinesins in the mouse brain. *Ann. N.Y. Acad. Sci.* **1074,** 97–103.

Yi, J. J., and Ehlers, M. D. (2005). Ubiquitin and protein turnover in synapse function. *Neuron* **47,** 629–632.

Yoshida, H., Ohno, M., and Watanabe, S. (1995). Roles of dopamine D1 receptors in striatal fos protein induction associated with methamphetamine behavioral sensitization in rats. *Brain Res. Bull.* **38,** 393–397.

Youdim, M. B. (1974). Heterogeneity of rat brain mitochondrial monoamine oxidase. *Adv. Biochem. Psychopharmacol.* **11,** 59–63.

Youdim, M. B., Edmondson, D., and Tipton, K. F. (2006). The therapeutic potential of monoamine oxidase inhibitors. *Nat. Rev. Neurosci.* **7,** 295–309.

Yui, K., Goto, K., Ikemoto, S., Ishiguro, T., Angrist, B., Duncan, G. E., Sheitman, B. B., Lieberman, J. A., Bracha, S. H., and Ali, S. F. (1999). Neurobiological basis of relapse prediction in stimulant-induced psychosis and schizophrenia: the role of sensitization. *Mol. Psychiatry* **4,** 512–523.

Yui, K., Goto, K., and Ikemoto, S. (2004). The role of noradrenergic and dopaminergic hyperactivity in the development of spontaneous recurrence of methamphetamine psychosis and susceptibility to episode recurrence. *Ann. N.Y. Acad. Sci.* **1025,** 296–306.

Zachariou, V., Bolanos, C. A., Selley, D. E., Theobald, D., Cassidy, M. P., Kelz, M. B., Shaw-Lutchman, T., Berton, O., Sim-Selley, L. J., Dileone, R. J., Kumar, A., and Nestler, E. J. (2006). An essential role for DeltaFosB in the nucleus accumbens in morphine action. *Nat. Neurosci.* **9,** 205–211.

Zhang, X., Lee, T. H., Xiong, X., Chen, Q., Davidson, C., Wetsel, W. C., and Ellinwood, E. H. (2006a). Methamphetamine induces long-term changes in GABAA receptor alpha2 subunit and GAD67 expression. *Biochem. Biophys. Res. Commun.* **351,** 300–305.

Zhang, J., Zhang, L., Jiao, H., Zhang, Q., Zhang, D., Lou, D., Katz, J. L., and Xu, M. (2006b). c-Fos facilitates the acquisition and extinction of cocaine-induced persistent changes. *J. Neurosci.* **26,** 13287–13296.

Zuddas, A., Oberto, G., Vaglini, F., Fascetti, F., Fornai, F., and Corsini, G. U. (1992). MK-801 prevents 1-methyl-4-phenyl-1,2,3,6-tetrahydropyridine-induced parkinsonism in primates. *J. Neurochem.* **59,** 733–739.

THE ROLE OF THE SOMATOTROPHIC AXIS IN NEUROPROTECTION AND NEUROREGENERATION OF THE ADDICTIVE BRAIN

Fred Nyberg

Department of Pharmaceutical Biosciences, Division of Biological Research
on Drug Dependence, Uppsala University, S-75124 Uppsala, Sweden

Early studies have shown that the abuse of alcohol, central stimulants, and opiates such as heroin destroys brain cells, reducing attention span and memory. However, new research has suggested that there may be a way to regain some of the lost attention and recall. It has recently been shown that brain cells targeted for early death by continued opiate use can be salvaged by injections of synthetic human growth hormone (GH). GH is a polypeptide hormone, normally secreted by the anterior pituitary gland, which stimulates cell growth and controls body metabolism. Recombinant human GH is currently used in replacement therapy to alleviate the symptoms of adults and children with GH deficiency syndrome. The recent observation that GH can reverse morphine-induced cell damage could open the door to new ways of treating and preventing damage from the abuse of opiates in addicts and also of treating cell damage induced by alcohol and central stimulants. This article reviews current knowledge of the somatotrophic axis, including GH and insulin-like growth factor-1 (IGF-1), in the brain and also discusses the potential use of GH/IGF-1 as agents for treatment of brain pathology in addictive diseases.

INTERNATIONAL REVIEW OF
NEUROBIOLOGY, VOL. 88
DOI: 10.1016/S0074-7742(09)88014-8

I. Introduction

The impact of growth hormone (GH) on functions related to the central nervous system (CNS) has received increased interest among many investigators over the past few years. The effect of the hormone and its mediator, insulin-like growth factor-1 (IGF-1), on brain function has been evaluated in many clinical and preclinical laboratories. Current data suggest that both GH and IGF-1 can penetrate the blood–brain barrier (BBB) and induce profound effects on various CNS-related behaviors (Nyberg, 2007). For instance, GH replacement therapy in adult patients with GH deficiency (GHD) has been shown to improve several psychological parameters, leading to improved quality of life (Burman and Deijen, 1998; Nilsson et al., 2007; Stabler et al., 2001; Svensson et al., 2001; Webb and Badia, 2008). Significant improvements regarding attention and cognitive-perceptual performance have also been documented in children with GHD after treatment with GH. Recombinant GH therapy greatly improves all pathophysiologic manifestations seen in GHD. For example, impairments in cognitive function, particularly attention and memory, in adults with GHD are improved and functionality appears to be restored following GH treatment (Falleti et al., 2006; Laursen et al., 2008; van Nieuwpoort and Drent, 2008). Furthermore, both GH and IGF-1 produce neuroprotective effects which counteract the damage induced by various kinds of CNS trauma (Scheepens et al., 2000). The observations made in human subjects regarding the beneficial effects of GH and IGF-1 on CNS trauma, memory, and cognition have been confirmed in a number of experimental animal studies. Preclinical studies using animal models or cell lines have confirmed that both GH and IGF-1 can stimulate neurogenesis and prevent apoptosis (Aberg et al., 2006; Svensson et al., 2008). These observations seem to be of particular importance, since they suggest a possible therapeutic use for GH and IGF-1 in the treatment of brain damage caused by addictive drugs. In the following text, this article highlights and extends past and current data on these issues, with particular focus on the mechanism behind the effects of the GH/IGF-1 axis (the somatotrophic axis) on CNS-related functions.

II. The Somatotrophic Axis in Brain Function

The discovery of specific receptors for GH and IGF-1 in discrete areas of the brain and spinal cord added significantly to our understanding of the effects of these molecules on brain function. The receptors for GH are expressed in a variety of CNS regions, including those that are believed to represent the functional site for the various behaviors associated with the hormone that were mentioned above (Nyberg, 2000). IGF-1 receptors have been found in various regions of the mature brain (Laviola *et al.*, 2007). Experimental animal models have been used to explore the molecular mechanisms by which GH and IGF-1 induce their effects on CNS-related behavior. During the last decade, studies have revealed that stimulation of brain receptors for GH and IGF-1 can affect several signaling pathways in the brain, including those involving monoamines, excitatory amino acids (glutamate), and various neuropeptides (Nyberg, 2000, 2007). However, before these hormones can access their brain targets, they need to cross the BBB.

A. PERMEATION OF GH AND IGF-1 ACROSS THE BLOOD–BRAIN BARRIER

Several studies have confirmed the active transport of various polypeptide hormones over the BBB. The hormones studied include insulin (Laron, 2009), leptin (Kastin *et al.*, 2001), and IGF-1 (Pan and Kastin, 2000). Kastin and cow-orkers reported that IGF-1 enters the CNS via a saturable transport system at the BBB, and suggested that this functions in synchrony with IGF-binding proteins in the periphery to regulate the availability of the growth factor to the CNS (Pan and Kastin, 2000). Using experimental animals, they observed that IGF-1 was present in the brain 20 min after an intravenous injection. They also noticed that, in the spinal cord, uptake was fastest in the cervical region, followed by the lumbar spinal cord.

The potential for GH to cross the BBB has been a topic of discussion among scientists for years. As the molecular weight of GH (22,000 kD) is considered too large for easy penetration of the BBB, it has been hard to believe that it can find its way over this barrier. However, over the years, accumulating evidence supporting the existence of a possible pathway for GH to the brain has emerged. For example, immunoreactive GH has been detected in several brain regions of various mammals (Hojvat *et al.*, 1982; Mustafa *et al.*, 1994b; Noteborn *et al.*, 1993). Furthermore, the identification and measurement of GH in cerebrospinal fluid (CSF) is indicative of the ability of the hormone to penetrate this barrier (Burman *et al.*, 1996; Johansson *et al.*, 1995; Nyberg, 2006). Interestingly, a positive correlation was found between the dosage and the CSF concentration of recom-binant human GH in patients given GH replacement therapy in a clinical study (Burman *et al.*, 1996). In addition, the beneficial effects of GH during replacement therapy suggest that the hormone can find its way from the circulatory system into the CNS (Nyberg, 2000, 2007; Nyberg and Burman, 1996). Several routes have been suggested as the mechanism for the hormone penetrating the BBB. For example, GH could reach its responsive sites in the brain by circumventing the BBB through the median eminence of the hypothalamus. The median eminence is characterized as a circumventricular organ that allows hypothalamic polypeptides to leave the brain without disrupting the BBB and permits hormonal entities that do not cross this barrier to trigger changes in brain function (Ganong, 2000). Or, as has been suggested for the pituitary hormone prolactin (Walsh *et al.*, 1987), GH could cross the BBB to reach the brain via the CSF by a receptor-mediated mechanism through the choroid plexus (Coculescu, 1999). Alternatively, the hormone could find its brain targets by actively or passively crossing the brain parenchyma capillaries of the BBB. It has also recently been reported (Pan *et al.*, 2005) that GH significantly diffuses into the CNS.

B. GH AND IGF-1 RECEPTORS IN THE BRAIN

Evidence for the existence of GH receptors (GHRs) in brain-related structures emerged from studies on regulation mechanisms for GH secretion, around three decades ago, when it was shown that GH interacts with catecholamine pathways in the hypothalamus. Thus, physiologic concentrations of systemically adminis-tered rat growth hormone (rGH) interacted with the turnover of dopamine and noradrenaline in the median eminence of the rat hypothalamus (Andersson *et al.*, 1983). This provided support for the notion that GH inhibits its own secretion through a feedback mechanism, partly via reduction of dopamine synthesis and release leading to enhanced somatostatin release, and partly via attenuated noradrenaline synthesis and turnover in the median eminence leading to reduced secretion of the GH releasing factor GH–RH. It was obvious that the effect of GH on hypothalamic monoamines would involve interaction with particular sites, and identification of the specific receptor sites in this brain area was reported soon after. A receptor protein for GH in the human hypothalamus was characterized as an entity with high affinity for the hormone, distinct from the prolactin receptor (Lai *et al.*, 1992, 1993). Currently, it is known that specific sites for GH are present in several regions of the mammalian brain (for review, see Nyberg, 2000, 2007). The regional distribution of specific receptor sites for the hormone in the human brain (Lai *et al.*, 1993; Nyberg, 2000) is similar to that in the rat brain (Zhai *et al.*, 1994). Binding studies in mammals have revealed a high density of GH binding in the CNS, including the choroid plexus, pituitary, hypothalamus, and hippocam-pus but also the spinal cord (Lai *et al.*, 1991, 1993; Walsh *et al.*, 1990; Zhai *et al.*, 1994). In human (Lai *et al.*, 1991, 1993) and rat (Mustafa *et al.*, 1994a,b; Zhai *et al.*, 1994) brains, the density of GH binding is highest in the choroid plexus. In most areas of both human (Lai *et al.*, 1993) and rat (Zhai *et al.*, 1994) brains, GH binding decreases with increasing age (Fig. 1). Expression of the gene transcripts of the GHR has been confirmed in both human and rat brain tissues (Le Grevés *et al.*, 2006; Nyberg, 2000). The *in situ* hybridization technique was used to identify the GHR in a number of different brain areas in rat studies (Burton *et al.*, 1992). Similar studies on the brain GHR have been reported by other investigators

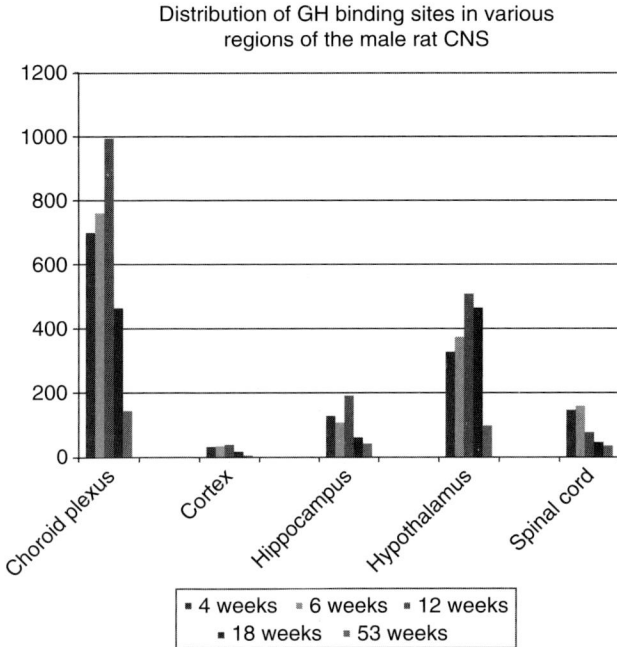

FIG. 1. Distribution of 125-I-labeled rat GH binding in various regions of the male rat CNS over age. The binding is expressed in fmol/mg protein with six animals in each group (data from Zhai *et al.*, 1994).

(Hasegawa *et al.*, 1993; Lobie *et al.*, 1993). Cells with high expression levels of GHR mRNA have been identified in the hypothalamus, the thalamus septal region, the hippocampus, the dentate gyrus, and the amygdala (Burton *et al.*, 1992). A GHR gene transcript of about 4.4 kb was found in an ovine choroid plexus cell line (Thornwall *et al.*, 1995) and the gene for the extracellular domain of GHR in the human choroid plexus has been cloned (Nyberg, 2000). The predicted amino acid sequence of this GHR domain was found to be homologous with that of cloned GHR in the human liver (Leung *et al.*, 1987). Furthermore, the nucleotide sequence for the GHR gene transcripts in the rat hippocampus and spinal cord has been evaluated (Thornwall-Le Grevés *et al.*, 2001), along with those identified in the rat hypothalamus and choroid plexus (Le Grevés, 2006). No deviation from the sequence for the GHR transcript determined for the rat liver receptor was found.

The IGF-1 receptor (IGF-1R) has also been identified in the mature brain. The early evidence for the existence of IGF-1-binding sites in the rat brain emerged from a study by Sara and coworkers in the early 1980s (Sara *et al.*,

1982). Further studies took place in the late 1980s (Bohannon *et al.*, 1988; Araujo *et al.*, 1989; Werther *et al.*, 1989) and investigations are continuing (e.g., Gualco *et al.*, 2009; Han, 1995; Melmed *et al.*, 1996). In peripheral tissues, the effects of IGF-1 are mediated through a receptor comprised of two extracellular α-subunits, with IGF-1-binding sites, and two membrane-spanning β-subunits, which encode an intracellular tyrosine kinase. The receptor seems to have a similar structure and organization in the brain. IGF-1Rs are highly expressed in the hippocampus (Le Grevés *et al.*, 2005; Zheng and Quirion, 2009), where they seem to be involved in pathways essential for cognition (Creyghton *et al.*, 2004; Sonntag *et al.*, 2005). Bohannon *et al.* (1988) used *in vitro* quantitative autoradiography to locate IGF-1-binding sites in the rat brain. Binding of labeled IGF-1 by autoradiographic techniques revealed that high affinity IGF-1-binding sites are widely distributed in discrete anatomic regions of the rat brain microarchitecture. High-density specific binding sites for IGF-1 were found in the choroid plexus of the lateral and third ventricles. Specific binding of the hormone was also seen in regions of the olfactory, visual, auditory, visceral, and somatic sensory systems, particularly in the glomerular layer of the olfactory bulb, anterior olfactory nucleus, accessory olfactory bulb, primary olfactory cortex, lateral–dorsal geniculate, superior colliculus, medial geniculate, and spinal trigeminal nucleus. Bohannon and coworkers also found significant concentrations of IGF-1 binding throughout the thalamus and dense IGF-1 binding was revealed in the hippocampus (dentate gyrus, Ca1, Ca2, Ca3). Moderate binding was observed in the paraventricular, supraoptic, and suprachiasmatic nuclei of the hypothalamus, while extensive IGF-1 binding was seen in the median eminence. Very low levels of specific binding were found in the hypothalamic arcuate nucleus and the optic chiasm and white matter regions (Bohannon *et al.*, 1988). This finding was followed up by Werther *et al.* (1989), who characterized specific binding sites for IGF-1 in the rat brain and pituitary gland using the technique of *in vitro* autoradiography and computerized densitometry. Using these techniques, IGF-1Rs were mapped, characterized and quantified in coronal and sagittal brain sections. Werther *et al.* reported a discrete and characteristic distribution of IGF-1R binding with specific binding and, by applying binding kinetics, they revealed a single class of binding site with dissociation constants in the nanomolar range. They noted that, as later seen for GH (see above), the IGF-1R density was very high in the choroid plexus. High levels were also found in the median eminence and the subfornical organ and low levels were found in the organum vasculosum of the lamina terminalis. The pituitary gland showed very high binding density, in both the anterior and posterior lobes, as did the thalamus, while the IGF-1-binding density was low in the basal ganglia. These workers also found IGF-1 binding in the hippocampus and the adjacent amygdala and reported a high density of IGF-1Rs over extensive areas of the dendritic arborizations, which receive rich synaptic inputs, in the cerebellum, hippocampus, and olfactory bulb.

It is obvious that IGF-1Rs are widespread throughout the rat brain and in the pituitary gland. Many of the regions with high densities of IGF-1 binding coincide with areas of high levels of GHR.

C. Mechanisms of GH/IGF-1 Action

In previous articles, it has been speculated that GH could exert its action on the CNS through at least three possible routes. Early work suggested that the hormone could undergo enzymatic degradation in the circulatory system to yield bioactive fragments that subsequently gained access to the CNS to interact with peptide receptors (Nyberg *et al.*, 1989). Another possibility was that the hormone could release peripheral mediators, such as IGF-1, from peripheral tissues, and that these in turn could cross the BBB to target responsive sites in the CNS (Nyberg, 2000). Finally, it was speculated that GH itself could penetrate the BBB and activate its brain receptors. In the latter case, stimulation of the GHRs in the brain would lead to the release of brain-derived IGF-1, which would subsequently mediate the effect. However, as mentioned above, studies have indicated that GH can at least cross the blood–CSF barrier (BCB), as CSF levels of the hormone significantly increase following GH replacement therapy (Burman *et al.*, 1996; Johansson *et al.*, 1995). Concomitant enhancement of IGF-1 levels in the CSF of patients receiving GH has also been demonstrated (Johansson *et al.*, 1995). The significant positive correlation between the dose of GH and the concentration of the hormone in the CSF (Burman *et al.*, 1996) further strengthens the evidence for this route. These studies also showed that GH injections affect the levels of various transmitter substances in the CSF. For instance, GH increased CSF concentrations of the opioid peptide β-endorphin (Johansson *et al.*, 1995) and the CSF levels of a dopamine metabolite and some amino acids were also affected (Burman *et al.*, 1996). Animal studies have suggested that GH could at least passively diffuse over the BBB into the CNS (Pan *et al.*, 2005).

Previous studies have shown that GH can affect short-term plasticity and spatial memory in the aging rat by interacting with the GABA system in the hippocampus. Ramsey *et al.* measured the expression of GABA-A receptors and found a significant age-related decrease in the level of the GABA-A receptor α_1-subunit, which was attenuated by GH treatment (Ramsey *et al.*, 2004). Furthermore, experimental animal studies have suggested that peripheral administration of GH can affect glutamate transmission in various brain areas. Gene transcripts of the N-methyl-D-aspartic acid (NMDA) receptor subunits in the hippocampus were affected in a fashion compatible with GH-promoting effects on memory and cognitive capabilities (Le Grevés *et al.*, 2002). This effect of GH was mimicked by IGF-1, which induced a similar alteration in the expression of the NMDA subunit gene transcripts (Le Grevés *et al.*, 2005). Thus, it seems possible that at least a part of the GH effect on hippocampal formation could result from the release of IGF-1. The suggestion of a stimulatory effect on brain GHRs was derived from the observation that peripherally administered GH elicited an effect on the NMDA receptor subunit type 2B (NR2B) that correlated positively with expression of GHRs (Le Grevés *et al.*, 2002).

Our current understanding of the mechanism of GH signaling at its target sites in the brain is limited. However, it is anticipated that the hormone interacts with its brain receptors using a mechanism similar to that in peripheral cells (Shafiei *et al.*, 2006). From this, it can be hypothesized that following GH binding to the extracellular part of its receptor in the brain, the hormone induces receptor dimerization and/or a conformational change of predimerized receptor complexes. As a consequence of this event, the receptor-linked kinase, Janus kinase 2 (JAK2), is phosphorylated, leading to phosphorylation of the intracellular part of the GHR, which undergoes further tyrosine autophosphorylation. The formation of the JAK2 kinase–GHR complex results in a process that culminates with the activation of the signal transducers and activator of transcription (STAT) proteins, which are involved in the STAT pathway. STAT1, STAT3, and STAT5 are cytoplasmic proteins that form homodimers and heterodimers. They are translocated to the nucleus, where they bind to specific response elements on DNA to initiate transcription of genes. The interactions of the STAT proteins with their gene targets in the nucleus are believed to be essential for the actions of GH. It has been suggested that the JAKs/STAT pathway in the brain is involved in the mediation of events that produce a calcium influx through NMDA receptors (Orellana *et al.*, 2005). Chronic infusion of GH in male dwarf rats increased the expression of hypothalamic STAT5b, while a single injection of GH into similar rats induced the phosphorylation of STAT5 proteins (Bennett *et al.*, 2005). The STAT5b protein may be involved in the GH-induced production of IGF-1, which mediates many of the anabolic actions of GH.

Available data suggest that GH can have profound effects on cell proliferation and cell differentiation and, in this regard, it has been suggested that the hormone exerts its effects through IGF-1. In support of this, the proliferation and differentiation induced by GH is blocked by IGF-1 antiserum (Ajo *et al.*, 2003). Levels of IGF-binding protein-3 (IGFBP-3), the IGF-IR protein and the phosphorylation of the latter molecule are increased by GH. It was further demonstrated that GH promotes proliferation of neural precursors, neurogenesis, and gliogenesis during brain development (Ajo *et al.*, 2003). All these responses are thought to be mediated by the action of locally produced IGF-1. Furthermore, it is also shown that GH and IGF-1, when administered separately, produce similar effects on brain targets. For instance, both hormones induced an upregulation of GHR expression in the hippocampus (Fig. 2).

The GH-elicited activation of the JAK2–STAT system also induces the expression of suppressors of cytokine signaling (SOCS) and/or cytokine-inducible SH2 protein (CIS) and this effect attenuates GH signaling. GH targeting in the

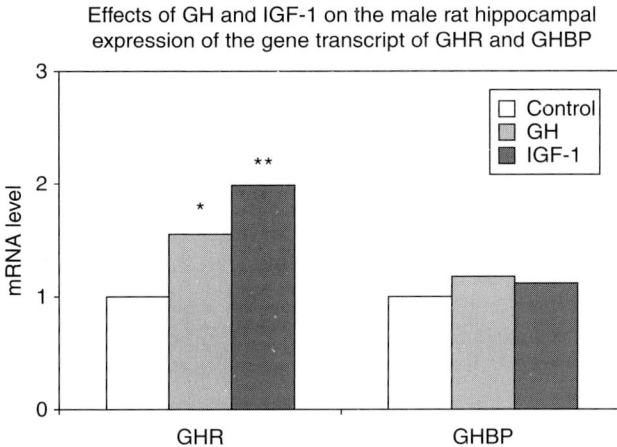

FIG. 2. The effects of GH and IGF-1 on the expression of the gene transcript of GHR and GHBP in the male rat hippocampus. The hormones induced a significant upregulation of GHR but had no significant effects on GHBP. $*p < 0.05$, $**p < 0.01$ (data from Le Grevés et al., 2002, 2005).

brain also seems to involve an interaction with the SOCS/CIS system. Studies of the rat hypothalamus have demonstrated that GH elicits an increase in the gene transcripts of SOCS3 and CIS in the arcuate nucleus as well as enhancing the SOCS3 message in the periventricular nucleus (Kasagi et al., 2004). Kasagi et al. concluded that GH acts directly on hypothalamic neurons, and enhances the gene expression of SOCS3 and CIS, which suggests that these proteins may be involved in the mechanism that regulates GH action in these cells. However, while GH did not affect expression of the suppressor of cytokine signaling-2 (SOCS2) in the hypothalamus (Kasagi et al., 2004), in studies using transgenic mice, SOCS2 abolished the effects of GH on neuronal differentiation (Ransome et al., 2004). From these studies, it was suggested that GH, possibly regulated by SOCS2, is involved in several processes in the development and maturation of the CNS, regulating the number and size of multiple neurons and even glial cells (Ransome et al., 2004; Scott et al., 2006).

The molecular mechanism underlying the effect of IGF-1 in the brain appears to be linked to the action of GH. The interaction between IGF-1 and its receptor has been extensively explored in several laboratories (for review, see Wine et al., 2009). Briefly, binding of the growth factor to IGF-1R results in the activation of a receptor kinase, subsequently inducing autophosphorylation and tyrosine phosphorylation of several substrates. Following these initial tyrosine phosphorylation reactions, signals that emerge from activation of the IGF-1R are transduced to a complex network of intracellular lipid and serine/threonine kinases, which are involved in the formation of the hormonal response, leading to effects such as cell proliferation, modulation of tissue differentiation, and protection from apoptosis. The IGF-1R and its endogenous ligand are both expressed in the brain. IGF-1 is highly expressed within the brain and is essential for normal brain development. It promotes growth of projecting neurons, arborization of dendrites and synaptogenesis and it is further known to act in an autocrine and/or paracrine manner in order to promote utilization of glucose. In this route, IGF-1 uses phosphatidylinositol 3 kinase (PI3K)/Akt or pMAPK pathways, similar to insulin signaling in peripheral tissues (Wine *et al.*, 2009).

IGF-1R is a transmembrane protein consisting of two extracellular α-subunits, which contain the IGF-1-binding site, and two transmembrane β-subunits, which have a cluster of three tyrosine residues which undergo phosphorylation and then activation following ligand binding (for review, see Philippou *et al.*, 2007). As mentioned above, the initial binding of the growth factor to its receptor induces its autophosphorylation, which results in the recruitment of specific cytoplasmic molecules containing the insulin receptor substrate proteins (IRSs). This recruitment is considered to be the critical point at which the actions of IGF-1 on proliferation and differentiation diverge. The IRSs are multisite docking proteins positioned immediately downstream of IGF-1R. In neuroendocrine pathways, IRS-1 and IRS-2 are major substrates of IGF-1 signaling. These two substrates are thought to be involved in the neuroprotective effects if IGF-1. A differential degradation of IRS-1 and IRS-2 contributes to their distinct modes of action and the increased neuroprotective effects of IRS-2 are thought to be due to the higher resistance of IRS-2 to caspase-mediated degradation (Kim and Feldman, 2009). At least six members of the IRS family (IRS-1 to IRS-6), which are expressed differentially depending on tissue and play specific and distinct but potentially overlapping roles in signal transduction, have been recognized in skeletal muscle. IRS-1 serves as a mediator and docking protein for other downstream signaling molecules in skeletal muscle. In studies involving muscle cell lines and satellite cell cultures, it has been found that activation of IGF-1R initiates intracellular signaling cascades involved in mitogenic and myogenic responses (Kim and Feldman, 2009). Depending on the timing and intracellular conditions, IGF-1 can stimulate both cellular proliferation and cellular differentiation. Recently, a considerable amount of progress has been made in the understanding of the specific signaling pathways downstream of IGF-1, mediating protein synthesis and hypertrophy. It has been proposed that at least two primary pathways are associated with the activation of the IGF-1R. One of these involves IGF-1-induced phosphorylation of IRS-1, leading to the activation of PI3K. Activation of PI3K is required for cellular processes such as protection from apoptosis via protein kinase Akt activation and alteration of intracellular calcium levels via the inositol phosphate cascade; it is also involved in increased translation. The initiation of translation occurs via alterations in the phosphorylation state of the p70 S6 kinase and of eukaryotic initiation factor 4 binding protein.

D. GH and IGF-1 in Aging and Neurodegeneration

The functional impairment of the somatotrophic axis in the brain during aging is an important example of the age-related changes that occur in humans but also in animals (Sonntag *et al.*, 2000a,b). This impairment results in a number of alterations in psychologic capabilities that are seen with increased age. Many of these changes are also included in the spectrum of potential disabilities related to GHD in young subjects. Common deficiencies related to this dysfunction are decreased energy and motivation, disabilities in cognitive function, and decreased mood and well-being; that is, an overall impairment in quality of life (Giordano *et al.*, 2008).

The decline in GH secretion that occurs with increasing age is generally reflected by a concomitant decrease in plasma levels of IGF-1, a reliable marker of GH status (Sherlock and Toogood, 2007). Studies have also demonstrated that the decline in pituitary secretion of GH is paralleled by an age-related decrease in the density of GHRs in the brain (Lai *et al.*, 1993; Nyberg, 2000). The positive influence of GH replacement therapy seen in GHD patients has raised the question of the potential clinical implications of GH replacement at advanced age, as a treatment approach exerting antiaging effects. It has been suggested that reduced IGF-1 support to the neurons is part of the pathological cascade during the degenerative process and contributes to neuronal demise in the majority of neurodegenerative diseases (Trejo *et al.*, 2004). Loss of IGF-1 stimulation of specific neuronal populations is considered to be the cause of certain neurodegenerative diseases and neurodegeneration following trauma to the CNS is shown to negatively affect the somatotrophic axis (Aimaretti and Ghigo, 2005; Behan *et al.*, 2008).

E. Neuroprotective Effects of GH and IGF-1

The concept of neuroprotection includes a variety of processes and components that may be involved in preserving CNS function. Evidence for the neuroprotective effects of GH and IGF-1 is well documented in the literature; knowledge generated in this research area may have an impact on the future treatment of brain or spinal cord trauma. Several review articles on this topic with particular focus on the GH/IGF-1 axis are available (Aberg *et al.*, 2006; Carro *et al.*, 2006; Kreitschmann-Andermahr *et al.*, 2008; Scheepens *et al.*, 2000; Smith 2003). A growing body of evidence suggests that the somatotrophic axis is integrally involved in the growth and development of the normal CNS. Some neuroprotective effects derive from locally produced IGF-1, whereas other effects originate from circulating IGF-1. Several of the neuroprotective effects attributed to GH and IGF-1 are the result of their effects on adult cell genesis (Aberg *et al.*, 2006). For instance, in the hippocampus, IGF-1 increases oligodendrocyte recruitment and newborn cells with the endothelial phenotype. The observed increase in endothelial cell phenotype produced by IGF-1 could explain the increase in cerebral arteriole density observed following GH treatment (Aberg *et al.*, 2006). All these findings seem to stress the role of IGF-1 being a putative regenerative agent in the CNS, whereas GH, which has been less studied in this context, is believed to produce similar effects.

Experimental data have shown (Scheepens *et al.*, 2000) that the IGF-1 gene transcript is induced, primarily within reactive microglia, as a consequence of brain injury, and suggest that the translation product, the IGF-1 protein, could act as a neurotrophic and antiapoptotic substance that acts directly on the stressed cells. In a parallel phase, IGF-1 could act as a prohormone for the generation of the tripeptide glycine–proline–glutamate (GPE) released from the IGF-1 N-terminal, and an additional N-terminal fragment of IGF-1, des-N-(1–3)-IGF-1. Both these IGF-1 fragments possess specific neuroprotective properties. Studies have also revealed that GH administered centrally to juvenile rats 2 h after a hypoxic–ischemic brain injury elicited significant neuroprotection which, in a spatiotemporal pattern, appeared distinct from the IGF-1-induced neuroprotective effect (Scheepens *et al.*, 2000).

Both GH and IGF-1 appear to reduce the outcomes of spinal cord trauma. Topical application of the hormones attenuated trauma-induced edema and cell damage in the rat (Nyberg, 2000; Nyberg and Sharma, 2002). Further, GH has the capacity to improve spinal cord conduction and attenuate edema formation and cell injury in the cord, indicating a potential therapeutic application for this peptide in spinal cord injuries (Winkler *et al.*, 2000).

In human subjects, injury to the CNS attenuates the levels of circulating GH and IGF-1 (for review see, e.g., Nyberg, 2000). However, as both GH and IGF-1 display a wide array of neuroprotective activities, as demonstrated in *in vitro* studies using various cell systems and in experimental animal studies, it is inviting to emphasize the potential therapeutic utility of both GH and IGF-1 in this context. As recent findings indicate that blood-born IGF-1 accounts for an important source of the molecule for brain cells (Carro *et al.*, 2006), and as GH is known to use this factor as a mediator of its effects, the importance of the somatotrophic axis in therapeutic strategies for treatment of various types of CNS injury should be considered for further clinical research.

It is plausible to believe that most of the neuroprotective effects of GH also involve the action of IGF-1. Indeed, many studies confirming a direct neuroprotective effect of IGF-1 have appeared in the literature. Multiple reports have confirmed a strong neuroprotective action of IGF-1 against different proapoptotic insults (for review, see Guan, 2008). Indeed, IGF-1 has been seen as a potential neuroprotective drug for the treatment of stroke and other forms of neural damage, and preclinical studies over the last decade have demonstrated that IGF-1 and some of its fragments can protect against neuronal and glial cell degeneration in animal models of stroke such as hypoxia–ischemia

(Smith, 2008). Also, fragments of IGF-1 have been demonstrated to be potent neuroprotective agents (Guan, 2008). For instance, the N-terminal tripeptide of IGF-1 (GPE) is neuroprotective after central administration, and intravenous infusion of GPE prevents brain injury and also improves long-term functional recovery, with a broad effective dose range in a wide ranged therapeutic window. Brain ischemia also induces the IGF-1 system in damaged regions, and exogenous administration of IGF-1 after injury is neuroprotective and improves long-term neurological function (Guan, 2008). It should also be mentioned that IGF-1 promotes neuronal survival during normal brain development, mainly in the hippocampal and olfactory systems, which depend on postnatal neurogenesis.

F. GH AND IGF-1 EFFECTS ON NEUROGENESIS

A substantial proportion of the impact of the somatotrophic axis on neuro-protection comes from its profound effects on neuroregeneration (neurogenesis) (Isgaard *et al.*, 2007). IGF-1 treatment has been shown to enhance cell genesis in the brains of adult GH- and IGF-1-deficient rodents (Aberg *et al.*, 2009; Anderson *et al.*, 2002). In the hippocampus, following 28 days of treatment with bovine GH (bGH), the number of BrdU/NeuN-positive cells increased proportionally to the increase in the number of BrdU-positive cells. *In vitro* incorporation of (3)H-labeled thymidine revealed that 24 h of bGH exposure was sufficient to increase cell proliferation in adult hippocampal progenitor cells. This finding shows for the first time that peripherally administered GH can increase the number of newborn cells in the adult brain and that the hormone can exert a direct proliferative effect on neuronal progenitor cells *in vitro* (Aberg *et al.*, 2009).

Earlier studies have also demonstrated positive effects of GH on neurogenesis. A study by Harvey and coworkers demonstrated that GH is expressed in the retinal ganglion cells of embryonic chicks, in which the hormone induces cell survival during neurogenesis. The underlying mechanism of this action was investigated in neural retina explants from 6- and 8-day-old embryos that were incubated for 48 h with GH to reduce the number of spontaneously apoptotic cells. This antiapoptotic action of the hormone was accompanied by a reduction in the expression of caspase-3 and, at embryonic day 8, by reduced expression of apoptosis-inducing factor-1, which is caspase independent. These actions were specific, since other components involved in apoptotic signaling (bcl-2, bcl-x, bid, and inhibitor of apoptosis protein-1) were unaffected. The results from this study were suggested to be indicative of caspase-dependent and caspase-independent pathways in GH-induced retinal cell survival (Harvey *et al.*, 2006).

Further studies have suggested that GH treatment of retina ganglion cells reduces Akt levels, while raising Akt-phos levels, consistent with a role for Akt signaling pathways in GH neuroprotective effects (Sanders *et al.*, 2009a). GH stimulation of GHRs involves downstream intracellular Trk pathways. The Akt and Trk pathways seem to depend on the activation of cAMP response element-binding protein (CREB), which can initiate transcription of pro- or antiapoptotic genes. From this it has been suggested that the neuroprotective effects of GH on embryonic retinal ganglion cells involve pathways common to those used by other neurotrophins, and that GH may act as a growth and differentiation factor in the development of the embryonic retina. Sanders and coworkers have also examined the relationship between the overlapping antiapoptotic effects of GH and IGF-1 and have observed that simultaneous immunoneutralization of GH and IGF-1 did not increase the level of apoptosis in the cultures above that were achieved by immunoneutralization of GH alone. On this basis, they concluded that the neuroprotective effects of GH on the developing retina are, to a significant extent, mediated through the action of IGF-1 (Sanders *et al.*, 2009b). In additional studies, these investigators examined the neuroprotective effects of GH using a quail embryo neural retina cell line (QNR/D) treated with GH siRNA to downregulate the local production of the hormone. They demonstrated that knock-down of GH by so-called gene silencing in cells of this cultured embryonic neural retina cell line correlated with the increased appearance of cells with apoptotic nuclear morphology in their cultures (Sanders *et al.*, 2009c).

Recent studies (for review, see Nyberg, 2000) have shown that GH targets many areas of the CNS, and GH deficiency has been associated with cognitive impairment, memory loss, and diminished overall well-being (Bengtsson *et al.*, 1993; Björk *et al.*, 1989; Burman and Deijen, 1998). GH replacement therapy ameliorates several of the adverse symptoms seen in patients with GHD (Bengtsson *et al.*, 1993; Björk *et al.*, 1989; Burman and Deijen, 1998; McMillan *et al.*, 2003). The hormone also prevents neuronal loss in the aged rat hippocampus, indicating a neuroprotective effect of GH on old animals (Azcoitia *et al.*, 2005). Levels of circulating GH and the density of GH-binding sites decrease with age in several areas of the human brain, including the hippocampus (Lai *et al.*, 1993; van Dam *et al.*, 2000). GH also increases the expression of the rat hippocampal gene transcript for the NR2B (Le Grevés *et al.*, 2002), which is known to enhance memory and cognitive capabilities in an age-dependent manner when overexpressed. Moreover, it was also demonstrated that GH replacement in hypophysectomized male rats improves spatial performance in the water maze test and increases the hippocampal gene transcript levels of NMDA receptor subunits and postsynaptic density protein 95 (Le Grevés *et al.*, 2006). Beneficial effects on animal performance have also been confirmed by studies using other experimental approaches (Fig. 3). Taken together, these findings suggest a link between lowered levels of GH and deterioration of cognitive function in the elderly, with a clear indication that the hormone could improve memory and cognitive capabilities and that this could be indicative of increased neurogenesis as a result of GH administration.

Effects of GH in neonatally hypophysis-ectomized male
rats on the time latency measures in the 8-armed maze

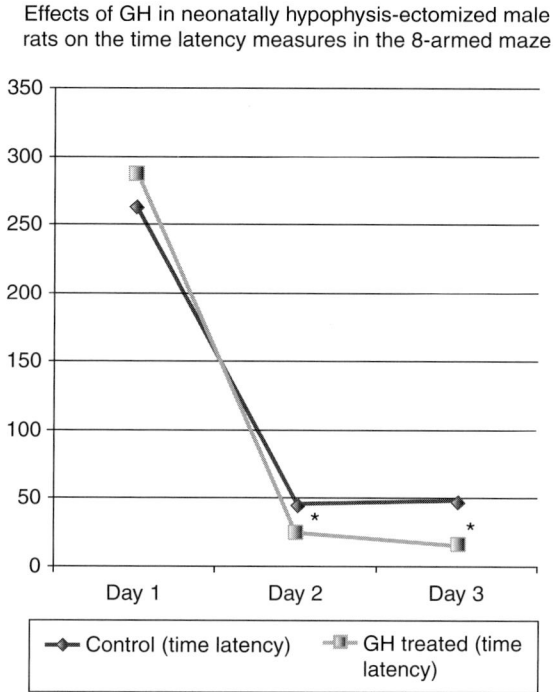

FIG. 3. Effects of human recombinant GH on memory function as recorded from studies using an eight-armed maze. Male rats neonatally hypophysis-ectomized and were treated with GH at an age of 52 weeks (data from Le Grevés *et al.*, 2000).

The details of the mechanism by which GH induces its beneficial effects on memory and cognition are still not clear. However, GH is known to promote nerve cell regeneration as well as gliogenesis during brain development in the fetal rat (Ajo *et al.*, 2003), presumably through local production of IGF-1. Peripherally administered GH reaching the CNS could induce the release of IGF-1 in the brain and IGF-1 could, in turn, account for mediation of the brain effects of GH. However, local production of GH and IGF-1 has also been suggested in certain brain regions, as mice with circulating GH and IGF-1 deficiency have normal levels of corresponding mRNAs in the hippocampus (Sun *et al.*, 2005). Also, GH is produced in the hippocampal formation, where it may be involved in functions associated with his region, such as learning and response to stress (Donahue *et al.*, 2006). Effects on these behaviors may be caused by the action of GH-induced release of IGF-1, as this mediator also affects hippocampal-related behavior. In fact, intracerebroventricular infusions of IGF-1 have been shown to reduce age-related decline in hippocampal neurogenesis in rats (Lichtenwalner *et al.*,

2001). Peripheral infusions of IGF-1 also induce neurogenesis in the adult rat hippocampus (Aberg *et al.*, 2000) and overexpression of IGF-1 promotes neurogenesis during postnatal development (O'Kusky *et al.*, 2000).

GH administration reduces circulating tumor necrosis factor-α (TNF-α) system and soluble apoptosis mediators in patients with idiopathic dilated cardiomyopathy (IDC). These GH-induced antiapoptotic effects may be associated with improvement in exercise capacity as well as with reversing left ventricular remodeling in patients with congestive heart failure and IDC (Parissis *et al.*, 2005).

G. NEURODEGENERATION IN THE ADDICTIVE BRAIN

Over the last decade, studies on the effects on the brain of abusing alcohol, central stimulants, and opiates have revealed adverse effects on many functions related to the CNS. While the extent and severity of the contribution of drugs of abuse to accelerated senescence remain uncertain, these putative aging effects add up to the dark side of drug addiction and will undoubtedly require a strong research effort in the near future (Carvalho, 2009). A variety of neuropathologic changes have been encountered in the brains of heroin abusers. These include complications such as hypoxic–ischemic changes with cerebral edema, ischemic neuronal damage and neuronal loss (Büttner *et al.*, 2000).

An observation that has received special attention during recent years is that chronic drug users display pronounced neuropsychologic impairment in the domains of executive and memory function (Ersche *et al.*, 2006). Emerging data from recent studies provide additional evidence for the cognitive impairment of alcohol-dependent patients with regard to tasks sensitive to frontal lobe function, and underline the importance of abstinence to allow these impairments to recover (Glass *et al.*, 2009; Loeber *et al.*, 2009).

Chronic exposure to opiates, such as heroin and morphine, impairs cognitive function (Eisch *et al.*, 2000). In fact, heroin is one of the most commonly abused illegal drugs, and abuse of this drug is linked to attention deficits and poor performance on memory tasks compared to controls (Guerra *et al.*, 1987). Chronic exposure to morphine also causes vigilance and attention to deteriorate in chronic pain patients (Mao *et al.*, 2002) and impairs acquisition of reference memory in rats (Spain and Newsom, 1991). These findings suggest an effect of chronic opiate stimulation on brain structures related to memory and learning, such as the hippocampus. The same appears true for several other drugs of abuse, including alcohol and central stimulants.

Earlier, drugs of abuse were believed only to induce nonspecific effects on the brain; however, it is now known that they produce selective neuroadaptations in particular brain regions. These neuroadaptations have been closely examined for clues to the development, maintenance, and treatment of addiction. The hippocampus is an area of particular interest, as it is central to many aspects of the addictive process, including relapse. A recently discovered hippocampal neuro-adaptation produced by drugs as diverse as opiates and psychostimulants involves decreased neurogenesis in the subgranular zone (SGZ). While the role of adult-generated neurons is not clear, their functional integration into hippocampal circuitry raises the possibility that decreased adult SGZ neurogenesis could alter hippocampal function in such a way as to maintain addictive behavior or contrib-ute to relapse (Arguello *et al.*, 2008). We have previously shown that a single dose of morphine affect the expression of GHR and GHBP in the rat hippocampus (Fig. 4). The gene transcripts were significantly attenuated 4 h following drug injection but was restored after 24 h. In rats subjected to chronic treatment with the opiate, a decrease in GH binding was observed during the acute phase but was restored when animals were tolerant to the drug (Zhai *et al.*, 1995).

Chronic morphine also decreases neurogenesis in the adult rat hippocampal granule cell layer and a similar effect has been seen in rats after chronic self-administration of heroin (Eisch *et al.*, 2000). Studies indicated that opiate regula-tion of nerve cell regeneration was not mediated by changes in circulating levels of glucocorticoids, as similar effects were observed in rats subjected to adrenalecto-my and subsequent corticosterone replacement. These findings suggest that opiate regulation of neurogenesis in the adult rat hippocampus may be one

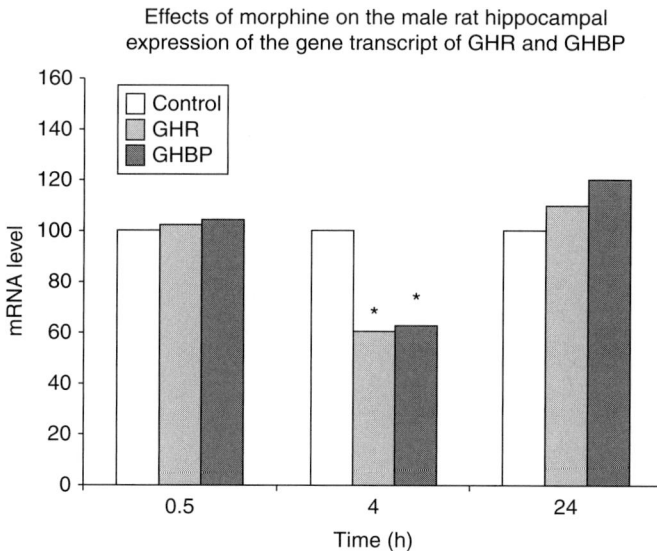

FIG. 4. Effects of a single dose of morphine on the expression of GHR and GHBP gene transcripts over time in the hippocampus of male rats. *$p < 0.05$ (data from Thornwall-Le Grevés *et al.*, 2001).

mechanism by which exposure to opioid drugs influences hippocampal function. The recent study by Arguello and coworkers mentioned above demonstrated that chronic morphine decreases SGZ neurogenesis by inhibiting dividing cells, particularly those in the S-phase, and progenitor cell progression to a more mature neuronal stage (Arguello *et al.*, 2008).

Moreover, a recent study has shown that chronic morphine significantly and dose-dependently reduces neuronal cell density in cultured hippocampal cells from the mouse fetus (Svensson *et al.*, 2008). The ability of morphine to inhibit cell growth and induce apoptosis is also known from earlier work. The opiate elicited apoptosis in human fetal microglia and neurons (Hu *et al.*, 2002), and apoptosis was also associated with morphine tolerance (Mao *et al.*, 2002). The apoptotic effect of morphine is blocked by the opioid receptor antagonist naloxone (Hu *et al.*, 2002), indicating an opioid receptor mechanism involved in this effect. The effect of morphine is known to be mediated mainly through the μ-opioid peptide (MOP) receptor although, at high concentrations, this opiate can also interact with both δ-opioid (DOP) and κ-opioid (KOP) receptors. Moreover, it seems that the various opioid receptor types (MOP, DOP, and KOP) may regulate different aspects of neuronal development (Hauser *et al.*, 2000). Evidence indicating that the MOP receptor could play an important role in regulating progenitor cell survival was recently described (Harburg *et al.*, 2007). In addition, morphine was earlier shown to promote abnormal programmed cell death by enhancing the expression of the proapoptotic Fas receptor protein and attenuating the expression of the antiapoptotic Bcl-2 oncoprotein through the sustained activation of opioid receptors (Boronat *et al.*, 2001). Studies also indicated that opiate-induced alteration of hippocampal function likely results from inhibited neurogenesis (Eisch and Harburg, 2006). Therefore, the recently observed decrease in neuronal cell density seen in the mouse hippocampal primary cell cultures (Svensson *et al.*, 2008) was expected, and a consequence of this reduction should be that markers of apoptosis, such as lactate dehydrogenase (LDH) and caspase-3 activity, will be affected; in fact, both of these enzymes have been found to be significantly enhanced (Svensson *et al.*, 2008) (Fig. 5).

This enhanced release of LDH in morphine-treated hippocampal cells provides further support that morphine can induce apoptosis in the cells of this brain area. LDH, a mitochondrial dehydrogenase, represents a critical component of the astrocyte–neuron lactate shuttle. It controls the formation and may regulate the turnover of lactate in the cells. Caspase-3 is another enzyme that serves as a marker of apoptosis; cleaved caspase-3 is an activated form of caspase-3 that acts as a lethal protease at the most distal stage of the apoptotic pathway (Kuribayashi *et al.*, 2006). This enzyme was also examined to find out whether the observed reduction in neuronal cell density involved elements related to apoptosis. It was noted that the level of cleaved caspase-3, as assessed by Western blot analysis, was significantly increased by morphine (Svensson *et al.*, 2008).

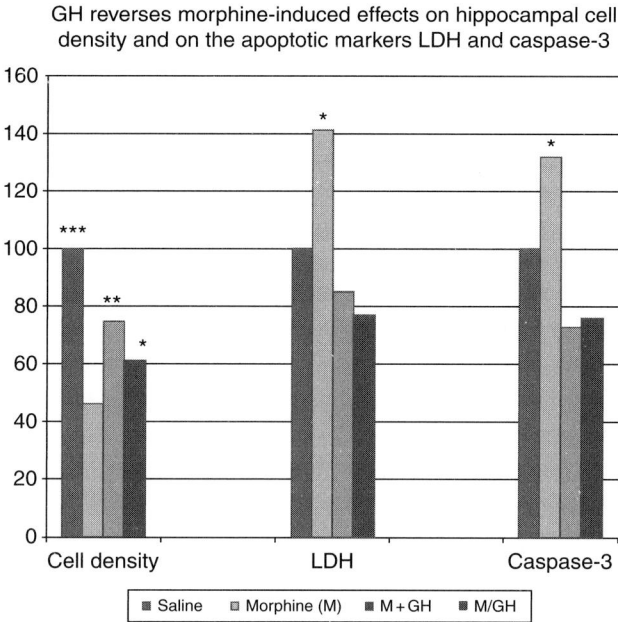

FIG. 5. Reversal by human recombinant GH of morphine-induced effects on hippocampal primary culture cells from mouse fetus. The cell density as well as the apoptotic markers lactodedy-drogenase (LDH) and caspase-3 activity are expressed in percentage of saline-treated control cells (data from Svensson et al., 2008).

The hippocampus is located in the limbic area of the brain. This area is well known as an important brain substrate required for the acquisition of declarative or explicit memory (Benfenati, 2007). From the literature cited above, it is evident that chronic administration of opiates can inhibit cell growth and trigger apoptosis, but it is also suggested that opiate-induced toxicity can include impaired regeneration of nerve cells (Eisch and Harburg, 2006; Hauser et al., 2000; Hu et al., 2002; Mao et al., 2002). This impact of adult-generated neurons on memory and learning has been suggested because training on associative learning tasks doubles these neurons in the rat dentate gyrus (Gould et al., 1999). Consequently, memory dysfunctions induced by opiate exposure could be caused by decreased adult neurogenesis as opiates can inhibit neurogenesis in the adult rat hippocampus (Eisch and Harburg, 2006; Eisch et al., 2000). The inhibition of neurogenesis could be related to decreased neural precursors; that is, increased apoptosis of newborn neurons. In recent years, several factors that may increase neurogenesis from preexisting neuronal precursors have been described. For instance, IGF-1 is essential for hippocampal neurogenesis (Aberg *et al.*, 2000). This factor is under the control of the somatotrophic axis, where GH has an important role as an activator and releaser of IGF-1 and its binding proteins.

III. Reversal of Drug-Induced Effects by GH

Studies of the reversal of the adverse effects induced by various drugs have shown that certain growth factors may be useful in attempts to counteract drug-induced cell damage and apoptosis. For instance, Gibson *et al.* (2002) demonstrated that stimulation of human embryonic kidney cells HEK 293 and the breast cancer cell line MDA MB 231 with epidermal growth factor (EGF) effectively and dose-dependently protected these cells from tumor necrosis factor-related apoptosis-inducing ligand (TRAIL)-induced apoptosis. This stimulation blocked apoptosis by inhibiting both TRAIL-mediated cytochrome *c* release from the mitochondria and caspase-3-like activation. It was further shown that the EGF survival response involved the activation of Akt. Expression of activated Akt was sufficient to block TRAIL-induced apoptosis, and kinase-inactive Akt expression blocked the EGF protective response. In contrast, inhibition of EGF stimulation of extracellular-regulated kinase (ERK) activity did not affect EGF protection. It was concluded from these findings that EGF receptor activation gives rise to a survival response against TRAIL-induced apoptosis by inhibiting mitochondrial cytochrome *c* release that is mediated by Akt activation in epithelial-derived cells. In addition, heparin-binding epidermal growth factor (HB-EGF), also a member of the EGF family of growth factors, has been reported to prevent apoptosis and differentiation and, in a very recent study, it was shown that stimulation with HB-EGF could reverse alcohol-induced apoptosis in human embryonic stem cells (Nash *et al.*, 2009). Another possibility for reversing alcohol-induced cell damage involves brain-derived neurotrophic factor (BDNF). BDNF signaling plays an important role in neural survival and differentiation and studies have shown that alcohol significantly reduces BDNF signaling in neuronal cells (Climent *et al.*, 2002). Also, the antiproliferative action of ethanol can be modulated by changing the sensitivity of the autophosphorylation of the IGF-1R to ethanol (Seiler *et al.*, 2000). This raised the question of whether IGF-1 could counteract the antiproliferative effects induced by alcohol. In fact, studies have shown that alcohol inhibits differentiation of the neural stem and that this effect is reduced by both IGF-1 and BDNF (Tateno *et al.*, 2004). These results suggest the possibility that stimulation of neurotrophic factor signaling can reverse apoptosis induced by alcohol exposure.

Attempts have also been made to use growth factors to reverse opioid-induced cell damage. HB-EGF also counteracts the opiate-induced inhibition of cell proliferation in brain cells and enhances DNA cell replication, suggesting a neuroprotective effect in opiate drug abuse (Opanashuk and Hauser, 1998).

Another study indicated that morphine-induced macrophage apoptosis is mediated through downstream signaling involving TGF-β and NO production (Bhat *et al.*, 2004). Moreover, NADPH oxidation activation involving phospholipase D and Ca^{2+} occurs, leading to the generation of superoxide. In *in vivo* studies, administration of *N*-acetyl cysteine and preinduction of heme oxygenase activity and epoetin-α prevented morphine-induced peritoneal macrophage apoptosis, thus further confirming the role of oxidative stress in morphine-induced macrophage apoptosis.

As mentioned above, opiates exert an inhibitory effect on neurogenesis, and previous studies have shown that morphine treatment decreases GH-binding site density in the rat hypothalamus in the acute phase of administration (Zhai *et al.*, 1995). A decrease in the levels of the GHR gene transcript levels in the rat hippocampus following a single dose of morphine (Fig. 4) has also been reported (Thornwall-Le Grevés *et al.*, 2001). Thus, to examine whether GH can counteract or reverse opiate-induced apoptosis or inhibition of neurogenesis, we investigated the effect of morphine and recombinant human GH (rhGH) on primary hippocampal neuronal cell cultures derived from fetal mice (Svensson *et al.*, 2008). The cells were treated with morphine and rhGH, and the neuronal cell density was recorded through microscopic studies. To investigate the effects of morphine and rhGH on cell viability, we also measured the release of LDH and caspase-3 activity, both well known as markers of cell death and apoptosis (Kajta *et al.*, 2005; Rami, 2003). We observed that rhGH significantly reversed the morphine-induced inhibition of neurite outgrowth and that cell density was restored after treatment with the hormone. We observed these effects of rhGH both when the hormone was added with morphine and when it was added after the opiate had induced its effect (Fig. 5). We further noted that the hormone was able to reverse the morphine-induced effects on the markers of apoptosis (LDH and caspase-3 activity) (Svensson *et al.*, 2008). Thus, combining these observations with the effects of GH on memory and spatial performance in rats and humans, it appears that this hormone may be of use for the reversal of the adverse effects of morphine or other opiates in humans.

IV. Conclusion

It is evident from this discussion that GH and its mediator IGF-1 can produce profound effects on brain function. The two hormones interact with the CNS at different levels. While they have many effects in common, the available data suggest that there are also some differences. These differences may be of more importance in short-term treatments. Further, GH exerts many effects through

IGF-1 but current data also suggest that it also elicits effects without its mediator. There are target tissues containing receptors for both GH and IGF-1 in a variety of brain regions. Both are important factors in the development and differentiation of the CNS and both seem to induce beneficial effects on memory and cognitive disabilities. Also, both GH and IGF-1 produce protective effects in neurodegeneration and trauma to the CNS. Recent studies suggest an important role for both GH and IGF-1 in the promotion of regeneration of nerve cells, both as exogenously administered agents and as essential components involved in mechanisms promoting stimulation of neurogenesis by other modes, such as exercise and enriched environments. The most important issue in this regard is to find routes for the implementation of GH and IGF-1 in therapeutic approaches to treat disorders related to impairment of the somatotrophic axis.

Acknowledgments

This work was supported by the Swedish Medical Research Council (grant 9459).

References

Aberg, M. A., Aberg, N. D., Hedbäcker, H., Oscarsson, J., and Eriksson, P. S. (2000). Peripheral infusion of IGF-I selectively induces neurogenesis in the adult rat hippocampus. *J. Neurosci.* **20**(8), 2896–2903.

Aberg, N. D., Brywe, K. G., and Isgaard, J. (2006). Aspects of growth hormone and insulin-like growth factor-I related to neuroprotection, regeneration, and functional plasticity in the adult brain. *Sci. World J.* **6,** 53–80 (Review).

Aberg, N. D., Johansson, I., Aberg, M. A., Lind, J., Johansson, U. E., Cooper-Kuhn, C. M., Kuhn, H. G., and Isgaard, J. (2009). Peripheral administration of GH induces cell proliferation in the brain of adult hypophysectomized rats. *J. Endocrinol.* **201**(1), 141–150.

Aimaretti, G., and Ghigo, E. (2005). Traumatic brain injury and hypopituitarism. *Sci. World J.* **5,** 777–781 (Review).

Ajo, R., Cacicedo, L., Navarro, C., and Sanchez-Franco, F. (2003). Growth hormone action on proliferation and differentiation of cerebral cortical cells from fetal rat. *Endocrinology* **144**(3), 1086–1097.

Anderson, M. F., Aberg, M. A., Nilsson, M., and Eriksson, P. S. (2002). Insulin-like growth factor-I and neurogenesis in the adult mammalian brain. *Brain Res. Dev. Brain Res.* **134**(1–2), 115–122.

Andersson, K., Fuxe, K., Eneroth, P., Isaksson, O., Nyberg, F., and Roos, P. (1983). Rat growth hormone and hypothalamic catecholamine nerve terminal systems. Evidence for rapid and discrete reductions in dopamine and noradrenaline levels and turnover in the median eminence of the hypophysectomized male rat. *Eur. J. Pharmacol.* **95**(3–4), 271–275.

Araujo, D. M., Lapchak, P. A., Collier, B., Chabot, J. G., and Quirion, R. (1989). Insulin-like growth factor-1 (somatomedin-C) receptors in the rat brain: distribution and interaction with the hippocampal cholinergic system. *Brain Res.* **484**(1–2), 130–138.

Arguello, A. A., Harburg, G. C., Schonborn, J. R., Mandyam, C. D., Yamaguchi, M., and Eisch, A. J. (2008). Time course of morphine's effects on adult hippocampal subgranular zone reveals preferential inhibition of cells in S phase of the cell cycle and a subpopulation of immature neurons. *Neuroscience* **157**(1), 70–79.

Azcoitia, I., Perez-Martin, M., Salazar, V., Castillo, C., Ariznavarreta, C., Garcia-Segura, L. M., and Tresguerres, J. A. (2005). Growth hormone prevents neuronal loss in the aged rat hippocampus. *Neurobiol. Aging* **26**(5), 697–703.

Behan, L. A., Phillips, J., Thompson, C. J., and Agha, A. (2008). Neuroendocrine disorders after traumatic brain injury. *J. Neurol. Neurosurg. Psychiatry* **79**(7), 753–759 (Review).

Benfenati, F. (2007). Synaptic plasticity and the neurobiology of learning and memory. *Acta Biomed.* **78**(Suppl. 1), 58–66.

Bengtsson, B. A., Edén, S., Lönn, L., Kvist, H., Stokland, A., Lindstedt, G., Bosaeus, I., Tölli, J., Sjöström, L., and Isaksson, O. G. (1993). Treatment of adults with growth hormone (GH) deficiency with recombinant human GH. *J. Clin. Endocrinol. Metab.* **76**(2), 309–317.

Bennett, E., McGuinness, L., Gevers, E. F., Thomas, G. B., Robinson, I. C., Davey, H. W., and Luckman, S. M. (2005). Hypothalamic STAT proteins: Regulation of somatostatin neurones by growth hormone via STAT5b. *J. Neuroendocrinol.* **17**(3), 186–194.

Bhat, R. S., Bhaskaran, M., Mongia, A., Hitosugi, N., and Singhal, P. C. (2004). Morphine-induced macrophage apoptosis: Oxidative stress and strategies for modulation. *J. Leukoc. Biol.* **75**(6), 1131–1138.

Björk, S., Jönsson, B., Westphal, O., and Levin, J. E. (1989). Quality of life of adults with growth hormone deficiency: A controlled study. *Acta Paediatr. Scand. Suppl.* **356,** 55–59 discussion 60, 73–74.

Bohannon, N. J., Corp, E. S., Wilcox, B. J., Figlewicz, D. P., Dorsa, D. M., and Baskin, D. G. (1988). Localization of binding sites for insulin-like growth factor-I (IGF-I) in the rat brain by quantitative autoradiography. *Brain Res.* **444**(2), 205–213.

Boronat, M. A., García-Fuster, M. J., and García-Sevilla, J. A. (2001). Chronic morphine induces up-regulation of the pro-apoptotic Fas receptor and down-regulation of the anti-apoptotic Bcl-2 oncoprotein in rat brain. *Br. J. Pharmacol.* **134**(6), 1263–1270.

Burman, P., Hetta, J., Wide, L., Månsson, J. E., Ekman, R., and Karlsson, F. A. (1996). Growth hormone treatment affects brain neurotransmitters and thyroxine [see comment]. *Clin. Endocrinol. (Oxf).* **44**(3), 319–324.

Burman, P., and Deijen, J. B. (1998). Quality of life and cognitive function in patients with pituitary insufficiency. *Psychother. Psychosom.* **67**(3), 154–167 (Review).

Burton, K. A., Kabigting, E. B., Clifton, D. K., and Steiner, R. A. (1992). Growth hormone receptor messenger ribonucleic acid distribution in the adult male rat brain and its colocalization in hypothalamic somatostatin neurons. *Endocrinology* **131**(2), 958–963.

Büttner, A., Mall, G., Penning, R., and Weis, S. (2000). The neuropathology of heroin abuse. *Forensic Sci. Int.* **113**(1–3), 435–442.

Carro, E., Trejo, J. L., Fernandez, S., Fernandez, A. M., and Torres-Aleman, I. (2006). Insulin-like growth factor-1 and neuroprotection. *In* "The Somatotrophic Axis in Brain Function" (F. Nyberg, Ed.), pp. 209–215. Elsevier Academic Press, San Diego, CA.

Carvalho, F. (2009). How bad is accelerated senescence in consumers of drugs of abuse? *Adicciones* **21**(2), 99–104.

Climent, E., Pascual, M., Renau-Piqueras, J., and Guerri, C. (2002). Ethanol exposure enhances cell death in the developing cerebral cortex: Role of brain-derived neurotrophic factor and its signaling pathways. *J. Neurosci. Res.* **68**(2), 213–225.

Coculescu, M. (1999). Blood–brain barrier for human growth hormone and insulin-like growth factor-I. *J. Pediatr. Endocrinol. Metab.* **12**(2), 113–124 (Review).

Creyghton, W. M., van Dam, P. S., and Koppeschaar, H. P. (2004). The role of the somatotrophic system in cognition and other cerebral functions. *Semin. Vasc. Med.* **4**(2), 167–172 (Review).

Donahue, C. P., Kosik, K. S., and Shors, T. J. (2006). Growth hormone is produced within the hippocampus where it responds to age, sex, and stress. *Proc. Natl. Acad. Sci. USA* **103**(15), 6031–6036.

Eisch, A. J., and Harburg, G. C. (2006). Opiates, psychostimulants, and adult hippocampal neurogenesis: Insights for addiction and stem cell biology. *Hippocampus* **16**(3), 271–286 (Review).

Eisch, A. J., Barrot, M., Schad, C. A., Self, D. W., and Nestler, E. J. (2000). Opiates inhibit neurogenesis in the adult rat hippocampus. *Proc. Natl. Acad. Sci. USA* **97**(13), 7579–7584.

Ersche, K. D., Clark, L., London, M., Robbins, T. W., and Sahakian, B. J. (2006). Profile of executive and memory function associated with amphetamine and opiate dependence. *Neuropsychopharmacology* **31**(5), 1036–1047.

Falleti, M. G., Maruff, P., Burman, P., and Harris, A. (2006). The effects of growth hormone (GH) deficiency and GH replacement on cognitive performance in adults: A meta-analysis of the current literature. *Psychoneuroendocrinology* **31**(6), 681–691.

Ganong, W. F. (2000). Circumventricular organs: Definition and role in the regulation of endocrine and autonomic function. *Clin. Exp. Pharmacol. Physiol.* **27**(5–6), 422–427 (Review).

Gibson, E. M., Henson, E. S., Haney, N., Villanueva, J., and Gibson, S. B. (2002). Epidermal growth factor protects epithelial-derived cells from tumor necrosis factor-related apoptosis-inducing ligand-induced apoptosis by inhibiting cytochrome *c* release. *Cancer Res.* **62**(2), 488–496.

Giordano, R., Bonelli, L., Marinazzo, E., Ghigo, E., and Arvat, E. (2008). Growth hormone treatment in human ageing: Benefits and risks. *Hormones (Athens)* **7**(2), 133–139 (Review).

Glass, J. M., Buu, A., Adams, K. M., Nigg, J. T., Puttler, L. I., Jester, J. M., and Zucker, R. A. (2009). Effects of alcoholism severity and smoking on executive neurocognitive function. *Addiction* **104**(1), 38–48.

Gould, E., Beylin, A., Tanapat, P., Reeves, A., and Shors, T. J. (1999). Learning enhances adult neurogenesis in the hippocampal formation. *Nat. Neurosci.* **2**(3), 260–265.

Gualco, E., Wang, J. Y., Del Valle, L., Urbanska, K., Peruzzi, F., Khalili, K., Amini, S., and Reiss, K. (2009). IGF-IR in neuroprotection and brain tumors. *Front. Biosci.* **14**, 352–375 (Review).

Guan, J. (2008). Insulin-like growth factor-1 and its derivatives: Potential pharmaceutical application for ischemic brain injury. *Recent Pat. CNS Drug. Discov.* **3**(2), 112–127 (Review).

Guerra, D., Sole, A., Cami, J., and Tobena, A. (1987). Neuropsychological performance in opiate addicts after rapid detoxification. *Drug Alcohol Depend.* **20**, 261–270.

Han, V. K. (1995). Is the central nervous system a target for growth hormone and insulin-like growth factors? *Acta Paediatr. Suppl.* **411**, 3–8 (Review).

Harburg, G. C., Hall, F. S., Harrist, A. V., Sora, I., Uhl, G. R., and Eisch, A. J. (2007). Knockout of the μ opioid receptor enhances the survival of adult-generated hippocampal granule cell neurons. *Neuroscience* **144**(1), 77–87.

Harvey, S., Baudet, M. L., and Sanders, E. J. (2006). Growth hormone and cell survival in the neural retina: Caspase dependence and independence. *Neuroreport* **17**(16), 1715–1718.

Hauser, K. F., Houdi, A. A., Turbek, C. S., Elde, R. P., and Maxson, W. III (2000). Opioids intrinsically inhibit the genesis of mouse cerebellar granule neuron precursors *in vitro*: Differential impact of mu and delta receptor activation on proliferation and neurite elongation. *Eur. J. Neurosci.* **12**(4), 1281–1293.

Hojvat, S., Baker, G., Kirsteins, L., and Lawrence, A. M. (1982). Growth hormone (GH) immunoreactivity in the rodent and primate CNS: Distribution, characterization and presence posthypophysectomy. *Brain Res.* **239**(2), 543–557.

Hu, S., Sheng, W. S., Lokensgard, J. R., and Peterson, P. K. (2002). Morphine induces apoptosis of human microglia and neurons. *Neuropharmacology* **42**, 829–836.

Isgaard, J., Aberg, D., and Nilsson, M. (2007). Protective and regenerative effects of the GH/IGF-I axis on the brain. *Minerva Endocrinol.* **32**(2), 103–113 (Review).

Johansson, J. O., Larson, G., Andersson, M., Elmgren, A., Hynsjo, L., Lindahl, A., Lundberg, P. A., Isaksson, O. G., Lindstedt, S., and Bengtsson, B. A. (1995). Treatment of growth hormone-deficient adults with recombinant human growth hormone increases the concentration of growth hormone in the cerebrospinal fluid and affects neurotransmitters. *Neuroendocrinology* **61**(1), 57–66.

Kajta, M., Trotter, A., Lasoń, W., and Beyer, C. (2005). Effect of NMDA on staurosporine-induced activation of caspase-3 and LDH release in mouse neocortical and hippocampal cells. *Brain Res. Dev. Brain Res.* **160**(1), 40–52.

Kasagi, Y., Tokita, R., Nakata, T., Imaki, T., and Minami, S. (2004). Human growth hormone induces SOCS3 and CIS mRNA increase in the hypothalamic neurons of hypophysectomized rats. *Endocr. J.* **51**(2), 145–154.

Kastin, A. J., Akerstrom, V., and Pan, W. (2001). Validity of multiple-time regression analysis in measurement of tritiated and iodinated leptin crossing the blood–brain barrier: Meaningful controls. *Peptides* **22**(12), 2127–2136.

Kim, B., and Feldman, E. L. (2009). Insulin receptor substrate (IRS)-2, not IRS-1, protects human neuroblastoma cells against apoptosis. *Apoptosis* **14**(5), 665–673.

Kreitschmann-Andermahr, I., Poll, E. M., Reineke, A., Gilsbach, J. M., Brabant, G., Buchfelder, M., Fassbender, W., Faust, M., Kann, P. H., and Wallaschofski, H. (2008). Growth hormone deficient patients after traumatic brain injury—Baseline characteristics and benefits after growth hormone replacement—An analysis of the German KIMS database. *Growth Horm. IGF Res.* **18**(6), 472–478.

Kuribayashi, K., Mayes, P. A., and El-Deiry, W. S. (2006). What are caspases 3 and 7 doing upstream of the mitochondria? *Cancer Biol. Ther.* **5**(7), 763–765 (Review).

Lai, Z., Emtner, M., Roos, P., and Nyberg, F. (1991). Characterization of putative growth hormone receptors in human choroid plexus. *Brain Res.* **546**(2), 222–226.

Lai, Z., Roos, P., Olsson, Y., Larsson, C., and Nyberg, F. (1992). Characterization of prolactin receptors in human choroid plexus. *Neuroendocrinology* **56**(2), 225–233.

Lai, Z., Roos, P., Zhai, O., Olsson, Y., Fholenhag, K., Larsson, C., and Nyberg, F. (1993). Age-related reduction of human growth hormone-binding sites in the human brain. *Brain Res.* **621**(2), 260–266.

Laron, Z. (2009). Insulin and the brain. *Arch. Physiol. Biochem.* **115**(2), 112–116.

Laursen, T., Jørgensen, J. O., and Christiansen, J. S. (2008). The management of adult growth hormone deficiency syndrome. *Expert Opin. Pharmacother.* **9**(14), 2435–2450 (Review).

Laviola, L., Natalicchio, A., and Giorgino, F. (2007). The IGF-I signaling pathway. *Curr. Pharm. Des.* **13**(7), 663–669 (Review).

Le Grevés, M. (2006). Growth hormone receptor message in the rat and human central nervous system. *In* "The Somatotrophic Axis in Brain Function" (F. Nyberg, Ed.), pp. 99–107. Elsevier Academic Press, San Diego, CA.

Le Grevés, M., Fhölenhag, K., Zhou, Q., Berg, M., Meyerson, B., and Nyberg, F. (2000). Effects of recombinant human growth hormone on learning and memory function. *Growth Horm. IGF Res.* **10**, 130.

Le Grevés, M., Steensland, P., Le Grevés, P., and Nyberg, F. (2002). Growth hormone induces age-dependent alteration in the expression of hippocampal growth hormone receptor and *N*-methyl-D-aspartate receptor subunits gene transcripts in male rats. *Proc. Natl. Acad. Sci. USA* **99**(10), 7119–7123.

Le Grevés, M., Le Grevés, P., and Nyberg, F. (2005). Age-related effects of IGF-1 on the NMDA-, GH- and IGF-1-receptor mRNA transcripts in the rat hippocampus. *Brain Res. Bull.* **65**(5), 369–374.

Le Greves, M., Zhou, Q., Berg, M., Le Greves, P., Fholenhag, K., Meyerson, B., and Nyberg, F. (2006). Growth hormone replacement in hypophysectomized rats affects spatial performance and hippocampal levels of NMDA receptor subunit and PSD-95 gene transcript levels. *Exp. Brain Res.* **173**(2), 267–273.

Lichtenwalner, R. J., Forbes, M. E., Bennett, S. A., Lynch, C. D., Sonntag, W. E., and Riddle, D. R. (2001). Intracerebroventricular infusion of insulin-like growth factor-I ameliorates the age-related decline in hippocampal neurogenesis. *Neuroscience* **107**(4), 603–613.

Lobie, P. E., García-Aragón, J., Lincoln, D. T., Barnard, R., Wilcox, J. N., and Waters, M. J. (1993). Localization and ontogeny of growth hormone receptor gene expression in the central nervous system. *Brain Res. Dev. Brain Res.* **74**(2), 225–233.

Mao, J., Sung, B., Ji, R. R., and Lim, G. (2002). Neuronal apoptosis associated with morphine tolerance: Evidence for an opioid-induced neurotoxic mechanism. *J. Neurosci.* **22**, 7650–7661.

McMillan, C. V., Bradley, C., Gibney, J., Healy, M. L., Russell-Jones, D. L., and Sönksen, P. H. (2003). Psychological effects of withdrawal of growth hormone therapy from adults with growth hormone deficiency. *Clin. Endocrinol. (Oxford)* **59**(4), 467–475.

Melmed, S., Yamashita, S., Yamasaki, H., Fagin, J., Namba, H., Yamamoto, H., Weber, M., Morita, S., Webster, J., and Prager, D. (1996). IGF-I receptor signalling: Lessons from the somatotroph. *Recent Prog. Horm. Res.* **51**, 189–215; discussion 215–216 (Review).

Mustafa, A., Adem, A., Roos, P., and Nyberg, F. (1994a). Sex differences in binding of human growth hormone to rat brain. *Neurosci. Res.* **19**(1), 93–99.

Mustafa, A., Nyberg, F., Bogdanovic, N., Islam, A., Roos, P., and Adem, A. (1994b). Somatogenic and lactogenic binding sites in rat brain and liver: Quantitative autoradiographic localization. *Neurosci. Res.* **20**(3), 257–263.

Nash, R. J., Heimburg-Molinaro, J., and Nash, R. J. (2009). Heparin binding epidermal growth factor-like growth factor reduces ethanol-induced apoptosis and differentiation in human embryonic stem cells. *Growth Factors.* Aug 4 **1** [Epub ahead of print].

Nilsson, A. G., Svensson, J., and Johannsson, G. (2007). Management of growth hormone deficiency in adults. *Growth Horm. IGF Res.* **17**(6), 441–462.

Noteborn, H. P., van Balen, P. P., van der Gugten, A. A., Hart, I. C., Ebels, I., and Salemink, C. A. (1993). Presence of immunoreactive growth hormone and prolactin in the ovine pineal gland. *J. Pineal Res.* **14**(1), 11–22.

Nyberg, F. (2000). Growth hormone in the brain: Characteristics of specific brain targets for the hormone and their functional significance. *Front. Neuroendocrinol.* **21**(4), 330–348.

Nyberg, F. (2006). Growth hormone and insulin-like growth factor-1 in human cerebrospinal fluid. *In* "The Somatotrophic Axis in Brain Function" (F. Nyberg, Ed.), pp. 69–74. Elsevier Academic Press, San Diego, CA.

Nyberg, F. (2007). Growth hormone and brain function. *In* "Growth Hormone Therapy in Pediatrics. 20 Years of KIGS" (M. B. Ranke, D. A. Price, and E. O. Reiter, Eds.), pp. 450–460. Karger, Basel.

Nyberg, F., and Burman, P. (1996). Growth hormone and its receptors in the central nervous system—Location and functional significance. *Horm. Res.* **45**(1–2), 18–22 (Review).

Nyberg, F., and Sharma, H. S. (2002). Repeated topical application of growth hormone attenuates blood–spinal cord barrier permeability and edema formation following spinal cord injury: An experimental study in the rat using Evans blue, ([125])I-sodium and lanthanum tracers. *Amino Acids* **23**(1–3), 231–239.

Nyberg, F., Nalén, B., Fhölenhag, K., Fryklund, L., and Albertsson-Wikland, K. (1989). Enzymatic release of peptide fragments from human growth hormone which displace (*H*)-dihydromorphine from rat brain opioid receptors. *J. Endocrinol. Invest.* **12**(Suppl. 2), 140.

O'Kusky, J. R., Ye, P., and D'Ercole, A. J. (2000). Insulin-like growth factor-I promotes neurogenesis and synaptogenesis in the hippocampal dentate gyrus during postnatal development. *J. Neurosci.* **20**(22), 8435–8442.

Opanashuk, L. A., and Hauser, K. F. (1998). Opposing actions of the EGF family and opioids: Heparin binding-epidermal growth factor (HB-EGF) protects mouse cerebellar neuroblasts against the antiproliferative effect of morphine. *Brain Res.* **804**(1), 87–94.

Orellana, D. I., Quintanilla, R. A., Gonzalez-Billault, C., and Maccioni, R. B. (2005). Role of the JAKs/STATs pathway in the intracellular calcium changes induced by interleukin-6 in hippocampal neurons. *Neurotox Res.* **8**(3–4), 295–304.

Pan, W., and Kastin, A. J. (2000). Interactions of IGF-1 with the blood–brain barrier *in vivo* and *in situ*. *Neuroendocrinology* **72**(3), 171–178.

Pan, W., Yu, Y., Cain, C. M., Nyberg, F., Couraud, P. O., and Kastin, A. J. (2005). Permeation of growth hormone across the blood-brain barrier. *Endocrinology* **146**(11), 4898–48904.

Parissis, J. T., Adamopoulos, S., Karatzas, D., Paraskevaidis, J., Livanis, E., and Kremastinos, D. (2005). Growth hormone-induced reduction of soluble apoptosis mediators is associated with reverse cardiac remodelling and improvement of exercise capacity in patients with idiopathic dilated cardiomyopathy. *Eur. J. Cardiovasc. Prev. Rehabil.* **12**(2), 164–168.

Philippou, A., Halapas, A., Maridaki, M., and Koutsilieris, M. (2007). Type I insulin-like growth factor receptor signaling in skeletal muscle regeneration and hypertrophy. *J. Musculoskelet. Neuronal. Interact.* **7**(3), 208–218 (Review).

Rami, A. (2003). Ischemic neuronal death in the rat hippocampus: The calpain–calpastatin–caspase hypothesis. *Neurobiol. Dis.* **13**(2), 75–88.

Ramsey, M. M., Weiner, J. L., Moore, T. P., Carter, C. S., and Sonntag, W. E. (2004). Growth hormone treatment attenuates age-related changes in hippocampal short-term plasticity and spatial learning. *Neuroscience* **129**(1), 119–127.

Ransome, M. I., Goldshmit, Y., Bartlett, P. F., Waters, M. J., and Turnley, A. M. (2004). Comparative analysis of CNS populations in knockout mice with altered growth hormone responsiveness. *Eur. J. Neurosci.* **19**(8), 2069–2079.

Sanders, E. J., Baudet, M. L., Parker, E., and Harvey, S. (2009a). Signaling mechanisms mediating local GH action in the neural retina of the chick embryo. *Gen. Comp. Endocrinol.* **163**(1–2), 63–69.

Sanders, E. J., Parker, E., and Harvey, S. (2009b). Endogenous growth hormone in human retinal ganglion cells correlates with cell survival. *Mol. Vis.* **15**, 920–926.

Sanders, E. J., Lin, W. Y., Parker, E., and Harvey, S. (2009c). Growth hormone expression and neuroprotective activity in a quail neural retina cell line. *Gen. Comp. Endocrinol.* June 17 [Epub ahead of print].

Sara, V. R., Hall, K., Von Holtz, H., Humbel, R., Sjögren, B., and Wetterberg, L. (1982). Evidence for the presence of specific receptors for insulin-like growth factors 1 (IGE-1) and 2 (IGF-2) and insulin throughout the adult human brain. *Neurosci. Lett.* **34**(1), 39–44.

Scheepens, A., Williams, C. E., Breier, B. H., Guan, J., and Gluckman, P. D. (2000). A role for the somatotrophic axis in neural development, injury and disease. *J. Pediatr. Endocrinol. Metab.* **13**(Suppl. 6), 1483–1491 (Review).

Scott, H. J., Stebbing, M. J., Walters, C. E., McLenachan, S., Ransome, M. I., Nichols, N. R., and Turnley, A. M. (2006). Differential effects of SOCS2 on neuronal differentiation and morphology. *Brain Res.* **1067**(1), 138–145.

Seiler, A. E., Ross, B. N., Green, J. S., and Rubin, R. (2000). Differential effects of ethanol on insulin-like growth factor-I receptor signaling. *Alcohol Clin. Exp. Res.* **24**(2), 140–148.

Shafiei, F., Herington, A. C., and Lobie, P. E. (2006). Mechanism of signal transduction utilized by growth hormone. *In* "The Somatotrophic Axis in Brain Function" (F. Nyberg, Ed.), pp. 39–49. Elsevier Academic Press, San Diego, CA.

Sherlock, M., and Toogood, A. A. (2007). Aging and the growth hormone/insulin like growth factor-1 axis. *Pituitary* **10**(2), 189–203.

Smith, P. F. (2003). Neuroprotection against hypoxia-ischemia by insulin-like growth factor-I (IGF-I). *IDrugs* **6**(12), 1173–1177 (Review).

Sonntag, W. E., Lynch, C., Thornton, P., Khan, A., Bennett, S., and Ingram, R. (2000a). The effects of growth hormone and IGF-1 deficiency on cerebrovascular and brain ageing. *J. Anat.* **197**(Pt. 4), 575–585 (Review).

Sonntag, W. E., Bennett, S. A., Khan, A. S., Thornton, P. L., Xu, X., Ingram, R. L., and Brunso-Bechtold, J. K. (2000b). Age and insulin-like growth factor-1 modulate N-methyl-D-aspartate receptor subtype expression in rats. *Brain Res. Bull.* **51**(4), 331–338.

Sonntag, W. E., Ramsey, M., and Carter, C. S. (2005). Growth hormone and insulin-like growth factor-1 (IGF-1) and their influence on cognitive aging. *Ageing Res. Rev.* **4**(2), 195–212 (Review).

Spain, J. W., and Newsom, G. C. (1991). Chronic opioids impair acquisition of both radial maze and Y-maze choice escape. *Psychopharmacology (Berl)* **105**(1), 101–106.

Stabler, B. (2001). Impact of growth hormone (GH) therapy on quality of life along the lifespan of GH-treated patients. *Horm. Res.* **56**(Suppl. 1), 55–58.

Sun, L. Y., Al-Regaiey, K., Masternak, M. M., Wang, J., and Bartke, A. (2005). Local expression of GH and IGF-1 in the hippocampus of GH-deficient long-lived mice. *Neurobiol. Aging* **26**(6), 929–937.

Svensson, J., Johansson, G., and Bengtsson, B. A. (2001). Body composition and quality of life as markers of the efficacy of growth hormone replacement therapy in adults. *Horm. Res.* **55**(Suppl. 2), 55–60 (Review).

Svensson, A. L., Bucht, N., Hallberg, M., and Nyberg, F. (2008). Reversal of opiate-induced apoptosis by human recombinant growth hormone in murine foetus primary hippocampal neuronal cell cultures. *Proc. Natl. Acad Sci. USA* **105**(20), 7304–7308.

Tateno, M., Ukai, W., Ozawa, H., Yamamoto, M., Toki, S., Ikeda, H., and Saito, T. (2004). Ethanol inhibition of neural stem cell differentiation is reduced by neurotrophic factors. *Alcohol Clin. Exp. Res.* **28**(8 Suppl. Proceedings), 134S–138S.

Thornwall, M., Chhajlani, V., Le Grevès, P., and Nyberg, F. (1995). Detection of growth hormone receptor mRNA in an ovine choroid plexus epithelium cell line. *Biochem. Biophys. Res. Commun.* **217**(1), 349–353.

Thornwall-Le Grevès, M., Zhou, Q., Lagerholm, S., Huang, W., Le Grevés, P., and Nyberg, F. (2001). Morphine decreases the levels of the gene transcripts of growth hormone receptor and growth hormone binding protein in the male rat hippocampus and spinal cord. *Neurosci. Lett.* **304**(1–2), 69–72.

Trejo, J. L., Carro, E., Garcia-Galloway, E., and Torres-Aleman, I. (2004). Role of insulin-like growth factor I signaling in neurodegenerative diseases. *J. Mol. Med.* **82**(3), 156–162 (Review).

van Dam, P. S., Aleman, A., de Vries, W. R., Deijen, J. B., van der Veen, E. A., de Haan, E. H., and Koppeschaar, H. P. (2000). Growth hormone, insulin-like growth factor I and cognitive function in adults. *Growth Horm. IGF Res.* **10**(Suppl. B), S69–S73 (Review).

van Nieuwpoort, I. C., and Drent, M. L. (2008). Cognition in the adult with childhood-onset GH deficiency. *Eur. J. Endocrinol.* **159**(Suppl. 1), S53–S57.

Walsh, R. J., Slaby, F. J., and Posner, B. I. (1987). A receptor-mediated mechanism for the transport of prolactin from blood to cerebrospinal fluid. *Endocrinology* **120**(5), 1846–1850.

Walsh, R. J., Mangurian, L. P., and Posner, B. I. (1990). The distribution of lactogen receptors in the mammalian hypothalamus: an *in vitro* autoradiographic analysis of the rabbit and rat. *Brain Res.* **530**(1), 1–11.

Webb, S. M., and Badia, X. (2008). Quality of life in growth hormone deficiency and acromegaly. *Endocrinol. Metab. Clin. North Am.* **36**(1), 221–232 (Review).

Werther, G. A., Hogg, A., Oldfield, B. J., McKinley, M. J., Figdor, R., and Mendelsohn, F. A. (1989). Localization and characterization of insulin-like growth factor-I receptors in rat brain and pituitary gland using *in vitro* autoradiography and computerized densitometry: A distinct distribution from insulin receptors. *J. Neuroendocrinol.* **1**(5), 369–377.

Wine, R. N., McPherson, C. A., and Harry, G. J. (2009). IGF-1 and pAKT signaling promote hippocampal CA1 neuronal survival following injury to dentate granule cells. *Neurotox Res.* **16**(3), 280–292.

Winkler, T., Sharma, H. S., Stålberg, E., Badgaiyan, R. D., Westman, J., and Nyberg, F. (2000). Growth hormone attenuates alterations in spinal cord evoked potentials and cell injury following trauma to the rat spinal cord. An experimental study using topical application of rat growth hormone. *Amino Acids* **19**(1), 363–371.

Zhai, Q., Lai, Z., Roos, P., and Nyberg, F. (1994). Characterization of growth hormone binding sites in rat brain. *Acta Paediatr. Suppl.* **406,** 92–95.

Zhai, Q. Z., Lai, Z., Yukhananov, R., Roos, P., and Nyberg, F. (1995). Decreased binding of growth hormone in the rat hypothalamus and choroid plexus following morphine treatment. *Neurosci. Lett.* **184**(2), 82–85.

Zheng, W. H., and Quirion, R. (2009). Glutamate acting on N-methyl-D-aspartate receptors attenuates insulin-like growth factor-1 receptor tyrosine phosphorylation and its survival signaling properties in rat hippocampal neurons. *J. Biol. Chem.* **284**(2), 855–861.

INDEX

CONTENTS OF RECENT VOLUMES

Control   Meth 23°C   Meth 29°C

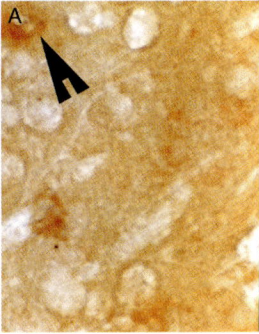 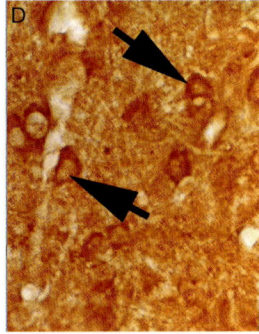 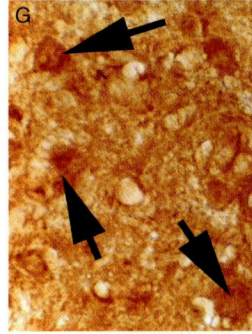

Albumin

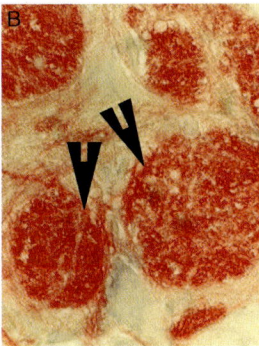 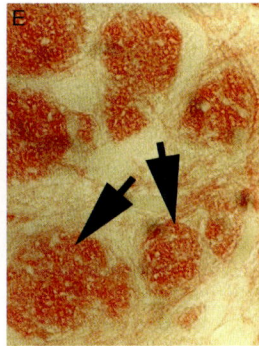 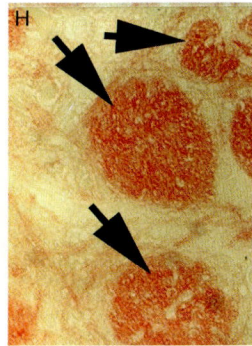

MBP

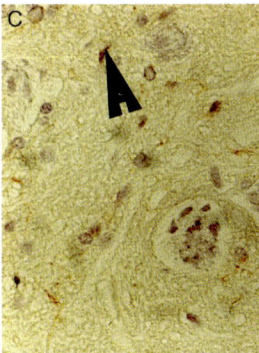 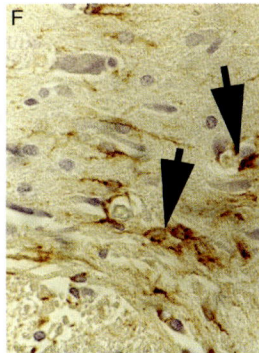 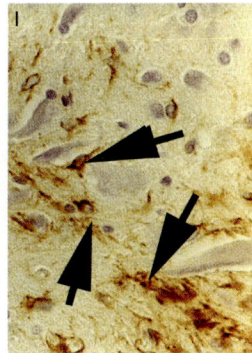

GFAP

KIYATKIN AND SHARMA, CHAPTER 4, FIG. 3. (See Legend in Text.)

Control

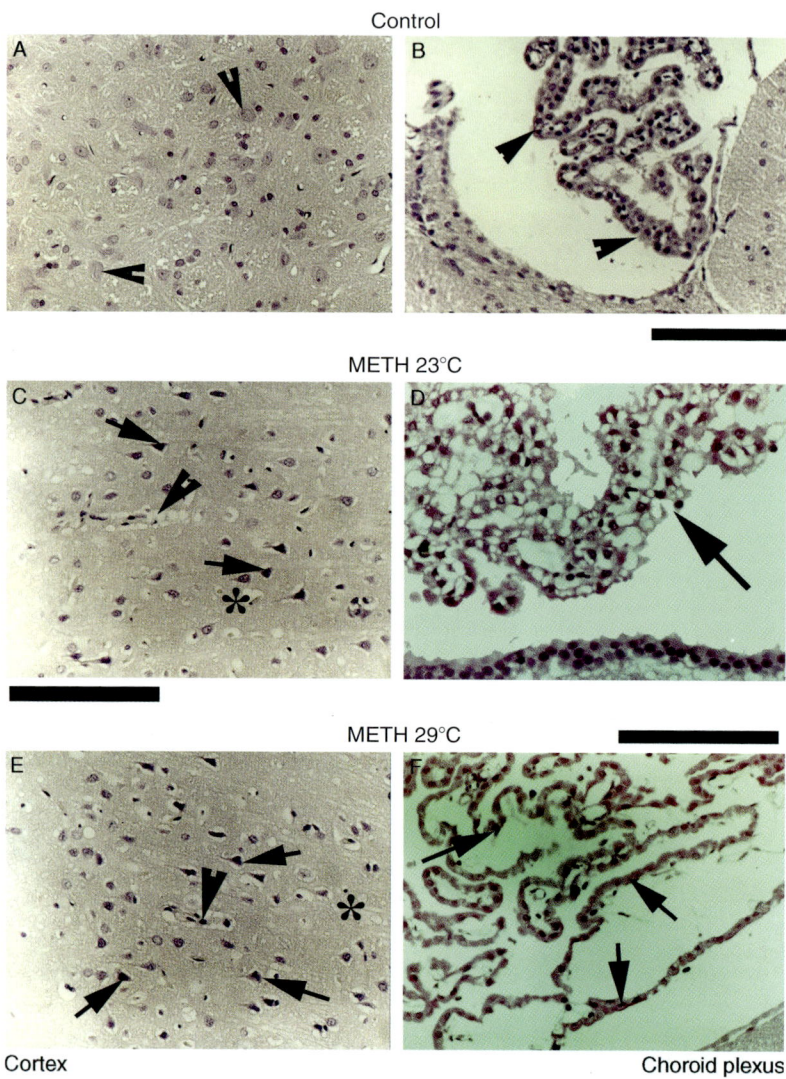

METH 23°C

METH 29°C

Cortex   Choroid plexus

KIYATKIN AND SHARMA, CHAPTER 4, FIG. 8. (See Legend in Text.)

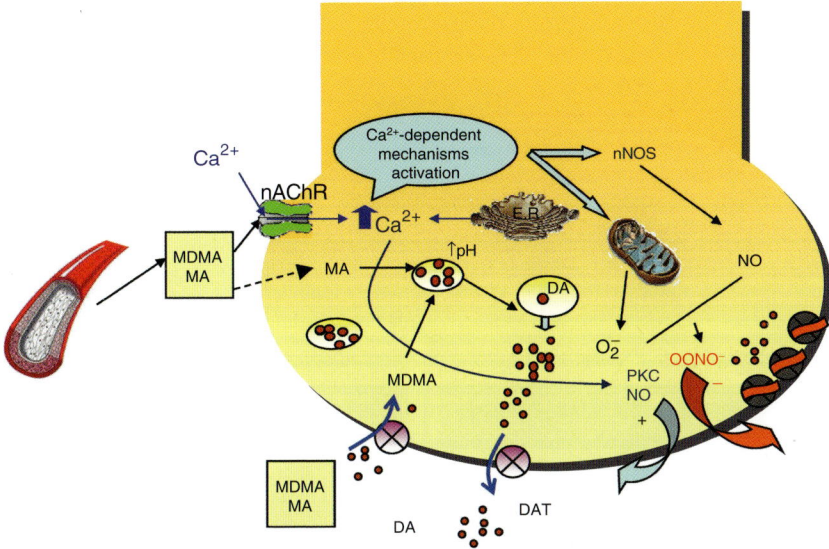

Escubedo et al., Chapter 6, Fig. 6. (See Legend in Text.)

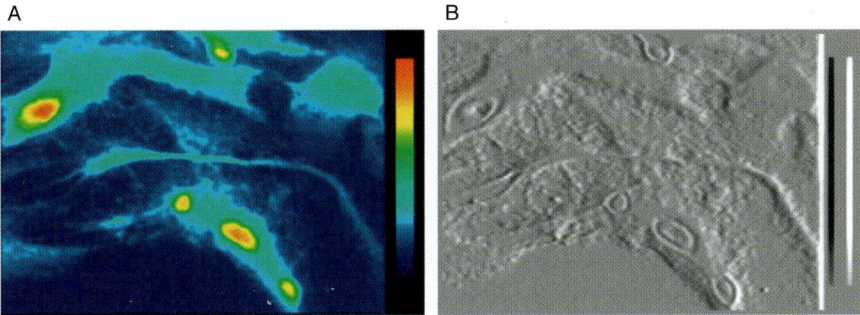

González and Salido, Chapter 7, Fig. 1. (See Legend in Text.)

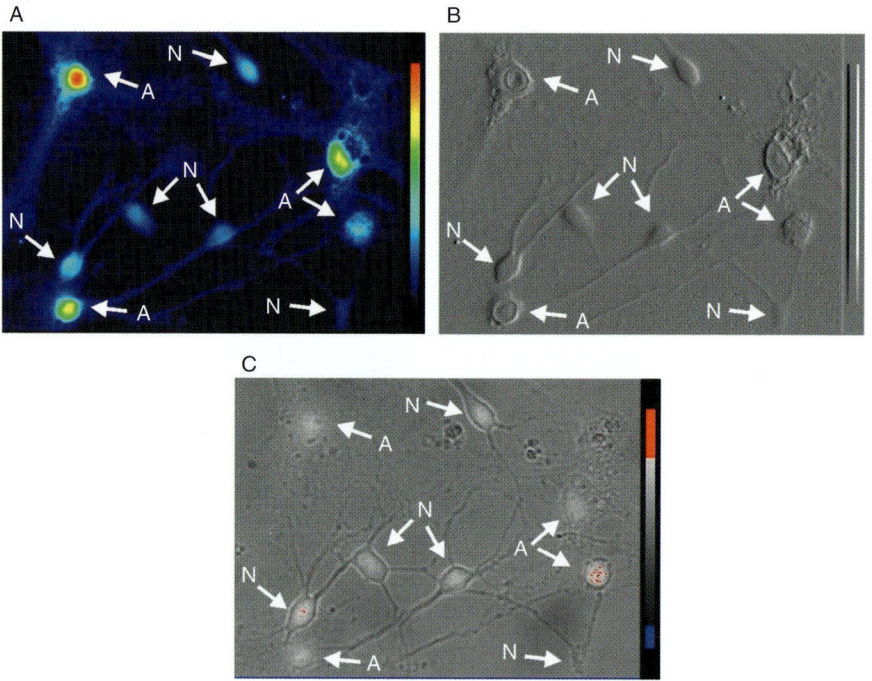

González and Salido, Chapter 7, Fig. 2. (See Legend in Text.)

A

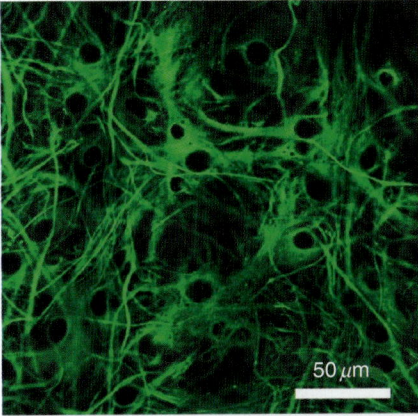

B

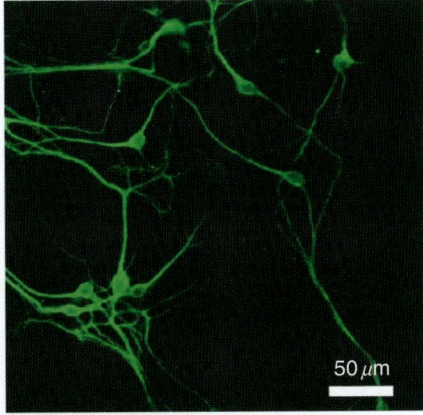

C

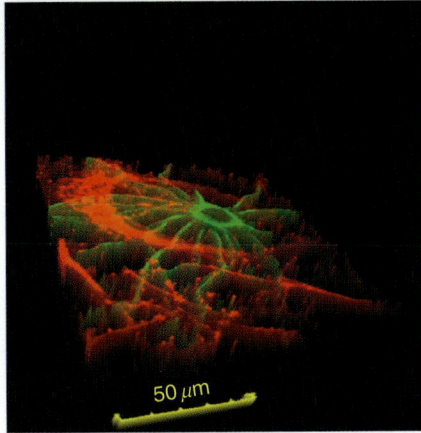

GONZÁLEZ AND SALIDO, CHAPTER 7, FIG. 3. (See Legend in Text.)

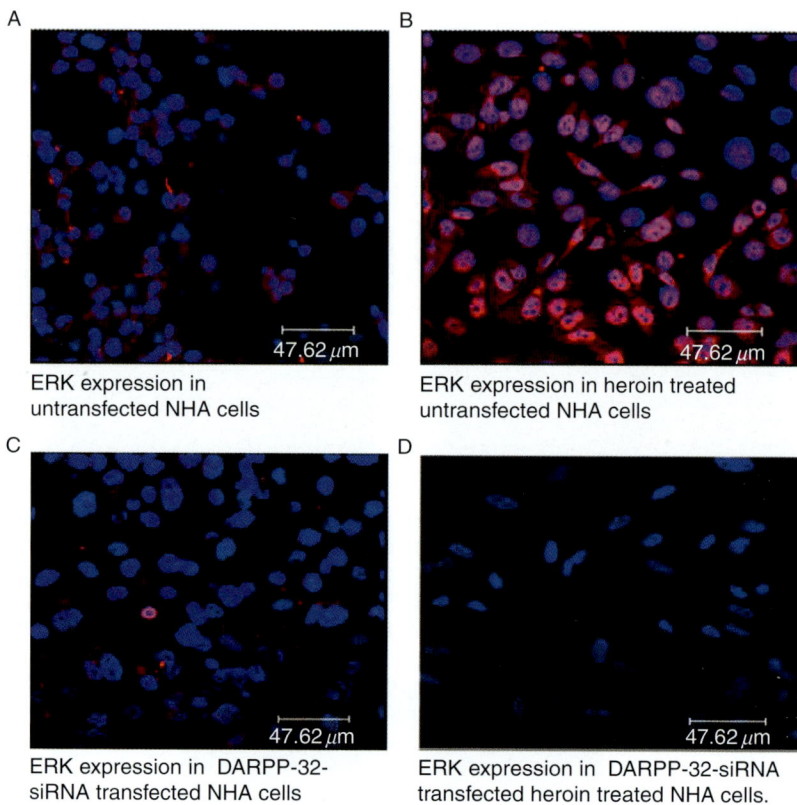

A

B

ERK expression in
untransfected NHA cells

ERK expression in heroin treated
untransfected NHA cells

C

D

ERK expression in DARPP-32-
siRNA transfected NHA cells

ERK expression in DARPP-32-siRNA
transfected heroin treated NHA cells.

Mahajan et al., Chapter 8, Fig. 7. (See Legend in Text.)

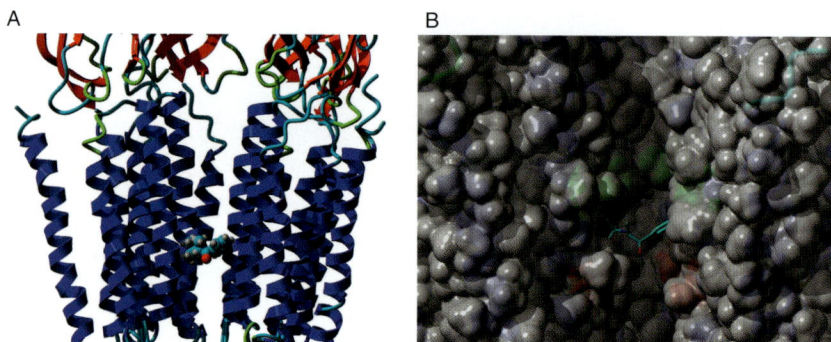

A

B

Arias et al., Chapter 9, Fig. 4. (See Legend in Text.)

Cocaine neurotoxicity

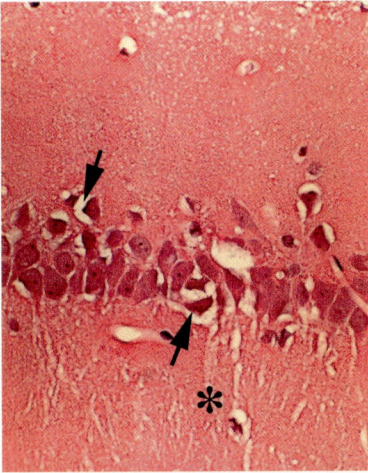

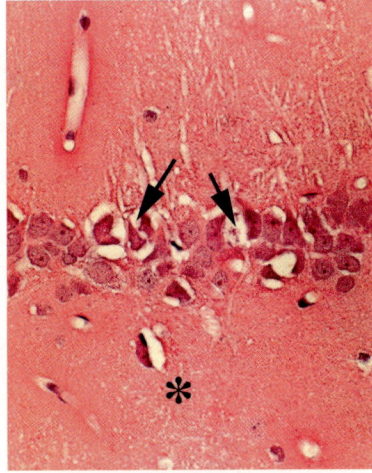

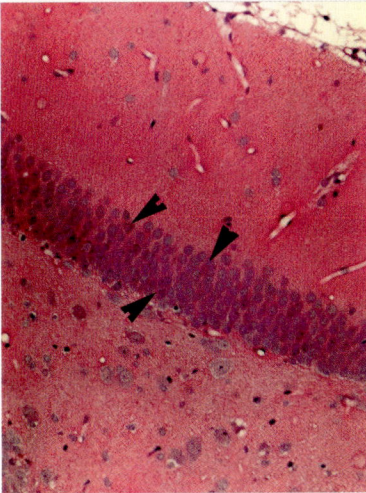

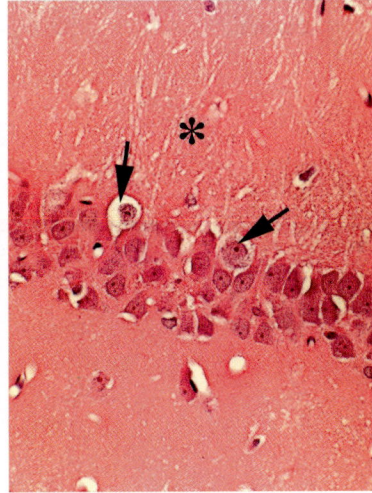

SHARMA ET AL., CHAPTER 11, FIG. 4. (See Legend in Text.)

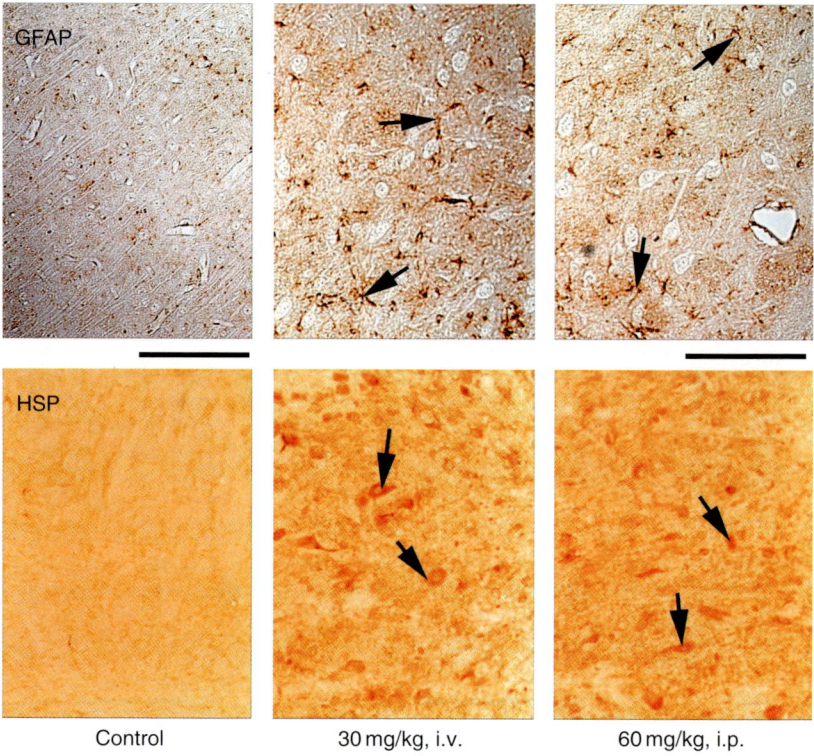

Cocaine neurotoxicity

GFAP

HSP

| Control | 30 mg/kg, i.v. | 60 mg/kg, i.p. |

Sᴍᴀʀᴍᴀ ᴇᴛ ᴀʟ., Cʜᴀᴘᴛᴇʀ 11, Fɪɢ. 5. (See Legend in Text.)

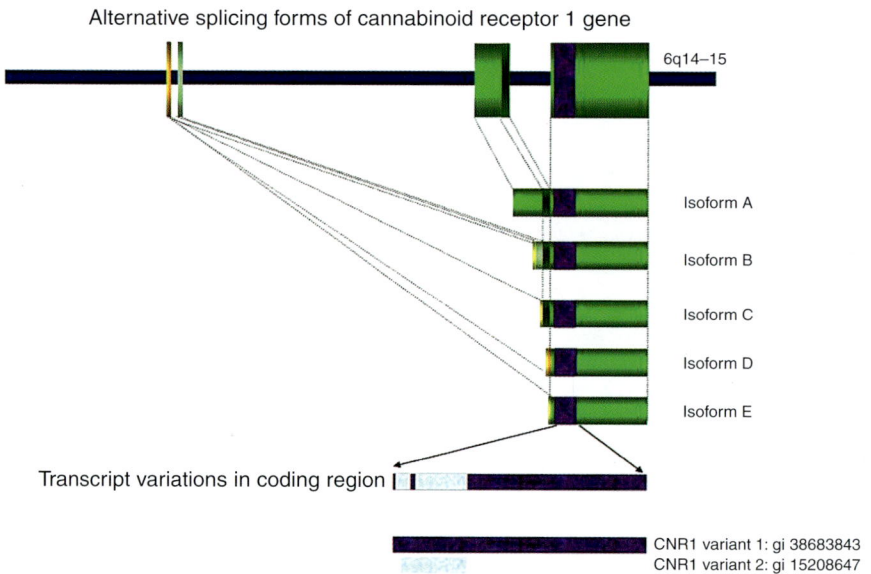

Alternative splicing forms of cannabinoid receptor 1 gene

6q14–15

Isoform A

Isoform B

Isoform C

Isoform D

Isoform E

Transcript variations in coding region

CNR1 variant 1: gi 38683843
CNR1 variant 2: gi 15208647

Oɴᴀɪᴠɪ, Cʜᴀᴘᴛᴇʀ 12, Fɪɢ. 1. (See Legend in Text.)

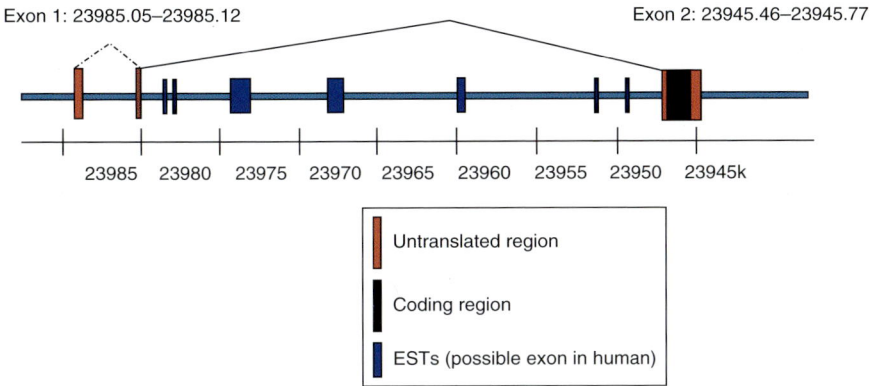

Cannabinoid *CB*2 receptor gene structure

Homo sapiens

Exon 1: 23985.05–23985.12

Exon 2: 23945.46–23945.77

23985 23980 23975 23970 23965 23960 23955 23950 23945k

Untranslated region

Coding region

ESTs (possible exon in human)

ONAIVI, CHAPTER 12, FIG. 2. (See Legend in Text.)

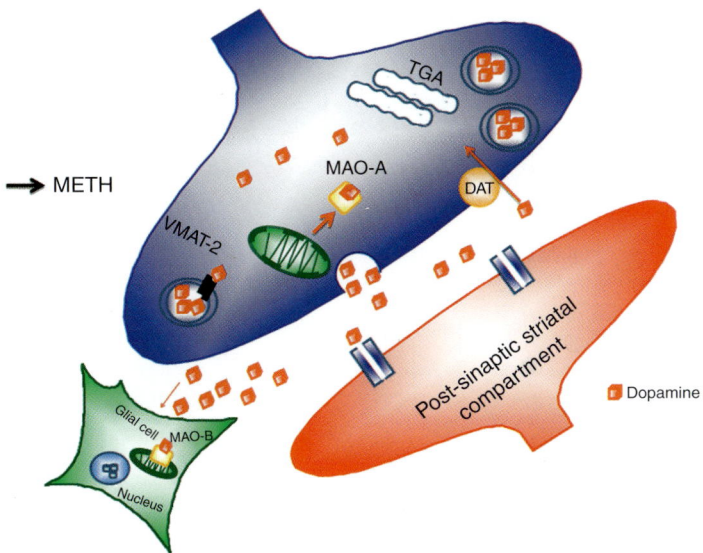

FORNAI ET AL., CHAPTER 13, FIG. 1. (See Legend in Text.)

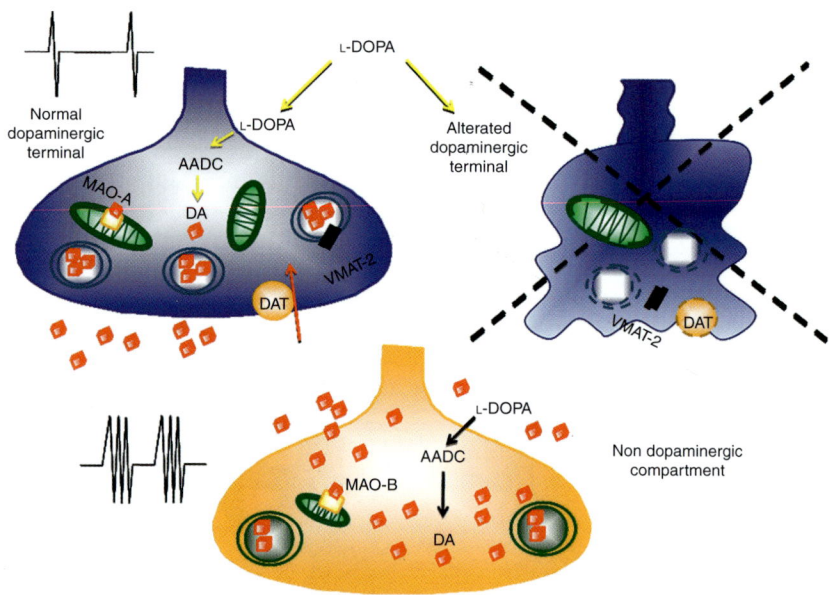

FORNAI ET AL., CHAPTER 13, FIG. 2. (See Legend in Text.)